Taurine 7

Advances in Experimental Medicine and Biology

Recent Volumes in this Series

Volume 635 GI MICROBIOTA AND REGULATION OF THE IMMUNE SYSTEM
Edited by Gary B. Huffnagle and Mairi Noverr

Volume 636 MOLECULAR MECHANISMS IN SPERMATOGENESIS
Edited by C. Yan Cheng

Volume 637 MOLECULAR MECHANISMS IN XERODERMA PIGMENTOSUM
Edited by Shamin Ahmad and Fumio Hanaoka

Volume 638 SOMITOGENESIS
Edited by Miguel Maroto and Neil Whittock

Volume 639 BREAST FEEDING: EARLY INFLUENCES ON LATER HEALTH
Edited by Gail Goldberg

Volume 640 MULTICHAIN IMMUNE RECOGNITION RECEPTOR SIGNALING
Edited by Alexander Sigalov

Volume 641 CELLULAR OSCILLATORY MECHANISMS
Edited by Miguel Maroto and Nick Monk

Volume 642 THE SARCOMERE AND SKELETAL MUSCLE DISEASE
Edited by Nigel G. Laing

Volume 643 TAURINE 7
Edited by Junichi Azuma, Stephen W. Schaffer, and Takashi Ito

---

Junichi Azuma · Stephen W. Schaffer ·
Takashi Ito
Editors

Taurine 7

Springer

Editors
Dr. Junichi Azuma
Osaka University
Department Clinical Evaluation
of Medicines and Therapeutics
1–6 Yamada Oka, Suita
Osaka
565-0871 Japan
azuma@phs.osaka-u.ac.jp

Dr. Stephen W. Schaffer
College of Medicine
University of South Alabama
Department of Pharmacology
307 University blvd., Mobile, Alabama
36688 USA
sschaffe@jagura1.usouthal.edu

Dr. Takashi Ito
College of Medicine
University of South Alabama
Department of Pharmacology
307 University blvd., Mobile, Alabama
36688 USA
takasy110@gmail.com

ISBN: 978-0-387-75680-6 e-ISBN: 978-0-387-75681-3
DOI: 10.1007/978-0-387-75681-3

Library of Congress Control Number: 2008937801

Printed on acid-free paper

springer.com

Preface

Taurine (2-ethanesulfonicacid) is a unique and mysterious compound. It is present in relatively high concentrations in a wide range of cells and tissues, but exists as a free amino acid in these cells without being utilized in protein synthesis. Taurine was first isolated more than 150 years ago from ox (Taurus) bile, where it was found in conjugation with bile acids through an amide linkage. Since that time, it has been reported to exist in particularly high concentrations in the cytoplasm of excitable tissues, such as certain parts of the brain, retina, skeletal muscle, myocardium and platelets. Some of its physiological functions have already been established, for example its role as an essential nutrient during development, an osmolyte and a neuromodulator. Recently, taurine has been advanced as a cytoprotective agent against certain pathological perturbations, but the mechanisms underlying its actions are still mostly a matter of speculation. Moreover, it is possible that other putative functions of taurine remain to be discovered.

The 16th International Taurine Meeting "Taurine for Future Healthcare" was held on September 2–5, 2007, in Shimoda, Shizuoka, Japan, with the site of the meeting being the Shimoda Central Hotel. Approximately 80 individuals from 11 nations, including newcomers as well as experts in taurine research, attended the scientific meeting. A total of 79 papers were presented as either oral or poster presentations. This meeting was multidisciplinary, with participants addressing multiple areas of the biological sciences. Typhoon "Fitow", which means "beautiful fragrant flower" in a Micronesian language, hit the Shimoda region with full force at the end of the meeting, but we were able to finish the scientific sessions and enjoy an excursion prior to the onslaught. The morning after Fitow's fury, the lingering scent of flowers reminded us that we are clearly at the dawn of a new era in taurine research.

The organizers of the taurine meeting would like to thank Taisho Pharmaceutical Co., Ltd., Tokyo, Japan, for their generous financial support and assistance in the organization of the meeting. We would also like to thank Dong-A Pharmaceutical Co, Korea, for their generous financial support. In addition, we would like to thank all participants of the meeting, especially the participants from the Osaka University and Taisho Pharmaceutical Company. The staff of the Shimoda Central Hotel was extremely helpful in making sure that all participants were comfortable and for ensuring the success of their planned events. Finally, we would like to express our appreciation for the untiring effort before, during and after the scientific sessions of Dr. Abe, who helped ensure the success of the meeting.

We are pleased to provide you with *Taurine 7*, which contains the proceedings of the 16th International Taurine Meeting consisting, of 54 original papers. This volume focuses on all aspects of taurine research, including topics of interest to today's scientists as well as future clinical applications.

Part I. Cardiovascular and Renal Effects of Taurine.
Part II. Effect of Taurine on Brain and Retina.
Part III. Effect of Taurine on Skeletal Muscle.
Part IV. Gastroenteric and Hepatic Effects of Taurine.
Part V. Effect of Taurine on Bone.
Part VI. Effect of Taurine on Diabetes and Obesity.
Part VII. Potential Therapeutic Effects of Taurine.
Part VII. Taurine as an Antioxidant; Role in Immune System and Other Tissues.
Part IX. Regulation of the Taurine Transporter.

Future interest in taurine will undoubtedly be robust. However, considerable work remains to develop and uncover key new facts regarding taurine. This book should provide insight into new avenues of investigation and help propel the field into the new era of taurine research. Finally, the organizers wish to thank all of the participants for their stimulating discussions, probing questions and written contributions that made the Shimoda taurine meeting an unmitigated success.

We are deeply thankful to all scientists who have an interest in taurine, and are looking forward to seeing the taurine family at the next Taurine meeting in Florida.

Alone we can do so little; together we can do so much.
Helen Keller

Osaka, Japan — Junichi Azuma, MD
USA — Stephen W. Schaffer, PhD
Osaka, Japan — Takashi Ito, PhD

Meeting Participants

Dr. Satoshi Abe
Taisho Pharmaceutical Co. Ltd.
JAPAN
s-abe@so.taisho.co.jp

Dr. Chang Soon Ahn
Ansan College
KOREA
csan@ansan.ac.kr

Mr. Takao Aketo
Taisho Pharmaceutical Co. Ltd.
JAPAN
t-aketo@so.taisho.co.jp

Dr. Medeya Akhalaya
M.V. Lomonosov Moscow State University
RUSSIA
AMedea@biophys.msu.ru

Dr. Junichi Azuma
Osaka University
JAPAN
azuma@phs.osaka-u.ac.jp

Dr. Kyung Ja Chang
Inha University
KOREA
kjchang@inha.ac.kr

Dr. Udom Chantharaksri
Mahidol University
THAILAND
scust@mahidol.ac.th

Dr. Russell W. Chesney
University of Tennessee
USA
rchesney@utmem.edu

Dr. Mi Ja Choi
Keimyung University
KOREA
choimj@kmu.ac.kr

Dr. Fili Fazzino
Instituto Venezolano de Investigaciones Científicas
VENESUELA
ffazzino2000@yahoo.com

Mr. Junichi Fukutome
Taisho Pharmaceutical Co. Ltd
JAPAN
j-fukudome@so.taisho.co.jp

Dr. Shoichiro Hamano
Hamano International Co., Ltd.
JAPAN
sho@hamanojp.com

Dr. Xiaobin Han
University of Tennessee
USA
xhan@utmem.edu

Dr. Takuzo Hano
Wakayama Medical University
JAPAN
hanotaku@wakayama-med.ac.jp

Dr. Abdeslem El Idrissi
City University of New York
USA
elidrissi@mail.csi.cuny.edu

Dr. Kenichi Imagawa
Pharmacogenome Tip Top Inc.
JAPAN
imagawak@otsuka.jp

Dr. Takashi Ito
Osaka University
JAPAN
takasy110@gmail.com

Dr. Jangmi Kang
Hyogo NCC college
JAPAN
kang@hyoei.ac.jp

Dr. Young Sook Kang
Sookmyung Women's University
KOREA
yskang@sookmyung.ac.kr

Dr. Byong Kak Kim
Seoul National University
KOREA
yunchil@naver.com

Dr. Ha Won Kim
University of Seoul
KOREA
hwkim@uos.ac.kr

Dr. Kyoung Soo Kim
Kyung Hee University
KOREA
labrea46@yahoo.co.kr

Dr. An Keun Kim
Sookmyung Women's University
KOREA
akkim@sookmyung.ac.kr

Dr. Chaekyun Kim
Inha University School of Medicine
KOREA
chaekyun@inha.ac.kr

Dr. Young Chul Kim
Seoul National University
KOREA
youckim@snu.ac.kr

Dr. Masayoshi Kuwahara
The University of Tokyo
JAPAN
akuwam@mail.ecc.u-tokyo.ac.jp

Dr. Cesar Lau-Cam
St. John's University
USA
claucam@usa.net

Dr. Dong-Hee Lee
University of Seoul
KOREA
leedh@uos.ac.kr

Dr. Yu Teh Li
Tulane University School of Medicine
USA
yli1@tulane.edu

Dr. Ning Ma
University of Tokushima
JAPAN
maning@basic.med.tokushima-u.ac.jp

Dr. Janusz Marcinkiewicz
Medical College Jagiellonian University
POLAND
mmmarcin@cyf-kr.edu.pl

Dr. Tadaomi A Miyamoto
Kokura Memorial Hospital
JAPAN
almiyamo@f3.dion.ne.jp

Dr. Teruo Miyazaki
George Washington University
USA
teruom@gwu.edu

Dr. Mari Mori
Mukogawa Women's University
JAPAN
mmori@mukogawa-u.ac.jp

Dr. Shigeru Murakami
Taisho Pharmaceutical Co. Ltd.
JAPAN
s-murakami@so.taisho.co.jp

Dr. Junichi Nakagawa
Tokyo University of Agriculture
JAPAN
j3nakaga@bioindustri.nodai.ac.jp

Dr. Seiichiro Nishida
Nara Medical University
JAPAN
s_nishida@indigo.plala.or.jp

Dr. Hideki Nishimura
Tokyo Womens Medical University
JAPAN
hidekigm@dnh.twmu.ac.jp

Dr. Naomichi Nishimura
Nayoro City University
JAPAN
nishimura@nayoro.ac.jp

Mr. Haruaki Nobori
Osaka University
JAPAN
nobori@phs.osaka-u.ac.jp

Dr. Conny Nordin
Lincoping University
SWEDEN
conny.nordin@inr.liu.se

Dr. Sonia Nusetti
Universidad de Oriente Cumana
VENESUELA
snusetti@yahoo.com

Dr. Tetsuya Ogino
Okayama University
JAPAN
togino@lapis.plala.or.jp

Dr. Simo S. Oja
University of Tampere
FINLAND
simo.oja@uta.fi

Mr. Yudhachai Rajatasereekul
Osotspa Co Ltd
THAILAND
yudhachai@osotspa.com

Dr. Sanya Roysommuti
Khon Kaen University
THAILAND
sanya@kku.ac.th

Dr. Martin Samuelsson
Lincoping University
SWEDEN
masam@inr.liu.se

Dr. Chaichan Sangdee
Chaing Mai University
THAILAND
csangdee@mail.med.cmu.ac.th

Dr. Pirjo Saransaari
University of Tampere
FINLAND
pirjo.saransaari@uta.fi

Dr. Hemanta K. Sarkar
BB Tech
USA
h202428@hotmail.com

Dr. Hiroyasu Satoh
Nara Medical University
JAPAN
hysat@naramed-u.ac.jp

Dr. Stephen W. Schaffer
University of South Alabama
USA
sschaffe@jaguar1.usouthal.edu

Dr. Yukio Seki
Chubu Rosai Hospital
JAPAN
seki.nes@chubuh.rofuku.go.jp

Dr. Makoto Shimizu
The University of Tokyo
JAPAN
ams316@mail.ecc.u-tokyo.ac.jp

Dr. Kyoko Takahashi
Osaka University
JAPAN
kyokot@phs.osaka-u.ac.jp

Mr. Toshio Takahashi
Taisho Pharmaceutical Co. Ltd
JAPAN
toshio.takahashi@po.rd.taisho.co.jp

Dr. Masaru Takahashi
Taisho Pharmaceutical Co. Ltd
JAPAN
m-takahashi@so.taisho.co.jp

Ms. Mika Takai
Osaka University
JAPAN
m-takai@phs.osaka-u.ac.jp

Ms. Atcharaporn Thaeomor
Khon Kaen University
THAILAND
thawmor@yahoo.com

Mr. Noppadol Tunglukmongkol
Osotspa Co Ltd
THAILAND
tnoppadol@osotspa.com

Mr. Shuichi Umeda
Taisho Pharmaceutical Co. Ltd
JAPAN
s-umeda@so.taisho.co.jp

Ms. Yoriko Uozumi
Taisho Pharmaceutical Co. Ltd
JAPAN
y-uozumi@po.rd.taisho.co.jp

Ms. Gaofeng Wu
Shen Yang Agricultural University
CHINA
gaofengwu@126.com

Dr. Jang Yen Wu
Florida Atlantic University
USA
jwu@fau.edu

Dr. Yukio Yamori
International Center for Research on
Primary Prevention of Cardiovascular
Diseases WHO Expert Advisory Panel
JAPAN
yamori@cardiacstudy.com

Dr. Yoshihisa Yatabe
Moriya Daiichi Hospital
JAPAN
yatabe@moriya.daiichi.or.jp

Contents

Part II Effect of Taurine on Brain and Retina

Part III Effect of Taurine on Skeletal Muscle

Part IV Gastroenteric and Hepatic Effects of Taurine

Part I
Cardiovascular and Renal Effects of Taurine

Chapter 1
Effect of Taurine on Protein Kinase C Isoforms: Role in Taurine's Actions?

Takashi Ito, Viktor Pastukh, Viktoriya Solodushko, Junichi Azuma, and Stephen W. Schaffer

Abstract Taurine is generally found to be cytoprotective, diminishing damage resulting from ischemia and from initiators of heart failure. Also linked to similar events in the heart is the protein kinase C (PKC) family, which consists of at least 12 different isoforms. Therefore, we proposed that PKC might contribute to the beneficial effects of taurine on cell viability and growth. One of the PKC isoforms that has been advanced as an important mediator of cytoprotection during ischemia is PKCε. In this study, we found that incubation of isolated cardiomyocytes with medium containing 20 mM taurine led to the translocation of PKCε into the membrane, an event commonly associated with the cardioprotective actions of the PKC isozyme. In addition, taurine promoted the upregulation of PKCα PKCβ2 and PKCζ. Because the effects of taurine and angiotensin II on PKC distribution were largely additive, PKC does not appear to contribute to the antagonism between taurine and angiotensin II. However, the upregulation of PKC by taurine is consistent with a role of taurine in normal cell growth. In the taurine deficient heart, cardiomyocyte size is reduced, an effect that is consistent with the effect of taurine on PKCε. In conclusion, the cytoprotective and pro-growth actions of taurine appears to be mediated in part by the activation of PKCε.

Abbreviations *PKC*, Protein kinase C; *Ang II*, Angiotensin II

1.1 Introduction

Taurine is the most abundant free amino acid in mammalian tissue, reaching concentrations as high as 5–20 μmol/g wet wt (Chapman et al. 1993; Chesney 1985). The relationship between intracellular taurine content and cardiac function remains unclear, largely because of the multiple functions of taurine. It is generally accepted that maintenance of intracellular taurine homeostasis is essential for normal cardiac function. Indeed, severe reductions in myocardial taurine content either through dietary taurine deficiency or genetic taurine transporter deficiency leads to the

T. Ito (✉)
Department of Pharmacology, University of South Alabama, College of Medicine, USA

J. Azuma et al. (eds.), *Taurine 7*, Advances in Experimental Medicine and Biology 643, DOI 10.1007/978-0-387-75681-3_1,

development of a cardiomyopathy (Novotny et al. 1991, 1994; Pion et al. 1987). In the failing heart, taurine levels rise, with the increase being directly associated with the severity of heart failure (Newman et al. 1977). The suggestion that the increase in taurine levels might represent an adaptation designed to re-establish normal function led the study of taurine therapy in various animal models of heart failure. These studies have generally shown a beneficial effect of taurine treatment. In the calcium sensitive cardiomyopathic hamster, oral taurine therapy reduced intracellular Ca^{2+} content and decreased the severity of myocardial lesions (Azari et al. 1980; McBroom and Welty 1977). Taurine therapy has also found to reduce mortality and improve contractile function in an aortic regurgitation model of congestive heart failure (Takihara et al., 1986), studies that led to clinical trials that established taurine as useful therapy in the treatment of congestive heart failure (Azuma et al. 1982). Among the factors implicated in these and other models of heart failure have been oxidative stress and calcium overload (Harada et al. 1990; Ohta et al. 1988). Significantly, taurine therapy prevented calcium overload and diminished the degree of oxidative stress in these models.

The pathophysiology of heart failure is complex, involving impaired contractile function, abnormal Ca^{2+} transport, elevations in neurohumoral agents, vascular resistance, diastolic dysfunction and ventricular remodeling. While the initial insult is a decrease in systolic function, the rise in sympathetic and angiotensin II (Ang II) activity triggers a constellation of events that lead to overt heart failure. Inhibition of the neurohumoral agents disrupts the progression of heart failure and reduces mortality, with inhibition of Ang II serving as the mainstay in the treatment of heart failure.

It has been proposed that taurine therapy may benefit the heart by preventing the actions of Ang II (Schaffer et al. 2000). This contention is largely based on the finding that incubation of isolated cardiomyocytes in medium containing 20 mM taurine prevents Ang II-mediated hypertrophy and cell death (Takahashi et al. 1997). Conversely, Ang II-mediated apoptosis is potentiated in taurine deficient cells (Schaffer et al. 2003). Since Ang II initiates signaling pathways that lead to enhanced oxidative stress, elevated $[Ca^{2+}]_i$ and cell death, taurine might act at an early step in Ang II signaling to protect the cardiomyocyte. The present study examines the effect of taurine treatment on the distribution of key protein kinase C (PKC) isoforms, enzymes involved in the pro-apoptotic and hypertrophic activities of angiotensin II.

1.2 Methods

1.2.1 Cell Culture

The care and treatment of animals were in accordance with the guidelines of the National Institute of Health and the procedures approved by the Institutional Care and Use Committee of the University of South Alabama. Rat neonatal cardiomyocytes were prepared as described previously (Pastukh et al. 2005). The cells were suspended in minimal essential medium containing 10% newborn calf serum and 0.1 mM 5-bromo-2-deoxyuridine and plated onto polystyrene treated Petri dishes at a density of 10×10^6 cells/dish (10 cm diameter). They were then placed in serum free medium containing either 0 (control) or 20 mM taurine, for a period of 3 days. The cells were then exposed to medium supplemented with either no addition (control) or 100 nM Ang II. The concentration of Ang II was chosen that induced apoptosis (Kajstura et al. 1997). At the appropriate time, the cells were used for Western blot analysis.

1.2.2 Western Blot Analyses

After the cells were detached from the dish with trypsin, they were washed in phosphate-buffered saline and then centrifuged for 5 min at 500 g at room temperature. Membrane and cytosolic fractions were prepared according to previous reports. Each sample was homogenized in ice-cold lysis buffer (pH 7.4) consisting of the following: 25 mM Tris-HCl; 2 mM EDTA; 5 mM EGTA; 100 mM NaF; protease inhibitors [a 1/100 dilution of protease inhibitor cocktail set III (Calbiochem) and 1% solutions of leupeptin and PMSF], 1 mM orthovanadate and 5 mM dithiothreitol. The samples were then centrifuged at 100,000 g for 60 min. The pellet represents the membrane-particulate fraction and particulate-free supernatant fraction was defined as the cytosolic fraction. The particulate fraction is resuspended in homogenizing buffer containing 0.5% Triton-X100 and centrifuged at 100,000 g for 60 min. The resulting detergent-treated supernatant was used in the Western blot analyses. The protein concentration of each sample was determined by the Bradford or Lowry assay. Cytosolic and membrane proteins were analyzed for PKC isoform content by electrophoresis using 8% SDS-polyacrylamide gels. Following electrophoresis the proteins were transferred to nitrocellulose membranes, where they were blocked. After incubation with the appropriate antibody, the membranes were washed and then incubated with a secondary antibody, goat anti-rabbit IgG. The Western blots were detected by the enhanced chemiluminescence reaction. All data were analyzed by densitometry using ChemiImage 4400 (Alpha Innotech).

1.2.3 Statistical Analysis

The statistical significance of the data was determined using either the Student's test for comparison with groups or ANOVA combined with Tukey's post hoc test for comparison between groups. Values of $P<0.05$ were considered statistically significant.

1.3 Results

1.3.1 Effect of Taurine Exposure on the Status of PKC Isoforms

One of the steps involved in the activation of specific PKC isoforms is their translocation within the cell (Churchill et al. 2008). As a measure of the activation state

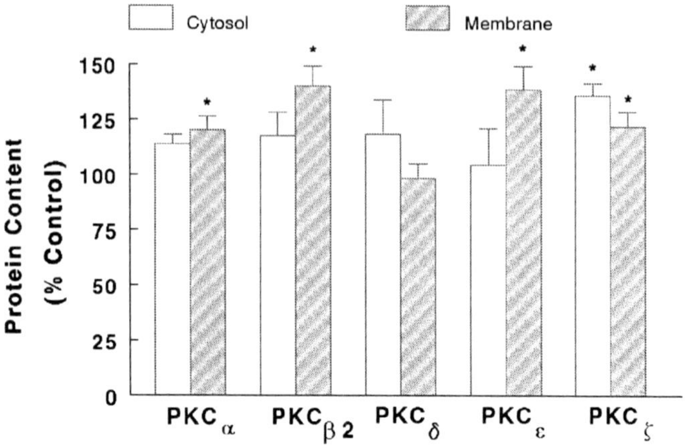

Fig. 1.1 Influence of taurine on the distribution of protein kinase C isoforms. Isolated cardiomyocytes were incubated for 3 days with medium containing either 0 or 20 mM taurine. After harvesting the cells, particulate and cytosolic fractions were obtained from the cell extract. Western blot analyses were performed using antibodies specific for the appropriate PKC isoform. The data are expressed as percent of the control, which is represented as 100%. All data denote means ± SEM of 5–6 different cellular preparations. *: $p<0.01$ vs. control

of the enzyme, the distribution of PKC in particulate and cytosolic fractions was determined. Figure 1.1 shows the protein levels of six PKC isoforms (α, β2, γ, δ, ε, ζ) in the cytosolic and membrane fractions of cardiomyocytes cultured in medium containing either 0 or 20 mM taurine for 3 days. Chronic taurine exposure led to a shift in the membrane/cytosol ratio for three PKC isoforms (α, β2 and ε) in favor of the membrane fraction. While the levels of PKCα and PKCζ in the membrane fraction were increased by taurine exposure, there was no shift in the membrane/cytosolic ratio associated with the activation event. Rather, taurine treatment led to a net increase in the content of the two isoforms in both the cytosolic and membrane fractions. Taurine exerted no influence on either the level or distribution of PKCδ.

1.3.2 Effect of Taurine on the Translocation of PKC by Angiotensin II

We have previously demonstrated that taurine exposure prevents Ang II-mediated hypertrophy of neonatal cardiomyocytes in culture (Azuma et al. 2000, Takahashi et al. 1997). One of the early events in the signaling pathway initiated by Ang II is the activation of PKC leading to the stimulation of NADPH oxidase (Ricci et al. 2008). Pastukh et al. (2005) previously showed that acute exposure of isolated cardiomyocytes to medium containing 1 nM Ang II led to an increase in the membrane/cytosol content ratio of PKCε and PKCδ without affecting that of PKCα,

PKCβ2 and PKCζ. Figure 1.2 shows that Ang II also increases the levels of PKCε and PKCδ in the membranes of cells exposed for 3 days to medium containing 20 mM taurine. Indeed, the effects of taurine and Ang II appear to be additive. While separate addition of Ang II and taurine to the medium elevated membrane levels of PKCε by 37% and 36%, respectively, addition of both Ang II and taurine to the medium increased membrane levels of PKCε by 76%. As expected, the levels of PKCδ were elevated to a similar degree in the presence of Ang II irrespective of taurine content; taurine alone had no effect on the distribution of PKCδ. Moreover, while taurine increased membrane levels of PKCα by 20%, Ang II reduced it by 17%. The combination of both effectors to the medium caused an inconsequential 4% increase in PKCα content.

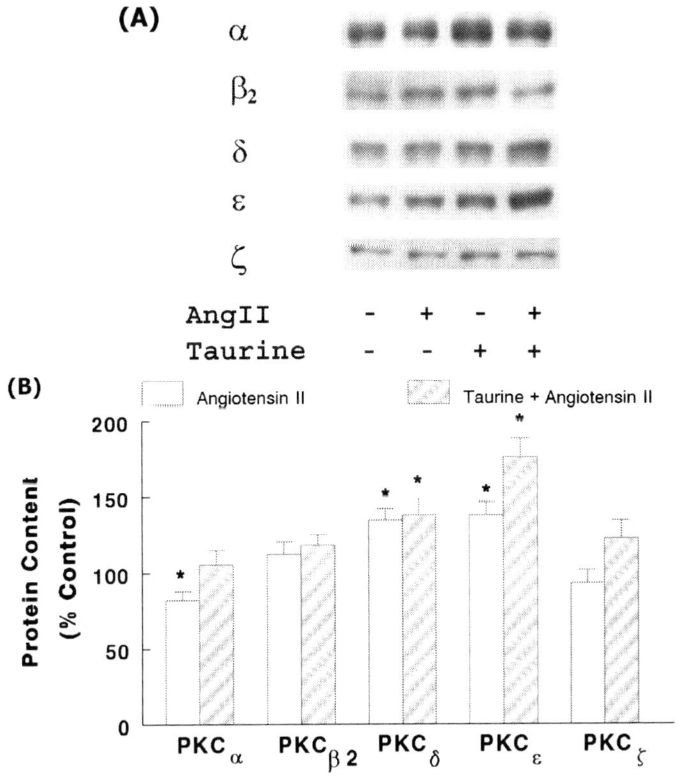

Fig. 1.2 Effect of taurine on AngII-mediated PKC translocation. Isolated cardiomyocytes incubated for 3 days with medium containing either 0 or 20 mM taurine were exposed for 5 min to 100 nM angiotensin II. The cellular particulate fraction was subjected to Western blot analyses using antibodies specific for the appropriate PKC isoform. (**A**) A representative gel showing membrane content of individual PKC isoforms of cells exposed to Ang II, taurine or the combination of Ang II and taurine. (**B**) The data are expressed as % control, with control = 100%. Data represent means ± SEM of 5–6 different cellular preparations. *:$p<0.01$ vs. control

1.4 Discussion

Upon activation individual PKC isoforms are translocated to a distinct subcellular site (Churchill et al. 2008). Mochly-Rosen and coworkers (2008) maintain that this site of translocation is defined by the location of the selective anchor protein, referred to as a RACK, to which the PKC isoform binds. The translocation of PKCδ and PKCε from the cytosol to the membrane fraction is a key step in the signaling pathway initiated by Ang II (Churchill et al. 2008). The involvement of the two PKC isoforms in Ang II signaling is a characteristic feature of pathways initiated by Gq proteins (Churchill et al. 2008).

Taurine also activates a number of PKC isoforms. In this study we showed that exposure of isolated cardiomyocytes to medium containing 20 mM taurine resulted in an increase in the membrane/cytosol ratio for PKCβ2 and PKCε. While the membrane levels of PKCα were significantly elevated in the taurine treated cells, cytosolic content of PKCα also tended to be elevated, leaving the membrane/cytosolic ratio unaffected. Taurine increased the cytosolic and membrane content of PKCζ, but the upregulation of PKCδ was not accompanied by a preferential association of PKCζ with the particulate fraction, ruling out a translocation event in the actions of taurine. In contrast to the other PKC isoforms, taurine had no effect on the levels of PKCδ associated with the membrane and cytosolic fractions.

The factor responsible for taurine-mediated modulation of PKC isoform status is a matter of conjecture. One means by which taurine could influence protein kinase C activity is through its osmoregulatory activity. Addition of 20 mM taurine to the extracellular medium has been shown to trigger cell shrinkage (unpublished data). In NIH/3T3 cells, hyperosmotic stress leads to an increase in diacylglycerol levels, resulting in the translocation of PKCα, PKCδ and PKCε from the cytosol to the membrane (Zhuang et al. 2000). Although membranes levels of PKCα, PKCδ and PKCε increase following exposure of the cardiomyocytes to medium containing 20 mM taurine (Fig. 1.1), the only protein kinase C isoform experiencing an elevation in the membrane/cytosol ratio, which is indicative of a translocation event, is PKCε. Together, these data suggest that either hyperosmotic stress is not responsible for the activation of protein kinase C in the taurine treated cells or that the osmotic stress experienced by the taurine treated cells is too mild to trigger the translocation of a significant number of protein kinase C isoforms. An alternative possibility is that taurine modulates the structure of the membrane, resulting in a change in the RACK-mediated translocation step. Hamaguchi et al. (1991) found that taurine interferes with phospholipid methyltransferase activity, thereby affecting the phosphatidylethanolamine/phosphatidylcholine ratio. This alteration is likely to affect the activity of membrane bound proteins, such as those associated with PKC translocation.

An important conclusion of the present study is that taurine-mediated modulation of PKC does not appear to play a role in the reversal of the myocardial actions of Ang II. Our initial hypothesis was that taurine should prevent Ang II-mediated activation of PKCδ and PKCε. However, Fig. 1.1 shows that taurine has no effect on PKCδ while increasing the translocation and activation of PKCε. Hence, combined treatment with Ang II and taurine leads to an elevation in both PKCδ and PKCε, in line with our conclusion that the two effects are additive. The other means by which taurine might reverse Ang II's actions is through a step downstream from the PKC activation step. In hyperosmotically stressed rat hepatocytes, a PKCζ activation step lies upstream from the activation of NADPH oxidase (Reinehr et al. 2006). Since Ang II signaling also leads to PKC-mediated activation of NADPH oxidase (Ricci et al. 2008), it is possible that taurine might influence Ang II signaling at the NADPH oxidase step. This possibility is worthy of consideration, as taurine serves as an indirect antioxidant. Moreover, NADPH oxidase-mediated generation of superoxide is a key step in Ang II-mediated apoptosis and cell hypertrophy (Ricci et al. 2008), events that are inhibited by taurine (Takahashi et al. 1997).

Although the present data rules out a role for PKC in the reversal of Ang II's actions by taurine, it raises the possibility that PKC might contribute to one of the other actions of taurine. Especially intriguing is the effect of taurine on PKCε. Because taurine promotes the translocation of PKCε from the cytosol to the membrane, it appears to initiate PKC signaling (Ping et al. 2001). PKCε has been the most widely studied PKC isoform in the heart. One of its recognized actions is as a mediator of ischemic preconditioning, a powerful strategy for protecting various tissues against ischemic injury. Ping et al. (1997) reported that ischemic preconditioning triggers a translocation of PKCε into the particulate fraction. This initiates a complex signaling response in which PKCε becomes associated with a large number of proteins (Ping et al. 2001). Selective inhibition of the interaction of PKCε with RACK abolishes the cardioprotection arising from ischemic or hypoxic preconditioning (Gray et al. 1997; Liu et al. 1999). Similarly, genetic PKCε deficiency has no effect on infarct size of the preconditioned heart (Saurin et al. 2002). Like ischemic preconditioning, taurine therapy has also been shown to protect the heart against an ischemic insult (Takahashi et al. 2003).

PKCε has also been implicated in hypertrophic growth (Ito et al. 2008). Not only is PKCε activated in response to hypertrophic stimuli but overexpression and activation of PKCε leads to myocardial hypertrophy (Churchill et al. 2008; Dorn and Force 2005). Although there are questions on the type of hypertrophy (maladaptive as in heart failure vs. adaptive as in development) mediated by PKCε, there is little question that it is a major component in the development of cardiac hypertrophy. Because taurine deficiency in the taurine transport knockout heart is associated with a reduction in cardiomyocyte size (Ito et al. 2008), taurine's ability to promote cell growth might be linked to its activation of PKCε.

1.5 Conclusion

The present data suggest that taurine promotes the translocation of PKCβ2 and PKCε into the particulate fraction. It also elevates the levels of PKCα and PKCζ. While these effects are not responsible for the interaction between angiotensin II and taurine, they may account for some of the cardioprotective effects of taurine. It may also explain the effect of taurine deficiency on cardiomyocyte size.

References

Azari J, Brumbaugh P, Huxtable R (1980) Prophylaxis by taurine in the hearts of cardiomyopathic hamsters. J Mol Cell Cardiol 12:1353–1366

Azuma J, Hasegawa H, Sawamura A, Awata N, Harada H, Ogura K, Kishimoto S (1982) Taurine for treatment of congestive heart failure. Int J Cardiol 2:303–304

Azuma M, Takahashi K, Fukuda T, Ohyabu Y, Yamamoto I, Kim S, Iwao H, Schaffer SW, Azuma J (2000) Taurine attenuates hypertrophy induced by angiotensin II in cultured neonatal rat cardiac myocytes. Eur J Pharmacol 403:181–188

Chapman RA, Suleiman MS, Earm YE (1993) Taurine and the heart. Cardiovasc Res 27:358–363

Chesney RW (1985) Taurine: its biological role and clinical implications. Adv Pediatr 32:1–42

Churchill E, Budas G, Vallentin A, Koyanagi T, Mochly-Rosen D (2008) PKC isozymes in chronic cardiac disease: possible therapeutic targets? Annu Rev Pharmacol Toxicol 48:569–599

Dorn GW, 2nd, Force T (2005) Protein kinase cascades in the regulation of cardiac hypertrophy. J Clin Invest 115:527–537

Gray MO, Karliner JS, Mochly-Rosen D (1997) A selective epsilon-protein kinase C antagonist inhibits protection of cardiac myocytes from hypoxia-induced cell death. J Biol Chem 272:30945–30951

Hamaguchi T, Azuma J, Schaffer S (1991) Interaction of taurine with methionine: inhibition of myocardial phospholipid methyltransferase. J Cardiovasc Pharmacol 18:224–230

Harada H, Cusack BJ, Olson RD, Stroo W, Azuma J, Hamaguchi T, Schaffer SW (1990) Taurine deficiency and doxorubicin: interaction with the cardiac sarcolemmal calcium pump. Biochem Pharmacol 39:745–751

Ito T, Kimura Y, Uozumi Y, Takai M, Muraoka S, Matsuda T, Ueki K, Yoshiyama M, Ikawa M, Okabe M, Schaffer SW, Fujio Y, Azuma J (2008) Taurine depletion caused by knocking out the taurine transporter gene leads to a cardiomyopathy, muscle atrophy and mitochondrial dysfunction. J Mol Cell Cardiol 44(5):927–37

Kajstura J, Cigola E, Malhotra A, Li P, Cheng W, Meggs LG, Anversa P (1997) Angiotensin II induces apoptosis of adult ventricular myocytes in vitro. J Mol Cell Cardiol 29:859–870

Liu GS, Cohen MV, Mochly-Rosen D, Downey JM (1999) Protein kinase C-epsilon is responsible for the protection of preconditioning in rabbit cardiomyocytes. J Mol Cell Cardiol 31: 1937–1948

McBroom MJ, Welty JD (1977) Effects of taurine on heart calcium in the cardiomyopathic hamster. J Mol Cell Cardiol 9:853–858

Newman WH, Frangakis CJ, Grosso DS, Bressler R (1977) A relation between myocardial taurine contest and pulmonary wedge pressure in dogs with heart failure. Physiol Chem Phys 9: 259–263

Novotny MJ, Hogan PM, Flannigan G (1994) Echocardiographic evidence for myocardial failure induced by taurine deficiency in domestic cats. Can J Vet Res 58:6–12

Novotny MJ, Hogan PM, Paley DM, Adams HR (1991) Systolic and diastolic dysfunction of the left ventricle induced by dietary taurine deficiency in cats. Am J Physiol 261:H121–H127

Ohta H, Azuma J, Awata N, Hamaguchi T, Tanaka Y, Sawamura A, Kishimoto S, Sperelakis N (1988) Mechanism of the protective action of taurine against isoprenaline induced myocardial damage. Cardiovasc Res 22:407–413

Pastukh V, Ricci C, Solodushko V, Mozaffari M, Schaffer SW (2005) Contribution of the PI 3-kinase/Akt survival pathway toward osmotic preconditioning. Mol Cell Biochem. 269(1-2):59-67

Ping P, Zhang J, Pierce WM Jr, Bolli R (2001) Functional proteomic analysis of protein kinase C epsilon signaling complexes in the normal heart and during cardioprotection. Circ Res 88: 59–62

Ping P, Zhang J, Qiu Y, Tang XL, Manchikalapudi S, Cao X, Bolli R (1997) Ischemic preconditioning induces selective translocation of protein kinase C isoforms epsilon and eta in the heart of conscious rabbits without subcellular redistribution of total protein kinase C activity. Circ Res 81:404–414

Pion PD, Kittleson MD, Rogers QR, Morris JG (1987) Myocardial failure in cats associated with low plasma taurine: a reversible cardiomyopathy. Science 237:764–768

Qi X, Inagaki K, Sobel RA, Mochly-Rosen D (2008) Sustained pharmacological inhibition of deltaPKC protects against hypertensive encephalopathy through J Clin Invest 118(1):173-82

Reinehr R, Becker S, Braun J, Eberle A, Grether-Beck S, Haussinger D (2006) Endosomal acidification and activation of NADPH oxidase isoforms are upstream events in hyperosmolarity-induced hepatocyte apoptosis. J Biol Chem 281:23150–23166

Ricci C, Pastukh V, Leonard J, Turrens J, Wilson G, Schaffer D, Schaffer SW (2008) Mitochondrial DNA damage triggers mitochondrial-superoxide generation and apoptosis. Am J Physiol Cell Physiol 294:C413–C422

Saurin AT, Pennington DJ, Raat NJ, Latchman DS, Owen MJ, Marber MS (2002) Targeted disruption of the protein kinase C epsilon gene abolishes the infarct size reduction that follows ischaemic preconditioning of isolated buffer-perfused mouse hearts. Cardiovasc Res 55: 672–680

Schaffer S, Solodushko V, Pastukh V, Ricci C, Azuma J (2003) Possible Cause of Taurine-deficient Cardiomyopathy: Potentiation of Angiotensin II Action. J Cardiovasc Pharmacol 41:751–759

Schaffer SW, Lombardini JB, Azuma J (2000) Interaction between the actions of taurine and angiotensin II. Amino Acids 18:305–318

Takahashi K, Azuma M, Taira K, Baba A, Yamamoto I, Schaffer SW, Azuma J (1997) Effect of taurine on angiotensin II-induced hypertrophy of neonatal rat cardiac cells. J Cardiovasc Pharmacol 30:725–730

Takahashi K, Ohyabu Y, Solodushko V, Takatani T, Itoh T, Schaffer SW, Azuma J (2003) Taurine renders the cell resistant to ischemia-induced injury in cultured neonatal rat cardiomyocytes. J Cardiovasc Pharmacol 41:726–733

Takihara K, Azuma J, Awata N, Ohta H, Hamaguchi T, Sawamura A, Tanaka Y, Kishimoto S, Sperelakis N(1986) Beneficial effect of taurine in rabbits with chronic congestive heart failure. Am Heart J 112(6):1278-84

Zhuang S, Hirai SI, Ohno S (2000) Hyperosmolality induces activation of cPKC and nPKC, a requirement for ERK1/2 activation in NIH/3T3 cells. Am J Physiol Cell Physiol 278: C102–C109

Chapter 2
Taurine as the Nutritional Factor for the Longevity of the Japanese Revealed by a World-Wide Epidemiological Survey

Yukio Yamori, Longjian Liu, Mari Mori, Miki Sagara, Shigeru Murakami, Yasuo Nara, and Shunsaku Mizushima

Abstract The initial observation that taurine (T) prevented stroke in stroke-prone spontaneously hypertensive rats (SHRSP) led us to study the effects of T on cardiovascular diseases (CVD), as well as the epidemiological association of T and mortality rates, by using the data from WHO-coordinated Cardiovascular Disease and Alimentary Comparison Study, which covered 61 populations in 25 countries. In this study, 24 hour urine (24-U) samples were examined along with biomarkers of CVD risk. The mortality rate from ischemic heart disease (IHD), which was lowest among the Japanese compared to the populations of other developed countries, was positively related to total serum cholesterol (TC) and inversely related to 24-U taurine excretion (24-UT), as well as the n-3 fatty acid to total phospholipids ratio of the plasma membrane, both biomarkers of seafood intake. Analysis of 5 diet-related factors revealed that TC and BMI were positively associated with IHD mortality in both genders while Mg and T were negatively associated with IHD mortality. TC and sodium (Na) were negatively and positively associated with stroke mortality, respectively. 24-UT was negatively associated with stroke mortality. These five diet-related factors explained 61 and 49% of IHD and stroke variances in male, 63 and 36% of IHD and stroke variances in female, respectively.

Abbreviations *T*, taurine; *Na*, sodium; *Mg*, magnesium; *CVD*, cardiovascular diseases; *IHD*, ischemic heart diseases

2.1 Introduction

Taurine (T) is abundant in the seafood consumed in large quantities by the Japanese, who are presently enjoy the longest life expectancy in the world, with a life expectancy of 86 among females and 79 among males, the latter which rates

Y. Yamori (✉)
Mukogawa Women's University, Japan

J. Azuma et al. (eds.), *Taurine 7*, Advances in Experimental Medicine and Biology 643,
DOI 10.1007/978-0-387-75681-3_2, © Springer Science+Business Media, LLC 2009

second in the world. Since various experimental studies, as well as epidemiological evidence from our world-wide cooperative study on nutrition and cardiovascular disease (CVD), suggest the importance of T in reducing the risk of CVD, it is possible that T is a food factor that contributes to Japanese longevity.

2.2 Basic Studies of T Effect on CVD Risks

2.2.1 T Effect on Hypertensive Models

Basic research on hypertension and stroke has made remarkable progress since the establishment of rat models of hypertension and stroke, the spontaneously hypertensive rat (SHR) (Okamoto and Aoki 1963) and the stroke-prone SHR (SMRSP) (Okamoto et al. 1974; Yamori 1984). These animal models develop severe hypertension and die from hemorrhagic and ischemic stroke, making these useful in studying the pathogenesis, prevention and treatment of hypertension and stroke (Yamori 1981; Yamori et al. 1987).

The effect of fish protein-rich diet on stroke prevention was first demonstrated by Yamori et al. (Yamori 1981; Yamori et al. 1987). When SHRSP were fed a normal or low protein diet maintained on drinking water containing 1% salt, they quickly developed severe hypertension and all died of stroke within a short period. Without excess salt, 80% of them died of stroke. In contrast, when fed a high protein fish diet with excess salt, the incidence of stroke was markedly reduced. And in the group fed a high protein fish diet without excess salt intake, the development of severe hypertension was attenuated and no stroke was observed. Thus, it can be concluded that a high protein fish diet attenuates the development of severe hypertension and counteracts the adverse effect of salt.

Of the amino acids in fish and protein that could attenuate the development of severe hypertension and counteract the adverse effect of salt in SHR and SHRSP were the sulfur amino acids, T and methionine (Yamori 1981; Yamori et al. 1987; Nara et al. 1978). T supplementation also prevented the development of hypertension in DOCA-salt hypertensive rats (Sato et al. 1991) and suppressed the elevation in plasma epinephrine and norepinephrine levels, which is likely one of the possible mechanisms underlying the anti-hypertensive actions of T.

2.2.2 Physiological Effect and Distribution of T

Although a simple sulfur amino acid, T has been experimentally found to exert various effects (Huxtable 1992, 2000), such as an antihypertensive effect through central suppression of sympathetic tone, a hypocholesterolemic effect through activation of hepatic 7α-hydroxylase activity to accelerate cholesterol excretion into bile acids and an antiatherogenic effect possibly through the scavenging of hypochlorous acid and formation of T chloramines (Jerlich et al. 2000).

There are species differences in T synthesis with synthesis being particularly poor in humans and cats (Huxtable 1992). In newborn humans, T is considered to be an essential amino acid, since the potential to synthesize T is limited (Huxtable 2000).

In men, T in mainly obtained from fish and seafood, which contain large amounts of T compared to meat (Tsuji and Yano 1984) and are eaten customarily by the Japanese.

2.3 Epidemiological Survey of T Effect on CVD Risks and Mortality

2.3.1 Food Culture and Lifespan in Various Populations

In order to prove whether or not dietary components, such as T, are important in preventing hypertension and atherosclerosis in humans, as well as in animal models, Yamori introduced the idea of performing a world-wide epidemiological study to WHO in 1982. The CARDIAC Study is the acronym of *Car*diovascular *Di*seases and *A*limentary *C*omparison Study and the study was designed to investigate the relation of biological markers of diet with hypertension in"Core Study"and with CVD mortalities in "Complete Study" (WHO and WHO Collaborating Centers 1986; Yamori, 1981, 1989, 2006; Yamori et al. 1990, 2006).

This epidemiological survey has been carried out over the past 20 years in 61 populations in 25 countries (Fig. 2.1). About 100 males and 100 females in ages ranging from 48 to 56 were randomly selected, with the number of participants being over 14,000 in all. Some study sites were revisited at 10 year intervals for a follow-up health survey (MONALISA study; *Mon*eo *Ali*mentationis *Sa*nae = Reminding healthy food), and we noted the populations marked with large clear dots in Fig. 2.1

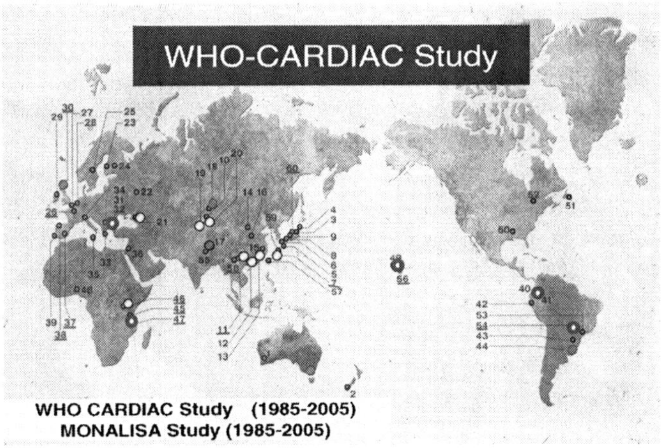

Fig. 2.1 Distribution of study sites in the world, 61 in total, for the WHO-coordinated Cardiovascular Diseases and Alimentary Comparison (WHO-CARDIAC) Study

had lesser risks of lifestyle-related diseases, as well as longer life-expectancy, such as Okinawans in the late 20th century. In contrast, the populations marked with large black dots, had higher risks and shorter life-expectancy. In some populations encircled with large black circles exhibited increased risk factors which shortened their life-expectancy due to the deterioration of their traditional food patterns, such as eating seafood.

2.3.2 Fish Intake and Urinary T Excretion

The CARDIAC study used the world-wide distribution of 24-UT excretion as an index of seafood intake (Yamori et al. 1996). T intake estimated by 24-UT (μmol/day) is high, particularly among the Japanese population (male, 1700–3300) and moderately high among the Mediterranean population (male, 1300–1700). The distribution among the Chinese population is variable, being lowest among Tibetans (male, 350) and among the Uygurs living in Oasis, but moderately high among the southern Chinese population living near the sea (Guangzhou, male, 1500).

The frequency analysis of fish consumption (x; times a week) was positively related with 24-UT excretion (y; μmol/day) [$y=277.3x+302.6$, $r=0.723$, $r^2=0.522$, $p<0.001$], indicating that T is a reliable bio-marker of fish intake. These data also indicate that Japanese and Mediterranean populations customarily eat seafood and consume considerable T. This may be associated with the epidemiological finding that both Japanese and Mediterranean foods are cardioprotective.

2.3.3 Low CVD Risks in Fish Eaters

Our CARDIAC Study populations can be classified into 2 groups, fish eaters and non-fish eaters according to the cut off level of 24-U taurine value of 1000 μmoles in males and 800 μmoles in females. We noted fish eaters had a lower prevalence of obesity, hypertension and hypercholesterolemia in comparison with that of non-eaters (data are not shown).

Furthermore, the indices of obesity, body mass index (BMI) (24.3 ± 0.1 in male (m), 24.4 ± 0.1 in female (f)), the grades of hypertension (systolic blood pressure (SBP) 124.3 ± 0.8 mmHg in m, 120.5 ± 0.8 mmHg in f, as well as hypercholesterolemia (total cholesterol (TC); 175.4 ± 1.5 mg/dl in m, 184.3 ± 1.5 mg/dl in f) were significantly lower in fish eaters than in non-fish eaters (BMI; 26.1 ± 0.1 in m, 26.7 ± 0.1 in f, SBP; 128.9 ± 0.6 in m, 128.7 ± 0.7 in f. TC; 201.3 ± 1.3 in m, 200.8 ± 1.5 in f).

2.3.4 T Effect on BP in Non-Fish Eaters

Since Tibetans were consuming too much salt and eating virtually no fish because of their strict religious discipline, the prevalence of hypertension was high, with

40% of the people aged 48 to 56 being hypertensive, which is nearly twice the world average of 20%. Moreover, the grades of hypertension were severe, many over 200 mmHg.

We, therefore, carried out an intervention study for Tibetans living in Namche Bazaar, which is located at the foot of Mt. Everest. After completing the WHO-CARDIAC health survey, volunteers with relatively high blood pressure were administered 3 g of T a day, 1 g of T per meal added to the tea. Their systolic and diastolic BP was significantly reduced in 2 months (Yamori et al. 1996).

2.3.5 *Low Fish Intake and CVD Risks in Japanese Brazilians*

Among the 61 populations examined were Okinawans living in Okinawa, Hawaii and Brazil in the 1990s. CVD risks were the lowest in Okinawans living in Okinawa, whose life expectancy was the longest in Japan and thus in the world (Yamori 2006). However, CVD mortality was higher and the life span was shorter in Japanese immigrants living in Brazil, indicating the importance of environmental factors in longevity (Mizushima et al. 1992). The most common Brazilian food is meat; Brazilians eat roasted meat seasoned with a lot of salt, with over 500 g of meat in one meal. Consequently, the prevalence of hypertension, hypercholesterolemia and obesity among Japanese immigrants living in Brazil were significantly higher than those living in Okinawa (Mizushima et al. 1992).

Among the Japanese immigrants living in Brazil, particularly in Campo Grande far from the sea coast, the frequency of fish intake was the lowest. They ate fish only once every 2 weeks and their life span was 17 years shorter than the Japanese living in Japan, an effect related to the very high IHD mortality rate. Low fish consumption and reduced T intake appeared to increase IHD.

2.4 T Effect on Hypercholesterolemia and Atherosclerosis

2.4.1 *Experiments in SHRSP*

The effects of T on hyperlipidemia and atherosclerosis were investigated experimentally in various animal models, particularly in SHRSP developing hypercholesterolemia and arterial fat deposits in response to high-fat cholesterol diet feeding (Yamori et al. 1975). Taurine supplementation attenuated the elevation of TC in SHRSP fed a high-fat-cholesterol diet (Murakami et al. 1996a) and decreased fat deposits in the mesenteric artery which were visible within a few weeks after high cholesterol feeding (Murakami et al. 1996b). Thus, the cholesterol-lowering and anti-atheroscle-rotic effects of T were confirmed in the hypertensive rat models.

2.4.2 Mechanisms of Hypocholesterolemic Effect

The mechanisms by which T lowers liver and serum TC levels appears to be linked in part to the acceleration in bile acid production and cholesterol 7α-hydroxylase activity, the rate-limiting enzyme of bile acid synthesis (Yokogoshi et al. 1999). Indeed, an inverse correlation exists between mRNA levels of the cholesterol 7α-hydroxylase (CYP7A1) gene and serum TC levels, indicating that T-mediated enhancement of cholesterol 7α-hydroxylase gene expression stimulates bile acid synthesis from cholesterol and lowers cholesterol levels (Yokogoshi et al. 1999).

The effect of T on serum TC levels and liver LDL receptors was studied in hamsters, because of the similarities between hamsters and humans relative to serum TC levels and cholesterol metabolism (Murakami et al. 2002a). T supplementation attenuated the rise of serum atherogenic LDL and VLDL cholesterol induced by the ingestion of a high-fat diet.

The effect of T on hepatic LDL receptor activity was evaluated by monitoring the disappearance of radio-labeled LDL from the blood. Although no difference was noted in non-receptor-mediated uptake of methylated LDL by liver cells, receptor-mediated LDL uptake was accelerated in T-treated hamsters (Murakami et al. 2002a).

Therefore, it is concluded that T stimulates bile acid synthesis from cholesterol through the enhancement of cholesterol 7α-hydroxylase activity, leading to depletion of the hepatic free cholesterol pool and resulting in enhanced LDL clearance.

2.4.3 Effect of T on Atherosclerotic Lesions

The effect of T on the development of atherosclerosis was further studied in genetically hyperlipidemic models, such as Watanabe heritable hyperlipidemic (WHHL) rabbits (Murakami et al. 2002b), in which T supplementation (1% in drinking water) suppresses the development of atherosclerotic lesion formation, such as lipid-rich macrophage accumulation in the aorta.

Thiobarbituric acid reactive substances (TBARS) in the serum and aorta were decreased in T-treated rabbits. Furthermore, ex-vivo experiments demonstrated that LDL isolated from T-treated rabbits was more resistant to copper-induced oxidation. Consequently, the formation of oxidized LDL was significantly suppressed in the T-treated rabbits (Murakami et al. 2002b).

The effects of T on cholesterol metabolism were different in various animal models, but the anti-atherosclerotic effects of T seemed to be universal (Kondo et al. 2001; Murakami et al. 1996a, b, 2002a, b; Yokogoshi et al. 1999), suggesting that the antioxidant actions of T might be involved in its anti-atherosclerotic effects. The important role of oxidized LDL in the pathogenesis and progression of atherosclerotic lesions has been well established (Penn and Chisoim 1994). Oxidized LDL itself is cytotoxic to endothelial cells, and accelerates the proliferation and the migration of smooth muscle cells. Moreover, excess uptake of oxidized LDL by macrophages results in the formation of lipid-laden foam cells. Oxidized

LDL also modulates the production of cytokines through the endothelial scavenger receptor, LOX-1 (Alexander 1995). Recently, T has been shown to suppress the upregulation of LOX-1 in the kidney of salt-loaded Dahl salt-sensitive rats and to normalize the salt-induced increase in 8-hydroxy-deoxy guanosine content (Chiba et al. 2002), confirming further the importance of the antioxidant effect of T against atherosclerosis.

2.5 Epidemiological Evidence for Anti-Atherosclerotic Nutrients

2.5.1 T as Putative Preventive Nutrient Against IHD

Our world-wide epidemiological study was the first to demonstrate a significant inverse relationship between 24-UT excretion and IHD in males (Fig. 2.2) and in females (Yamori 2006; Yamori et al. 1992, 1996, 2001, 2006) suggesting the importance of T found in the rich seafood diet in preventing IHD. The populations in Fig. 2.2 with the highest T excretion and the lowest IHD mortality are all Japanese populations who also enjoy the longest life expectancy. The Mediterranean populations were second to the Japanese in these parameters. From these data, the dietary goal for the prevention of IHD is the maintenance of sufficient T intake to yield a urinary T excretion rate of over 2000 μmole per day.

2.5.2 T Intake, IHD and Longevity

Japanese, particularly Japanese women enjoy the longest life span in the world, including the lowest mortality rate for IHD, all of which may be ascribed to the highest T intake in the form of seafood.

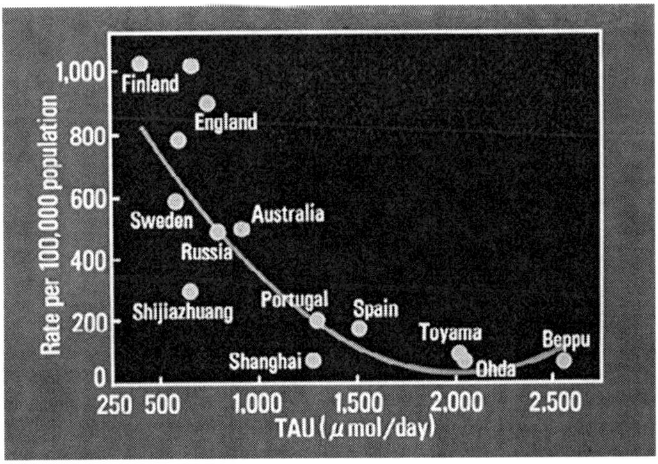

Fig. 2.2 Association between 24-hour urinary taurine (TAU) excretion (x) and age-adjusted mortality rates of IHD (y) in males ($y = 2.95 \times 10^{-4}\chi^2 - 1.21\chi + 1256$, $r = 0.759$, $p = 0.0011$)

Our CARDIAC Study indicates that the average life expectancy is inversely associated with the age-adjusted IHD mortality rate (Yamori 2006). The Japanese populations with the lowest IHD mortality rates had the longest life expectancy of the developed countries. The French population in Orlean had the second longest life expectancy and the second lowest IHD mortality rate.

Although serum TC levels are positively associated with the IHD mortality rate (Yamori 2006; Yamori et al. 2006) a large discrepancy in the association is noted among Scottish and French people. They exhibit a large variation in the intake of vegetables and fruit containing antioxidants, as well as the intake of T from seafood, the latter which is far greater in French than in English, particularly Scottish people (Fig. 2.2). These data indicate the importance of antioxidants in the prevention of IHD.

2.5.3 Diet-Related Factors and IHD

Serum TC levels and the intake of T are positively and inversely related with IHD mortality, respectively. Consequently, we applied structural equation modeling to the male data of the CARDIAC Study populations (Fig. 2.3) (Liu and Yamori 2007). The results help clarify the effects of 5 diet-related factors, namely, TC, BMI,

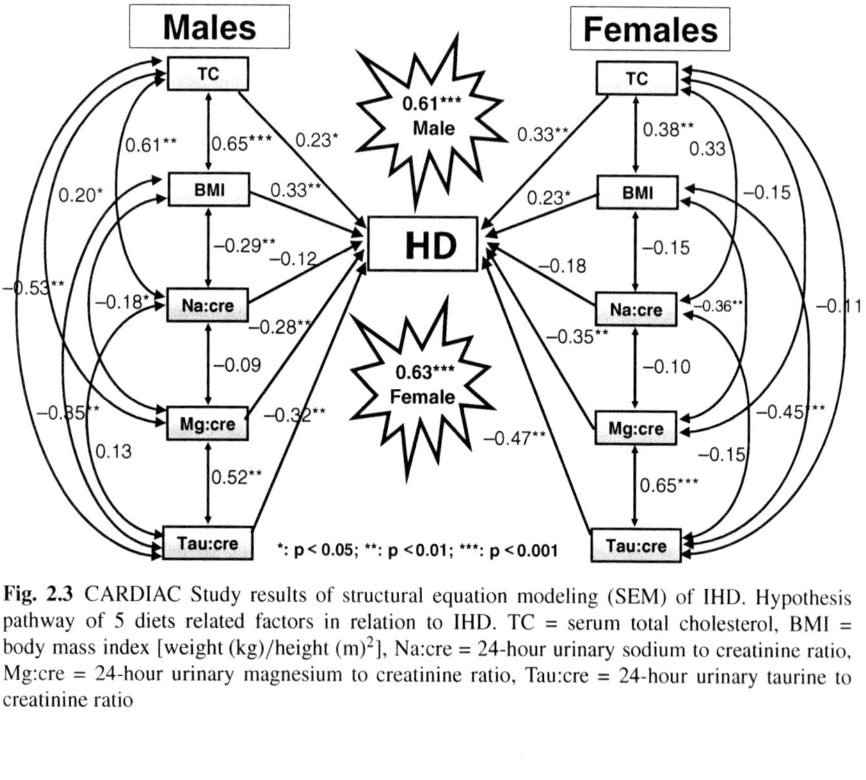

Fig. 2.3 CARDIAC Study results of structural equation modeling (SEM) of IHD. Hypothesis pathway of 5 diets related factors in relation to IHD. TC = serum total cholesterol, BMI = body mass index [weight (kg)/height (m)^2], Na:cre = 24-hour urinary sodium to creatinine ratio, Mg:cre = 24-hour urinary magnesium to creatinine ratio, Tau:cre = 24-hour urinary taurine to creatinine ratio

sodium (Na) to creatinine ratio, magnesium (Mg) to creatinine ratio and T to creatinine ratio in relation to IHD. Each regression coefficient indicates the effect of changing each factor on IHD mortality. For example, an increase in TC by one standard deviation, elevated IHD mortality by 0.33 while an increase in the T to creatinine ratio by one standard deviation reduces IHD mortality by 0.47 standard deviations in females. Thus, the T to creatinine ratio is inversely related with TC, the major risk factor of IHD and positively related with the Mg to creatinine ratio because seafood contains T, as well as Mg. Both were proven to be beneficial dietary factors against IHD. The squared multiple standard regression coefficient equals 0.61 and 0.63 for males and females respectively. It indicates that these 5 diet-related factors accountted for 61% and 63% of IHD variance in the male and female populations, respectively. Among the 5 diet-related factors T is the most potent factor in reducing IHD mortality. It's also interesting that the T to creatinine ratio is inversely related to BMI, confirming the above-mentioned finding that the populations with higher 24-UT excretion are less obese and on average showed lower BMI values.

Moreover, a recent Japanese study reported that T is produced in small-size adipocytes but its production decreases in fat-containing hypertrophic adipocytes (Tsuboyama-Kasaoka et al. 2006). Dietary T supplementation increases the mobilization of fat and basal energy expenditure. Thus, T helps prevent high fat diet-induced obesity. This experimental finding is consistent with our epidemiological data and suggests that T might attenuate obesity.

2.5.4 *T and Heart Rate*

Further analysis of the epidemiological data demonstrate that fish eaters, whose 24-UT excretions were over 1000 μmoles in males and 800 μmoles in females, had significantly lower BP and also significantly slower heart rates (68.1 ± 0.4 in m, 71.4 ± 0.4 in f) than non-fish eaters (70.0 ± 0.4 in m, 74.1 ± 0.4 in f). Population studies in the US (Dyer et al. 1980; Palatini and Julius 1997), as well as an extensive cohort study in Japan (Okamura et al. 2004) showed that heart rates are inversely related with life expectancy in the elderly. Therefore, more T intake may contribute to longevity by slowing down heart rate.

2.5.5 *Five Diet-related Factors and Stroke*

We first noted the beneficial effect of T against stroke in SHRSP (Yamori 1981, 1984; Yamori et al. 1987). The results of our structural equation modeling for males and females relating to stroke among CARDIAC populations demonstrated the clear adverse effect of Na on stroke (Fig. 2.4). As Na goes up one standard deviation, it is estimated that stroke mortality goes up 0.49 and 0.35 standard deviations in males and females, respectively. In contrast to Na, when the T to creatinine ratio

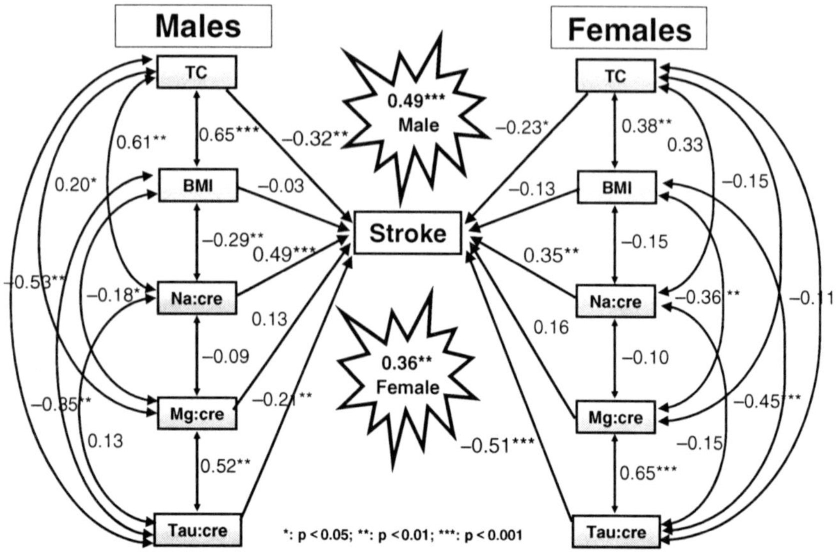

Fig. 2.4 CARDIAC Study results of the structural equation modeling (SEM) for stroke. Hypothesis pathway of 5 diet-related factors in relation to stroke. Abbreviations are explained in the legend for Fig 2.3

goes up one standard deviation, it is estimated that stroke mortality goes down 0.21 and 0.51 standard deviations in males and females, respectively. These five diet-related factors explain 49 and 36% of the variability in stroke among males and females, respectively. These data allow us to speculate that T is protective against both IHD and stroke, the mortality rates of which are inversely related to average life expectancy. Therefore, we conclude that T is a food factor that contributes to longevity.

The role of T in the prevention of stroke may be ascribed to its anti-inflammatory action (Schuller-Levis and Park 2004). Recent epidemiological studies indicate that C-reactive protein (CRP), the marker of inflammation, is closely related not only to IHD but also to stroke. Moreover, since there is early involvement of macrophages in the development of cerebrovascular lesions in SHRSP and humans (Tagami et al. 1993), CRP may be regarded as a predictive marker for stroke (Ballantyne 2005). T may be beneficial by producing T chloramines and attenuating oxidative stress arising from the inflammatory process.

2.6 Conclusion

T has various physiological and pharmacological effects on lifestyle-related diseases (Jacobsen and Smith 1968). Such beneficial effects can be attributed to its basic physiological actions, including osmoregulation, antioxidant/membrane stabilization and calcium regulation, for which there has been abundant research done internationally.

Our data based on experimental and epidemiological studies support the hypothesis that T intake beneficially affects IHD and stroke, the major outcome of life style-related diseases, through physiological effects on cell viability and function.

Therefore, an adequate level of T inside the body may be important for reducing the risk of IHD and stroke and for the prevention of lifestyle-related diseases.

Acknowledgments The CARDIAC Study was partially supported by grants-in-aid from the Japanese government and a WHO donation through the Japan Heart Foundation, Tokyo, Japan from Otsuka Pharmaceutical Co., Tokyo, Taisho Pharmaceutical Co., Ltd., Tokyo and other companies and also from about 300,000 individuals in total since 1983 through Japan Vascular Disease Research Foundation, Kyoto, Japan to International (former WHO-Collaborating) Center for Research on Primary Prevention of Cardiovascular Diseases, Kyoto, Japan.

We are grateful for the research fund from International Medical Center of Japan, and for the cooperation of medical professionals, public health workers and researchers as well as volunteer participants in the health survey at 61 CARDIAC Study sites in the world.

References

Alexander RW (1995) Theodore Cooper Memorial Lecture: Hypertension and the pathogenesis of atherosclerosis, oxidative stress and the mediation of arterial inflammatory response:a new perspective. Hypertension 25:155–161

Ballantyne CM, Hoogeveen RC, Bang H, Coresh J, Folsom AR, Chambless LE, Myerson M, Wu KK, Sharrett AR, Boerwinkle E (2005) Lipoprotein-associated phospholipase A^2, high-sensitivity c-reactive protein, and risk for incident ischemic stroke in middle-aged men and women in the atherosclerosis risk in communities (ARIC) study. Arch Intern Med 165: 2479–2484

Chiba Y, Ando K, Fujita T (2002) The protective effects of taurine against renal damage by salt loading in Dahl salt-sensitive rats. J Hypertens 20:2269–2274

Dyer AR, Persky V, Stamler J, Paul O, Shekelle RB, Berkson DM, Lepper M, Schoenberger JA, Lindberg HA (1980) Heart rate as a prognostic factor for coronary heart disease and mortality: findings in three Chicago epidemiologic studies. Am J Epidemiol 112:736–749

Huxtable RJ (1992) Physiological actions of taurine. Physiol Rev 72:101–163

Huxtable RJ (2000) Expanding the circle 1975–1999:Sulfur biochemistry and insights on the biological functions of taurine. Adv Exp Med Biol 483:1–25

Jacobsen JG, Smith LH (1968) Biochemistry and physiology of taurine and taurine derivatives. Physiol Rev 48:424–511

Jerlich A, Fritz G, Kharrazi H, Hammel M, Tschabuschnig S, Glatter O, Schaur RJ (2000) Comparison of HOCl traps with myeloperoxidase inhibitors in prevention of low density lipoprotein oxidation. Biochem Biophys Acta 481:109–118

Kondo Y, Toda Y, Kitajima H, Oda H, Nagate T, Kameo K, Murakami S (2001) Taurine inhibits development of atherosclerotic lesions in apolipoprotein E-deficient mice. Clin Exp Pharmacol Physiol 28:809–815

Liu L, Yamori Y (2007) Gender difference of cardiovascular diseases as a clue for healthier again: Asian food. Food Culture: Development and Education UNESCO:78–83

Mizushima S, Moriguchi EH, Nakada Y, Gonzalez B, Nara Y, Murakami K, Horie R, Moriguchi Y, Mimura G, Yamori Y (1992) The relationship of dietary factors to cardiovascular diseases among Japanese in Okinawa and Japanese immigrants, originally from Okinawa, in Brazil. Hypertens Res 15:45–55

Murakami S, Yamagishi I, Asami Y, Ohta Y, Toda Y, Nara Y, Yamori Y (1996a) Hypolipidemic effect of taurine in stroke-prone spontaneously hypertensive rats. Pharmacology 52:303–313
Murakami S, Nara Y, Yamori Y (1996b) Taurine accelerates the regression of hypercholesterolemia in stroke-prone spontaneously hypertensive rats. Life Sci 58:1643–1651
Murakami S, Kondo Y, Toda Y, Kitajima H, Kameo K, Sakono M, Fukuda N (2002a) Effect of taurine on cholesterol metabolism in hamsters: Up-regulation of low density lipoprotein (LDL) receptor by taurine. Life Sci 70:2355–2366
Murakami S, Kondo Y, Sakurai T, Kitajima H, Nagate T (2002b) Taurine suppresses development of atherosclerosis in Watanabe heritable hyperlipidemic (WHHL) rabbits. Atheroschlerosis 163:79–87
Nara Y, Yamori Y, Lovenberg W (1978) Effects of dietary taurine on blood pressure in spontaneously hypertensive rats. Biochem Pharmacol 27:2689–2692
Okamoto K, Aoki K (1963) Development of a strain of spontaneously hypertensive rats. Jpn Circ J 27:202–293
Okamoto K, Yamori Y, Nagaoka A (1974) Establishment of the stroke-prone spontaneously hypertensive rat (SHR). Circ Res 34/35:143–153
Okamura T, Hayakawa T, Kadowaki T, Kita Y, Okayama A, Elliott P, Ueshima H, NIPPON-DATA80 Research Group (2004) Resting heart rate and cause-specific death in a 16.5-year cohort study of the Japanese general population. Am Heat J 147:1024–1032
Palatini P, Julius S (1997) Heart rate and the cardiovascular risk. J Hypertens 15:3–17
Penn MS, Chisoim GM (1994) Oxidized lipoproteins, altered cell function and atherosclerosis. Atherosclerosis 108:21–29
Sato Y, Ogata E, Fujita T (1991) Hypotensive action of taurine in DOCA-salt rats-involve-ment of sympathoadrenal inhibition and endogenous opiate. Jpn Circ J 55:500–508
Schuller-Levis GB, Park E (2004) Taurine and its chloramine: Modulators of immunity. Neurochem Res 29:117–126
Tagami M, Tsukada T, Kubota A, Nara Y, Yamori Y (1993) Immunocytochemical study of cerebral perforating arteries in patients with cerebral infactions. Acta Histochem Cytoshem 26: 109–115
Tsuboyama-Kasaoka N, Shozawa C, Sano K, Kamei Y, Kasaoka S, Hosokawa Y, Ezaki O (2006) Taurine (2-aminoethanesulfonic acid) deficiency creates a vicious circle promoting obesity. Endocrinology 147:3276–3284
Tsuji K, Yano S (1984) Taurine/cholesterol ratio of well-consumed animal foods. Sulfur Amino Acids 7:249–255
WHO and WHO Collaborating Centers (1986) CARDIAC (Cardiovascular Diseases and Alimentary Comparison) Study Protocol. Shimane, Geneva
Yamori Y (1981) Environmental influences on the development of hypertensive vascular diseases in SHR and related models, and their relation to human disease. In: Worcel M et al. (eds). New Trends in Arterial Hypertension (INSERM Symposium No. 17). Elsevier, Amsterdam: 305–320
Yamori Y (1984) The stroke-prone spontaneously hypertensive rat: Contribution to risk factor analysis and prevention of hypertensive diseases. In: de Jong W (ed). Handbook of Hypertension. Elsevier, Amsterdam:240–255
Yamori Y (1989) Hypertension and biological dietary markers in urine and blood: A progress report from the CARDIAC study group. In: Yamori Y and Strasser T (eds). New Horizons in Preventing Cardiovascular Diseases. Elsevier, Amsterdam:111–126
Yamori Y (2006) Food factors for atherosclerosis prevention. Asian perspective derived from analysis of world wide dietary biomarkers. Exp Clin Cardiol 11:94–98
Yamori Y, Hamashima Y, Horie R, Handa H, Sato M (1975) Pathogenesis of acute arterial fat deposition in spontaneously hypertensive rats. Jpn Circ J 39:601–609
Yamori Y, Horie R, Nara Y, Tagami M, Kihara M, Mano M, Ishino H (1987) Pathogenesis and dietary prevention of cerebrovascular diseases in animal models and epidemiological evidence for the applicability in man. In: Yamori Y and Lenfant C (eds). Prevention of Cardiovascular Diseases, an Approach to Active Long Life. Elsevier, Amsterdam:163–177

Yamori Y, Nara Y, Mizushima S, Mano M, Sawamura M, Kihara M, Horie R, Hatano S (1990) International cooperative study on the relationship between dietary factors and blood pressure: A report from the cardiovascular diseases and alimentary comparison study. J Cardiovas Pharmacol 16:S43–S47

Yamori Y, Nara Y, Mizushima S, Murakami S, Ikeda K, Sawamura M, Nabika T, Horie R (1992) Gene-environment interaction in hypertension, stroke and atherosclerosis in experimental models and supportive findings from a world-wide cross-sectional epidemiological survey: A WHO-CARDIAC Study. Clin Exp Pharmacol Physiol 20:43–52

Yamori Y, Nara Y, Ikeda K, Mizushima S (1996) Is taurine a preventive nutritional factor of cardiovascular diseases or just a biological marker of nutrition. Adv Exp Med Biol 403:623–629

Yamori Y, Liu L, Ikeda K, Miura A, Mizushima S, Miki T, Nara Y (2001) Distribution of twenty-four hour urinary taurine excretion and association with ischemic heart disease mortality in 24 populations of 16 countries: Results from the WHO-CARDIAC Study. Hypertens Res 24: 453–457

Yamori Y, Liu L, Mizushima S, Ikeda K, Nara Y and CARDIAC Study Group (2006) Male cardiovascular mortality and dietary markers in 25 population samples of 16 countries. J Hypertens 24:1499–1505

Yokogoshi H, Mochizuki H, Nanami K, Hida Y, Miyachi F, Oda H (1999) Dietary taurine enhances cholesterol degradation and reduces serum and liver cholesterol concentrations in rats fed a high-cholesterol diet. J Nutr 129:1705–1712

Chapter 3
Taurine-Mediated Cardioprotection is Greater When Administered upon Reperfusion than Prior to Ischemia

Tadaomi-Alfonso Miyamoto, Takayuki Ueno, Yoshihumi Iguro, Goichi Yotsumoto, Yoshihiro Fukumoto, Kazuo Nakamura, and Ryuzo Sakata

Abstract Taurine (TA) administered exogenously before the induction of myocardial ischemia decreases lactic acid production and increases pyruvic acid production during ischemia. It also preserves the activity of GOT, GPT, LDH and CPK during ischemia and enhances recovery of CKMB synthesis as early as 5 minutes after onset of reperfusion. The aim of the study was to determine the optimal conditions for administering TA in order to reduce myocardial ischemia-reperfusion injury. Left ventricular (LV) function, creatine kinase (CK) and lipid peroxide products (LPOP=oxidant stress), as well as the area at risk (AAR), and infarct size (IS) after reperfusion were studied in 3 groups of isolated rat hearts perfused with Krebs Henseleit Buffer (KHB)-stabilized isolated rat hearts that were subjected to 20 minutes(') of global ischemia at 37°C followed by 60′ of reperfusion with KHB: Hearts were perfused with TA containing KHB for 10′ just prior to ischemia or during the first 10' of reperfusion. *Conclusion:* Taurine before ischemia or during reperfusion was equally effective in preventing infarction; however, when administered at reperfusion, taurine reduced lipid peroxidation and myocardial injury more, thereby providing improved early recovery of function.

Abbreviations TA, taurine; *AAR*, area at risk; *IS*, infarct size; *LPO*, lipid peroxide; *CK*, creatine kinase

3.1 Introduction

The fundamental molecular mechanisms underlying taurine's cardioprotection have been largely uncovered using isolated cardiomyocytes subjected to hypoxia, where

---

T.-A. Miyamoto (✉)
Kokura Memorial Hospital, Kitakyushu, Fukuoka, Japan

J. Azuma et al. (eds.), *Taurine 7*, Advances in Experimental Medicine and Biology 643,
DOI 10.1007/978-0-387-75681-3_3, © Springer Science+Business Media, LLC 2009

taurine prevents apoptosis (Takahashi et al. 2003; Takatani et al. 2004a, b). Isolated rat hearts perfused with taurine containing buffer during the pre-ischemic phase and then subjected to normothermic ischemia exhibited minimal lactic acid production during ischemia, allowing recovery of aerobic metabolism as early as 5 minutes after onset of reperfusion (Oriyanhan et al. 2006a) In isolated rat hearts subjected to hypothermic ischemia for 360 minutes, taurine pretreatment completely prevented oxidative DNA damage, restored CKMB production as early as 5 minutes after onset of reperfusion and significantly improved LV functional recovery (Oriyanhan et al. 2006b). The free radical scavenging effects of taurine are well known (Franconi et al. 1985; Huxtable 1992; Kaplan et al. 1993; Milei et al. 1992; Öz et al. 1999; Wright et al. 1986), but which of these effects plays a role in myocardial protection is not known. Because taurine is extruded from the intracellular to the interstitial space by adenosine during CNS ischemia (Benveniste et al. 1984; Miyamoto and Miyamoto 1999), we hypothesized that taurine's antioxidant actions upon reperfusion might be greater than its role in preserving energy production or preventing DNA damage before or during ischemia. Calcium overload occurs primarily during reperfusion (Bagchi et al. 1997; Grace 1994; Lemasters et al. 1997) and induces alterations in mitochondrial function, enzyme release, disruption of the cell membrane, necrosis of tissue, reductions in contractility and increases in diastolic pressure (Akizuki et al. 1985; Lemasters et al. 1997).

Taurine also exerts anti-arrhythmic actions, which enables a fast return in normal electrical and mechanical activity of the hypothermically preserved heart (Oriyanhan et al. 2006b). Furthermore, it decreases the accumulation of calcium in the reoxygenated myocardium (Franconi et al. 1985), and protects against lipid peroxidation of the myocardial cell membrane (Huxtable 1992; Kaplan et al. 1993; Milei et al. 1992; Öz et al. 1999).

In spite of its cytoprotective effects, it is unclear whether taurine administered prior to ischemia or upon onset of reperfusion is most effective. The present study was aimed at elucidating the most effective condition of providing maximum cardioprotection.

3.2 Methods

Details are as reported elsewhere (Oriyanhan et al. 2006a; Ueno et al. 2007). The standard Langendorff rat heart preparation was used. Rat hearts were randomly assigned to one of three (ABC) groups after normothermic stabilization for 20 minutes with plain Krebs Henseleit Buffer (KHB) solution:

(A) *Global ischemia with no treatment (Control group; KHB $n = 8$).* Hearts were subjected to a 20-minute period of global ischemia and a 60-minute period of reperfusion with plain KHB.

(B) *Preischemic administration of taurine (Pre-TA group; KHB-preT $n = 8$).* Hearts were treated with continuous perfusion of KHB containing 10 mmol/L taurine for the last 10 minutes of stabilization just prior to induction of global ischemia, followed by a 20-minute period of global ischemia and a 60-minute period of reperfusion with plain KHB.

(C) *Postischemic administration of taurine (Post-TA group; KHB-post T $n = 8$).* Stabilized hearts were subjected to a 20-minute period of global ischemia, reperfused for the initial 10 minutes with KHB containing 10 mmol/L taurine, and then reperfused for the following 50 minutes with plain KHB.

3.2.1 Isovolumic Left Ventricular Function

A left ventricular balloon filled with fluid (to produce an end-diastolic pressure of 10 mmHg at the end of the stabilization period) was connected to a pressure transducer. The balloon was emptied during ischemia, but the removed fluid was reintroduced during reperfusion.

3.2.2 Biochemical Markers

Creatine kinase (CK) = myocardial damage Total CK leakage (expressed as IU/g dry heart weight) was assessed with a commercially available kit (Sigma-Aldrich, Poole, Dorset, UK) by spectrophotometric analysis of enzyme activity in the coronary effluent collected during reperfusion.

Lipid peroxide (LPO) = oxidative stress Total LPO leakage (expressed as nmol/g dry heart weight) was assessed in the coronary effluent collected during reperfusion with a commercially available kit (Determiner LPO, Kyowa Medicus, Tokyo, Japan) using chemiluminescence.

3.2.3 Myocardial Injury and Viability

Myocardial Area at Risk (AAR); At the end of the reperfusion period, 0.2 ml of 10% Evans blue dye was injected into the aortic cannula until the heart turned blue. Each heart was immediately removed from the Langendorff perfusion apparatus, weighed, frozen with liquid nitrogen, and stored in a freezer. The frozen hearts were then cut from the apex to the base into 4 transverse slices of equal thickness (2 mm). The AAR was defined as the region lacking Evans blue staining.

Infarct Size (IS); Slices were then incubated for 20 minutes at 37°C in a 1% solution of triphenyltetrazolium chloride (TTC) dissolved in isotonic phosphate buffer solution (pH 7.4). Slices were subsequently fixed in 10% formaldehyde for 6 hours. The dark-red stained viable myocardial tissue was easily distinguished from the pale unstained necrotic tissue. Infarct size and AAR were determined by computer morphometry using the Scion Image Beta 4.02 for Windows software (Scion Corporation, Frederick, MD). AAR and IS volumes were calculated by multiplying the area by the slice thickness. The AAR was shown as a percentage of the left ventricular (LV) area and the IS as a percentage of the AAR.

3.2.4 *Statistical Analysis*

Data are presented as means ± standard deviation. Between-group differences at baseline and 60 minutes after reperfusion were analyzed by oneway analysis of variance (ANOVA) followed by Newman-Keul,s and/or Tukey's test for multiple comparisons. Statistical significance was set at p less than 0.05(*) (** <0.01; *** <0.001). All statistical analyses were performed with PRISM GRAPH PAD ver 3.0 for MAC.

3.3 Results

Baseline ventricular functional parameters of the 3 experimental groups: mean left ventricular end-diastolic pressure (LVEDP), LV developed pressure (LVDP), heart rate (HR), rate pressure product (RPP), and LV maximal rate of ± pressure change ($\pm dP/dt_{max}$) were similar.

3.3.1 *Postischemic Ventricular Function*

Postischemic LVEDP measured 60 minutes after reperfusion was elevated in all hearts. However, the postischemic LVEDP of Pre-TA and Post-TA groups were significantly lower than that in the control group, but the differences between the Pre-TA group and Post-TA group were not significant (Fig. 3.1A). Postischemic LVDP: recovery of LVDP was significantly greater in the Post-TA group than in the control group or the Pre-TA group.

Pre-TA treatment provided no protection (Fig. 3.1B). Postischemic HR was similar in all 3 groups. Recovery of Postischemic RPP: the recovery of RPP was significantly greater in the Post-TA group than in the Pre-TA group; differences between the control group and Pre-TA group were not significant (Fig. 3.1C). Recovery of Postischemic ± dP/dtmax (Fig. 3.1D,E): recovery of ± dP/dtmax was significantly greater in the Post-TA group than in the control group and the Pre-TA group; differences were not significant between the control group and the Pre-TA group.

3.3.2 *Postischemic Biochemical Markers*

CK Leakage. Total CK leakage in the Post-TA group was significantly less than in the Pre-TA group and control group. Differences between the control group and Pre-TA group were also significant (Fig. 3.2A). *LPO Products.* Total LPO product

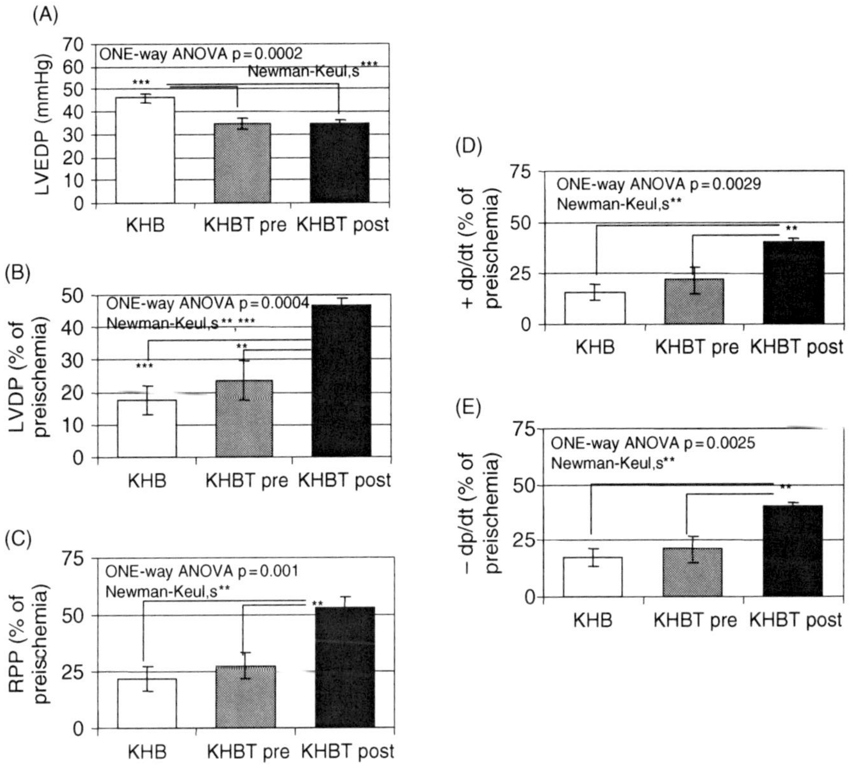

Fig. 3.1 Left ventricular function: (**A**) LVEDP: KHB>(KHBT-pre = KHBT-post); (**B**) LVDP: KHBT-post>KHBT-pre and KHB; (**C**) Rate Pressure Product: KHBTpost>KHBT-pre or KHB. (**D**) +dp/dt; (**E**) - dp/dt; Recovery of the rates of contraction and relaxation were greater in the KHBT-post group than in the KHBT-pre or KHB hearts. Values shown represent means ± SEM

in the Post-TA group was minimal and markedly lower than in the control group and the Pre-TA group (Fig. 3.2B).

3.3.3 Postischemic Myocardial Viability

Myocardial AAR, an index of myocardial ischemia, was significantly smaller in both TA groups than in the control group. However, myocardial AAR of the Post-TA group myocardial AAR was significantly smaller than that of the Pre-TA group (Fig. 3.2C). Myocardial IS, an index of irreversible myocardial injury, was significantly smaller in both TA groups than in the control group. Myocardial IS of the post-TA group was smaller than that of the Pre-TA group; however, when normalized relative to the AAR, the IS/AAR was similar between the two TA groups.

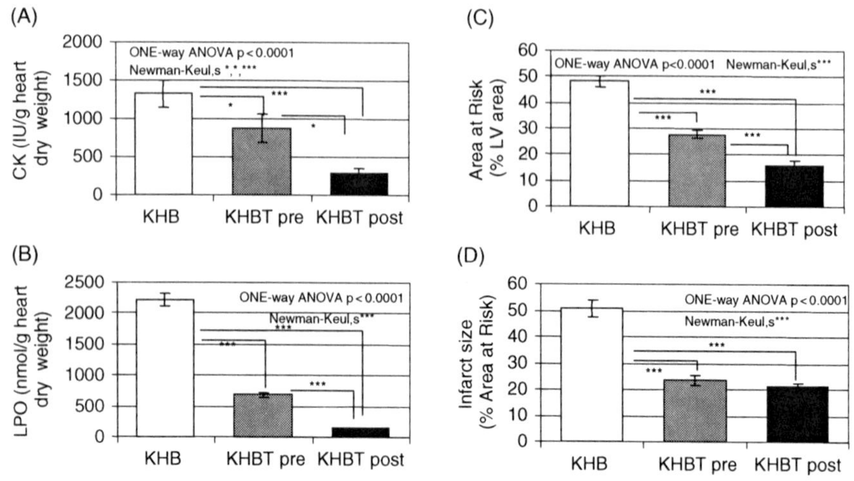

Fig. 3.2 Biochemical markers of injury and oxidative stress: (**A**) CK and (**B**) LPO of KHBT-post<(KHBT-pre<KHB) hearts after 60' reperfusion. Myocardial viability: (**C**) AAR of KHBT-post <(KHBT-pre <KHB); (**D**) IS of (KHBT-post=KHBT-pre) <KHB hearts. Values represent means ± SEM

3.4 Discussion

Reperfusion itself has been shown to initiate a chain of events that worsens damage; the combined damage has been termed ischemia/reperfusion (I/R) injury. The relative roles played by ischemia and reperfusion in determining this injury is not known, but our results suggest a greater detrimental effect of reperfusion. I/R injury is characterized by a unique histological pattern, with the formation of contraction bands; calcific granules within the mitochondria, and cell swelling with disruption of sarcoplasmic and mitochondrial membrane (Whalen et al. 1974). During ischemia, the increase in calcium in the cytosol activates proteases, which convert xanthine dehydrogenase to xanthine oxidase. Xanthine oxidase uses molecular oxygen available upon reperfusion to convert hypoxanthine to xanthine releasing superoxide in the process (Grace 1994; Grisham and Granger 1989). Reactive oxygen species cause lipid peroxidation of membrane polyunsaturated fatty acids especially arachidonic acid (Poli et al. 1987). Malondialdehyde (MDA) is a relatively stable end product of lipid peroxidation that is used to assess the degree of lipid peroxidation (Kaplan et al. 1993). Taurine deficiency increases MDA levels (Harada et al. 1989), but in taurine supplemented animals MDA levels are decreased (Kaplan et al. 1993; Öz et al. 1999). The extent of myocardial oxidative damage was evaluated by determining lipid peroxide (LPO) levels in the coronary effluent obtained during the reperfusion period and showed that taurine administration at the onset of reperfusion yielded lesser myocardial oxidative damage than taurine administration before ischemia.

The following sequence of events is proposed: Ischemia/reperfusion generates peroxynitrites, which convert the active octamer of mitochondrial creatine kinase (mitCK) into an inactive dimer (Beutner et al. 1998; Grisham and Granger 1989; Koufen et al. 1999; Stachowiak et al. 1998; Wallimann et al. 1998). In the presence of Ca^{++} the mitochondrial permeability transition pores open and oxidative phosphorylation ceases (Beutner et al. 1998; Grisham and Granger 1989; Koufen et al. 1999; Lemasters et al. 1997; Li et al. 1997, 1999; Stachowiak et al. 1998; Wallimann et al. 1998). Phosphotransfer from mitochondria to the nucleus is catalyzed by mitCK/AK3 (adenylate kinase) that relies mostly on AT∼P supplied by mitochondrial oxidative phosphorylation. However, when oxidative supplied phosrylation cases (Dzeja et al. 2002) AT∼P supplied by glycolysis is not sufficient to sustain nuclear integrity for extended periods and endonucleases are activated (Li et al. 1999). This leads to activation of p53 the apoptosis inducing molecule (Takahashi et al. 2003) which together with activated endonucleases and other cytosolic proteases that increase expression of the pro-apoptotic molecule BAD eventually causing DNA nicking and ultimately fragmentation (apoptosis) (Thatte et al. 2004). Tatsumi et al. (2003) demonstrated that the presence of AT∼P generated through glycolysis determines the extent of apoptosis in hypoxic cultured cardiomyocytes, with little necrosis occurring in the presence of AT∼P. Apoptosis in the liver is also an energy dependent phenomenon (Jaskille et al. 2006), and support the well-known fact that nuclear integrity depends on oxidative phosphorylation, as well as mitCK/AK3, which catalyzes phosphotransfer from the mitochondria to the nucleus. Köhler et al. (1999) demonstrated that in apoptotic cells AK2 that is localized to the mitochondrial intermembrane space, rather than the mitochondrial matrix is translocated into the cytosol concomitantly with cytochrome c in Jurkat leukemic cells.

Taurine is concentrated in the mitochondrial matrix (Hansen et al. 2006), as is the AK3 (Köhler et al. 1999) suggesting that could be directly involved in the regulation of glucose metabolism, as indicated by the interaction between taurine and pyruvate dehydrogenase phosphatase (Lombardini 1996), as well as higher GPT and pyruvic acid production (Oriyanhan et al. 2006a, b). Thus, it is hypothesized that taurine plays an important role in mitochondrial production of energy and transfer from the mitochondria to the nucleus, a process dependent on creatine-kinase (CK) activity (Beutner et al. 1998). When added before ischemia, TA preserves oxidative metabolism, prevents oxidative DNA damage (Oriyanhan et al. 2006a), attenuates myocardial apoptosis and preserves function.

In cultured cardiomyocytes taurine added to the culture-medium before induction of anoxia prevents apoptosis through several pathways: (a) inhibition of ischemia-induced activation of p53, an apoptosis initiator (Takahashi et al. 2003); (b) elevation in expression of the antiapoptotic protein, Bcl-2, and decrease in the proapoptotic protein, BAD (Takahashi et al. 2003), and (c) inhibition in caspase-3 activation (Takatani et al. 2004a,b) through a reduction in the formation of the Apaf-1/caspase-9 apoptosome (Takatani et al. 2004a). Even though taurine administration before or after ischemia reduces infarct size almost to the same extent, the fact that hearts recover more function when taurine was administered during reperfusion points to the improvement in lipid peroxidation and a reduction in oxidative stress (Wright et al. 1986). This suggests that reperfusion might be the major cause of I/R injury. The potent antioxidant property of taurine may be linked to its modulation of glutathione levels (Biasetti and Dawson 2002; Huxtable 1992; Oudit et al. 2004; Sevier and Kaiser 2002; Woo et al. 2003). Reduced glutathione (GSH) plays an important role in the cellular defenses against oxidative stress. GSH depletion and accumulation of oxidized glutathione (GSSG) occur in the heart during oxidative stress caused by increased cellular demand and lead to impaired cell function because of the shift in the redox state (Forgione et al. 2002; Hill and Singal 1997; Li et al. 2003). Although not determined, we surmise taurine administered early after reperfusion must prevent the decline in GSH levels and preserved the GSH/GSSG redox.

Because the protective mechanisms of taurine administered before ischemia or after reperfusion are different, we intuitively assume that the combined systemic administration of taurine, before ischemia to the donor, and at reperfusion to the recipient, will result in maximal protection during transplant of multi organs.

3.4.1 Study Limitations

The model lacks the detrimental effects of leukocytes. However, taurine exerts a powerful anti-inflammatory effect against hypochlorous acid generation by the leukocyte. Larger whole animal in-situ blood perfused models would be ideal, but because we have only amplified the natural taurine defense mechanisms against ischemia/reperfusion injury, similar results would be anticipated in larger animals (Miyamoto et al. 2006).

3.5 Conclusion

Administration of taurine during onset of reperfusion period hastens recovery of postischemic pump function of the isolated rat heart more than taurine administered prior to ischemia, as evidenced by the decrease in infarct size. Taurine-mediated myocardial protection seems to be secondary to the marked prevention of oxidative damage.

References

Akizuki S, Yoshida S, Chambers DE, Eddy LJ, Parmley LF, Yellon DM, Downey JM (1985) Infarct size limitation by the xanthine oxidase inhibitor, allopurinol, in closed chest dogs with small infarcts. Cardiovasc Res 19:686–692

Bagchi D, Wetscher GJ, Bagchi M, Hinder PR, Perdikis G, Stohs SJ, Hinder RA, Das DK (1997) Interrelationship between cellular calcium homeostasis and free radical generation in myocardial reperfusion injury. Chem Biol Interact 104:65–85

Benveniste, H, Drejer J, Schousboe A, Diemer NH (1984) Elevation of extracellular concentrations of glutamate and aspartate in rat hippocampus during transient cerebral ischemia monitored by intracerebral microdialysis. J Neurochem 43:1369–1374

Beutner G, Rück A, Riede B, Brdiczka D (1998) Complexes between porin, hexokinase, mitochondrial creatine kinase and adenylate translocator display properties of the permeability transition pore. Implication for regulation of permeability transition by the kinases. Biochim Biophys Acta 1368:7–18

Biasetti M, Dawson R Jr (2002) Effects of sulfur containing amino acids on iron and nitric oxide stimulated catecholamine oxidation. Amino Acids 22:351–368

Dzeja PP, Bortolon R, Perez-Terzic C, Holmuhamedov EL, Terzic A (2002) Energetic communication between mitochondria and nucleus directed by catalyzed phosphotransfer. Proc Natl Acad Sci USA 99:10156–10161

Forgione MA, Cap A, Liao R, Moldovan NI, Eberhardt RT, Lim CC, Jones J, Goldschmidt-Clermont PJ, Loscalzo J (2002) Heterozygous cellular glutathione peroxidase deficiency in the mouse: abnormalities in vascular and cardiac functionand structure. Circulation 106: 1154–1158

Franconi F, Stendardi I, Failli P, Matucci R, Baccaro C, Montorsi L, BandinelliR, Giotti A (1985) The protective effects of taurine on hypoxia (performed in the absence of glucose) and on reoxygenation (in the presence of glucose) in guineapig heart. Biochem Pharmacol 34: 2611–2615

Grace PA (1994) Ischaemia-reperfusion injury. Br J Surg 81:637–647

Grisham MB, Granger DN (1989) Metabolic sources of reactive oxygen metabolites during oxidant stress and ischemia with reperfusion. Clin Chest Med 10:71–81

Hansen SH, Andersen ML, Birkedal H, Cornett C, Wibrand F (2006) The important role of taurine in oxidative metabolism. Adv Exp Med Biol 583:129–135

Harada H, Cusack BJ, Olson RD, Stroo W, Azuma J, Hamaguchi T, Schaffer SW (1990) Taurine deficiency and doxorubicin: interaction with the cardiac sarcolemmal calcium pump. Biochem Pharmacol 39:745–751

Hill MF, Singal PK (1997) Right and left myocardial antioxidant responses during heart failure subsequent to myocardial infarction Circulation 96:2414–2420

Huxtable RJ (1992) Physiological actions of taurine. Physiol Rev 72:101–163

Jaskille A, Koustova E, Rhee P, Britten-Webb J, Chen H, Valeri CR, Kirkpatrick JR, Alam HB (2006). Hepatic apoptosis after hemorrhagic shock inrats can be reduced through modifications of conventional Ringer's solution. J Am Coll Surg 202:25–35

Kaplan B, Aricioglu A, Erbas D, Erbas S, Turkozkan N (1993) The effects of taurine on perfused heart muscle malondialdehyde levels. Gen Pharmacol 24:1411–1413

Köhler C, Gahm A, Noma T, Nakazawa A, Orrenius S, Zhivotovsky B (1999) Release of adenylate kinase 2 from the mitochondrial intermembrane space during apoptosis. FEBS Lett 447:10–12

Koufen P, Rück A, Brdiczka D, Wendt S, Wallimann T, Stark G (1999) Free radical-induced inactivation of creatine kinase: influence on the octameric and dimeric states of the mitochondrial enzyme (Mib-CK). Biochem J 344 Pt 2:413–417

Lemasters JJ, Nieminen AL, Qian T, Trost LC, Herman B (1997) The mitochondrial permeability transition in toxic, hypoxic and reperfusion injury. Mol Cell Biochem 174:159–165

Li PA, Liu GJ, He QP, Floyd RA, Siesjö BK (1999) Production of hydroxyl free radical by brain tissues in hyperglycemic rats subjected to transient forebrain ischemia. Free Radic Biol Med 27:1033–1040

Li PA, Uchino H, Elmér E, Siesjö BK (1997) Amelioration by cyclosporin A of brain damage following 5 or 10 min of ischemia in rats subjected to preischemic hyperglycemia. Brain Res 753:133–140

Li S, Li X, Rozanski GJ (2003) Regulation of glutathione in cardiac myocytes. J Mol Cell Cardiol 35:1145–1152

Lombardini JB (1996) Taurine depletion in the intact animal stimulates in vitro phosphorylation of an approximately 44-kDa protein present in the mitochondrial fraction of the rat heart. J Mol Cell Cardiol 28:1957–1961

Milei J, Ferreira R, Llesuy S, Forcada P, Covarrubias J, Boveris A (1992) Reduction of reperfusion injury with preoperative rapid intravenous infusion of taurine during myocardial revascularization. Am Heart J 123:339–345

Miyamoto TA, Miyamoto KJ (1999) Does adenosine release taurine in the A1 receptor-rich hippocampus? J Anesth 13:94–98
Miyamoto TA, Miyamoto KJ, Miyamoto MR (2006) Systemically administered taurine Part III. Pharmacologically activated mechanisms. Adv Exp Med Biol 583:335–351
W, Miyamoto TA, Yamazaki K, Miwa S, Takaba K, Ikeda T, Komeda M (2006a) Regionally perfused taurine. Part I. Minimizes lactic acidosis and preserves CKMB and myocardial contractility after ischemia/reperfusion. Adv Exp Med Biol 583:271–278
Oriyanhan W, Miyamoto TA, Yamazaki K, Miwa S, Takaba K, Ikeda T, Komeda M (2006b) Regionally perfused taurine. Part II Taurine addition to St Thomas solution prevents DNA oxidative stress and maintains contractile function. Adv Exp Med Biol 583:279–288
Oudit GY, Trivieri MG, Khaper N, Husain T, Wilson GJ, Liu P, Sole MJ, Backx PH (2004) Taurine supplementation reduces oxidative stress and improves cardiovascular function in an iron-overload murine model. Circulation 109:1877–1885
Öz E, Erbas D, Gelir E, Aricioglu A (1999) Taurine and calcium interaction in protection of myocardium exposed to ischemic reperfusion injury. Gen Pharmacol 33(2):137–141
Poli G, Albano E, Dianzani MU (1987) The role of lipid peroxidation in liver damage. Chem Phys Lipids 45:117–142
Sevier CS, Kaiser CA (2002) Formation and transfer of disulphide bonds in living cells. Nat Rev Mol Cell Biol 3:836–847
Stachowiak O, Dolder M, Wallimann T, Richter C (1998) Mitochondrial creatine kinase is a prime target of peroxynitrite-induced modification and inactivation. J Biol Chem 273:16694–16699
Takahashi K, Ohyabu Y, Takahashi K, Solodushko V, Takatani T, Itoh T, Schaffer SW, Azuma J (2003) Taurine renders the cell resistant to ischemiainduced injury in cultured neonatal rat cardiomyocytes. J Cardiovasc Pharmacol 41:726–733
Takatani T, Takahashi K, Uozumi Y, Shikata E, Yamamoto Y, Ito T, Matsuda T, Schaffer SW, Fujio Y, Azuma J (2004a) Taurine inhibits apoptosis by preventing formation of the Apaf-1/caspase-9 apoptosome. Am J Physiol Cell Physiol 287:C949–953
Takatani T, Takahashi K, Uozumi Y, Matsuda T, Ito T, Schaffer SW, Fujio Y, Azuma J (2004b) Taurine prevents the ischemia-induced apoptosis in cultured neonatal rat cardiomyocytes through Akt/caspase-9 pathway. Biochem Biophys Res Commun 316:484–489
Tatsumi T, Shiraishi J, Keira N, Akashi K, Mano A, Yamanaka S, Matoba S, Fushiki S, Fliss H, Nakagawa M (2003) Intracellular ATP is required for mitochondrial apoptotic pathways in isolated hypoxic rat cardiac myocytes. Cardiovasc Res 59:428–440
Thatte HS, Rhee JH, Zagarins SE, Treanor PR, Birjiniuk V, Crittenden MD, Khuri SF (2004) Acidosis-induced apoptosis in human and porcine heart. Ann Thorac Surg 77:1376–1383
Ueno T, Iguro Y, Yotsumoto G, Fukumoto Y, Nakamura K, Miyamoto TA, Sakata R (2007) Taurine at early reperfusion significantly reduces myocardial damage and preserves cardiac function in the isolated rat heart. Resuscitation 73:287–295
Wallimann T, Dolder M, Schlattner U, Eder M, Hornemann T, O'Gorman E, Rück A, Brdiczka D (1998) Some new aspects of creatine kinase (CK): compartmentation, structure, function and regulation for cellular and mitochondrial bioenergetics and physiology. Biofactors 8:229–234
Whalen DA Jr, Hamilton DG, Ganote CE, Jennings RB (1974) Effect of a transient period of ischemia on myocardial cells. I. Effects on cell volume regulation. Am J Pathol 74:381–397
Woo HA, Chae HZ, Hwang SC, Yang KS, Kang SW, Kim K, Rhee SG (2003) Reversing the inactivation of peroxiredoxins caused by cysteine sulfinic acid formation. Science 300:653–656
Wright CE, Tallan HH, Lin YY, Gaull GE (1986) Taurine: biological update. Annu Rev Biochem 55:427–453

Chapter 4
Vascular Modulation of Rat Aorta by Taurine

Seiichiro Nishida and Hiroyasu Satoh

Abstract Taurine is found in high concentration in smooth muscle and heart muscle (approximately 10–20 mM). We found that taurine affects NE- and KCl induced vasoconstriction. The mechanisms regulating these vasoconstrictions mostly involve Ca^{2+} channels and EDRF(NO). Taurine exerted either a vasodilation or vasoconstriction depending on cellular Ca^{2+} concentration. When vascular tone was excessively low, taurine promoted vasoconstriction allowing the maintenance of blood pressure. On the other hand, taurine dilates vessels to increase blood flow during ischemia or hypoxia. Thus, taurine modulates vascular wall tone to maintain blood flow. These results indicate that taurine plays an important homeostatic function on vascular smooth muscles as well as cardiac muscle.

Abbreviations $[Ca^{2+}]_o$, extracellular Ca^{2+} concentration; $[Ca^{2+}]_i$, intracellular Ca^{2+} concentration; *CHF*, congestive heart failure; *EDHF*, endothelium-derived hyperpolarizing factors; *EDRF*, endothelium-derived releasing factor; *EGTA*, ethylene glycol-O,O'-bis (2-aminocthyl) N,N,N',N'-tetraacetic acid; *HOCl*, hypochlorous acid; *LDL*, low density lipoprotein; *L-NAME*, Nω-nitro-L-arginine methyl ester; *NE*, norepinephrine; *SPC*, sphingosylphosphorylcholine; *TEA*, tetraethylammonium; *VLDL*, very low density lipoprotein

4.1 Introduction

The sulfur amino acid, taurine, plays an important role in the maintenance of cardiac function during hypoxia, ischemia and cardiac failure. Patients in congestive heart failure (CHF) with less than 50% ejection fraction were treated with 3 g taurine a day (Azuma et al. 1992). After 6 weeks, a significant improvement in systolic function was observed. This improvement is thought to be caused the modulation of cardiac ion channels. At low intracellular Ca^{2+} concentration ($[Ca^{2+}]_i$), taurine enhanced L-type Ca^{2+} channel current (I_{Ca}) but inhibited I_{Ca} at high $[Ca^{2+}]_i$ (Satoh

H. Satoh (✉)
Department of Pharmacology, Nara Medical University, Nara, Japan

J. Azuma et al. (eds.), *Taurine 7*, Advances in Experimental Medicine and Biology 643,
DOI 10.1007/978-0-387-75681-3_4,

and Sperelakis 1993, 1998; Satoh and Horie 1997). Taurine also inhibited the fast Na^+ current (TTX-sensitive), an effect which might contribute to the antiarrhythmic activity of taurine (Satoh and Sperelakis 1992; Satoh 1994d). Thus, taurine can regulate $[Ca^{2+}]_i$ through the modulation of ionic channels (such as Ca^{2+}, K^+ and Na^+ channels) and secondly through the modulation of Na/Ca exchange (Satoh et al. 1992; Sperelakis and Satoh et al. 1993; Satoh and Sperelakis 1998). Therefore, taurine acts to maintain homeostasis in the presence of low $[Ca^{2+}]_i$ as well as Ca^{2+} overload.

Similarly, taurine could also modulate ionic channels of vascular smooth muscle cells. Not only does the aorta maintain high levels of taurine (126 μg/g wet) (Song et al. 1998; Satoh et al. 2002) but it dilates rabbit ear arteries in the presence of high-K^+ medium (Franconi et al. 1982) and has a similar effect on rat aorta induced to contract in the presence of either high K^+ or NE (Risori and Verdetti 1991). Thus, taurine mediated dilation of vascular smooth muscle involves either the inhibition of Ca^{2+} channels or some other undetermined mechanisms.

The basis underlying the treatment of heart failure consists of (1) redution in the workload of the heart, (2) protection of the cardiomyocyte, and (3) restriction of plasma volume and sodium. In order to reduce both pre-load and afterload, the dilation of arterioles and veins is strongly recommended when filling pressure is elevated. Therefore, taurine could serve as a therapeutic agent to not only modulate ionic channel function of the heart but also to regulate vascular tone. The effects of taurine on the vasculature are not mediated by α- and β-adrenoceptors and muscarinic receptors. Furthermore, unlike the heart (Satoh 1994a, b, c, 1995a, b, c 1966, 1998a, b, c, 1999) taurine's actions are independent of the endothelium and extracellular Ca^{2+} concentration ($[Ca^{2+}]_o$) (Risori and Verdetti 1991). Because few studies have examined the effects of taurine on vascular smooth muscle the present study was initiated to elucidate the pharmacological effects of taurine on vascular smooth muscle.

4.2 Materials and Methods

All experiments were carried out according to the guidelines laid down by the Nara Medical University Animal Welfare Committee, and also under the terms of the Declaration of Helsinki.

Wistar male rats (5–14 weeks old) were anesthetized with ether and euthanized by exsanguinations using methods similar to those described previously (Nishida and Satoh 2003, 2004). The thoracic aorta was quickly removed and the isolated aorta was cut into 3-mm rings in length. The rings were suspended between two triangular-shaped stainless steel stirrups in a jacketed organ chamber filled with 20 ml modified Krebs-Henseleit solution. The modified solution was, in mM: 118 NaCl, 4.6 KCl, 1.2 $MgSO_4$, 1.2 KH_2PO_4, 11.1 glucose, 27.2 $NaHCO_3$, 0.03 ethylene glycol-O,O'-bis (2-aminoethyl)-N,N,N',N'-tetraacetic acid (EGTA), and 1.8 $CaCl_2$. The chamber solution was kept at 36.5°C and oxygenated with 95% O_2 and 5% CO_2. The lower stirrup was anchored and the upper stirrup was attached to a force-displacement transducer (TB-652T, Nihon Kohden, Tokyo, Japan) to record the isometric force. All rings were stretched to generate a resting tension of 1.2 g.

After a 40 min-rest period, 1 μM norepinephrine (NE), different concentrations of KCl (30 to 60 mM), or 30 nM sphingosylphosphorylcholine (SPC, a Rho-kinase activator) was added to the bath to induce vasoconstriction. After a steady state contractile response was achieved, taurine was added to the bath. The responses were measured 6–10 min later. The relaxation response was analyzed as a percent decrease in maximal contraction induced by NE. Pretreatment with L-NAME (a NO synthase inhibitor), nicardipine and TEA (an inhibitor of Ca^{2+}-activated K^+ channel) were carried out to evaluate the role of key modulators of contractile function.

4.3 Results

4.3.1 Effects on NE-Induced Vasoconstriction

Taurine dilated aortic strips which were induced to contract with NE (1 μM). However, the effect of taurine depended both $[Ca^{2+}]_o$ and taurine concentration. In normal Krebs solution (1.8 mM Ca^{2+}), taurine had no effect on NE-induced vasoconstriction, but at high $[Ca^{2+}]_o$ (3.6 and 5.4 mM), it mediated a vasodilatory response. At the high $[Ca^{2+}]_o$ of 5.4 mM, the vasodilating effect of taurine was weaker than that seen at 3.6 mM $[Ca^{2+}]_o$ (Fig. 4.1)

4.3.2 Effects in the Presence of Some Inhibitors

To examine the involvement of the voltage-dependent Ca^{2+} channels, pretreatment with nicardipine was carried out. In the presence of the inhibitor, the effect of taurine on the NE-induced vasoconstriction was examined at 3.6 mM $[Ca^{2+}]_o$.

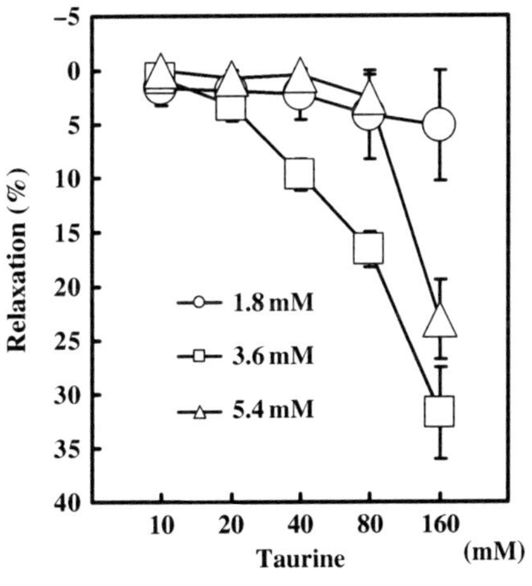

Fig. 4.1 Effect of taurine on NE-induced vasoconstriction at different $[Ca^{2+}]_o$ (1.8 to 5.4 mM)

Table 4.1 Modulation by inhibitors of the vasodilation induced by taurine

		Taurine (mM)				
	n	10	20	40	80	160
Control	6	0.7±0.7	3.2±1.4	9.6±1.5	16.6±1.6	31.7±4.7
Nicardipine 0.1 μM	6	0±0	0±0	2.5±1.7[a]	4.7±2.2[a]	10.8±2.7[b]
L-NAME 100 μM	6	1.4±1.2	4.3±1.9	6.3±2.3	10.5±2.7[a]	30.2±2.1
TEA 10 μM	6	1.1±0.8	7.1±3.3	10.5±2.7	20.8±2.3	32.7±3.8

Values shown represent % taurine-mediated vasorelaxation of aortic strips exposed to NE 3.6 mM Ca^{2+} in the presence of the indicated inhibitor.
[a]: $P<0.05$, [b]: $P<0.01$.

Taurine-mediated vasodilation was attenuated by nicardipine (0.1 μM); it fell from 31.7 ± 4.6 to 10.8 ± 2.7% ($n = 6$, $p<0.01$) at 160 mM taurine.

L-NAME (100 μM) pretreatment, which affects the endothelium, inhibited vasorelaxation induced by 80 mM taurine. TEA (10 μM), which examines the effect of Ca^{2+}-activated K^+ channels, did not affect taurine-induced relaxation (Table 4.1).

4.3.3 Effects on KCl-Induced Vasoconstriction

The effects of taurine on KCl-induced vasoconstriction were also investigated. In normal Krebs solution, taurine (10 to 160 mM) did not dilate KCl (30 mM)-induced vasoconstriction, but rather accentuated the degree of vasoconstriction (Fig. 4.2A). The taurine-mediated vasoconstriction was blocked by nicardipine (0.1 μM), but not by phentolamine (10 μM). On the other hand, taurine reduced the degree of vasoconstriction triggered by higher KCl (45 or 60 mM). Taurine mediated the most vasorelaxation at 45 mM, although it was not significant (Fig. 4.3A).

By contrast, at a medium concentration of 3.6 mM $[Ca^{2+}]_o$, taurine (10–80 mM) dilated KCl (30 mM)-induced vasoconstriction (Fig. 4.3B). However, even in the presence of 3.6 mM $[Ca^{2+}]_o$, 160 mM taurine triggered a vasoconstricting response. At a high KCl concentration (45 and 60 mM), taurine mediated vasodilation in contracted aortic slices.

When $[Ca^{2+}]_o$ was elevated to 5.4 mM, taurine similarly exhibited a biphasic response (Fig. 4.3C). Taurine (10 to 160 mM) potentiated 30 mM KCl-induced

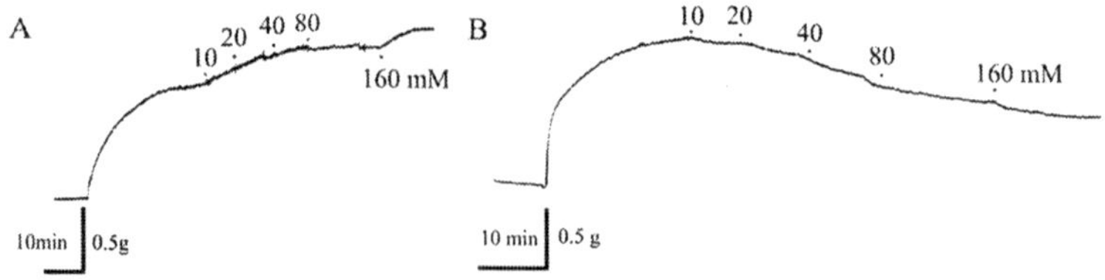

Fig. 4.2 Effect of taurine on KCl-induced vasoconstriction. (**A**) Vasoconstriction induced by taurine of aorta undergoing 30 mM KCl-induced vasoconstriction at 1.8 mM Ca^{2+}. (**B**) Vasorelaxation induced by taurine of aorta undergoing 60 mM KCl-induced vasoconstriction at 3.6 mM Ca^{2+}. In both experiments, KCl was added at time 0

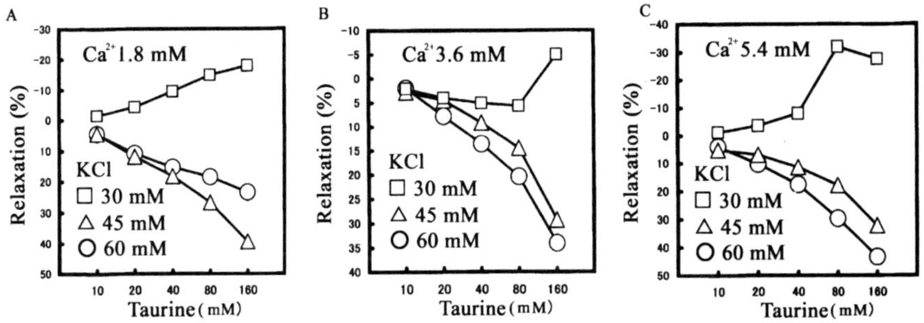

Fig. 4.3 Taurine-induced vasoresponses to KCl (30–60 mM)-induced vasoconstriction at different Ca^{2+} concentrations (1.8–5.4 mM)

vasoconstriction (Fig. 4.2B), but dilated vasoconstriction mediated by a high concentration of KCl (45 and 60 mM). Taurine 160 mM exerted a marked vasodilation (43.5 ± 8.6% ($n = 6$, $P < 0.01$) of aortic strips made to contract with 60 mM KCl at 5.4 mM $[Ca^{2+}]_o$.

4.3.4 Effects on Ca2+-independent constriction

For a Ca2+-independent vasoconstriction, the effect of taurine on SPC (a Rho-kinase activator)-induced vasoconstriction was examined. SPC alone caused a week vasoconstriction. But taurine did not dilate SPC-induced vasoconstriction but rather constricted it further (Fig. 4.4).

In addition, high concentrations of taurine (160 mM) constricted resting aorta (lacking pretreatment with drugs). This vasoconstriction was attenuated by nicardipine (0.1 μM), but not by phentolamine (10 μM) (Fig. 4.5)

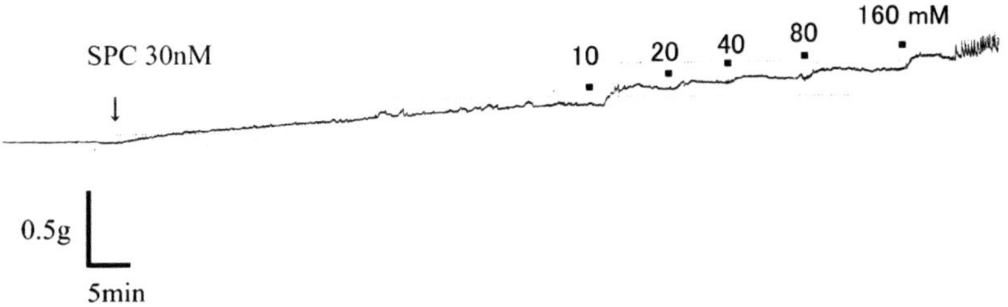

Fig. 4.4 Effect of taurine on SPC-induced vasoconstriction

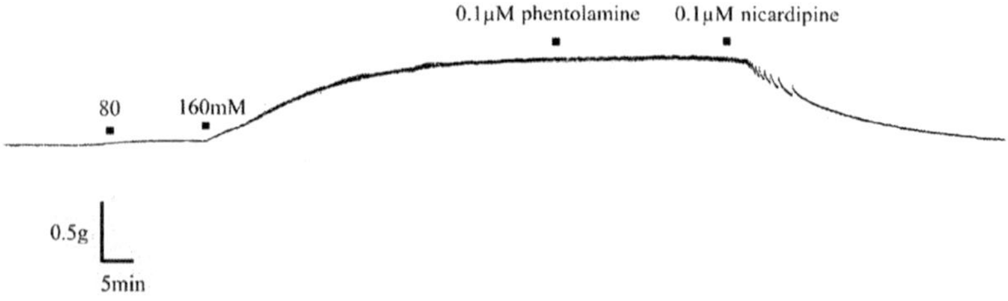

Fig. 4.5 Taurine-induced vasoconstriction of resting aorta

4.4 Discussion

4.4.1 Effects on NE-Induced Vasoconstriction

Taurine reduced the degree of NE-induced vasoconstriction in a concentration-dependent manner. Aorta contains about ten times the taurine content (10–20 mM) of cardiac muscle (Song et al. 1998; Satoh et al. 2002). Thus, the higher concentrations might be required to produce an effective response. The vasorelaxation was Ca^{2+}-dependent. At normal $[Ca^{2+}]_o$ (1.8 mM), taurine elicited a slight vasoconstriction, but at higher $[Ca^{2+}]_o$ it caused a marked vasorelaxation.

4.4.2 Ca^{2+}-Dependency of Taurine's Actions

Taurine's actions were modified by $[Ca^{2+}]_o$. In normal Krebs (1.8 mM Ca^{2+}), taurine did not dilate NE-induced vasoconstriction, but at a higher $[Ca^{2+}]$ (3.6 mM) it produces vasodilation, but the vasodilation at 5.4 mM Ca^{2+} was weaker than that at 3.6 mM Ca^{2+}. In KCl-induced vasoconstricted aorta, taurine mediated a Ca^{2+}-dependent vasorelaxation. In aorta constricted at 60 mM KCl taurine's vasorelaxation was clearly dependent on $[Ca^{2+}]_o$. With an increase in $[Ca^{2+}]_o$ from 1.8 to 5.4 mM, taurine-induced vasorelaxation became more obvious, but in 30 mM KCl-induced vasoconstricted aorta taurine-dependent relaxation was not observed at all $[Ca^{2+}]_o$.

Thus, taurine-induced vasorelaxation is only observed at high $[Ca^{2+}]_o$ and high KCl (45–60 mM). When vasoconstriction was induced by lower KCl (30 mM), taurine mediated a further constriction at 1.8 to 5.4 mM $[Ca^{2+}]_o$. Taurine also constricted the resting aorta. This constriction was reduced by nicardipine but not by phentolamine. Therefore, these results indicate that a major action of taurine is exerted by the modulation of Ca^{2+} channels. The present experiments show that taurine is strongly dependent on $[Ca^{2+}]_i$ and $[Ca^{2+}]_o$, different from the report of Risori and Verdetti (1991). However, the present results are consistent with the response of cardiac myocytes (Satoh and Horie 1997; Satoh 2001, 2003). Taurine activates the

Ca^{2+} channel at low $[Ca^{2+}]_i$, but inhibits it at high $[Ca^{2+}]_i$. The Ca^{2+} dependence should contribute to the regulation of vascular wall tone.

4.4.3 Taurine-Induced Vasorelaxation and Vasoconstriction

Taurine dilates KCl-induced vasoconstriction. The vasoconstriction induced by taurine was inhibited by nicardipine. Since TEA did not affect taurine-induced vasorelaxation, the taurine effect does not appear to involve Ca^{2+}-activated K^+ channels. L-NAME partially modified taurine-induced vasodilation. Therefore, these results indicate that the vascular mechanisms are mainly caused by the inhibition of Ca^{2+} channels of smooth muscle cells, although they also involve endothelium-dependent relaxation, another effect that is inconsistent with the results of Risori and Verdetti (1991). The data suggest that taurine might affect endothelium-derived releasing factor (EDRF) and endothelium-derived hyperpolarizing factors (EDHF).

SPC activates Rho-kinase and constricts vascular muscle without an increase in $[Ca^{2+}]_i$, an effect likely involving the enhancement of Ca^{2+} sensitization (Shirao et al. 2002; Hirano 2007). In the present experiments, taurine did not dilate SPC-induced vasoconstriction, but rather caused further vasoconstriction in a concentration-dependent manner. Taurine also constricted resting aorta and this constriction was blocked by Ca^{2+} antagonists (Fig. 4.5). Thus, consistent with results in cardiac muscle, taurine-induced vasoconstriction in smooth muscle requires the activation of Ca^{2+} channels (Satoh 2001, 2003; Satoh and Sperelakis 1993; Satoh and Horie 1997). Taurine constricts vascular smooth muscles when $[Ca^{2+}]_i$ is not elevated, as in SPC-treated aorta or untreated resting aorta (under the low $[Ca^{2+}]_i$) (Fig. 4.6). On the other hand, taurine administration in the presence of high $[Ca^{2+}]_i$ dilates smooth muscle.

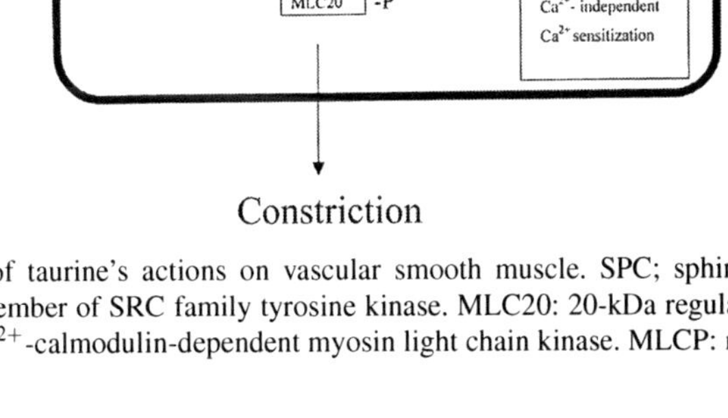

Fig. 4.6 Sheme of taurine's actions on vascular smooth muscle. SPC; sphingosylphosphorylcholine, Fyn; a member of SRC family tyrosine kinase. MLC20: 20-kDa regulatory myosin light chain. MLCK: Ca^{2+}-calmodulin-dependent myosin light chain kinase. MLCP: myosin light chain phosphatase

4.4.4 Vascular Regulation by Taurine

Induction of aortic regurgitation in rabbits leads to the development of CHF and a mortality rate of 53% after 8 weeks. However, daily administration of taurine (p.o. 100 mg/kg) reduced the mortality rate after 8 weeks to 10%. Cardiac function was maintained in the taurine-treated rabbits, but was depressed in the untreated rabbits (Takihara et al. 1986). Thus, taurine dilates vessels to maintain blood flow, improving cardiac function under these conditions.

Taurine treatment also reduces serum low density lipoprotein (LDL) and very low density lipoprotein (VLDL) by 44% in mice fed a high fat diet. Hypochlorous acid (HOCl) produced by myeloperoxidase in neutrophils and macrophages oxidizes LDL (Jerlich et al. 2000). By scavenging HOCl, taurine exerts a cytoprotective action (Kearns and Dawson 2000). Thus, taurine prevents atherosclerosis by inhibiting the oxidization of LDL. Simultaneously taurine prevents endothelial dysfunction. Taurine facilitates ACh-induced relaxation of the aorta in cholesterol-fed and streptozotocin-induced diabetic mice (Kamata et al. 1996). Therefore, taurine reduces vascular wall tone, and may act as an anti-atherosclerotic agent (Murakami et al. 1999).

4.5 Conclusion

Taurine has the potential to modulate vascular wall tone to maintain blood flow. If vascular tone is excessively low, as occurs in hypotension and bacterial shock (low $[Ca^{2+}]_i$ level), taurine can constrict vessels to maintain blood pressure. On the other hand, taurine dilates vessels to increase blood flow during ischemia or hypoxia (calcium overload). Therefore, taurine possesses homeostatic actions on vascular smooth muscles as well as cardiac muscle.

References

Azuma J, Sawamura A, Awata N (1992) Usefulness of taurine in chronic heart failure and its prospective application. Jpn Circ J 56:95–99

Franconi F, Glotti A, Manzini S, Martini F, Stendardi I, Zilletti L (1982) The effects of taurine on high potassium- and noradrenalline- induced contraction in rabbit ear artery. Br J Pharmacol 75:605–612

Hirano K (2007) Current topics in the regulatory mechanism underlying the Ca^{2+} sensitization of the contractile apparatus in vascular smooth muscle. J Pharmacol Sci 104:109–115

Jerlich A, Fritz G, Kharrazi H, Hammel M, Tschabuschnig S, Glatter O, Schaur RJ (2000) Comparizon of HOCl traps with myeloperoxidase inhibitors in prevention of low density lipoprotein oxidation. Biochim Biophys Acta 80:27–32

Kamata K, Sugiura M, Kojima S, Kasuya Y (1996) Restoration of endothelium-dependent relaxation in both hypercholesterolemia and diabetes by chronic taurine. Eur J Pharmacol 303:47–53

Kearns S, Dawson R Jr (2000) Cytoprotective effect of taurine against hypochlorous acid toxicity to PC12 cells. Adv Exp Med Biol 483:563–570

Murakami S, Kondo Y, Tomisawa K, Nagate T (1999) Prevention of atherosclerotic lesion development in mice by taurine. Drug Exp Clin Res 25:227–234
Nishida S, Satoh H (2003) Mechanisms for vasodilations induced by Ginkgo biloba extract and its main constitute bilobalide in rat aorta. Life Sci 72:2659–2667
Nishida S, Satoh H (2004) Comparative vasodilating actions among terpenoids and flavonoids contained in Ginkgo biloba extract. Clinica Chimica Acta 339:129–133
Risori MT, Verdetti J (1991) Effects of taurine on rat aorta in vitro. Fundam Clin Pharmcol 5: 245–258
Satoh H (1994a) Antagonistic actions of taurine on Ca^{2+} -induced responses in cardiac muscle cells. Jpn Heart J 35:457–458
Satoh H (1994b) Cardioprotective actions of taurine against intracellular and extracellular Ca^{2+}-induced effects. In: Huxtable RJ, Michalk D (eds) Taurine in health and disease. Plenum Press, New York, pp 181–196
Satoh H (1994c) Taurine-induced hyperpolarizing shift of the reversal potential for the fast Na^+ current in embryonic chick cardiomyocytes. Gen Pharmacol 26:517–521
Satoh H (1994d) Regulation of the action potential configuration by taurine in guinea-pig ventricular muscle. Gen Pharmacol 25:47–52
Satoh H (1995a) Regulation by taurine of the spontaneous activity in young embryonic chick cardiomyocytes. J Cardiovasc Pharmacol 25:3–8
Satoh H (1995b) A dual actions of taurine on the delayed rectifier K^+ current in young embryonic chick cardiomyocytes. Amino Acids 9:235–246
Satoh H (1995c) Electropysiological actions of taurine on spontaneously beating rabbit sino-atrial nodal cells. Jpn J Pharmacol 67:29–34
Satoh H (1996) Direct inhibition by taurine of the ATP-sensitive K^+ channel in guinea pig ventricular cardiomyocytes. Gen Pharmacol 27:625–627
Satoh H (1998a) Modulation by taurine of the spontaneous action potentials in right atrial muscles of rat. Gen Pharmacol 30:209–212
Satoh H (1998b) Inhibition by taurine of the inwardly rectifying K^+ current in guinea pig ventricular cardiomyocytes. Eur J Pharmacol 346:309–313
Satoh H (1998c) Inhibition of the fast Na^+ current by taurine in guinea pig ventricular myocytes. Gen Pharmacol 31:155–158
Satoh H (1999) Taurine modulates I_{K_1} but not I_{K_s} in guinea pig ventricular cardiomyocytes. Br J Pharmacol 126:87–92
Satoh H (2001) $[Ca^{2+}]_i$-dependent actions of taurine in spontaneously beating rabbit sino-atrial nodal cells. Eur J Pharmacol 424:19–25
Satoh H (2003) Electropharmacology of taurine on the hyperpolarization-activated inward current and the sustained inward current in spontaneous beating rat sino-atrial nodal cells. J Pharmacol Sci 91:229–238
Satoh H, Horie M (1997) Actioin of taurine on the L-type Ca^{2+} channel current in the guinea pig ventricular cardiomyocytes. J Cardiovasc Pharmacol 30:711–716
Satoh H, Sperelakis N (1992) Taurine inhibition of Na^+ current in embryonic chick ventricular myocytes. Eur J Pharmocol 218:83–89
Satoh H, Sperelakis N (1993) Taurine effects on Ca^{2+} currents in young embryonic chick cardiomyocytes. Eur J Pharmacol 231:443–449
Satoh H, Sperelakis N (1998) Review of some actions of taurine on ion channels of cardiac muscle cells and others. Gen Pharmacol 30:451–463
Satoh H, Nakatani A, Tanaka T, Haga T (2002) Cardiac functions and taurine's actions at different extracellular calcium concentrations in forced swimming stress-loaded rats. Biol Trac Ele Res 87:171–182
Shirao S, Kashiwagi S, Sato M, Miwa S, Nakao F, Kurokawa T, Todoroki-Ikeda N, Mogami K, Mizulami Y, Kuriyama S, Haze K, Suzuki M, Kobayashi S (2002) Sphingosylphosphorylcholine is a novel messenger for Rho-kinase-mediated Ca^{2+} sensitization in the bovine cerebral artery: unimportant role for protein kinase C. Circ Res 91:112–119

Song D, O'Regan MH, Phillis JW (1998) Mechanisms of amino acid release from the isolated anoxic/reperfused rat heart. Eur J Pharmacol 351:313–322

Sperelakis N, Satoh H (1993) Taurine effects on ion channels of cardiac muscle. In: Noble D, Earn Y (eds) Ionic channels and effect of taurine on the heart. Kluwer Academic Publishers, Boston, pp 93–118

Sperelakis N, Satoh H, Bkaily G (1992) Taurine's effects on ionic current myocardial cells. In: Schaffer SW, Lombardim B (eds) Taurine: new dimensions on its mechanisms and actions, pp 129–143

Takihara K, Azuma J, Awata N, Ohta H, Hamaguchi T, Sakamura A, Tanaka Y, Kishimoto S, Sperelakis N (1986) Beneficial effects of taurine in rabbits with chronic congestive heart failure. Am Heart J 112:1278–1284

Chapter 5
Modulation by Taurine of Human Arterial Stiffness and Wave Reflection

Hiroyasu Satoh and Jangmi Kang

Abstract Effects of taurine (1000–2000 mg) on hemodynamic function and the arterial pulse wave were investigated for 102 healthy medical and paramedical students. The vascular parameters were generally dependent on aging, with the arterial stiffness parameters, such as baPWV, ABI and AI, are considered the indicators of "vascular aging". Acute administration of taurine decreased BP and HR and attenuated the stiffness parameters derived from the pulse waveform. Thus, taurine can cause significant changes in the cardiovascular system and the arterial pulse wave. However, approximately 5% of the students were non-responders. This may be related to the notion that taurine would be expected to exert greater effects on the vascular functions of unhealthy individuals. Based our previous experiments, therefore, taurine plays a role in the regulation of the cardiac and vascular function.

Abbreviations *ABI*, ankle and brachial pressure index; *AI*, augmentation index; *baPWV*, brachial to ankle pulse wave velocity; *BP*, blood pressure; *CBP*, central arterial blood pressure; *CHF*, congestive heart failure; DBP, diastolic blood pressure; $[Ca^{2+}]_o$, extracellular Ca^{2+} concentration; $[Ca^{2+}]_i$, intracellular Ca^{2+} concentration; *HR*, heart rate; *MBP*, mean blood pressure; *SBP*, systolic blood pressure

5.1 Introduction

Taurine (a sulfur amino acid) is present in high concentration (around 10 mM) in myocardial cells, but is found at relatively low levels in the plasma. Taurine has been reported to be involved in numerous physiological functions, such as osmoregulation, antioxidant action, and Ca^{2+} modulation. Taurine has been found to produce many electrical and mechanical actions on cardiac muscle cells (Huxtable 1992; Sperelakis and Satoh 1993; Satoh and Sperelakis 1998; Satoh 1998b, 1999; Sperelakis et al. 1992).

H. Satoh (✉)
Department of Pharmacology, Nara Medical University,
Nara, and Hyogo NCC College, Hyogo, Japan

J. Azuma et al. (eds.), *Taurine 7*, Advances in Experimental Medicine and Biology 643,
DOI 10.1007/978-0-387-75681-3_5, © Springer Science+Business Media, LLC 2009

The intracellular taurine content of myocardial cells is reduced during ischemia and hypoxia, leading to calcium overload (an excess of $[Ca^{2+}]_i$ level) (Crass and Lombardini 1978; Schaffer et al. 1980; Satoh 1994a, 1994b, 1994c, 1996). Under those conditions, administration of taurine produces a beneficial effect, as the physiological and pharmacological actions induced by taurine are dependent on $[Ca^{2+}]_i$ and $[Ca^{2+}]_o$ (Satoh 1994a 1994b 1994c). For example, taurine exhibits normalizing activity, which is dependent on $[Ca^{2+}]_i$. Thus, taurine acts to maintain cellular homeostasis via its actions on ion channels, signal transduction pathways and transport systems. At a high $[Ca^{2+}]_o$, taurine inhibits cardiac function, whereas at low $[Ca^{2+}]_o$, taurine enhances it.

In arteries, the pulse wave velocity (PWV) depends on regional large artery stiffness. A high PWV allows the reflected wave to return to the aortic root sooner and to increase systolic pressure of the heart. The arterial pulse wave is formed by combining the ejection and reflection pulses and is altered in atheriosclerosis and hypertension. An augmentation index (AI) depends not only on systemic arterial elasticity but also on arterial geometry and tone (Nichols and O'Rourke 1998; Pannier et al. 2002; Davies and Struthers 2003).

In rat aorta, taurine modulates contractile function in a $[Ca^{2+}]_i$-dependent manner (unpublished data). Like cardiac cells, modulation of aortic function by taurine depends upon modulation of automatic activity, ionic channels and contractile force (Franconi et al. 1982; Satoh 1995a, 1995b 1996, 1998a, 1998b, 1998c, 1999, 2001; Satoh et al. 2002; Satoh and Sperelakis 1998). Thus, clinically, taurine may regulate vascular wall tone, mediating in the process declines in central arterial pressure provided that the tissue is exposed to high $[Ca^{2+}]_i$ with an opposite effect seen at lower $[Ca^{2+}]_i$. The aim of the present study is to investigate the arterial pulse wave. The effect of taurine on hemodynamic properties and on arterial stiffness was examined.

5.2 Materials and Methods

The study was approved by the Ethical Committee of Nara Medical University Hospital, and an informed consent was obtained from all students.

5.2.1 Taurine Intake

This study investigated the effects of tauirne administration on the hemodynamic properties of 102 healthy paramedical and medical students (averaged 21.5 ± 2.1 years old). The test was performed about 2 h after lunch, and the students took 1000-2000 mg taurine (Waken Co., Kyoto, Japan) with a cup of water. To avoid confounding influences, the students were told not to consume large amounts of beverage at lunch time before the test.

Measurements were obtained every 20 min following consumption of taurine. Significant differences were assessed with ANOVA and Student's *t*-test for paired data followed by gaussian distribution. Values are presented as means ± S.E.M.

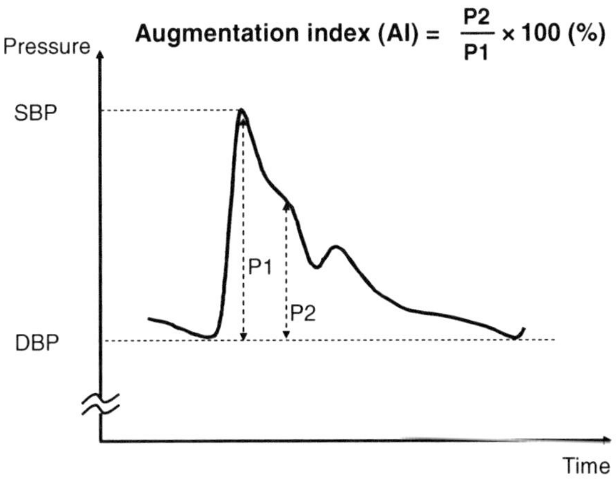

Fig. 5.1 Radial arterial pressure contour. Augmentation index is calculated from the equation; AI = P2/P1 × 100. P1: ejection wave. P2: reflection wave

5.2.2 Measurements of baPWV and ABI

Brachial to ankle pulse wave velocity (baPWV) was measured separately on the right and left sides, using a validated non-invasive device (BP-203RPEII, Omron Healthcare, Kyoto, Japan), with pressure transducers placed at the base of the brachial artery and in the inguinal region on assess the femoral artery. Simultaneously ankle and brachial pressure index (ABI), the ratio of the ankle to the brachial blood pressure, was also recorded on both right and left sides. The value of baPWV was calculated by dividing the distance to the distal site by the transient time for conducted wave.

5.2.3 Measurement of AI

An augmentation index (AI) has been established as a reflection coefficient, a ratio of reflection pressure (P2)/an ejection pressure (P1) (Fig. 5.1). AI is determined by a computer algorithm developed from invasive pressure and flow data (Kelly et al. 1989). AI was measured using automated applanation tonometry (HEM-9000AI, Omron Healthcare, Kyoto, Japan). Simultaneously left brachial HR and BP were recorded. In addition, the central arterial blood pressure (CBP) was estimated from the reflection pulse using automated applanation tonometry.

5.3 Results

Effects of taurine on arterial stiffness and wave reflection were investigated after administration of taurine (1000–2000 mg). The baPWV, the ABI, and the AI, the latter an index of wave reflection, were analyzed using a validated non-invasive device and applanation tonometry. No students complained of any symptoms following taurine administration.

5.3.1 Blood Pressure

The hemodynamic properties of 79 students were examined before and following taurine administration. The responses were time-dependent and were apparent almost 40–60 min post-administration. Prior to taurine administration, systolic blood pressure (SBP) of brachial and ankle arteries were 114.1 ± 1.0 mmHg and 126.8 ± 1.5 mmHg on the right side, and 115.2 ± 1.3 mmHg and 116.9 ± 1.4 mmHg on the left side, respectively. Brachial and ankle diastolic blood pressure (DBP) on the left and right sides were 65.2 ± 1.5 and 64.1 ± 1.2 mmHg, and 64.8 ± 1.5 and 66.0 ± 1.1 mmHg, respectively. Right and left mean BP (MBP) were 82.3 ± 2.1 and 81.7 ± 1.6 mmHg, respectively, while HR was 73.9 ± 1.7 beats/min before taurine administration.

Approximately 40–60 min after administration of 1000 mg taurine, both SBP and DBP tended to decrease although it was not significant (by approximately 2.5 to 5%). Taurine also tended to decrease HR in a time dependent manner. However, 2000 mg taurine mediated a significant decline in both SBP and DBP (Fig. 5.2).

5.3.2 baPWV and ABI

Prior to taurine exposure, the baPWV values for the right and left sides of the 79 students were 1125.9 ± 23.2 and 1167.5 ± 20.1, respectively, while the corresponding measurements of ABI were 1.1 ± 0.2 and 1.2 ± 0.2, respectively. Taurine (2000 mg) reduced right and left baPWV by 4.5 ± 0.3% ($P > 0.05$) and 5.2 ± 0.1% ($P < 0.05$),

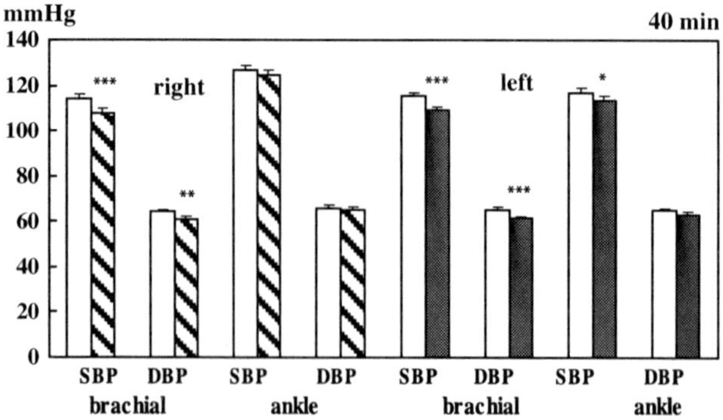

Fig. 5.2 Effects of taurine on SBP and DBP of brachial and ankle on right and left sides (n = 79). White columns: control. Shadow and black columns: 40 min after taking 2000 mg taurine. ∗: $P < 0.05$, **: $P < 0.01$, ∗∗∗: $P < 0.001$, with respect to control value

Table 5.1 Depressions of baPWV and ABI before and after taurine administration

	n	baPWV		ABI
		right	left	
1000 mg	32	2.8 ± 0.1	3.1 ± 0.3	2.9 ± 0.1
2000 mg	79	4.5 ± 0.3	5.2 ± 0.1[a]	3.1 ± 0.1

Values (%) are represented as means ± SEM. baPWV: a pulse wave velocity of brachial to ankle arteries. ABI: a ratio of ankle vs. brachial BP. [a]: $P < 0.05$, with respect to control value.

respectively, but failed to affect ABI (Table 5.1). At 1000 mg, taurine exerted no significant effect.

5.3.3 AI and CBP

The AI and CBP from radial artery were also measured, and their average values were 57.9 ± 3.1% and 113.8 ± 11.2 mmHg, respectively. Following administration of 1000 mg taurine ($n = 32$), AI decreased by 3.2 ± 0.3% within 40 min and by 2.9 ± 0.3% within 60 min. By comparison, 40 and 60 min after taurine administration CBP decreased insignificantly by 4.3 ± 0.3% and 4.4 ± 0.3%, respectively. After administration of 2000 mg taurine ($n = 63$), BP and HR were markedly reduced, with SBP falling 12.7 ± 0.3% ($P < 0.01$), DBP 11.2 ± 0.4% ($P < 0.01$) and HR 10.5 ± 0.8% ($P < 0.05$). Other stiffness parameters were also significantly affected as summarized in Table 5.2.

However, there were non-responders to taurine application for almost 5% students. Taurine did not affect or enhanced the parameters (but not markedly).

5.3.4 Age-Dependent Responses to Taurine

The vascular parameters are generally dependent on aging. The students were divided into three groups: teens, twenties, and thirties. The hemodynamic actions of taurine in the 3 groups were age-dependent. After administration of taurine (2000 mg), the stiffness parameters were reduced. But the coefficients of correlation with aging were not markedly affected, presumably because elder persons were excluded from the study. Taurine decreased the correlation coefficient of AI with respect to aging from 0.365 in the control to 0.283 after 40 min of taurine

Table 5.2 Decreases in stiffness parameters before and after taurine administration

	n	SBP	DBP	HR	AI	CBP
1000 mg	32	4.7 ± 0.2	2.5 ± 0.2	2.4 ± 0.4	3.2 ± 0.3	4.4 ± 0.3
2000 mg	63	12.7 ± 0.3[b]	11.2 ± 0.4[b]	10.5 ± 0.8[a]	10.4 ± 0.4[a]	11.9 ± 0.3[a]

Values (%) are represented as means ± SEM. SBP: systolic blood pressure. DBP: diastolic blood pressure. HR: heart rate. AI: an augmentation index from radial artery. CBP: central arterial blood pressure. [a]: $P < 0.05$, [b]: $P < 0.01$, with respect to control value.

exposure. Also the correlation coefficient of CBP with respect to aging was reduced. The effects on other parameters were not age-dependent.

5.4 Discussion

Taurine exhibits numerous physiological and pharmacological actions in various tissues. Taurine administration mediates relaxation of smooth muscle, stimulation of skeletal muscle, and diuresis (Rall 1990) Furthermore, in isolated hearts, taurine modulates the ionic currents and the action potentials, and affects the developmental tension and sinus rhythm (Satoh 1994d, 1995a, 1995c, 1998a, 2001, 2003; Satoh and Sperelakis, 1998; Satoh et al. 2002). The intracellular taurine level (10-20 mM) of myocardial cells is reduced in the ischemic heart and hypoxia, leading to a calcium overload (an excess of $[Ca^{2+}]_i$ level) (Suleiman et al. 1997; Song et al. 1998; Satoh et al. 2002). Calcium overload in turn elicits triggered activity and provokes arrhythmias. Under those conditions, taurine administration would be expected to improve cardiac function. Indeed, taurine treatment actually mediates a beneficial effect in congestive heart failure (CHF). On the other hand, taurine stimulates cardiac function under low $[Ca^{2+}]_i$ conditions. Thus, the actions of taurine have been demonstrated to be dependent on $[Ca^{2+}]_i$ and $[Ca^{2+}]_o$11 (Satoh and Sperelakis 1992; 1993; Satoh and Horie 1997).

Pulse waveform is generated by the superposition of the reflected backward wave on the incident forward wave (Nichols and O'Rourke 1998). Pulse waveform depends on two factors; arterial stiffness with increased PWV and a progressively earlier wave reflection. The pressure wave-contour analysis is a key indicator of arterial elastic properties and a prognosticator of cardiovascular risk.

The baPWV is a modality to assess arterial stiffness non-invasively, and also expresses the tone of the peripheral muscular arteries (Asmer et al. 1995; Munakata et al. 2004). Normal PWV value is approximately 950 to 1200 individuals ranging in age from teenagers to the thirties. Values of PWV increase with arterial stiffness induced by aging, hypertension, diabetes, smoking and stress. On the other hand, normal ABI values range from 0.9 to 1.3. Higher ABI values are predictive of arterial occlusive diseases, such as arteriosclerosis obliterans and Raynoud's disease.

The augmentation index has been recently established as a reflection coefficient (Kelly et al. 1989). It is determined by a computer algorithm developed from invasive pressure and flow data. Central aortic augmentation has been evaluated non-invasively by mathematically transforming the radial artery pulse waveform to the aortic pulse waveform (Chen et al. 1997; Gallagher et al. 2004). Recent technical progress makes non-invasive measurements of AI possible. The normal AI value is approximately 50% in teens, 55% in twenties, and 62% in thirties. Reflected pressure waves are also responsible for raising CBP (Karamanoglu et al. 1994). Cardiac afterload depends on the central aortic BP (or CBP) rather than on the peripheral brachial BP. Central arterial stiffening causes the pulse pressure (systolic-diastolic) to widen and leads to the syndrome of isolated systolic hypertension (Wang and Parker 2004).

After administration of taurine, a significant decrease in BP occurs within 40–60 min. In our laboratory, in rat aorta, taurine dilates the NE-induced constriction, due to mainly Ca^{2+} channel inhibition. As a result, taurine can reduce vascular wall tone and might therefore decrease baPWV and AI. But ABI may be unaffected, since taurine reduces both the brachial and ankle BP to almost a similar extent.

In the clinical treatment of CHF, the aims are to achieve several endpoints: (1) reducing workload of the heart, (2) protection of the cardiomyocyte, and (3) restriction and control of volume and sodium. In order to reduce both preload and afterload in cases of elevated filling pressure, arterioles and veins need to be dilated and cardiac output needs to decrease. Taurine might be clinically beneficial through its ability to modulate ion channels of cardiac cells and regulate blood vessel tension.

In general, the values of BP and the stiffness parameters are age-dependent (Marchais et al. 1993). In this study, the values increased with age. Therefore, the stiffness parameters, such as baPWV, ABI and AI, are considered indicators of "vascular aging". In 102 students, acute administration of taurine (1000–2000 mg) produced hemodynamic effects. The AI value was reduced by taurine, but the reduction was not marked because only young people were studied. Taurine concentration in the blood reached a peak ~40–60 min post-administration. In rat, the serum taurine level is 36.3 ± 1.0 μg/g wet ($n = 20$) (Satoh et al. 2002). Taurine content in the aorta is usually high (~120–130 μg/g wet) (Song et al. 1998). This high concentration is likely required for its physiological function in arterial vessels. The present finding illustrates the response to pharmacological doses of taurine.

In this study, some students did not respond to taurine, which is consistent with pharmacological non-responders in clinical treatment using herbal medicine (Kampo Medicine). The present results indicate that taurine (at the relatively high concentration of 2000 mg) reduces vascular wall tone, slows baPWV and leads to declines in AI and CBP.

5.5 Conclusion

Administration of taurine (1000–2000 mg) mediated significant hemodynamic changes in healthy students. Thus, even acute administration of taurine can cause beneficial effects; namely, a decline in BP and alterations in the pulse waveforms. Based on our previous findings, taurine would be expected to exert greater effects on vascular function of unhealthy individuals, with greater depressions occurring in the presence of high $[Ca^{2+}]_i$, and an enhancement under low $[Ca^{2+}]_i$ conditions. This hypothesis is supported by the existence of non-responders. This pattern would lead to cytoprotection against cardiovascular diseases. Irrespective, taurine exerts hemodynamic actions, resulting in effective actions on the heart and vascular tissue.

References

Asmer R, Benetos A, Topouchian J, Laurent P, Pannier B, Brisac AM, Target R, Levy B (1995) Assessment of arterial distensibility by automatic pulse wave velocity measurement:validation and clinical application studies. Hypertension 26:485–490

Chen CH, Nevo E, Fetics B, Pak PH, Yin FCR, Maughan WL, Kass DA (1997) Estimation of central aortic pressure waveform by mathematical transformation of radial tomometry pressure Validation of generalized transfer function. Circulation 95:1827–1836
Crass MF, Lombardini JB (1978) Release of tissue taurine from the oxygen-deficient perfused rat heart. Proc Soc Exp Biol Med 157:486–488
Davies JI, Struthers AD (2003) Pulse wave analysis and pulse wave velocity:a critical review of their strengths and weakness. J Hyertens 21:463–472
Franconi F, Martini, Stendari I, Matucci R, Zilleti L, Giotti A (1982) Effect of taurine on calcium level and contractility in guinea pig ventricular strips. Biochem Pharmacol 31:3181–3185
Gallagher D, Adji A, O'Rourke MF (2004) Validation of the transfer function technique for generating central from peripheral upper limb pressure waveform. Am J Hypertens 17:1059–1067
Huxtable RJ (1992) Physiological actions of taurine. Physiol Rev 72:101–163
Karamanoglu M, Gallagher DE, Avolio AP, O'Rourke MF (1994) Functional origin of reflected pressure waves in a multibranched model of the human arterial system. Am J Physiol 267:H1681–H1688
Kelly R, Hayward C, Avolio A, O'Rourke M (1989) Noninvasive determinstion of age-related changes in the human arterial pulse. Circulation 80:1652–1659
Marchais SJ, Guerin AP, Pannier BM, Levy BI, Safar ME, London GM (1993) Wave reflections and cardiac hypertrophy in chronic uremia Influence of body size. Hypertension 22:876–883
Munakata M, Nagasaki A, Nunokawa T, Sakuma T, Kato H, Yoshinaga K, Toyota T (2004) Effects of varsartan and nifedipine coatcore on systemic arterial stiffness in hypertensive patients. AJH 17:1050–1055
Nichols WW, O'Rourke MF (1998) McDonald's blood flow in arteries: theoretical experimental and clinical principles. Edward Arnold, London
Pannier BM, Avolio AP, Hoeks A, Mancia G, Takazawa K (2002) Methods and devices for measuring arterial compliance in human. Am J Hypertens 15:743–753
Rall JA (1990) Sixty years of investigation into the foundamental nature of muscle contraction. Prog Clin Biol Res 327:1–15
Satoh H (1994a) Antagonistic actions of taurine on Ca^{2+-}induced responses in cardiac muscle cells. Jpn Heart J 35:457–458
Satoh H (1994b) Cardioprotective actions of taurine against intracellular and extracellular Ca^{2+}-induced effects. In: Huxtable RJ, Michalk D (eds) Taurine in health and disease. Plenum Press, New York, pp 181–196
Satoh H (1994c) Taurine-induced hyperpolarizing shift of the reversal potential for the fast Na^{+} current in embryonic chick cardiomyocytes. Gen Pharmacol 26:517–521
Satoh H (1994d) Regulation of the action potential configuration by taurine in guinea-pig ventricular muscle. Gen Pharmacol 25:47–52
Satoh H (1995a) Regulation by taurine of the spontaneous activity in young embryonic chick cardiomyocytes. J Cardiovasc Pharmacol 25:3–8
Satoh H (1995b) A dual actions of taurine on the delayed rectifier K^{+} current in young embryonic chick cardiomyocytes. Amino Acids 9:235–246
Satoh H (1995c) Electropysiological actions of taurine on spontaneously beating rabbit sino-atrial nodal cells. Jpn J Pharmacol 67:29–34
Satoh H (1996) Direct inhibition by taurine of the ATP-sensitive K^{+} channel in guinea pig ventricular cardiomyocytes. Gen Pharmacol 27:625–627
Satoh H (1998a) Modulation by taurine of the spontaneous action potentials in right atrial muscles of rat. Gen Pharmacol 30:209–212
Satoh H (1998b) Inhibition by taurine of the inwardly rectifying K^{+} current in guinea pig ventricular cardiomyocytes. Eur J Pharmacol 346:309–313
Satoh (1998c) Inhibition of the fast Na^{+} current by taurine in guinea pig ventricular myocytes. Gen Pharmacol 31:155–158
Satoh (1999) Taurine modulates I_{Kr} but not I_{Ks} in guinea pig ventricular cardiomyocytes. Br J Pharmacol 126:87–92

Satoh H (2001) $[Ca^{2+}]_i$-dependent actions of taurine in spontaneously beating rabbit sino-atrial nodal cells. Eur J Pharmacol 424:19–25
Satoh H (2003) Electropharmacology of taurine on the hyperpolarization-activated inward current and the sustained inward current in spontaneous beating rat sino-atrial nodal cells. J Pharmacol Sci 91:229–238
Satoh H, Horie M (1997) Actions of taurine on the L-type Ca^{2+} channel current in guinea pig ventricular cardiomyocytes. J Cardiovasc Pharmacol 30:711–716
Satoh H, Sperelakis N (1992) Taurine inhibition of Na^{+}current in embryonic chick ventricular myocytes. Eur J Pharmacol 218:83–89
Satoh H, Sperelakis N (1993) Taurine effects on Ca^{2+}currents in young embryonic chick cardiomyocytes. Eur J Pharmacol 231:443–449
Satoh and Sperelakis N (1998) Review of some actions of taurine on ion channels of cardiac muscle cells and others. Gen Pharmacol 30:451–463
Satoh H, Nakatani A, Tanaka T, Haga T (2002) Cardiac functions and taurine's actions at different extracellular calcium concentrations in forced swimming stress-loaded rats. Biol Trac Ele Res 87:171–182
Schaffer SW, Kramer J, Chovan JP (1980) Regulation of calcium homeostasis in the heart by taurine. Fed Proc 39:2691–2694
Song D, O'Regan MH, Phillis JW (1998) Mechanisms of amino acid release from the isolated anoxic/reperfused rat heart. Eur J Pharmacol 351:313–322
Sperelakis N, Satoh H (1993) Taurine effects on ion channels of cardiac muscle. In: Noble D, Earn Y (eds) Ionic channels and effect of taurine on the heart. Kluwer Academic Publishers, Boston, pp 93–118
Sperelakis N, Satoh H, Bkaily G (1992) Taurine's effects on ionic current myocardial cells. In: Schaffer SW, Lombardim B (eds) Taurine: new dimensions on its mechanisms and actions, pp 129–143
Suleiman MS, Dihmis WC, Caputo M, Angelini GD, Bryan AJ (1997) Changes in myocardial concentration of glutamine and aspartate during coronary artery surgery. Am J Physiol 272:H1063–H1069
Wang JJ, Parker KH (2004) Wave propagation in a model of the arterial circulation. J Biomech 37:457–470

Chapter 6
Taurine Suppresses Pressor Response Through the Inhibition of Sympathetic Nerve Activity and the Improvement in Baro-Reflex Sensitivity of Spontaneously Hypertensive Rats

Takuzo Hano, Miki Kasano, Hiromi Tomari, and Naomi Iwane

Abstract To investigate the effect of taurine on the sympathetic nervous system, I observed pressor responses of perfused mesenteric arteries by electrical stimulation and baro-reflex sensitivity of spontaneously hypertensive rats (SHR/Izm) and control Wistar-Kyoto rats (WKY/Izm) treated with taurine. Taurine added to the perfusate suppressed norepinephrine (NE) overflow and the pressor response mediated by electrical stimulation of isolated perfused mesenteric arteries. Taurine showed a greater suppressive effect on NE overflow and the pressor response mediated by electrical stimulation in SHR/Izm rats than in WKY/Izm rats. Increment of renal sympathetic nerve activity induced by jet-air stress was suppressed in SHR/Izm rats treated with 3% taurine. The arterial baro-reflex sensitivity was more sensitive in SHR/Izm rats treated with taurine than in those exposed to no taurine. These data suggest that taurine lowers blood pressure directly via the suppression of NE release from the peripheral sympathetic nerve and the improvement of impaired baro-reflex sensitivity.

Abbreviations *SHR*, spontaneously hypertensive rat; *NE*, norepinephrine

6.1 Introduction

It has been reported in previous studies that spontaneously hypertensive rats (SHR) contain elevated plasma norepinephrine (NE) levels and increased hyper-reactivity to physical stress. We showed that arterial boro-reflex sensitivity (Kasamatsu 1996) and NE overflow from nerve ending was increased in mesenteric arteries of young SHRs (Hano and Rho 1989). Although taurine, a sulfur-containing amino-acid, has been shown to lower blood pressure in SHRs (Yamamoto et al. 1985; Trachtman

T. Hano (✉)
Wakayama Medical University, Postgraduate School of Medicine, Medical Education and Population-based Medicine, Wakayama, Japan

J. Azuma et al. (eds.), *Taurine 7*, Advances in Experimental Medicine and Biology 643,
DOI 10.1007/978-0-387-75681-3_6, © Springer Science+Business Media, LLC 2009

et al. 1986) and DOCA-salt rats (Fujita and Sato 1986) in part through suppression of plasma norepinephrine and stress-induced over-response of sympathetic activity in hypertensive models, the precise mechanism of taurine action is not clear.

We have previously detected pressor responses of perfused mesenteric arteries of spontaneously hypertensive rats (Izumo strain; SHR/Izm) as well as control Wistar-Kyoto rats (WKY/Izm) and renal sympathetic nerve activity of SHRs. In the present study, we will investigate the effects of taurine on sympathetic nervous system activity and the development of hypertension. Taurine was applied acutely to isolated mesenteric artery to investigate norepinephrine (NE) release from peripheral nerve. To clarify the effect of taurine on baro-reflex sensitivity in hypertensive rats, the rodents were chronically maintained on drinking water containing 3% taurine.

6.2 Methods

Six-week-old Wistar rats, SHR/Izm and age matched WKY/Izm rats were used in the present study. Male SHR/Izm strain rats were obtained from Disease Model Cooperative Research Association and were housed in 12-hour light/dark cycles. Animals were handled according to the institutional guidelines for animal research of Wakayama Medical University. In the chronic experiment, SHR/Izm rats were maintained for 3 weeks on tap water containing 3% taurine.

The isolated mesenteric arterial preparation was made as previously reported (Hano T and Rho J 1989). Polyethylene tubes were inserted into the superior mesenteric artery and the mesenteric artery-intestine loop preparation was perfused with Krebs-Henseleit solution. Platinum electrode was attached to the main trunk of the mesenteric artery to stimulate the peripheral sympathetic nerve and to measure NE in the effluent. Perfusion pressure was measured by a pressure transducer (TR-200 Gulton Ind. Inc.). NE was measured by HPLC using an electrochemical detector.

Arterial baro-reflex sensitivity was examined in animals allowed free movement. Stainless steel electrodes were attached around the renal nerve and fixed by Wacker–Sil Gel 604. Twenty four hours later, renal nerve activity was measured using a polygraph (AVB11-A Nihon-Kohden Co. Ltd.). The signal was magnified two thousand times as it passed through a band-pass filter (low cut 100Hz, high cut 1000Hz) and integral amplifier (EL601-G Nihon-Kohden Co. Ltd. Ltd.). Blood pressure and heart rate were measured by a pressure transducer (P23XL Vigo-Spectramed Inc.) which was connected to a polyethylene tube (PE-5 Intramedics Inc.) that was inserted into the femoral artery. Renal nerve activity was analyzed by non-linear regression of blood pressure and renal sympathetic activity after infusion of phenylephrine.

6.3 Results

6.3.1 Acute Experiment

Taurine added acutely to the perfusion solution suppressed NE overflow and the pressor response mediated by electrical stimulation of isolated mesenteric arteries of Wistar rats. However, taurine had no effect on vascular responsiveness caused by the administration of exogenous NE Fig. 6.1.

The suppressive effect of taurine on peripheral sympathetic nerve was more evident in the SHR/Izm rat than in the control WKY/Izm rat Fig. 6.2. Taurine did not suppress pressor responsiveness mediated by exogenous NE in either the SHR/Izm or the WKY/Izm rat Fig. 6.3

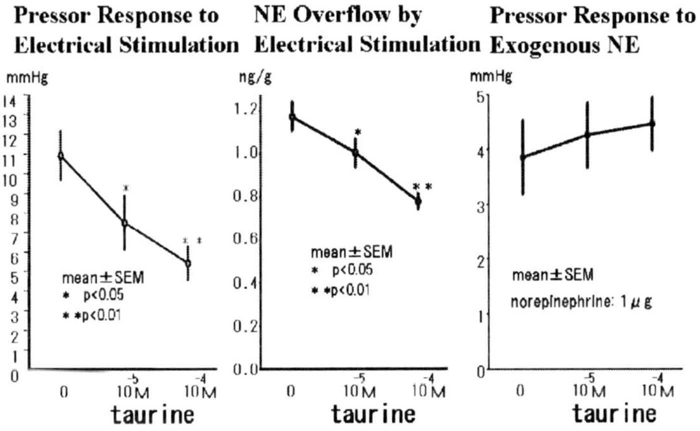

Fig. 6.1 Effects of taurine on pressor response to electrical stimulation (*left*), NE overflow by electrical stimulation (*middle*) and pressor response to exogenous NE (*right*) in Wistar rats. Taurine suppressed NE overflow and the pressor response mediated by electrical stimulation of isolated perfused mesenteric arteries of Wistar rats

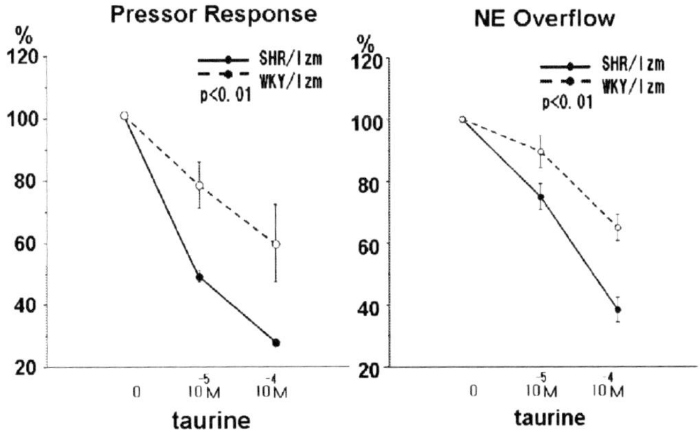

Fig. 6.2 Effect of taurine on pressor response (*left*) and NE overflow (*right*) mediated by electrical stimulation in SHR/Izm rats. Taurine suppressed NE overflow and the pressor response mediated by electrical stimulation of isolated perfused mesenteric arteries of SHR/Izm and WKY/Izm rats

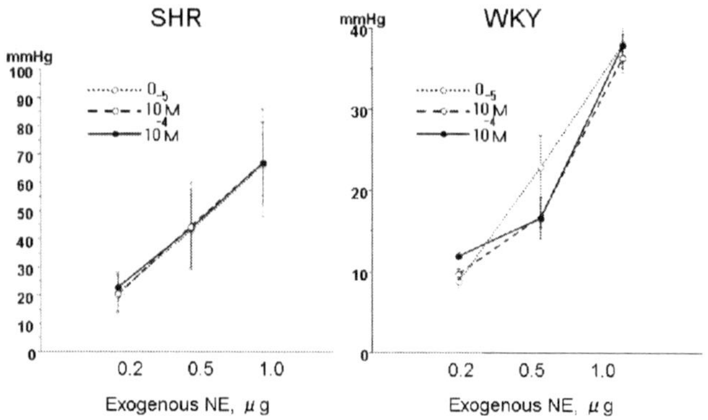

Fig. 6.3 Effect of taurine on pressor response to exogenous NE in SHR/Izm WKY/Izm rats. *Left panel* reveals pressor response in SHRs and *right panel* shows response in WKYs. Taurine had no significant effect on exogenous NE-induced vascular contraction in either SHR or WKY rats

6.3.2 Chronic Experiment

As shown in Table 6.1 chronic treatment with taurine blunted blood pressure elevation more in the SHR/Izm rat than in the control SHR/Izm rats.

Mesenteric arteries from SHRs that drank 3% taurine for 3 weeks were isolated and perfused with taurine free Krebs-Henseleit solution. The treated mesenteric arteries showed no significant difference in the pressor response to electrical stimulation. The pressor response to exogenous NE tended to be lower in SHR/Izm rats maintained on 3% taurine than in control rats maintained on tap-water (Fig. 6.4).

Kasamatsu (1996) reported that the magnitude of increased renal sympathetic nerve activity induced by jet-air stress was enhanced and baro-reflex sensitivity was blunted more in SHR/Izm than WKY/Izm rats. In the present study we found that arterial baro-reflex sensitivity was improved in SHRs maintained on 3% taurine (Fig. 6.5). The rise in blood pressure and renal nerve activity induced by jet-air stress was suppressed in the taurine group (Figs. 6.6, 6.7. Arterial baro-reflex sensitivity was examined by infusion of phenylephrine and analyzed by non-linear regression;

Table 6.1 Body weight (BW), systolic blood pressure (SBP) and heart rate (HR) of SHR/Izm rats

	6 week		9 week	
SHR/Izm	Taurine (–)	Taurine (+)	Taurine (–)	Taurine (+)
BW (g)	191.7 ± 1.9	189.9 ± 1.5	272.3 ± 2.3	262.3 ± 1.9
SBP (mmHg)	139.1 ± 3.8	136.4 ± 4.5	240.0 ± 4.4	209.0 ± 3.7
HR (/min)	411.6 ± 5.8	408.8 ± 7.9	410.0 ± 6.5	411.1 ± 7.6

Data shown are mean ± SEM Taurine (+): SHR/Izm rats maintained on 3% taurine for 3 weeks, Taurine (-): SHR/Izm rats maintained on tap-water. Chronic taurine treatment reduced blood pressure in SHR/Izm rats.

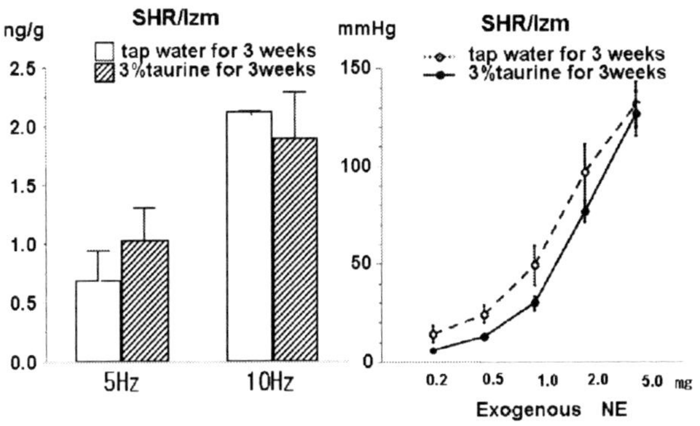

Fig. 6.4 Effect of chronic taurine treatment on NE overflow (*left*) and pressor response to exogenous NE (*right*) in SHR/Izm rats. Chronic taurine treatment showed no significant effect on pressor response and NE overflow in isolated mesenteric arteries of SHR/Izm rats perfused with Krebs-Henseleit solution lacking taurine

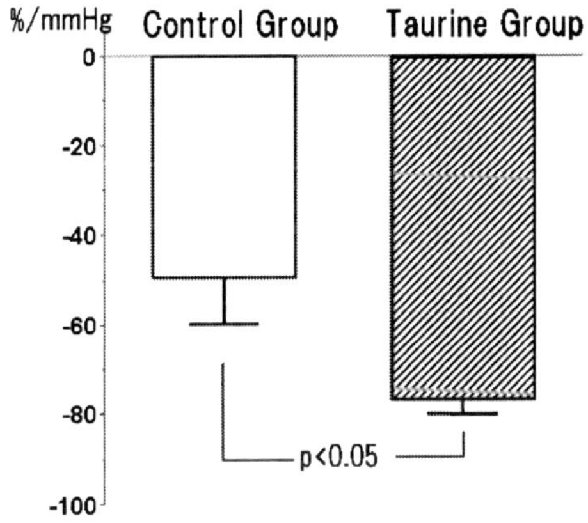

Fig. 6.5 Effect of taurine on baro-reflex sensitivity in SHR/Izm rats. Arterial baro-reflex sensitivity, examined by infusion of phenylephrine and analyzed by non-linear regression, was more sensitive in SHR/Izm rats with taurine (Taurine group) than in rats lacking taurine treatment (Control group)

these revealed greater sensitivity in SHRs treated with taurine than in those maintained on taurine free tap water.

6.4 Discussion

Several reports have shown that basal plasma NE and stress-evoked plasma NE elevations are increased in SHRs. We have observed enhanced norepinephrine overflow from peripheral sympathetic nerve terminals of isolated mesenteric arteries of young SHRs (Hano et al. 1986). Arterial baro-reflex sensitivity was blunted in our

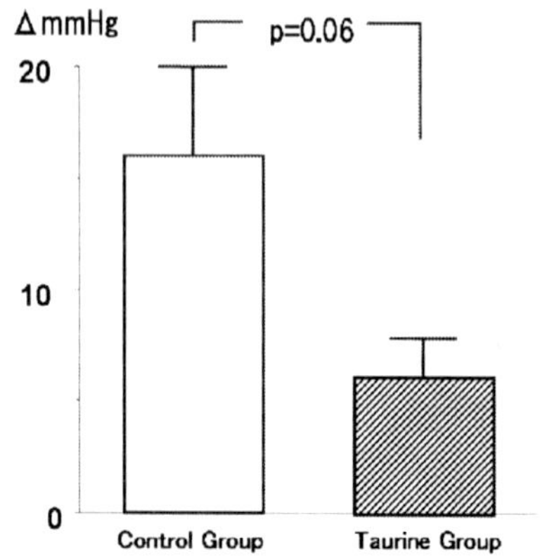

Fig. 6.6 Blood pressure elevation mediated by jet-air blowing in SHR/Izm rats. Taurine treatment tended to suppress blood pressure elevation mediated by jet-air stress ($p = 0.06$)

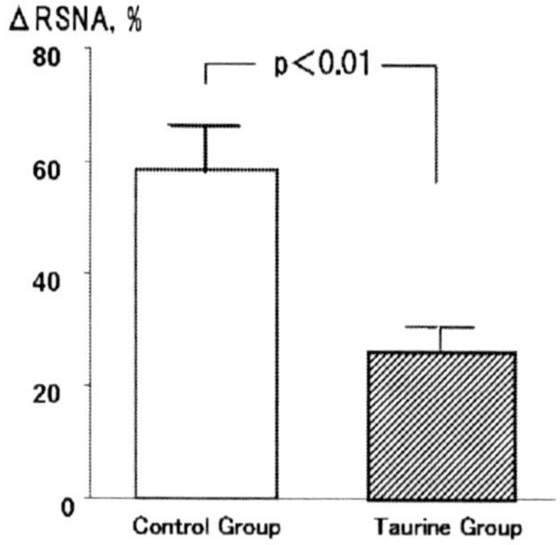

Fig. 6.7 Renal nerve activation by jet-air blowing in SHR/Izm rats. Enhanced renal sympathetic nerve activity derived from jet-air stress was markedly suppressed in the SHR/Izm rats treated with 3% taurine (Taurine group) compared with control SHR rats (Control group)

data and this mechanism may contribute to altered sympathetic activity of the SHR rats (Kasamatsu1996).

Although taurine is present in high amounts in the brain and heart, its physiological role is unclear. It has been reported that plasma and cerebral levels of taurine are reduced in SHRs (Kuriyama et al. 1984). Taurine supplementation suppresses blood pressure, stress-induced blood pressure and plasma catecholamine changes in SHRs (Yamamoto et al. 1985) as well as blood pressure in DOCA-salt rats (Fujita and Sato 1986). Sympatho-adrenal tone is increased in young borderline hypertensive patients but oral administration of taurine attenuates the elevated sympathetic tone and the elevated blood pressure (Fujita et al. 1987).

In the present study, taurine can act on peripheral sympathetic nerve terminals to suppress NE overflow. Taurine also suppresses blood pressure elevation and sympatho-inhibitory action in SHR/Izm rats, an effect that is less in control WKY/Izm rats. Although the mechanism underlying the inhibitory effects of taurine on sympathetic nerve activity is unclear, alterations in membrane potential or calcium influx may contribute to these effects (Satoh 1998). Chronic treatment with taurine improves arterial baro-reflex sensitivity in SHRs, which exhibit impaired baro-reflex function. Blood pressure elevation and renal nerve activation mediated by jet-air stress was blunted more in taurine-treated SHR/Izm rats than in control SHR/Izm rats.

6.5 Conclusion

In summary, these data suggest that taurine suppresses NE release from the peripheral sympathetic nerves, an effect of taurine that is both direct and dependent upon central mechanisms that regulate baro-reflex sensitivity and stress-induced activation. These mechanisms may counter the consequences arising from reduced plasma and central taurine content.

Acknowledgments This study was funded by a grant from Taisho Co. Ltd.

References

Hano T, Rho J (1989) Norepinephrine overflow in perfused mesenteric arteries of spontaneously hypertensive rats. Hypertension 14:44–53

Yamamoto J, Akabane S, Yoshimi H et al. (1985) Effects of taurine on stress-evoked hemodynamic and plasma catecholamine changes in spontaneously hypertensive rats. Hypertension 7: 913–922

Trachtman H, Del Pizzo R, Rao P et al (1986) Taurine lowers blood pressure in the spontaneously hypertensive rats by a catecholamine independent mechanism. Am J Hypertens 2:909–912

Fujita T, Sato Y (1986) Changes in blood pressure and extra cellular fluid with taurine in DOCA-salt rats. Am J Physiol 250:R1014–R1020

Kasamatsu K (1996) The relationship between the reactivity of sympathetic nervous system and the development of hypertension in experimental hypertension. J Wakayama Med Soc 47:267–273

K, Ida S, Ohkuma S (1984) Alteration of cerebral taurine biosynthesis in spontaneously hypertensive rats. J Neurochem 42:1600–1606

Fujita T, Ando K, Noda H et al. (1987) Effects of increased adrenomedullary activity and taurine in young patients with borderline hypertension. Circulation 75:525–532

Satoh H (1998) Cardiac actions of taurine as a modulator of the ion channels. Adv Exp Med Biol 442:121–128

Chapter 7
Beneficial Effect of Taurine Treatment Against Doxorubicin-Induced Cardiotoxicity in Mice

Takashi Ito, Satoko Muraoka, Kyoko Takahashi, Yasushi Fujio, Stephen W. Schaffer, and Junichi Azuma

Abstract Though the administration of taurine is clinically efficacious against heart failure, the mechanism underlying its cardioprotection remains to be established. To provide information on the mechanism, we examined the effects of taurine on doxorubicin (DOX)-induced cardiotoxicity, with an emphasis on ROS generation and cardiac gene inhibition. Oral administration of taurine (3% w/v in tap water) dramatically reduced the mortality rate in both the acute or sub-acute toxic models of DOX toxicity. It was shown that taurine prevented DOX-induced oxidative stress as determined from cardiac glutathione content. Interestingly, Northern blot analysis revealed that DOX altered cardiac gene expression, including that of α-myosin heavy chain, ventricular myosin light chain-2 isoform and brain natriuretic peptide, an effect partially ameliorated by taurine treatment. In conclusion, taurine suppresses ROS generation and regulates gene expression in the DOX treated heart.

Abbreviations *DOX*, doxorubicin; *ROS*, reactive oxygen species; *GSH*, glutathione; α*MHC*, α Myosin heavy chain; *MLC2v*, myosin light chain type 2v; *BNP*, brain natriuretic peptide

7.1 Introduction

Doxorubicin (DOX) is an antineoplastic agent used against a wide variety of malignancies. However, the clinical use of the drug is limited largely because of its cardiotoxicity that leads to the development of a cardiomyopathy and eventually to overt heart failure (Minnotti et al. 2004; Singal and Iliskovic 1998). Although the mechanisms underlying the development of irreversible myocardial damage remain unclear, it has been shown that the generation of reactive oxygen species (ROS) is one of the critical events in the onset of the cardiomyopathy. ROS generation is closely related with the impairment in mitochondrial function and calcium handling. Importantly, the DOX-induced cardiomyopathy is associated with the

T. Ito (✉)
Department of Clinical Pharmacology and Pharmacogenomics, Graduate School of Pharmaceutical Sciences, Osaka University, Japan; Department of Pharmacology, University of South Alabama, College of Medicine, USA

J. Azuma et al. (eds.), *Taurine 7*, Advances in Experimental Medicine and Biology 643, DOI 10.1007/978-0-387-75681-3_7,

downregulation of cardiac muscle-specific gene expression, such as α-myosin heavy chain (αMHC), myosin light chain ventricular type (MLC2v), brain natriuretic peptide (BNP) and sarcoplasmic reticulum proteins (Arai et al. 1998; Ito et al. 1990), resulting in myofibrillar loss and cardiac dysfunction.

Taurine (2-aminoethylsulfonic acid), a sulfur-containing amino acid, is found in millimolar concentrations in most mammalian tissues, being especially high in the heart. Taurine mediates many physiological functions, such as calcium handling, osmoregulation, membrane stabilization and detoxication (Huxtable 1992; Schaffer et al. 2000a). In cultured cardiomyocytes, taurine promotes cell survival during ischemia (Takatani 2004a, b). Collectively, taurine contributes to the maintenance of cellular homeostasis and is therefore clinically useful in treating certain pathological conditions (Azuma et al. 1982). However, in spite of the pharmacological benefits of taurine, detailed mechanisms of its cardioprotection have not been clarified.

In present study, we tested the effects of taurine on mortality following DOX treatment, focusing on changes in ROS generation and cardiac gene expression. This study provides the biological basis for the treatment of the DOX-induced cardiomyopathy with taurine.

7.2 Methods

7.2.1 Animals and Treatment

The experimental procedures conformed to the guidelines of the Institutional Animal Care and Use Committee of Osaka University. Six-week-old male C57BL6 mice were used in the present study. In the taurine treated group, animals were maintained on tap water containing 3% (w/v) taurine starting 1 week before DOX administration. To produce the acute toxicity model, DOX (Kyowa Hakko, Japan) was administered to mice as a single injection (15 mg/kg i.p.) (Kunisada et al. 2000). In the sub-acute toxicity model, DOX was administered for 6 weeks (5 mg/kg/week i.p.) (Taniyama and Walsh 2002). The survival rate was monitored for either 1 month (acute) or 2 months (sub-acute).

7.2.2 Measurement of Glutathione

The level of total glutathione (GSH+GSSG) was measured by the glutathione reductase (GR)/ 5,5′-dithibis(2-aminobenzoic acid) (DTNB) recycling assay (Anderson 1985). Tissues were homogenized in 0.1 M phosphate buffer, pH 7.5, containing 5% sulfosalicylic acid and were centrifuged at 10000 xg. Aliquots of supernatant were added to phosphate buffer (30°C) containing 0.27 mM NADPH, DNTB and 0.8 U GR. The reaction was monitored at 412 nm for 30 min.

7.2.3 Northern Blotting

Total RNA was isolated from control and DOX treated hearts and Northern blots were obtained as previously described (Ita et al. 2004). cDNA probes for BNP (Nakaoka et al. 2003), and GAPDH were labeled with Megaprime DNA Labeling System (Amersham Bioscience, USA) according to the protocol. The following oligonucleotides (Zhang et al. 2001) (Invitrogen, USA) were labeled with $\gamma-{}^{32}$P ATP by using T4 polynucleotide kinase (TOYOBO, Japan) according to the protocol. The following probes were used:

α-MHC: 5'-CGA ACG TTT ATG TTT ATT GTG GAT TGG CCA CAG CGA GGG TCT GCT GGA GAG GTT ATT CCT CGT C-3'.
β-MHC: 5'-GAG GGC TTC ACG GGC ACC CTT AGA GCT GGG TAG CAC AAG ATC TAC TCC TCA TTC AGG CC-3'.
MLC2v: 5'-CAC AGC CCT GGG ATG GAG AGT GGG CTG TGG GTC ACC TGA GGC TGT GGT TCA G-3'.
Sarcoplasmic reticulum Ca^{2+}-ATPase 2v (SERCA2a): 5'-TCA GTC ATG CAG AGG GCT GGT AGA TGT GTT GCT AAC AAC GCA CAT GCA CGC ACC CGA ACA-3'.

7.2.4 Statistical Analysis

Each value was expressed as the mean ± SEM. Statistical significance was determined by the Student's t-test. Survival data were analyzed by the Kaplan Meier method. Differences were considered statistically significant when the calculated P value was less than 0.05.

7.3 Results

7.3.1 Taurine Improved the Survival Rate After DOX Treatment

Taurine promotes cardiomyocyte survival in models of DOX toxicity. Therefore, we examined the effect of taurine treatment on the development of the in acute and sub-acute models of DOX cardiotoxic. As seen in Fig 7.1, DOX treated mice maintained on tap water containing 3% taurine observed greater rates of survival than mice maintained on taurine free tap water. In the acute DOX model, mice that were maintained on normal tap water exhibited a mortality rate of 70% 28 days following DOX injection while mice maintained on tap water supplemented with 3% taurine exhibited a mortality rate of only 10%. Similarly, in the sub-acute DOX model, 100% of the mice without taurine treatment were dead 43 days after the initial DOX injection, however, 50% of the mice treated with taurine survived at least 60 days, at which time the experiment was terminated. Thus, treatment with taurine significantly attenuated DOX-induced mortality.

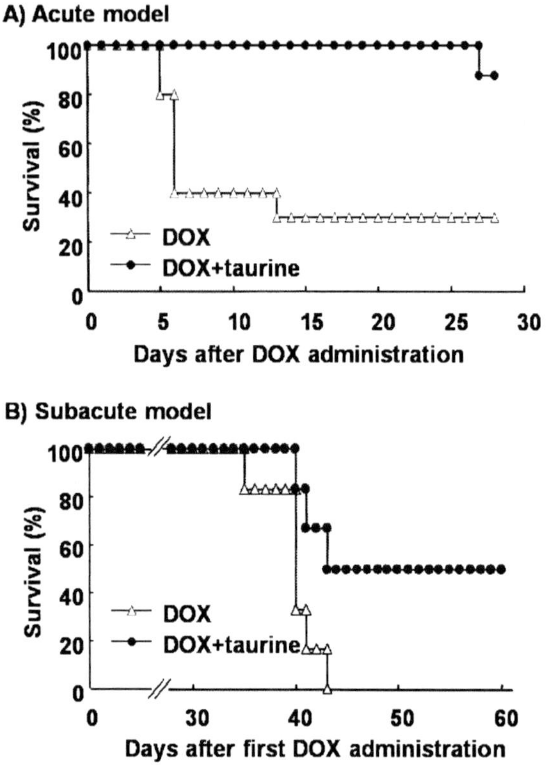

Fig. 7.1 The effects of taurine treatment on survival in acute (A) and sub-acute (B) models of DOX-induced cardiomyopathy. (**A**) Ten mice in each group were subjected to a single injection of 15 mg/kg i.p. of DOX and were monitored for an additional 4 weeks. (**B**) Six mice in each group were subjected to 6 injections of 5 mg/kg i.p. of DOX once a week and were monitored for an additional 8 weeks. *P* values were 0.011 (A) and 0.022 (B)

Cardiac enlargement is a common phenotype of cardiac remodeling seen following pathological stress. In order to examine the effects of taurine on DOX-induced cardiac enlargement, we analyzed the heart weight to body weight (HW/BW) ratio 28 days following DOX injection (15 mg/g BW) (Fig. 7.2). DOX increased the HW/BW ratio by 125%, an effect suppressed by taurine administration.

7.3.2 *DOX-Mediated Oxidant Generation was Inhibited by Taurine Treatment*

DOX-induced free radical production plays an important role in the progression of the DOX cardiomyopathy. To investigate the effect of taurine on DOX-induced

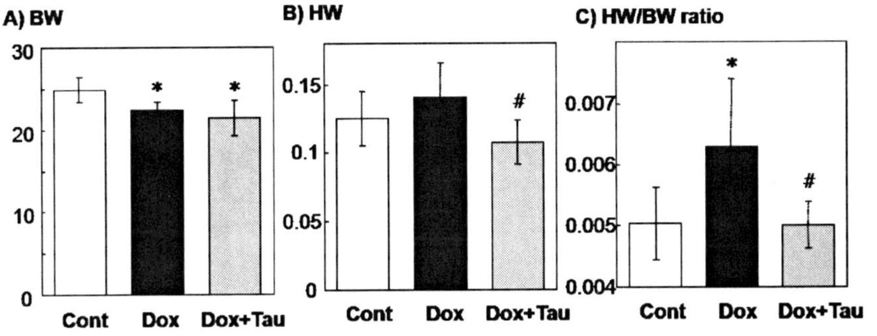

Fig. 7.2 The effect of taurine on DOX-induced cardiac remodeling. In the acute DOX model, body (**A**) and heart (**B**) weight were measured 28 days following initial DOX administration and the heart weight/body weight ratio was calculated (**C**). Data represent means ± S.D. *; p<0.05 vs. control group (Cont), #; p<0.05 vs. DOX

ROS generation in the heart, glutathione (GSH) was measured, which is known to be inversely associated with ROS production, as GSH is intrinsically reduced by oxidants (Zhou et al. 2001). The GSH content of the heart was reduced by DOX treatment, an effect attenuated by taurine treatment (Fig. 7.3). This result indicates that taurine suppresses DOX-induced ROS generation.

7.3.3 Dox-Induced Alterations in Cardiac Gene Expression were Inhibited by Taurine Treatment

The downregulation of specific genes of the heart is characteristic of the DOX-induced cardiomyopathy (Ito et al. 1990). To examine the effect of taurine treatment on the expression of cardiac genes of mice injected with DOX, Northern blots of

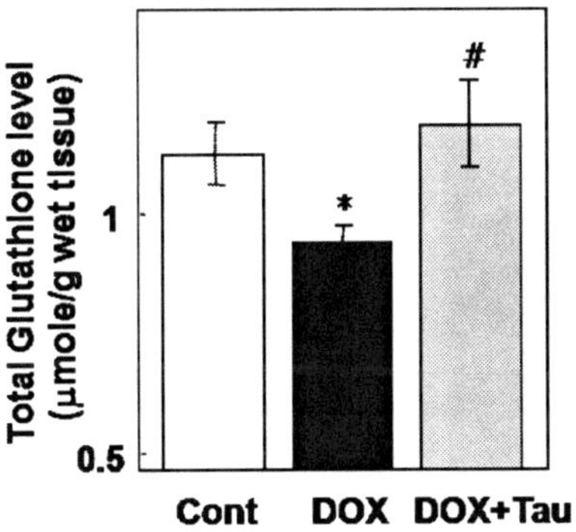

Fig. 7.3 The effect of taurine on the generation of ROS in cardiac tissues after DOX administration. Hearts of mice were removed 2 days after DOX injection and were immediately homogenized. Total glutathione content was then measured. Assays were repeated twice with similar results. Data represent means ± S.D., n=3. *; p< 0.05 vs. control group (Cont), #; p< 0.05 vs. DOX

αMHC, MLC2v, BNP and SERCA2a were obtained (Fig. 7.4). All of these genes were downregulated 2 days after the administration of DOX, an effect suppressed by taurine treatment. βMHC mRNA was increased in DOX-treated mice, but the change was only partially attenuated by taurine treatment (Data not shown). These results suggest that taurine may ameliorate cardiac remodeling through the suppression of the DOX-induced phenotype.

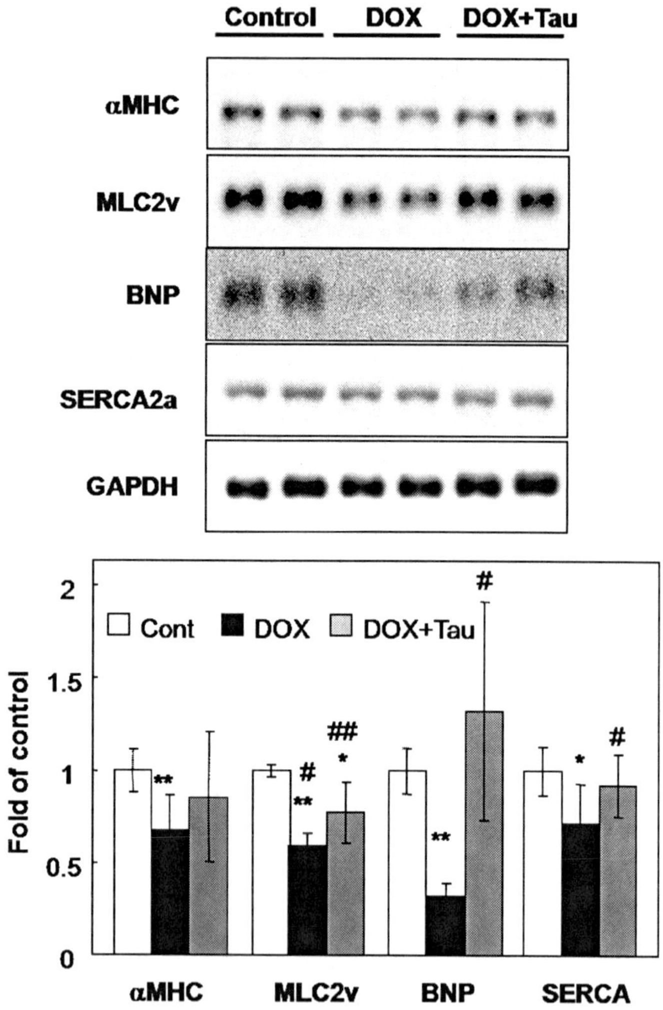

Fig. 7.4 The effects of taurine on cardiac gene expression in DOX-treated mice. Total RNA was prepared from hearts 2 days after DOX injection and then subjected to Northern blot analysis. Representative autoradiograms from 2–3 independent experiments involving 5–12 mice in each group are shown. Data are mean ± S.D. *; $p < 0.05$, **; $p < 0.01$ vs. control (Cont), $^{\#}$; $p < 0.05$, $^{\#\#}$; $p < 0.01$ vs. DOX

7.4 Discussion

In the present study, we demonstrated that taurine significantly reduced mortality in both the acute and sub-acute models of DOX cardiotoxicity. Treatment with taurine attenuated the progression of cardiac dysfunction, including the generation of ROS and the alteration in myocardial gene expression. While taurine has been implicated in various functions of the heart, including modulation of ion current, antioxidation and osmoregulation (Huxtable 1992; Schaffer et al. 2000b), the critical cytoprotective role of taurine remains to be clarified. Consistent with the present study, it has been demonstrated that treatment with taurine attenuates the degree of myocardialROS generation caused by various toxins or pathological stimulants (Oudit et al. 2004), an effect unlikely to be related to free radical scavening since taurine chemically reacts with ROS to only a limited extent (Cunningham et al. 1998). Treatment with antioxidant agents, such as probucol (Siveski-Iliskovic et al. 1995) and N-acetylcysteine (Villani et al. 1990), or overexpression of antioxidant proteins, such as Mn-SOD (Yen et al. 1996) and catalase (Kang et al. 1996), attenuate DOX-induced cardiotoxicity. Thus, the antioxidant effect of taurine may contribute to the suppression of DOX-induced cardiac remodeling.

DOX downregulates specific muscle proteins, which contribute to alterations in Z-band structure and to disarray of the thin filaments (Ito et al. 1990). The transcription of these genes is mediated by regulatory factors Nkx2.5 (Lints et al. 1993; Tanaka et al. 1999), MEF2c (Edmondson et al. 1994) and GATA4 (Molkentin et al. 1994, Thuerauf et al. 1994), as well as transcription co-factor p300 (Poizat et al. 2000). Importantly, overexpression of Nkx2.5 or p300 in mice not only blocks DOX-induced inhibition of myocardial gene expression but also increases survival (Kawamura et al. 2004; Toko et al. 2002). Thus, the control of cardiac genes by taurine may be the critical step in the cytoprotective actions of taurine against DOX-induced cardiotoxicity and cardiac remodeling. To test this hypothesis, we measured the levels of the transcriptional factors in our animal model. It was revealed that the levels of neither Nkx2.5 nor MEF2c were influenced by DOX injection *in vivo* (data not shown), although they were reportedly downregulated by DOX in vitro (Poizat et al. 2000). Yet in agreement with a study by Aries et al. (2004), we found that the levels of GATA4 were downregulated by DOX administration. However, taurine treatment did not attenuate the response (data not shown). Thus, although we cannot completely exclude the possibility that alterations in the expression of these transcriptional factors may be partially caused by DOX-induced downregulation of cardiac genes, such as αMHC, MLC2v, BNP and SERCA2a, causality between the downregulation of transcriptional factors and that of cardiac specific genes seems unlikely. However, further studies examining the influence of DOX on myocardial gene regulation, especially in vivo, could be beneficial in understanding the molecular mechanisms underlying the cardioprotective actions of taurine.

Several studies have addressed the relationship between DOX-induced ROS generation and the downregulation of myocardial genes. It has been demonstrated that anti-oxidants, such as N-acetylcysteine and catalase, do not prevent DOX-induced inhibition of gene expression in cultured cardiomyocytes (Torti et al. 1998).

Moreover, while it has been reported that heart-specific overexpression or upregulation of metallothionein in mice protects against DOX-induced cardiac injury through the suppression of ROS generation (Kang et al. 1997; Wang et al. 2001; Oshima et al. 2002), elevated metallothionein did not prevent the downregulation of cardiac genes, including αMHC and MLC2v (our unpublished data). These data suggest that the myocardial genes are downregulated by DOX injection independent of oxidative stress. Collectively, our results indicate that taurine protects cardiac tissue through diverse mechanisms and these pharmacological effects of taurine might explain the remarkable improvement in survival following DOX administration.

7.5 Conclusion

Treatment with taurine confers resistance against DOX-induced cardiotoxicity. The protective effect correlates with the suppression of ROS generation and amelioration of impaired myocardial gene expression. These findings suggest that taurine might be clinically useful in the treatment of the DOX-induced cardiomyopathy. It raises the interesting possibility that overexpression of taurine-linked proteins, such as the taurine transporter, might represent a novel therapeutic approach in reducing the severity of the DOX cardiomyopathy.

Acknowledgments We thank Yasuko Murao for her secretary work. The work was supported by a grant from Taisho Pharmaceutical Co. Ltd and a Grant-in-Aid for Scientific Research from Ministry of Education, Science, Sports, and Culture of Japan. We acknowledge Kyowa Hakko for the donation of DOX.

References

Anderson ME (1985) Determination of glutathione and glutathione disulfide in biological samples. Methods Enzymol 113:548–555

Arai M, Tomaru K, Takizawa T, Sekiguchi K, Yokoyama T, Suzuki T, Nagai R (1998) Sarcoplasmic reticulum genes are selectively down-regulated in cardiomyopathy produced by doxorubicin in rabbits. J Mol Cell Cardiol 30:243–254

Aries A, Paradis P, Lefebvre C, Schwartz RJ, Nemer M (2004) Essential role of GATA-4 in cell survival and drug-induced cardiotoxicity. Proc Natl Acad Sci USA 101:6975–6980

Azuma J, Hasegawa H, Sawamura A, Awata N, Harada H, Ogura K, Kishimoto S (1982) Taurine for treatment of congestive heart failure. Int J Cardiol 2:303–304

Cunningham C, Tipton KF, Dixon HB (1998) Conversion of taurine into N-chlorotaurine (taurine chloramine) and sulphoacetaldehyde in response to oxidative stress. Biochem J 330(Pt 2): 939–945

Edmondson DG, Lyons GE, Martin JF, Olson EN (1994) Mef2 gene ex-pression marks the cardiac and skeletal muscle lineages during mouse embryogenesis. Development 120:1251–1263

Huxtable RJ (1992) Physiological actions of taurine. Physiol Rev 72:101–163

Ito H, Miller SC, Billingham ME, Akimoto H, Torti SV, Wade R, Gahlmann R, Lyons G, Kedes L, Torti FM (1990) Doxorubicin selectively inhibits muscle gene expression in cardiac muscle cells in vivo and in vitro. Proc Natl Acad Sci USA 87:4275–4279

Ito T, Fujio Y, Hirata M, Takatani T, Matsuda T, Muraoka S, Takahashi K, Azuma J (2004) Expression of taurine transporter is regulated through the TonE (tonicity-responsive element)/TonEBP (TonE-binding protein) pathway and contributes to cytoprotection in HepG2 cells. Biochem J 382:177–182

Kang YJ, Chen Y, Epstein PN (1996) Suppression of doxorubicin cardiotoxicity by overexpression of catalase in the heart of transgenic mice. J Biol Chem 271:12610–12616

Kang YJ, Chen Y, Yu A, Voss-McCowan M, Epstein PN (1997) Overexpression of metallothionein in the heart of transgenic mice suppresses doxorubicin cardiotoxicity. J Clin Invest 100: 1501–1506

Kawamura T, Hasegawa K, Morimoto T, Iwai-Kanai E, Miyamoto S, Kawase Y, Ono K, Wada H, Akao M, Kita T (2004) Expression of p300 pro-tects cardiac myocytes from apoptosis in vivo. Biochem Biophys Res Commun 315:733–738

Kunisada K, Negoro S, Tone E, Funamoto M, Osugi T, Yamada S, Okabe M, Kishimoto T, Yamauchi-Takihara K (2000) Signal transducer and activator of transcription 3 in the heart transduces not only a hypertrophic sig-nal but a protective signal against doxorubicin-induced cardiomyopathy. Proc Natl Acad Sci USA 97:315–319

Lints TJ, Parsons LM, Hartley L, Lyons I, Harvey RP (1993) Nkx-2.5: a novel murine homeobox gene expressed in early heart progenitor cells and their myogenic descendants. Development 119:419–431

Minotti G, Menna P, Salvatorelli E, Cairo G, Gianni L (2004) Anthracyclines: molecular advances and pharmacologic developments in antitumor activity and cardiotoxicity. Pharmacol Rev 56:185–229

Molkentin JD, Kalvakolanu DV, Markham BE (1994) Transcription factor GATA-4 regulates cardiac muscle-specific expression of the alpha-myosin heavy-chain gene. Mol Cell Biol 14: 4947–4957

Nakaoka Y, Nishida K, Fujio Y, Izumi M, Terai K, Oshima Y, Sugiyama S, Matsuda S, Koyasu S, Yamauchi-Takihara K, Hirano T, Kawase I, Hirota H (2003) Activation of gp130 transduces hypertrophic signal through interaction of scaffolding/docking protein Gab1 with tyrosine phosphatase SHP2 in cardiomyocytes. Circ Res 93:221–229

Oshima Y, Fujio Y, Funamoto M, Negoro S, Izumi M, Nakaoka Y, Hirota H, Yamauchi-Takihara K, Kawase I (2002) Aldosterone augments endothelin-1-induced cardiac myocyte hypertrophy with the reinforcement of the JNK pathway. FEBS Lett 524:123–126

Oudit GY, Trivieri MG, Khaper N, Husain T, Wilson GJ, Liu P, Sole MJ, Backx PH (2004) Taurine supplementation reduces oxidative stress and improves cardiovascular function in an iron-overload murine model. Circulation 109:1877–1885

Poizat C, Sartorelli V, Chung G, Kloner RA, Kedes L (2000) Proteasome-mediated degradation of the coactivator p300 impairs cardiac transcription. Mol Cell Biol 20:8643–8654

Schaffer S, Solodushko V, Azuma J (2000a) Taurine-deficient cardiomyopathy: role of phospholipids, calcium and osmotic stress. Adv Exp Med Biol 483:57–69

Schaffer S, Takahashi K, Azuma J (2000b) Role of osmoregulation in the actions of taurine. Amino Acids 19:527–546

Singal PK, Iliskovic N (1998) Doxorubicin-induced cardiomyopathy. N Engl J Med 339:900–905

Siveski-Iliskovic N, Hill M, Chow DA, Singal PK (1995) Probucol protects against adriamycin cardiomyopathy without interfering with its antitumor effect. Circulation 91:10–15

Takatani T, Takahashi K, Uozumi Y, Matsuda T, Ito T, Schaffer SW, Fujio Y, Azuma J (2004a) Taurine prevents the ischemia-induced apoptosis in cultured neonatal rat cardiomyocytes through Akt/caspase-9 pathway. Bio-chem Biophys Res Commun 316:484–489

Takatani T, Takahashi K, Uozumi Y, Shikata E, Yamamoto Y, Ito T, Ma-tsuda T, Schaffer SW, Fujio Y, Azuma J (2004b) Taurine inhibits apoptosis by preventing formation of the Apaf-1/caspase-9 apoptosome. Am J Physiol Cell Physiol 287:C949–C953

Tanaka M, Chen Z, Bartunkova S, Yamasaki N, Izumo S (1999) The cardiac homeobox gene Csx/Nkx2.5 lies genetically upstream of multiple genes essential for heart development. Development 126:1269–1280

Taniyama Y, Walsh K (2002) Elevated myocardial Akt signaling amelio-rates doxorubicin-induced congestive heart failure and promotes heart growth. J Mol Cell Cardiol 34:1241–1247
Thuerauf DJ, Hanford DS, Glembotski CC (1994) Regulation of rat brain natriuretic peptide transcription. A potential role for GATA-related transcription factors in myocardial cell gene expression. J Biol Chem 269:17772–17775
Toko H, Zhu W, Takimoto E, Shiojima I, Hiroi Y, Zou Y, Oka T, Aka-zawa H, Mizukami M, Sakamoto M, Terasaki F, Kitaura Y, Takano H, Nagai T, Nagai R, Komuro I (2002) Csx/Nkx2-5 is required for homeosta-sis and survival of cardiac myocytes in the adult heart. J Biol Chem 277:24735–24743
Torti SV, Akimoto H, Lin K, Billingham ME, Torti FM (1998) Selective inhibition of muscle gene expression by oxidative stress in cardiac cells. J Mol Cell Cardiol 30:1173–1180
Villani F, Galimberti M, Monti E, Piccinini F, Lanza E, Rozza A, Favalli L, Poggi P, Zunino F (1990) Effect of glutathione and N-acetylcysteine on in vitro and in vivo cardiac toxicity of doxorubicin. Free Radic Res Com-mun 11:145–151
Wang GW, Klein JB, Kang YJ (2001) Metallothionein inhibits doxorubicin-induced mitochondrial cytochrome c release and caspase-3 activation in cardiomyocytes. J Pharmacol Exp Ther 298:461–468
Yen HC, Oberley TD, Vichitbandha S, Ho YS, St Clair DK (1996) The protective role of manganese superoxide dismutase against adriamycin-induced acute cardiac toxicity in transgenic mice. J Clin Invest 98:1253–1260
Zhang X, Azhar G, Chai J, Sheridan P, Nagano K, Brown T, Yang J, Khrapko K, Borras AM, Lawitts J, Misra RP, Wei JY (2001) Cardiomyopathy in transgenic mice with cardiac-specific overexpression of serum response factor. Am J Physiol Heart Circ Physiol 280:H1782–H1792
Zhou S, Palmeira CM, Wallace KB (2001) Doxorubicin-induced persistent oxidative stress to cardiac myocytes. Toxicol Lett 121:151–157

Chapter 8
Antihypertensive Effect of Taurine in Rat

Jianmin Hu, Xingli Xu, Jiancheng Yang, Gaofeng Wu, Changmian Sun, and Qiufeng Lv

Abstract To investigate the effect of taurine on hypertension, a rat model of hypertension was produced by administering N-nitro-L-arginine methylester (L-NAME) to reduce the levels of the vasodilator, nitric oxide. At the same time that L-NAME was administered, taurine treated animals received either 1% or 2% taurine in the drinking water. As a control, 1% taurine was added to the water without L-NAME administration in order to investigate the effects of taurine on blood pressure of normal rats. The results showed that taurine increased serum levels of nitric oxide and nitric oxide synthase, inhibited the elevation of blood pressure, interfered with the activity of the renin-angiotensin-aldosterone system and minimized the elevation in serum cytokine, endothelin, neuropeptide Y and thromboxane B_2. It also reduced oxygen derived free radical generation, upregulated the antioxidant defenses and inhibited the proliferation of vascular smooth muscle cells. These data indicate that taurine benefits hypertensive rats, with 2% taurine mediating greater improvement than 1% taurine.

Abbreviations *VSMC*, vascular smooth muscle cells; *ET*, endothelins; *NPY*, neuropeptide Y; *AngI*, angiotensin I; *AngII*, angiotensin II; *ACE*, angiotensin converting enzyme; *TXB2*, thromboxane B2; *ANP*, atrial natriuretic polypeptide; *MDA*, malondialdehyde; *SOD*, superoxide dismutase; *NO*, nitric oxide; *NOS*, nitric oxide synthase

8.1 Introduction

Hypertension is a common cardiovascular disease throughout the world and is associated with complications in heart, brain, kidney and other tissues that seriously endanger human health.

J. Hu (✉)
College of Animal Science and Veterinary Medicine, Shenyang Agricultural University, Shenyang, China

J. Azuma et al. (eds.), *Taurine 7*, Advances in Experimental Medicine and Biology 643, DOI 10.1007/978-0-387-75681-3_8, © Springer Science+Business Media, LLC 2009

Taurine is a free amino acid present in high concentrations in tissues of most mammals. It has been reported that taurine plays an important role in neuro-modulation, regulation of calcium-dependent processes, osmoregulation and thermoregulation (Schaffer and Azuma 1992). There is evidence that taurine mediates a number of cardiovascular functions, including the regulation of blood pressure (Trachtman et al. 1998). It has been reported that taurine administration lowers blood pressure in various hypertensive animal models, including spontaneous hypertension (Paakkari et al. 1983), deoxycorticosterone acetate (DOCA) -salt induced hypertension (Fujita et al. 1987), insulin resistant hypertension (Yu et al. 2000) and renal hypertension (Zheng et al. 2003), as well as hypertension in human beings. However, there are few reports concerning the beneficial effect of taurine therapy on the severity of hypertension. The present study was initiated to elucidate the mechanism of taurine action, which could ultimately prove valuable in the treatment of hypertension.

8.2 Methods and Materials

8.2.1 Experimental Animals

Male SD rats weighing 120–140 g were used and were acclimatized for 7 days before onset of the experiment. Rats were maintained under a controlled environment of light (12 hr of light, 12 hr of dark) and temperature (23±2°C), and were given free access to commercial standard rat chow and water.

8.2.2 Chemicals

L-NAME and taurine were purchased from Sigma Chemical Company. (St. Louis, MO, USA.) Reagent kits of malondialdehyde (MDA), superoxide dismutase (SOD), nitric oxide (NO) and nitric oxide synthase (NOS) were purchased from Nanjing Jiancheng Bioengineering Institute. Reagent kit of Angiotensin-Converting (ACE) was purchased from navy general hospital of People's Liberation Army. endothelins (ET), neuropeptide Y (NPY), angiotensin I (AI), angiotensin II (AII), thromboxane B_2(TXB_2) and atrial natriuretic polypeptide (ANP) radioimmunoassay were purchased from Beijing Chemclin Biotech Co., Ltd. High-density lipoprotein (HDL) assay kit was purchased from Zhejiang Dongou Biochemical Engineering Co., Ltd. Total cholesterol (T-CHO) assay kit was purchased from Shanghai Rongsheng Biotech Co., Ltd.

8.2.3 Experimental Design

The rats were randomly divided into 5 groups (*n*=20). The rats in the control group (C) were given free access to tap water. The rats in the model group (M) were given water containing L-NAME 70 mg/100 ml. The rats in the preventive groups (I, II) were administered taurine 1% or 2% in the drinking water at the same time as the L-NAME administration. Rats in the 1% taurine alone group (A) were given water containing 1% taurine only. Systolic arterial pressures (SBP) of arteria caudilis were measured every two days. Rats were euthanatized after 2–3 weeks. Blood and abdominal aortas were collected for biochemical analyses and histological tests.

8.2.4 *Biochemical Determinations*

8.2.4.1 Blood Pressure Measurements

SBP of arteria caudilis was measured using the ALC-NIBP non-invasive analytical system for caudal arterial SBP.

8.2.4.2 Serum Biochemical Index

Serum was separated by centrifugation (+4°C, 1500 r/min, 15 min). Serum concentrations of NO, NOS, SOD, MDA were determined by a colorimetry method according to reagent kits. Serum content of ACE was determined by ultraviolet spectrometry according to the reagent kit. Serum concentrations of T-CHO and HDL were determined by a biochemical immunoenzyme technique according to the reagent kits.

8.2.4.3 Plasma Biochemical Index

Blood was collected in tubes that were treated with anticoagulant. Plasma was separated by centrifugation (+4°C, 3000 r/min,10 min). Plasma content of AngII, AngI, ET, ANP, NPY and TXB2 were determined by a radioimmunoassay according to the reagent kits.

8.2.5 *Histological Tests*

Abdominal aortas were fixed in 10% (v/v) phosphate buffered formalin solution pH 7.0 and were embedded in paraffin wax. Sections were cut into 5μm and stained with haematoxylin and eosin HE. Proliferation of VSMCs were observed under an optical microscope.

8.2.6 *Statistical Analysis*

Data were presented as means ± SE and significant differences were determined by the Duncan's multiple range test using SPSS 11.5 statistical analysis software. P values less than 0.05 were considered significant.

8.3 Results

8.3.1 Blood Pressure Measurement

Table 8.1 reveals that the blood pressure of the model group was significantly elevated after L-NAME administration, but no significant differences were observed between 1% taurine preventive group and the model group at the end of the first week. At the end of the third week, the blood pressure in the 1% and 2% taurine preventive groups were significantly lower than that of the model group ($P<0.05$); 2% taurine attenuated the L-NAME mediated blood pressure rise more than 1% taurine. There was no significant difference between the 1% taurine alone group and the control group.

8.3.2 Biochemical Index

8.3.2.1 Serum Levels of NO and NOS

As shown in Table 8.2, the serum levels of NO and NOS were significantly decreased in the model group compared with that of the control group ($P<0.05$). The values were significantly increased in the taurine preventive groups compared with that of the model group, and 2% taurine better effect than 1% taurine ($P<0.05$).

Table 8.1 Changes in blood pressure after L-NAME administration

	Weeks after administration			
Group	0	1	2	3
C	102.68±9.657	107.82±9.72^{a}	111.64±7.35^{a}	115.05±9.97^{a}
M	101.86±8.10	124.29±8.22^{b}	145.51±7.26^{b}	165.73±9.22^{b}
I	103.38±9.01	120.04±10.82^{b,c}	137.64±6.57^{c}	150.35±8.67^{c}
II	102.60±13.63	114.13±10.56^{a,c}	125.58±6.88^{d}	131.53±11.22^{d}
A	99.88±3.44	107.41±10.94^{a}	109.64±8.60^{a}	112.05±6.13^{a}

Results represent means ± SE (n = 5). Superscripts represent significant differences between two groups (P < 0.05).

Table 8.2 Effects of taurine on serum NO and NOS

Group	NO (μmol/l)	NOS (U/ml)
C	31.26±2.25^{a}	19.78±2.63^{a}
M	9.91±1.96^{b}	8.91±1.73^{b}
I	18.76±1.59^{c}	12.49±2.25^{c}
II	25.59±1.60^{d}	16.19±1.92^{d}
A	31.22±1.91^{a}	19.45±1.92^{a}

NO: nitric oxide, *NOS*: nitric oxide synthase. Results are presented as mean ± SE (n=5). Different letters indicate significant difference between two groups (P <0.05).

There was no significant difference between the 1% taurine alone group and the control group.

8.3.2.2 Plasma Content of ACE, AngI and Ang II

As shown in Table 8.3, plasma content of ACE in the model group was significantly higher than that of the taurine preventive groups ($P< 0.05$). There were no significant differences between the 1% taurine alone group and the control group. Plasma content of AngI and AngII in the model group were significantly higher than those of the control group ($P<0.05$). Taurine administered preventively significantly lowered the plasma levels of AngI and AngII ($P<0.05$). There was no significant difference between the 1% taurine alone group and the control group.

8.3.2.3 Plasma Cytokine Content

Plasma content of ANP, ET, NPY and TXB2 was significantly increased after L-NAME administration in the model group ($P< 0.05$). When administered preventively, taurine significantly decreased the plasma levels of the cytokines, with 2% taurine being more effective ($P< 0.05$). Plasma content of the cytokines in the 1% taurine alone group was similar to the control value (Table 8.4).

Table 8.3 Effect of taurine on plasma content of ACE, AngI, AngII

Group	ACE (nM/ml/min)	AngI (ng/ml)	AngII (pg/ml)
C	60.92±8.93[a]	3.11±0.45[a]	316.91±43.16[a]
M	93.39±7.13[b]	4.92±0.74[b]	585.87±62.37[b]
I	77.55±5.39[c]	4.12±0.80[c]	519.57±68.66[c]
II	67.17±6.21[d]	3.85±0.53[c]	416.01±36.39[d]
A	59.72±5.65[a]	3.25±0.52[a]	320.99±62.78[a]

ACE: angiotensin-converting enzyme, *Ang I*: angiotensin I, *Ang II*: angiotensin II. Results represent means ± SE (n=5). Superscripts represent significant differences between two groups ($P<0.05$).

Table 8.4 Effect of taurine on plasma content of ANP, ET, NPY and TXB_2

Group	ANP (pg/ml)	ET (pg/ml)	NPY (ng/ml)	TXB2 (pg/ml)
C	312.51±67.50[a]	137.40±11.54[a]	42.50±3.66[a]	375.54±79.24[a]
M	529.24±54.19[b]	200.83±26.06[b]	65.26±9.40[b]	806.57±119.86[b]
I	446.52±59.98[c]	176.10±17.35[c]	56.02±7.67[c]	674.08±82.60[c]
II	392.39±53.61[a,c]	155.46±15.67[a,c]	47.35±7.01[a,c]	546.46±74.69[d]
A	323.65±68.72[a]	138.07±11.31[a]	41.02±4.69[a]	367.97±76.75[a]

ANP:atrial natriuretic polypeptide, *ET*:endothelins, *NPY*:neuropepitide Y, TXB_2: thromboxane B_2. Results represent means ± SE (n=5). Superscripts represent significant differences between two groups ($P<0.05$).

Table 8.5 Effect of taurine on serum levels of TC, HDL, MDA and SOD

Group	TC (mmol/L)	HDL (mmol/L)	MDA(nmol/ml)	SOD (U/ml)
C	1.62 ± 0.42^{a}	1.42 ± 0.24^{a}	4.53 ± 0.63^{a}	332.97 ± 38.73^{a}
M	2.96 ± 0.63^{b}	0.90 ± 0.19^{b}	6.69 ± 1.00^{b}	245.57 ± 35.12^{b}
I	2.32 ± 0.55^{c}	$1.09\pm0.21^{b,c}$	$5.81\pm0.60^{b,c}$	$289.44\pm43.90^{a,b}$
II	$1.88\pm0.42^{a,c}$	$1.29\pm0.19^{a,c}$	$5.24\pm0.50^{a,c}$	309.30 ± 38.60^{a}
A	1.60 ± 0.34^{a}	1.42 ± 0.26^{a}	4.49 ± 0.68^{a}	335.48 ± 50.80^{a}

TC: total cholesterol, *HDL*: high-density lipoprotein, *MDA*: malondialdehyde, *SOD*: superoxide dismutase. Results represent means ± SE (n=5). Superscripts represent significant differences between two groups ($P<0.05$).

8.3.2.4 Serum Concentration of Lipometabolic Index

Serum content of TC was significantly increased by L-NAME treatment in the model group ($P<0.05$). However, TC levels were significantly lower in the taurine preventive groups than in the model group ($P<0.05$). There were no significant differences in the TC levels of the 1% taurine alone group and the control group. However, serum content of HDL was significantly lower in the model group than in the control group ($P<0.05$). Taurine administered preventively significantly increased serum content of HDL, with 2% taurine being more effective. Serum HDL content of rats treated with only taurine was similar to the values of the control group (Table 8.5).

8.3.2.5 Serum Content of MDA and SOD

Serum content of MDA was significantly higher in the model group than in the control group ($P< 0.05$). However, the content of MDA was significantly decreased in the 2% preventive group. Serum content of SOD was significantly decreased by L-NAME treatment in the model group. Taurine administered preventively inhibited the decrease in serum levels of SOD in rats treated with L-NAME. No significant difference was observed in serum levels of MDA and SOD between the 1% taurine alone group and the control group (Table 8.5).

8.3.3 Histological Analysis

Histological analysis revealed that vascular smooth muscle cells VSMC proliferate and migrate into the endangium after exposure to L-NAME (Fig. 8.1a, b). There was a subsequent thickening of the vessel walls and conversion of the endothelial cells into a cube shaped phenotype (Fig. 8.1a). Significant proliferation of VSMC in the subendothelial layer of blood vessels was seen in the model group (Fig 8.1b). Blood vessel endothelium was exfoliated and the endomembrane swelled (Fig. 8.1c). The degree of VSMC proliferation and vessel wall thickening was diminished in the 1% taurine preventive group (Fig 8.1e) and the 2% taurine preventive group (Fig. 8.1f). The number of VSMCs was nearly identical in the 1% taurine alone

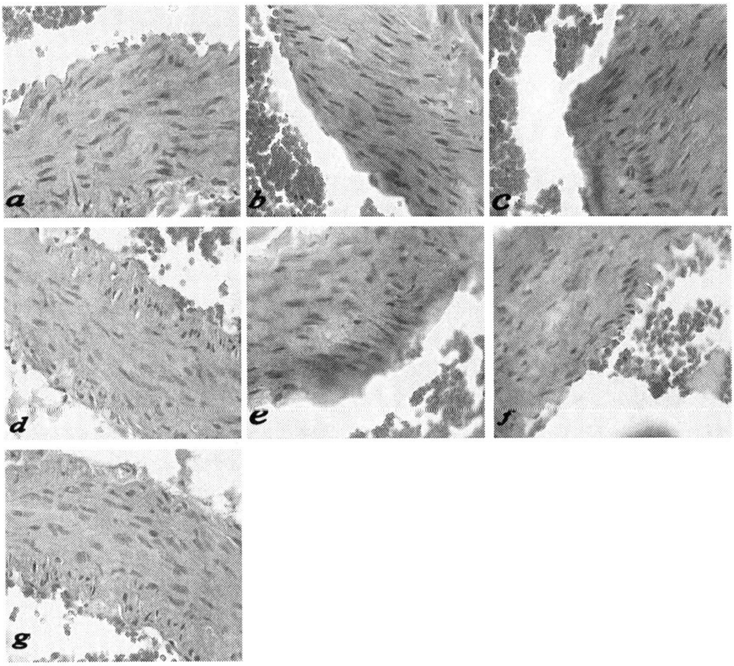

Fig. 8.1 Effect of taurine on the proliferation of vascular smooth muscle cells VSMC of rat abdominal aorta. (**a**, **b** and **c**): model group. (**d**): control group. (**e**): 1% taurine + L-NAME group. (**f**): 2% taurine + L-NAME group. (**g**): 1% taurine alone group. HE stain. × 400

group (Fig. 8.1g) and the control group (Fig. 8.1d). The VSMCs lined up and endothelial cells assumed an applanate shape (Fig. 8.1d, g).

8.4 Discussion

Significant elevation in blood pressure and decreases in serum concentrations of NO and NOS were observed in rats treated with L-NAME. A number of studies have shown that L-NAME administration can induce hypertension in rats (Allen et al. 1986; Lahera et al. 1992; Böger et al. 1994; Dunn et al. 1995). It is known that NO, which is synthesized via the L-Arg-NO pathway, plays an important role in the regulation of angiotasis and blood pressure. Importantly, the production of NO is affected by both the content of the nitric oxide precursor, L-arginine, and the activity of NOS. L-NAME, which is a NOS inhibitor, reduces the synthesis of NO by inhibiting the activity of eNOS that is locolized in vascular endothelial cells. In the taurine treated animals, the decrease in blood pressure may be related to the increase in NOS activity and the production of NO. On the other hand, taurine may decrease blood pressure by diminishing intracellular production of superoxide anions, which oxidizes NO after it is synthesized. Taurine administration 1% had no significant effect on the blood pressure of normal rats, suggesting that it exerts no influence on serum NOS and NO content under normal conditions.

The renin-angiotensin-aldosterone system (RAAS) is an important pressure regulating system, which is a key regulator of vasoconstriction. It consists of angiotensinogen, feritin, AngI, AngII, AngIII, ACE and the catabolic enzymes of angiotenin. AngII is the most important biological active component of the RAAS and is the most effective smooth muscle contracting substance. It had been reported that serum ACE is significantly increased in spontaneously hypertensive rats (Ke and Qu 2003). In the present study, serum concentrations of ACE, AngII and AngI were significantly increased in the model group. In the preventive groups, taurine significantly inhibited the activity of the RAAS. Administration of low levels of taurine for a short period of time had no effect on the activity of the RAAS. However, at higher concentrations, taurine inhibited the activity of serum ACE, thereby reducing the conversion of AngI to AngII and diminishing L-NAME mediated elevations in blood pressure. It has been proposed that taurine might reduce hypertension by reducing the secretion of aldosterone and thus inhibit sodium reabsorption by the kidney and lower blood volume. Moreover, taurine could also diminishh blood pressure by acting as a neuromodulator to inhibit sympathetic activity.

ANP has a significant effect on vascular tone and therefore blood pressure (Li et al. 2002). As one of the most effective endogenous vasoactive peptides, ET is a potent vasoconstrictor, stimulator of myocardial contractile force MCF, promoter of neuroendocrine function and regulator of blood pressure. In this study, taurine administration significantly inhibited the L-NAME mediated increase in plasma ANP and ET in the hypertensive rats. Several mechanisms may be involved in this action, including reductions in ET and ANP synthesis and release, acceleration in ET clearance, reduction in atrioventricular pressure and attenuation of the rise in sympathetic nervous system SNS activity. Alternatively, taurine could enhance the activity of NOS and increase serum NO, thereby reducing plasma ET and ANP content. Taurine could also reduce the release of ANP and ET through a decline in AngII levels.

The neurotransmitter NPY elevates blood pressure by promoting vasoconstriction, stimulating the release of contract reaction factors from blood vessel endothelium, enhancing VSMC proliferation, increasing vessel wall proliferation, elevating peripheral resistance and upregulating endogenous vasostimulants. It has also been reported that NPY elevates blood pressure by increasing volume load (Liu et al. 2003). In this study, taurine administration significantly decreased the secretion of plasma NPY in the hypertension rats, an observation consistent with a previous study showing that taurine inhibits the secretion of plasma NPY in stress trained rats (Jin et al. 2001). The results show that taurine lowers plasma TXB_2 levels, which may be another mechanism by which taurine could prevent hypertension.

This study shows that taurine significantly improved lipid metabolism in the hypertensive rats, lowering serum TC and elevating serum HDL.

Taurine therapy 2% significantly reduced serum MDA levels and elevated SOD levels, partially restoring content that had been altered by L-NAME. It is known that SOD is one of the most important antioxidants, while MDA is a end product of lipid peroxidation. Since taurine exhibits antioxidant activity (Hanna et al. 2004), it is possible that taurine could reduce blood pressure by reducing oxidative stress and in the process increase the levels of the vasodilator, nitric oxide.

L-NAME mediated hypertension involves the migration of VSMC from the tunica media to the subintimal space, the thickening of the blood vessel and the narrowing of the lumen of the blood vessel, the increase in peripheral resistance and the elevation in blood pressure. Structural alternations of the vessel wall are common pathological features of the hypertensive condition and include VSMC proliferation, hypertrophy of the arterial wall and thickening of the blood vessel wall. It has been reported that taurine inhibits the proliferation of VSMCs by preventing large amounts of $Ca^{2+}$from entering the cell (Wu JM et al. 1991; Zhang et al. 1999).

8.5 Conclusion

In summary, taurine administration attenuated L-NAME mediated hypertension by inhibiting the activity of the RAAS, reducing serum levels of ANP, ET, NPY and TXB2, lowering blood fat levels and reducing the degree of oxidative stress. The above results provide important insight into the control of hypertension and potential for future therapy.

Acknowledgments This study was supported by the National Nature Science Foundation of China (Grant No. 30471263). The authors thank Dr. Masugi Nishihara, Tokyo University (Japan) for reviewing the manuscript and giving valuable advice.

References

Allen JC, Navran SS, Kahn AM (1986) Na^+-K^+-ATPase in vascular smooth muscle. Am J Physiol 250:C536–C539

Böger, RH, Bode-Böger SM, Gerecke U, Frölich JC (1994) Long term administration of L-arginine, L-NAME and the exogeneous NO donor molsidomine modulate urinary nitrate and cyclic GMP excretion in rats-correlation which vascular reactivity. Cardiovas Res 28:494–499

Dunn WR, Gardiner SM (1995) NO evidence for vascular remodelling during hypertension induced by chronic inhibition of nitric oxide synthase in Brattleboro rats. Hypertension 13: 849–857

Fujita T, Ando K, Noda H, Ito Y, Sato Y (1987) Effects of increased adrenomedullary activity and taurine in young patients with borderline hypertension. Circulation 75:525–532

Hanna J, Chahine R, Aftimos G, Nader M, Mounayar A, Esseily F, Chamat S (2004) Protective effect of taurine against free radicals damage in the rat myocardium. Exp Toxicol Pathol 56:189–194

Jin QG, Deng RH, Li NC (2001) Effects of taurine on secretion of cardiovascular systematic regulatory peptide in heavy stress training rats. Chinese J Sports Med 20:126–131

Ke YH, Qu SB (2003) Effects of mediodespidine caps on AngII and Ald in renal vascular hypertensive rats. Chinese Exp Pharmacol J Trad Chinese Med Formulae 9:51–52

Lahera V, Salazar J, Salon MG, Romer JC (1992) Deficient production of nitric oxide induces volume dependent hypertension. Hypertension 10(Suppl 7):173–177

Li J, Hua Q, Liu RK (2002) Sodium diuresis polypeptides and hypertension. Chinese J Med Guide 4:295–297

Liu RM, Xia YF, Chen SJ, Zhai HX, Tong WR (2003) Variations of plasma NPY and neurotensin in hypertensive patients and its significance. Chinese Heart J 15:126–128

Paakkari P, Paakkari I, karppanen H, Paasonen MK (1983) Mechanisms of inhibitory cardiovascular effects of taurine and homotaurine. Acta Med Scand 677:134–137

Schaffer SW, Azuma J (1992) Review: myocardial physiological effects of taurine and their significance. Adv Exp Biol Med 315:105–120

Trachtman H, Pizzo RD, Rao P, Rujikarn N, Sturman JA (1998) Taurine lowers blood pressure in the spontaneously hypertensive rats by a catecholamine independent mechanism. Am J Hypertents 2:909–912

Wu JM, Cheng T, Tang CS (1991) Effects of Mg^{2+} and taurine on proliferation effects of endothelin on VSMC. J Peking Univ Heal Sci 5:351–352

Yu WJ, Yao XH and Hou YJ, Guo LF, Zhang LK, Tang CY (2000) Effects of taurine on L-Arg/NO pathway of akaryocytes in insulin resistant rats caused by fructose. Chinese Pathophysiol J 16:1202–1206

Zhang X, Tenner TE Jr, Lombardini JB (1999) Inhibition of rat vascular smooth muscle cell proliferation by taurine and taurine analogues. Biochem Pharmacol 11:1331–1339

Zheng M, Wu LJ, Li ZL, Yin SH, Bao CY (2003) Effects of taurine on serum and cardiac muscle Agt II and aldosterone in renal hypertensive rats. Chinese J Angiocarpy 8:313

Chapter 9
Attenuating Action of Taurine and Labetalol on Cardiovascular Alterations by Pyridoxal-Isoproterenol, a Vitamin-Drug Interaction with Cardiopathologic Significance

Miteshkumar Acharya and Cesar A. Lau-Cam

Abstract Taking into account the ability of vitamin B_6 vitamers (i.e., pyridoxal, pyridoxine, pyridoxamine) to stimulate adrenomedullary catecholamine outflow and biochemical and cardiovascular changes consistent with adrenoceptor stimulation, this study was undertaken in mice to verify the possibility that pyridoxal (PL), the most potent of the B_6 vitamers, could enter into an adverse vitamin-drug interaction when coadministered with a cardioactive drug, and that compounds like taurine (TAU) and labetalol (LAB) could be protective. To this end, mice were treated with PL (200 mg/kg) and isoproterenol (ISO 200 mg/kg), with and without a pretreatment with TAU (300 mg/kg), LAB (30 mg/kg), or TAU plus LAB, and monitored for heart rate, electrocardiographic (ECG) and myocardial electrolyte changes. PL plus ISO lowered the LD_{50} of ISO and magnified the changes in heart rate, ECG and myocardial electrolytes seen with ISO alone. Pretreating rats with TAU or TAU plus LAB attenuated the cardiovascular changes induced by PL plus ISO to a greater extent than LAB alone. The attenuating actions of TAU and LAB probably reflect independent and complementary mechanisms of action.

Abbreviations *TAU*, taurine; *PL*, pyridoxal; *ISO*, isoproterenol; *LAB*, labetalol

9.1 Introduction

In rats, a treatment with a single, 100–300 mg/kg, intraperitoneal (i.p.) dose of a vitamin B_6 vitamer, i.e., pyridoxine (PN), pyridoxamine (PA) or PL, was found to stimulate the secretion of adrenomedullary catecholamines (CATs) in a dose-dependent manner (Lau-Cam et al. 1991). At the dose of 300 mg/kg, PL increased

M. Acharya (✉)
Department of Pharmaceutical Sciences, St. John's University, College of Pharmacy and Allied Health Professions, Jamaica, NY, USA

J. Azuma et al. (eds.), *Taurine 7*, Advances in Experimental Medicine and Biology 643,
DOI 10.1007/978-0-387-75681-3_9,

the plasma total CAT levels to a greater extent (74%) than either PA (29%) or PN (5%) (Lau-Cam et al. 1991) and, above this dose, PL elevated the levels of $Na^{+}$and Ca^{++} and lowered the K^{+} and Mg^{++} in the myocardium, and induced myocardial changes characterized by preventricular contractions, coupled beats, ventricular tachycardia, R on T phenomenon, and ventricular fibrillation (Saber 1987). Since adrenalectomy, splanchnectomy and drugs with anticholinergic and ganglionic action attenuated the elevation in plasma CATs associated with a vitamin B_6, it was concluded that this vitamin stimulated adrenal CATs secretion by a mechanism of central origin and utilizing the influence of the splanchnic innervation to adrenomedullary chromaffin granules (Lau-Cam et al. 1991; Saber 1987). On the other hand, the reported ability of TAU to suppress the mobilization of adrenomedullary epinephrine and hepatic glycogen in rats exposed to immobilized cold stress has been explained as being the result of a membrane-stabilizing effect (Nakagawa and Kuriyama 1975). This laboratory has not only confirmed this view but also has found TAU to display a dose-dependent inhibitory action on the adrenal gland and accompanying metabolic alterations in rats treated with a vitamin B_6 vitamer (Patel and Lau-Cam 2006).

The purpose of the present investigation was to determine whether a high dose ($\geq$ 100 mg/day) of vitamin B_6, like that prescribed for the treatment of adult attention-deficit hyperactivity disorder (Arnold 2001) and autism (Kleijnen and Knipschild 1991), can have potential for causing an adverse interaction with medication possessing intrinsic cardiovascular actions. To this effect, mice were treated with isoproterenol (ISO), a cardioactive agent, in the presence and absence of a vitamin B_6 vitamer such as PL. In addition, the mice also received TAU and LAB, singly or in combination, to gauge the contribution of the adrenal medulla and of cardiac adrenoceptors to a vitamin B_6-ISO interaction. The cardiac effects of the various experimental treatments were monitored electrocardiographically and by assessing myocardial electrolyte levels.

9.2 Materials and Methods

9.2.1 Animals

All experiments were conducted on male Swiss-Webster mice, 30–35 g in weight. The animals were housed in a room maintained at a constant temperature of 21.1 ± 3°C and a normal 12-hour light-dark cycle. During a 3 day acclimation period the animals had free access to a commercial animal diet (Purina Lab Chow®, Ralston Purina Co., St. Louis, MO) and filtered tap water.

9.2.2 Treatment Solutions and Treatments

The treatment solutions were prepared in distilled water. The doses (mg/kg/0.5 mL) were: PL 200, ISO 200, TAU 300 (in two divided doses, given 30 min apart), and

LAB 30. Control animals received 0.5 mL of physiological saline (PHS) in place of a treatment solution. All treatments were by the intraperitoneal (i.p.) route, at 30 min intervals from each other, and in the order TAU, LAB, PL, ISO.

9.2.3 Lethal Dose (LD)$_{50}$ Determinations

Groups of 6 mice each were treated with increasing doses of PL (100–700 mg/kg) and ISO (200–550 mg/kg), and the LD_{50} was obtained from a plot of number of deaths vs. treatment doses. The results of these studies suggested a dose of PL or ISO equal to 200 mg/kg for use in subsequent experiments.

9.2.4 Electrocardiographic Studies

Electrocardiograms (ECGs) were recorded every 5 min for 30 min or until the time of death, starting from the time of a treatment or treatment combination, on a model RPS7C8B electrophysiograph (Grass Instruments, Quincy, MA). QT intervals (in msec) and T wave amplitude (in mm) were measured from 9-beat panels, and the values were averaged. The values of heart rates, QT intervals, and T wave amplitudes prior to a given treatment were taken as the baseline (control) values. Changes due to a treatment were calculated as a percentage (%) of the control value. Corrected heart rates (QT_c) values (in msec) were calculated from the QT intervals using the formula reported by (Mitchell et al. 1998), i.e., $QTc = QT/\sqrt{RR/100}$, where the RR interval is in msec. For the sake of brevity, only the 10 min and 15 min are discussed here.

9.2.5 Collection of Heart Samples and Preparation of Heart Homogenates

The hearts were surgically removed from a dead animal or from a surviving animal that had been sacrificed with a high i.p. dose of phenobarbital sodium (80 mg/kg), 60 min after the start of an experiment. The ventricles were removed, washed with distilled water, blotted dry, cut into small pieces, and dried at 85°C for 12 hr in a in a vacuum oven. A 30 mg portion of dry heart was homogenized with about 0.8 mL of 3N HCl. The suspension was incubated at 110°C for 4 hr, cooled to ambient temperature, vacuum filtered through a 0.45 μm filter paper into a polyethylene microtube, and made to 1 mL with 3N HCl.

9.2.6 Assay of Heart Electrolytes

Each heart homogenate was analyzed for Ca^{++}, Mg^{++} and K^{+} using assay kits obtained from Advanced Diagnostics, South Plainfield, NJ, USA. Concentrations were expressed in mEq/g of wet tissue.

9.2.7 *Statistical Analysis of Data*

The experimental results are expressed as the mean ± SEM for $n = 6$. Statistical differences between control (PHS) group and treatment groups were determined by Student's t-test, and were regarded as being significant at $p < 0.05$.

9.3 Results

9.3.1 LD_{50} *Values for PL and ISO*

The LD_{50} of PL was 575 mg/kg and that of ISO 400 mg/kg. Co-administering PL (200 mg/kg) with ISO (400 mg/kg) lowered the LD_{50} of ISO from 400 to 200 mg/kg. Hence, in studies in which PL was concurrently used with ISO, the dose of each of these compounds was 200 mg/kg.

9.3.2 *Cardiac Changes Due to PL-ISO*

From the ECG traces (not shown) and corresponding changes in heart rate, QT and QT_c intervals, and T wave amplitude values (Table 9.1), it was inferred that PL (200 mg/kg) ISO accentuated the effect of ISO (200 mg/kg) on the heart rate by 2.25- to 3.5-fold, and shortened the onset of the maximum effect by 10 min. Similarly, PL enhanced the effects of ISO on the amplitude of the QT and QT_c intervals (up by ~25% and 41%, respectively), which occurred faster (by 5 min and 20 min, respectively) than with ISO alone. In addition, a treatment with PL plus ISO significantly increased the T wave amplitude relative to control values, but not as much as with PL alone (1.27-fold vs. 2.44-fold, respectively, relative to control).

PL magnified the changes in cardiac electrolytes due to ISO alone (Table 9.2). Relative to control values, the Ca^{++} level was increased by 22% ($p < 0.05$), and the Mg^{++} and K^{+} levels were decreased by 29% ($p < 0.01$) and 15%, ($p < 0.05$), respectively. Raising the dose of ISO to 400 mg/kg led to cardiac electrolytes changes that were not markedly different from those observed at one-half this dose (Ca^{++} up by 24%, Mg^{++} down by 29% and K^{+} down by 17%, all at $p < 0.05$).

9.3.3 *Cardiac Changes Due to PL-ISO in the Presence of TAU*

A pretreatment with TAU offset the lethality of PL-ISO by shifting the LD_{50} of PL-ISO upwardly from 200 mg/kg to 300 mg/kg. From the data summarized in Table 9.1, it was evident that TAU attenuated the effect of PL-ISO on the heart rate at 10 and 15 min post-PL-ISO, with the average increase being about 58% lower than the value seen with PL-ISO alone ($p < 0.001$). Likewise, TAU attenuated the average increases in QT and QT_c values seen with PL plus ISO by 87% ($p < 0.001$) and 92% ($p < 0.001$),

Table 9.1 Effect of PL + ISO, in the presence or absence of TAU and/or LAB, on heart rate, QT interval, QT_c interval and T wave amplitude at 15 and 30 min after their administration[a,b,c,d]

	Heart rate		QT interval		QTc interval		T wave amplitude	
Treatment	10 min	15 min	10 min	30 min	15 min	30 min	15 min	30 min
PHS	6.70 ±0.70	7.28 ±0.09	6.66 ±0.20	3.33 ±0.72	4.82 ±1.70	3.57 ±0.05	25.00 ±7.56	25.00 ±5.50
PL	20.00 ±0.50***	15.00 ±2.35**	20.00 ±0.05**	18.06 ±3.95***	31.00 ±5.95***	25.10 ±9.10**	120.00 ±17.00***	75.00 ±9.00***
ISO	15.20 ±1.40*	32.20 ±2.85***	7.25 ±3.25	59.60 ±0.30***	18.60 ±6.00**	23.86 ±4.38**	18.06 ±1.84	25.00 ±2.10
PL+ISO	55.60 ±2.50***	61.30 ±6.10***	29.20 ±8.00***	5.26 ±0.20	59.60 ±4.30***	37.80 ±0.32***	35.00 ±8.2	ND
PL+ISO+ TAU	27.00 ±5.25**	17.30 ±0.54***	5.20 ±1.40	3.33 ±0.50	12.50 ±3.70**	8.32 ±0.31*	32.65 ±7.35	30.13 ±8.47
PL+ISO+ LAB	6.52 ±0.30	3.34 ±8.65*	18.50 ±2.50**	12.30 ±2.89*	20.30 ±6.87**	14.70 ±0.07**	25.00 ±8.26	25.25 ±10.23
PL+ISO+ TAU+LAB	7.00 ±1.79	7.40 ±1.00	7.00 ±1.79	6.73 ±4.00	7.00 ±1.79	6.73 ±0.40	22.50 ±7.50	22.00 ±8.33

[a]Doses (mg/kg): PL 200, ISO 200, TAU 300 (two divided equal doses), LAB 30.
[b]Results are reported as the mean ± SEM for $n = 6$.
[c]Statistical comparisons were vs. PHS at $^{*}p < 0.05$, $^{**}p < 0.01$ and $^{***}p < 0.001$.
[d]ND = not done (due to animal death).

respectively. While TAU showed no effect on the change in T wave amplitude at 10 min after PL-ISO, it had a small lowering effect at 15 min (~36%).

TAU was also able to reverse the changes in electrolytes seen with PL plus ISO (Table 9.2). Thus, in comparison to the control group the Ca^{++} was decreased (by 20%, $p < 0.05$) rather than increased, and the Mg^{++} and K^{+} levels were, respectively, equal to and slightly below (by 8%) the control value.

9.3.4 Cardiac Changes Due to PL-ISO in the Presence of LAB

LAB was more effective than TAU in offsetting the lethality of PL-ISO since it caused a greater upward shift in the LD_{50} of PL-ISO (300 mg/kg vs. 325 mg/kg, respectively, from 200 mg/kg). From the data derived from an ECG of a mouse receiving LAB prior to ISO plus PL and summarized in Table 9.1 it was determined that LAB was more effective than TAU in attenuating the increase in heart rate (average decrease of ~90% vs. average decrease of 53%, respectively) and increase in T wave amplitude (average decrease of 20% vs. average increase of 26%, respectively) seen at 10 and 15 min post-PL plus ISO. During the same intervals, TAU was more protective against the increases in QT (average decrease of ~87% vs. 53%, respectively) and QT_c values (average decrease of ~93% vs. 67%, respectively) due to PL plus ISO.

Table 9.2 Effect of PL + ISO, in the presence or absence of TAU and/or LAB, on myocardial electrolytes[a,b,c]

	Electrolyte content, mEq/g of wet tissue (mean ± SEM, $n = 6$), % of control		
Treatment	Ca^{++}	Mg^{++}	K^{+}
PHS	0.506 ± 0.019	0.126 ± 0.007	0.156 ± 0.005
PL	0.534 ± 0.046*	0.097 ± 0.007*	0.139 ± 0.005*
ISO	0.581 ± 0.046*	0.098 ± 0.007*	0.140 ± 0.007*
PL + ISO	0.616 ± 0.055*	0.085 ± 0.007**	0.127 ± 0.005*
PL + ISO + TAU	0.440 ± 0.034^{+}	0.122 ± 0.006^{++}	0.159 ± 0.005^{+}
PL + ISO + TAU + LAB	0.477 ± 0.028^{+}	0.103 ± 0.007^{+}	0.160 ± 0.010^{+}

[a]Doses (mg/kg): PL 200, ISO 200, TAU 300 (two divided equal doses), LA 30.
[b]Statistical comparisons were vs. PHS at $*p < 0.05$ and $**p < 0.01$; and vs. PL + ISO at $^{+}p < 0.05$ and $^{++}p < .001$
[c]Animals required sacrifice with a high (80 mg/kg) i.p. dose of phenobarbital sodium.

By analogy to TAU, LAB was able to reverse the changes in electrolytes caused by PL-ISO, although to a slightly lesser extent. Thus, all three electrolytes were decreased by LAB plus PL-ISO to values that were not significantly different from those of the control group (Ca^{++} by 10%, $p < 0.05$; Mg^{++} by 18%, $p < 0.05$; and K^{+} by 6%) (Table 9.2). However, in comparison to TAU, LAB was slightly less effective against changes in Ca^{++} and Mg^{++} and about equipotent against changes in K^{+} induced by PL-ISO.

9.3.5 Cardiac Changes Due to PL-ISO in the Presence of TAU-LAB

A combined pretreatment with TAU plus LAB raised the LD_{50} of PL-ISO from 200 mg/kg to > 400 mg/kg, and attenuated the increase in cardiac rate seen 10 and 15 min after the administration of PL-ISO to a value approximating that attained with LAB alone (i.e., average decrease of ~87%, $p < 0.001$) and about equal to the control value (Table 9.1). Furthermore, while TAU and LAB were more effective in attenuating the increases in QT (average decrease of ~79%) and QT_c (average decrease of 89%) intervals seen at 10 and 15 min after PL-ISO when given together than separately, their combination increased the T wave amplitude to a value similar to that seen with LAB alone (by ~18% vs. ~20%, respectively). A pretreatment with TAU-LAB precluded the changes in cardiac electrolytes induced by PL-ISO in a manner and to an extent similar to that by LAB alone. Thus, relative to control values, giving TAU plus LAB before PL-ISO resulted in a small decrease in cardiac Ca^{++} (by 6%) and Mg^{++} (by 18%, $p < 0.05$) and a minimal increase in K^{+} (~2%). In contrast, PL-ISO increased Ca^{++} and decreased Mg^{++} and K^{+} in the myocardium (Table 9.2, $p < 0.05$ vs. control values).

9.4 Discussion

ISO is a synthetic CAT which, like its natural counterparts epinephrine and norepinephrine, can stimulate myocardial β-adrenoceptors in a manner analogous to exercise, namely to reduce after load and end-diastolic and end-systolic dimensions, and to increase myocardial contractility, cardiac index, heart rate, and ventricular force-velocity relation (Manger 1982). However, at variant with epinephrine and norepinephrine, ISO is a strong nonselective β-adrenergic agonist with low affinity for α-adrenoceptors, possessing a longer duration of action and demonstrating a weaker hyperglycemic effect (Hoffman and Lefkowitz 1995). In addition to tachycardia, ISO can also induce cardiac ischemia (Hoffman and Lefkowitz 1995), cardiac necrosis (Barletta et al. 1978) and arrhythmias (Barletta et al. 1978; Guideri et al. 1975). In this study, ISO exhibited a greater toxicity than PL (i.p. LD_{50} 400 vs. 575 mg/kg, respectively).

The ECG changes elicited by ISO in doses ranging from 200 to 550 mg/kg were consistent with those reported in the literature (Furushima et al. 2001; Urao et al. 2004). However, in contrast to PL, ISO made the heart rate faster and stronger, had a greater effect on the QT_c than on the QT interval, and more rapidly and transiently increased the amplitude of the T wave. At a dose equal to 300 mg/kg, this sympathomimetic agent induced sinus bradycardia, tachycardia, arrhythmia, and eventual cardiac failure. The myocardial electrolytes changes observed in the present study for sublethal (200 mg/kg) and lethal (400 g/kg) doses of ISO were somewhat dose-related, with the lower dose altering Ca^{++} and Mg^{++} but not K^{+}, and the higher one affecting all three electrolytes. According to Guideri et al. (1974), a cardioactive compound such as ISO may interfere with energy-producing mechanisms in the cardiac cell to cause Mg^{++} and K^{+} loss and the accumulation of Na^{+} and Ca^{++}, especially at doses above 200 mg/kg.

When coadministered in sublethal equidoses (200 mg/kg), ISO and PL caused the LD_{50} of ISO to become the LD_{100}. An additional consequence of this vitamin-drug interaction was the accentuation of the ECG changes seen only with ISO and which, in turn, led to tachycardia and myocardial infarction. These developments probably reflect an increase in contractile myocardial activity along with an increased demand for oxygen (Braunwald 1971). Interestingly, however, the changes in myocardial electrolytes seen with PL-ISO were not substantially different from those derived using ISO alone. This finding could imply that either compound had affected the stores of myocardial electrolytes maximally when administered singly or, alternatively, that the dose of ISO used in the experiment had not been sufficiently high. This question was clarified by an experiment in which the dose of PL was kept constant and that of ISO was doubled (i.e. to 400 mg/kg). The similarity of the electrolyte values derived with the lower and higher doses of ISO supported the validity of former assumption and confirmed the appropriateness of the doses of PL and ISO chosen for the study.

An insight into the mechanism underlying the cardiac effects of PL and ISO was provided by experiments in which TAU, an inhibitor of adrenomedullary CAT release, and LAB, a α- and β-adrenoceptor blocker, were used as pretreatments.

TAU may be lowering adrenomedullary CAT outflow by acting as a membrane stabilizer on adrenomedullary chromaffin granules (Nakagawa and Kuriyama 1975). In addition, this sulfur compound may display a regulatory, although nonspecific, role on myocardial electrolytes, especially on Ca^{++} accumulation, as a result of ISO administration (Satoh and Sperelakis, 1998; Welty et al. 1982). On the other hand, LAB was selected taking into account the presence of both α- and β-adrenoceptors in cardiac cells and its known antagonizing actions on the log dose response curves, diastolic pressure changes, tachycardia, Ca^{++} accumulation, and vasodepressor responses by ISO (Drew et al. 1979). Additional considerations were its reported ability to completely block hypokalemia and QT interval prolongation (Grubb 1991), and to prevent the development of ectopic rhythms and premature ventricular contractions (Read and Welty, 1963) following acute dosing with epinephrine. The differences in potency between TAU and LAB in terms of their effects on the LD_{50} of PL plus ISO could be related to contribution by CATs of nonadrenal origin or by extracardiac effects inherent to PL. However, as it relates solely to the heart, such additional contributions are probably not significant enough to offset the blockade of cardiac adrenoceptors by LAB in the presence of TAU. Not surprisingly, dosing mice with TAU plus LAB prior to PL plus ISO resulted in ECG and myocardial electrolyte changes that were not very different from those observed with TAU as the only pretreatment.

9.5 Conclusions

PL can enhance the intrinsic cardioactivity of ISO to a significant extent when both compounds are co-administered in sublethal doses. TAU and LAB can effectively antagonize this pharmacological interaction when given either singly or in combination. Based on differences in intrinsic potency between PN, the nutritionally most recognizable form of vitamin B_6, and PL, the aldehyde analog, to raise plasma CAT levels, it is safe to assume that under ordinary circumstances a vitamin B_6-cardioactive agent interaction leading to exaggerated cardiovascular responses such as those described here will be unlikely at daily doses of PN below 1 g.

Acknowledgments We thank Dr. Michael A. Barletta, Department of Pharmaceutical Sciences, St. John's University, for his help and advice during the electrocardiographic studies.

References

Arnold LE (2001) Alternative treatments for adults with attention-deficit hyperactivity disorder (ADHD). Ann NY Acad Sci 931:310–341

Barletta MA, Lehr D, Guideri G (1978) Hemodynamic alterations in isoproterenol-induced myocardial cardiac arrhythmias in corticoid-treated rats. J Pharm Sci 67:1390–1394

Braunwald E (1971) Control of myocardial oxygen consumption: physiologic and clinical considerations. Am J Cardiol 27:416–432

Drew GM, Hilditch A, Levy GP (1979) The relationship between the cardiovascular effects, alpha- and beta-adrenoreceptor blocking actions and plasma concentration of labetalol in doca hypertensive rats. Clin Exp Hypertens 1:597–611

Furushima H, Chinushi M, Washizuka T, Aizawa Y (2001) Role of α-blockade in congenital long QT syndrome investigation by exercise. Jpn Circ J 65:654–658

Grubb BP (1991) The use of oral labetalol in the treatment of arrhythmias associated with the long QT syndrome. Chest 100:1724–1725

Guideri G, Barletta M, Lehr D (1974) Extraordinary potentiation of isoproterenol cardiotoxicity by corticoid pretreatment. Cardiovasc Res 8:775–786

Guideri G, Barletta M, Chau R, Green M, Lehr D (1975) Method for the production of severe ventricular dysrhythmias in small laboratory animals. Recent Adv Stud Cardiac Struct Metab 10:661–679

Hoffman B, Lefkowitz R (1995) Catecholamines, sympathomimetic drugs, and adrenergic receptor antagonists. In: Hardman JG, Limbird LE (eds) Goodman & Gilman's the pharmacological basis of therapeutics, McGraw-Hill, New York, pp 212–214

Kleijnen J, Knipschild P (1991) Niacin and vitamin B6 in mental functioning: a review of controlled trials in humans. Biol Psychiatry 29:931–941

Lau-Cam CA, Thadikonda KP, Kendall, BF (1991) Stimulation of rat liver glycogenolysis by vitamin B_6: a role for adrenal catecholamines. Res Commun Chem Pathol Pharmacol 73:197–207

Manger WM (1982) Catecholamines in normal and abnormal cardiac function. In: Kellerman JJ, Hashomer T (eds) Advances in cardiology, Karger Publishers, New York, pp 24–27

Mitchell GF, Jeron A, Koren G (1998) Measurement of heart rate and Q-T interval in the conscious mouse. Am J Physiol 274 (Pt 2):H747–H751

Nakagawa K, Kuriyama K (1975) Effect of taurine on alteration in adrenal functions induced by stress. Jpn J Pharmacol 25:737–746

Patel JP, Lau-Cam CA (2006) Taurine attenuates pyridoxal-induced adrenomedullary catecholamine release and glycogenolysis in the rat. In: Oja SS, Saransaari P (eds) Taurine 6, Springer, New York, pp 147–156

Read WO, Welty JD (1963) Effect of taurine on epinephrine and digoxin induced irregularities of the dog heart. J Pharmacol Exp Ther 139:283–289

Saber S (1987) Elucidation of the mechanism of cardiotoxicity of vitamin B6 in rats, PhD Thesis, St John's University, Jamaica, New York

Satoh H, Sperelakis N (1998) Review of some actions of taurine on ion channels of cardiac muscle cells and others. Gen Pharmacol 30:451–463

Urao N, Shiraishi H, Ishibashi K, Hyogo M, Tsukamoto M, Keira N, Hirasaki S, Shirayama T, Nakagawa M (2004) Idiopathic long QT syndrome with early afterdepolarization induced by epinephrine. A case report. Circ J 68:587–591

Welty MC, Welty JD, McBroom MJ (1982) Effect of isoproterenol and taurine on heart calcium in normal and cardiomyopathic hamsters. J Mol Cell Cardiol 14:353–357

Chapter 10
Tool from Traditional Medicines is Useful for Health-Medication: Bezoar Bovis and Taurine

Kyoko Takahashi, Yuko Azuma, Shizu Kobayashi, Junichi Azuma, Koichi Takahashi, Stephen W. Schaffer, Masao Hattori, and Tsuneo Namba

Abstract Bezoar Bovis (BB:dried cattle gallbladder stones) has been used empirically in Asia for over 3000 years to treat heart and liver disorders. Yet its therapeutic potential remains unexplored by Western researchers. The aim of this study has been to clarify the actions of BB on cultured cardiomyocytes and to identify its active component(s). BB is a component of 98.7% of the Japanese over the counter (OTC) cardioactive drugs. The water-extract of BB exhibits protection action against arrhythmias produced by low Ca^{2+} and high Ca^{2+} in the medium. On the other hand, the Ca^{2+}-antagonist, verapamil, did not suppress arrhythmias that developed in cell culture. Rather, it aggravated the beating status of the cardiomyocytes. The major constituents of the BB extract are bile salts (cholate, deoxycholate, taurocholate) and amino acids (taurine, cysteine, leucine, isoleucine). Most cells incubated with bile salts developed morphological damage. However, one of the major constituents of the BB extract, taurine, was effective in protecting against the abnormal beating pattern induced by high Ca^{2+}. Since beta-alanine, an inhibitor of taurine transport, antagonized the protective effects of both BB and taurine, it is likely that the effect of BB is partly mediated by taurine.

Abbreviations *BB*, Bezoar Bovis; *OTC*, Over the counter drug; *Ver*, verapamil; *CA*, cholate; *DCA*, deoxycholate; *TCA*, taurocholate; *ACR*, arrhythmic cell rate

10.1 Introduction

Traditional medicine has received considerable attention as an alternative to clinical therapy and demand for these remedies has currently increased. About the 6th Century, Traditional Chinese Medicine was gradually introduced from China to Japan, where it developed as an independent form of medical care (Ishibashi 2005). This medicine, which has been called "Kampo", was officially integrated into the Japanese healthcare system. Many Japanese medical doctors utilize Kampo-

K. Takahashi (✉)
Graduate School of Pharmaceutical Sciences, Osaka University, Osaka, Japan

J. Azuma et al. (eds.), *Taurine 7*, Advances in Experimental Medicine and Biology 643,
DOI 10.1007/978-0-387-75681-3_10,

formulas in their daily practice either as the sole source of therapy or in combination with Western drugs.

Bezoar Bovis (BB), a dried gallbladder stone of Bos taurus L. var. domesticus GMELIN, is called “Go-o” in Japanese and is one of the expensive drugs in Kampo medicine. It has been used empirically in Asia for over 3000 years to treat heart and liver related disorders (Mori 1971, Matsumoto et al. 1995). However, its therapeutic potential has not been studied using the Western research approach. The present study was designed to test the idea that BB may act to modulate cardiac disorder and to achieve better therapeutic efficacy. In particular, we focused on taurine and bile acids, which are found in gallbladder and gallstones.

10.2 Methods

10.2.1 Cell Culture

Cardiomyocytes were isolated from the heart of 14–16-day-old mouse embryos (ICR strain) as described previously (Takahashi et al. 1989). In brief, cardiac ventricles were minced and dissociated with 0.125% trypsin (Invitrogen) and 0.025% collagenase type IV (Sigma Aldrich). Dispersed cells were plated and incubated for 1 hr at 37°C. Nonattached myocytes were collected and cultured in Eagle’s MEM medium containing 10% newborn calf serum for 2 days.

10.2.2 Measurement of Beating

The beating status of cultured myocardial cells was monitored with an inverted phase-contrast microscope equipped with a video camera, and a video monitor in a chamber controlled at 37°C. The shape and location of each cell in the dish were recorded before initiating the experiments. The beating properties of the same cells were monitored following the chosen perturbation. Data were expressed as percent of initial values. A cell was considered arrhythmic if it exhibited at least one irregular beat/min; the number of such cells/100 cells is referred to as the arrhythmic cell rate (ACR).

10.2.3 Preparation of BB-Extract and Taurine Measurement

Australian BB (Lot. AMG 86003, Tochimoto Tenkaido Co. Fig. 10.1A-a,b) contains a high concentration of bile acid, which adversely affect cells by disrupting their cell membrane. To minimize the extraction of these bile acids, water was used as the solvent to extract the desired components. Four hundred mg of well-pulverized crude drug was extracted with 2 ml of water for 0.5 hr at room temperature. The extract was centrifuged at 12000 × g for 20 min at 4°C. The taurine content of

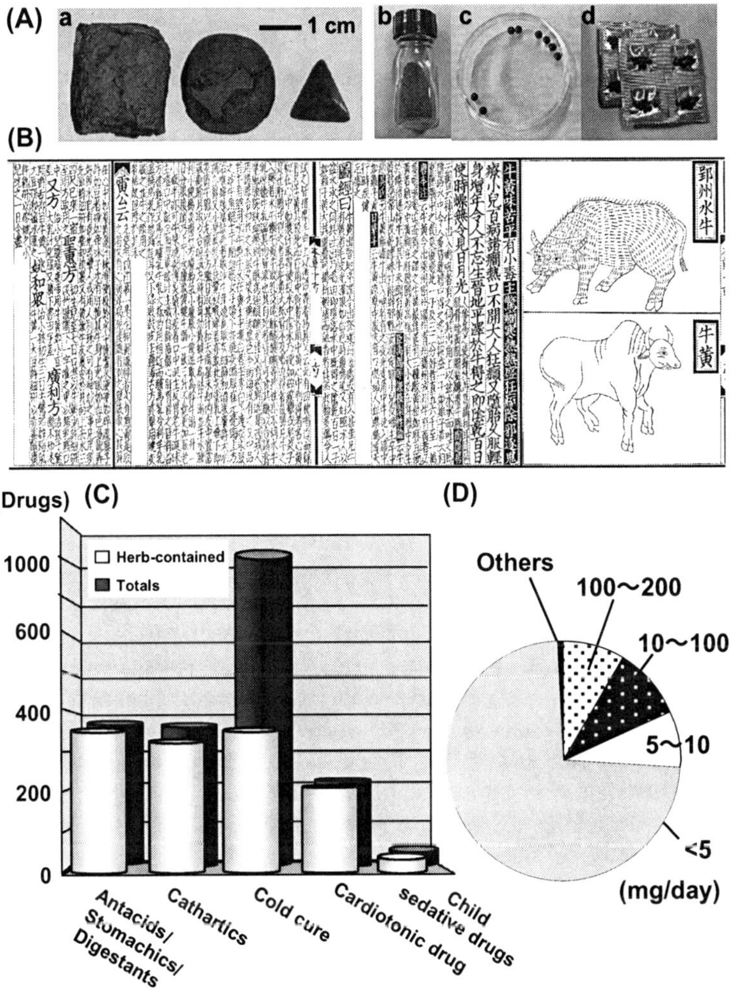

Fig. 10.1 Traditional herbal medicines. **A-a,b**: Australian BB. **A-c,d**: Typical herbal medicines (Kampo drugs) in OTC Japanese drugs. **B**: Chinese herbal classic "Shouruihonzo (http://ethmed.u-toyama.ac.jp/honzou/)" indicates how BB should be used. **C**: Ratio of herbal drugs in OTC Japanese drugs. **D**: Content of BB in the 230 cardioactive agents of OTC Japanese drugs

pharmaceutical products (Fig. 10.1A-c: Kyusin, http://www.kyushin.co.jp/about/index.html; Fig. 10.1A-d: Yutangan, http://www.tsumura.co.jp/products/otc/ yutangan/) was measured by the procedure Jones and Gilligan (1983) using high-performance liquid chromatography (HPLC) system (JASCO880-PU) equipped with a HITACHI F1000 fluorescence detector and HITACHI D-2500 integrator.

10.2.4 Statistic Analysis

Statistical significance was determined by the Student's t-test or χ^2-test. Each value was expressed as the mean ± S.E. Differences were considered statistically significant when the calculated P value was less than 0.05.

10.3 Results

10.3.1 Clinical Application of Traditional Medicine Containing Crude Animal Drugs

Of the 13000 over the counter (OTC) Japanese drugs, the majority of them contain herbal products (Drugs in Japan forum 2006). The ratio of Kampo drugs to the total number of OTC drugs is about 20%. Figure 10.1C shows the ratio of herbal containing articles in stomachic, cathartic or cold cure medicines etc. Eighty-nine % articles in 370 cathartics contain herbal ingredients.

We analyzed the ratio of OTC drugs containing BB. A Chinese review of herbal classic "Shouruihonzo" (AD1108, the monograph on medicinal materials in China, http://ethmed.u-toyama.ac.jp/honzou/) indicated how the drugs should be used (Fig. 10.1B). Namely, it was used mainly as a cardiotonic, an antifebrile, an antispasmodic and an antidotal remedy. In child sedative medicines, almost all drugs contained crude animal drugs, such as BB, dried ox gallstone and Moschus. Notably, of the 230 cardioactive types of OTC Japanese drugs (Drugs in Japan forum 2006), 228 (98.7%) contained BB at concentrations varying from 5 mg to 200 mg per day (Fig. 10.1D). Of the 228 cardioactive drugs, 20 drugs, 22 drugs and 18 drugs contained 100–200 mg, 10–100 mg and 5–10 mg of BB, respectively. The others 167 drugs contained less than 5 mg BB/day. Figure 10.1A shows the number of OTC Japanese drugs containing crude animal drugs. The main drug form is a pill, a tablet or a granule. We measured taurine concentration by HPLC and by amino acid analysis. Taurine content of the OTC drugs (Fig. 10.1A-c,d) varied widely (Kyusin: 0.83 ± 0.01 microg/a pill=15.8 mg; Yutangan: 1.8 ± 0.08 microg/a pill=17 mg).

10.3.2 The Approach to the In Vitro Evaluation of Contractile Beating Pattern and Morphology of Cultured Cardiomyocytes

In order to examine the cardioactive effect of BB, we tested the ability of BB to alter the beating pattern of spontaneously contracting cultured cardiomyocytes following changes in extracellular Ca^{2+}.

Cells were incubated in medium containing 1–2 mM Ca^{2+} and allowed to contract normally for more than 3 hr. However, soon after modifying the medium Ca^{2+} concentration, the beating status of the cultured cardiomyocytes began to change.

After a 1 hr incubation in medium containing either low (0.4 mM) or high (20 mM) Ca^{2+}, the number of beating cells decreased from 98% to 67% and from 98% to 79%, respectively. Addition of 20 mg BB to the medium containing low Ca^{2+} increased the number of beating cells from 67% to 89% (Takahashi et al. 1989). Whereas exposure of myocytes to the Ca^{2+} antagonist, verapamil, decreased the

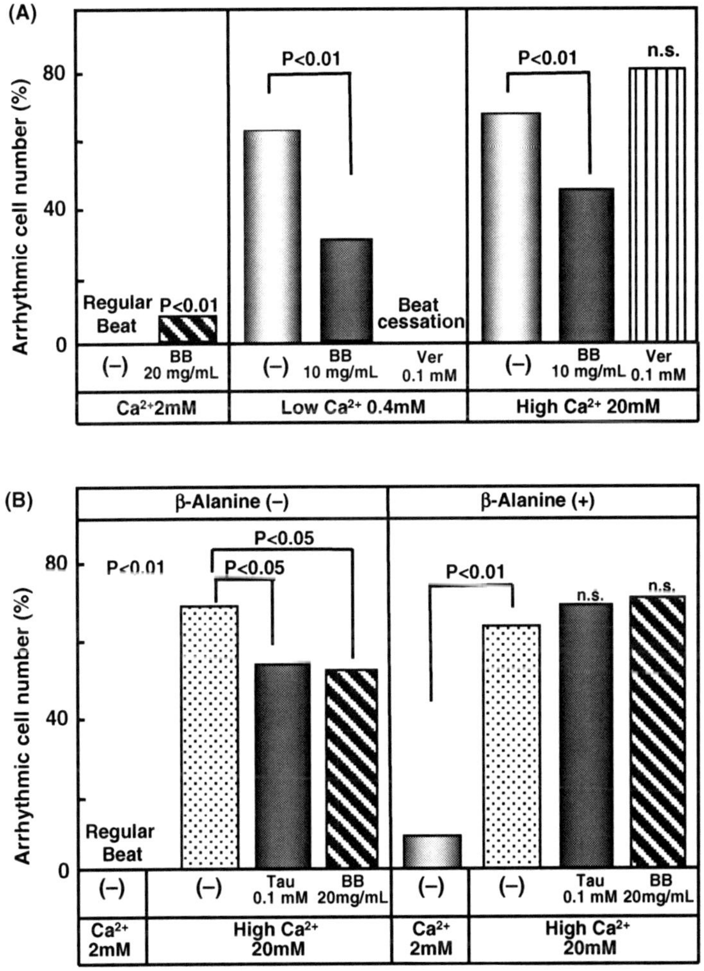

Fig. 10.2 Relationship between the antiarrhythmic effects of BB and taurine. **A**: The effect of BB on beating abnormalities induced by high and low calcium medium. Cardiomyocytes were incubated for 1 hr in each medium. **B**: Influence of beta-alanine pretreatment on the antiarrythmic actions of BB and taurine. Cardiomyocytes were preincubated for 1 hr with 0.1 mM beta-alanine prior to exposure to either BB or taurine. Following the addition of either BB or taurine, the cardiomyocytes were incubated for additional 1 hr, at which point the cellular beating pattern was measured

number of beating cells by 2–3%, verapamil did not exert cardiotonic actions on cells incubated in normal and low Ca^{2+} medium (not data shown).

Figure 10.2 summarizes the effect of changes in medium Ca^{2+} on the frequency of arrhythmias. Cells incubated in medium containing either low or high Ca^{2+} exhibited an increased frequency of arrhythmias. However, BB-extract decreased ACR at low and high medium Ca^{2+} from 62% to 32% and from 66% to 45%, respectively. By comparison, verapamil did not exhibit an antiarrhythmic effect. Rather, it aggravated the beating status of the cardiomyocytes.

The next question we asked was: What is the effective compound in Bezor Bovis? To address this question, the cardioactive effects of several soluble substances were examined using the same culture conditions. The BB extract contained very low levels of the cations Na^+, K^+, Mg^{2+} and Ca^{2+}. The major constituents of BB-extract were bile salts (cholate:CA, 7500; deoxycholate:DCA,1000; taurocholate : TCA, 6700 nmol/mg BB weight/ml) and several amino acids Most cells incubated with concentrations of CA, DCA or TCA ranging from 1∼10 mM underwent marked morphological changes and in some cases lifted from the dish after 1 hr (data not shown).

What substance is responsible for the antiarrhythmic effect? Taurine (0.1 mM) as well as the BB-extract exhibited antiarrhythmic effects. Addition of 0.1 mM taurine to medium containing 20 mM Ca^{2+} decreased ACR from 68 % to 52 % ($p < 0.05$). However, a low dose of bile salts (0.1 mM) did not exert an antiarrhythmic effect. Indeed, TCA (87%, $p < 0.01$) and DCA (85 %, $p < 0.05$) further aggravated the incidence of arrhythmias (Fig. 10.3).

Taurine, cysteine, leucine and isoleucine were detectable in the preparation. Taurine content was 7.1 nmol/mg BB weight/ml, which is extremely high compared with that of the other amino acids (cycteine:Cys, 0.6; leucine:Leu, 0.4; isoleucine:Isoleu, 0.4 nmol/mg BB weight/ml). High levels of taurine (20 mM)

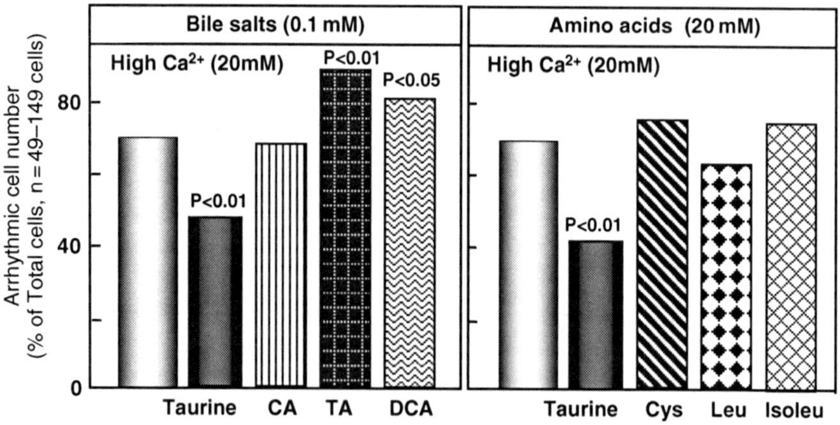

Fig. 10.3 What is the effective compound in the water extract of BB? Cells were incubated with medium containing 20 mM Ca^{2+} and supplemented with bile salts (0.1 mM) or amino acids (20 mM) for 1 hr, at which point the cellular beating pattern was determined

decreased the incidence of arrhythmias from 68% to 42% without causing morphological changes. The same concentration (20 mM) of Cys, Leu and Isoleu did not exert an antiarrhythmic effect (Fig. 10.3).

Beta-alanine is an inhibitor of taurine transport. Addition of 0.1 mM beta-alanine alone did not affect the beating status of myocytes at 20 mM Ca^{2+} (Fig. 10.2B). Although both BB (20 mg/ml) and taurine (0.1 mM) protected against high Ca^{2+}-induced arrhythmias in cells that were not pretreated with beta-alanine pretreatment, when the cells were preincubated with beta-alanine for 1 hr, addition of either BB or taurine did not improve the frequency of arrhythmias.

10.4 Discussion

The results suggest an efficacy of BB as a traditional knowledge-product. The mechanism underlying the action of BB still remains unclear. Based on several lines of evidence it has been postulated that a relationship exists between BB, Ca^{2+} and taurine. First, the water extract of BB prevents arrhythmias produced by exposure of isolated cardiomyocytes to medium containing either low Ca^{2+} or high Ca^{2+}. By contrast, the Ca^{2+}-antagonist, verapamil, did not exhibit an antiarrhythmic effect and in fact aggravated the beating status of the cardiomyocytes incubated in medium containing a low concentration of Ca^{2+}. Therefore, although both BB and verapamil alter Ca^{2+} metabolism, their effects are mediated by different mechanisms.

Second, we have attempted to identify the cardioactive component(s) in the BB extract. Although BB contains many constituents, including bile acids, bile pigments, bile salts and amino acids (Chen et al. 2002; Namba et al. 1982; Hashimoto et al. 1994), the present study has focused on the cardioactive activity of the water soluble fraction of BB. The BB extract contained low levels of certain cations, such as Na^+, K^+, Ca^{2+} and Mg^{2+}. The major constituents of the BB extract are bile salts (CA, DCA, TCA) and amino acids (taurine, Cys, Leu, Isoleu). Most of the cells that were incubated with bile salts (0.1–1 mM) developed morphological damage. Even at the lowest dosages examined, bile salts (0.1 mM) failed to exert an antiarrhythmic effect. In the case of DCA, there was a further aggravation in the incidence of arrhythmias. Addition of Cys, Leu and Isoleu to the medium at a concentration of 20 mM did not exert an antiarrhythmic effect. Thus, the actions of BB appear to be unrelated to the standard amino acid composition of BB.

Third, taurine was effective in protecting against the abnormal beating pattern induced by high Ca^{2+}. Because beta-alanine, an inhibitor of taurine transport, antagonized the beneficial effects of both BB and taurine, it is likely that the effect of BB is partly mediated by taurine. Taurine, which is found in millimolar concentrations in all mammalian tissues, exerts several cytoprotective effects, such as modulation of calcium handling, osmoregulation, antioxidation and detoxification (Huxtable 1992; Schaffer et al. 2000). Cardiac tissue is especially sensitive to the cardioprotective actions of taurine (Satoh and Sperelakis 1998; Xu et al. 2006).

Biliary products, including BB and bear bile, have been used in China and other Asian countries as traditional therapy for thousands of years. The present study provides substantial evidence for the early introduction of medicine from overseas into Japan as Shosoin drugs. Interestingly, one of the Shosoin drugs, which was identified as "Jyutan" gallbladder, contained remarkable amounts of taurine (Yoneda 2000). Hashimoto et al. (1994) measured the taurine content of several crude animal drugs and showed that BB and Kaiba contained high concentrations of taurine. We have also found detectable levels of taurine in BB and some OTC drugs from animal sources. Ancient and modern Asian pharmacopoeias classified biliary products as a medicine that exerted the effects of sedation, anti-hyperspasmia, relieving fever, diminishing inflammation and normalizing function of the gallbladder (Matsumoto et al. 1995; Chen et al. 2002). The therapeutic effect of BB is exactly the same as that of taurine, and was used empirically even before taurine was discovered.

In the countries with an aging or impoverished society, the role of traditional medicine seems to be much more important than ever before. As BB and bear bile are obtained from animals, production is limited and the price is extremely high. For practical reasons and for the purpose of conservation, it is desirable to find an acceptable substitute for the animal biliary products used in traditional medicine. Alternatively, identification of the active agents in biliary products and their pharmacological properties may eliminate the need for an animal source. Because the global market for herbal products is booming, the exhaustion of medicinal sources is critical to the continued existence of traditional medicine. Thus, it is necessary to make use of modern scientific tools to establish verifiable synthetic substitutes as sustainable replacements for herbal sources.

10.5 Conclusion

It has been shown that taurine is a major active component of Kampo medicine and formulae in which biliary products are an ingredient. Presently, the standards in most pharmacopoeias do not provide information on the efficacy or safety of the crude drug. Therefore modern scientific tools are very important in the evaluation of traditional herbs.

Acknowledgments We thank Ms. Y Murao for her secretarial work. This study was funded by a grant from Tochimoto Tenkaido Co. (Osaka Japan).

References

Chen X, Mellon RD, Yang L, Dong H, Oppenheim JJ, Howard OMZ (2002) Regulatory effects of deoxycholic acid, a component of the anti-inflammatory traditional Chinese medicine Niuhung, on human leukocyte response to chemooattractants. Biochem Pharmacol 63:533–541

Drugs in Japan forum (2006) Drugs in Japan: OTC drugs 2007–08. Jiho Inc, Tokyo.

Hashimoto A, Yamasaki K, Kokusenya Y, Miyamoto T, Sato T (1994) Investigation of "signal" constituents for the evaluation of animal crude drugs. I. Free amino acids and total amino acids. Chem Pharm Bull 42:1636–1641
Huxtable RJ (1992) Physiological actions of taurine. Physiol Rev 72:101–163
Ishibashi A, Kosoto H, Ohno S, Sakaguchi H, Yamada T, Matsuda K (2005) Genaral introduction to Kampo. In: The Japan society for oriental medicine (eds) Introduction to KAMPO: Japanese Traditional medicine, Elsevier Japan, Tokyo, pp 2–8
Jones BN, Gilligan JP (1983) O-phthaldehyde precolumn derivatization and reversed-phase high-performance liquid chromatography of polypeptide hydrolysates and physiological fluids. J Chromatogr 266:471–482
Matsumoto N, Nakashita T, Kashima K (1995) Effectiveness of bovine gallstone (Goou) and bear gall powder (Yutan) on chronic liver diseases: a preliminary report. Tokai J Exp Clin Med 20:9–16
Mori T (1971) Goo. In: Mori T (ed) Shen-Nong-Ben-Cao-jing, Seibundo, Tokyo, p 14
Namba T, Nunome S, Hattori M, Higashidate S, Maekubo T (1982) Fundamental stidies on the evaluation crude drug, VII. On animal gall (1). Yakugaku Zasshi 102:760–767
Satoh H, Sperelakis N (1998) Review of some actions of taurine on ion channels of cardiac muscle cells and others. Gen Pharmacol 130:451–463
Schaffer SW, Lombardini JB, Azuma J (2000) Interaction between the actions of taurine and angiotensin II. Amino Acids. 18:305–318
Takahashi K, Azuma J, Park S, Awata N, Kishimoto S, Namba T, Shaffer SW (1989) Pharmacological study of a traditional Chinese medicine: effect of Bezoar Bovis on the irregular beating pattern of cultured mouse myocardial cells. Res Commun Chem Phathol Pharmacol 63: 317–330
Xu Y, Saini HK, Zhang M, Elimban V, Dhalla NS (2006) MAPK activation and apoptotic alterations in hearts subjected to calcium paradox are attenuated by taurine. Cardiovasc Res 72: 163–174
Yoneda K (2000) Jutan. In: Office of Shosoin Treasure-House, Imperial Household Agency (ed) Shosoin Medicaments Chuokoron-shinsha, Tokyo, pp 136–138

Chapter 11
Mechanism of TauT in Protecting Against Cisplatin-Induced Kidney Injury (AKI)

Xiaobin Han and Russell W. Chesney

Abstract Acute renal failure (ARF) results from ischemic or toxic renal injury and is a common disorder with an overall mortality rate of about 50%. Despite the advent of dialysis and increasing knowledge regarding the causes and the effects of ARF, this trend has hardly changed for several decades. Cisplatin is a broadly used anti-cancer drug and is widely known to induce acute renal failure as a result of renal tubular injury. In this study, we have demonstrated that: (1) *TauT* is down-regulated by the p53 tumor suppressor gene in renal proximal tubule LLC-PK1 cells; (2) cisplatin down-regulates the expression of *TauT* in renal proximal tubule cells through, at least in part, a p53-dependent pathway; and (3) forced overexpression of *TauT* can attenuate cisplatin-induced apoptosis of renal proximal tubule cells.

Abbreviations *TauT* , taurine transporter gene; *AKI,* acute kidney injury

11.1 Introduction

Cisplatin is a commonly used chemotherapeutic agent that has a major limitation because of its nephrotoxicity. Taurine is an important osmolyte that has been found to protect against cisplatin-induced apoptosis in renal cells *in vitro* (Han and Chesney 2006). Elevated levels of the tumor suppressor gene p53 have been found in the kidneys of animal models of acute renal failure induced by cisplatin administration (Miyaji et al. 2001). A recent study has demonstrated that p53 activation is an early signal in cisplatin-induced apoptosis in renal tubular cells (Jiang et al. 2004). Our studies have shown that *TauT* is negatively regulated by p53 in renal cells (Han et al. 2002). Notably, the progressive renal failure seen in p53 transgenic mice is similar to that previously observed in the offspring of taurine-deficient cats, which showed ongoing kidney damage and abnormal renal and retinal development (Han and Chesney 2006), suggesting that the taurine transporter gene is an important target of p53 during kidney development and renal injury.

X. Han (✉)
Department of Pediatrics, University of Tennessee Health Science Center, and the Children's Foundation Research Center at Le Bonheur Children's Medical Center, Memphis, TN, USA

J. Azuma et al. (eds.), *Taurine 7*, Advances in Experimental Medicine and Biology 643,
DOI 10.1007/978-0-387-75681-3_11,

11.2 Materials and Methods

11.2.1 Construction of the Reporter Gene

In this study, we generated a pGL963-reporter gene construct by using a 1.1 kb fragment of the *TauT* promoter region DNA as the template for PCR (GenBank accession number AR151716) and the PCR fragment was cloned into the promoter-less luciferase vector pGL3-Basic or SV40/pGL-3 control vector (Promega, Madison, WI). The conditions used were 30 cycles of 1 min of denaturation at 94°C, 1 min of annealing at 58°C, and 1 min of elongation at 72°C. The sense primer (5′-GGGGTACCTTAC-TGAAGGTCACACAGC-3′) designed for PCR contained a unique site for *KpnI*, and the antisense primer (5′-AAGATCTTGGCACGGGAGTT CA-3′) contained a unique site for *BgI II*. PCR products were digested with *KpnI* and *BgI II* and re-ligated into *KpnI* and *BgI II* sites of pGL3-Basic or pGL3-control vectors.

11.2.2 Cell Culture

Wild-type and *TauT*–overexpressing LLC-PK1 cells were cultured according to ATCC (American Type Culture Collection) guidelines. Briefly, cells are grown as confluent monolayers in 10 cm diameter tissue culture plates in media specific for each cell line with 10% fetal calf serum at 37°C in the presence of 5% CO_2 in a humidified incubator. For experiments, 1×10^6 cells were seeded and cultured overnight, and 80% confluent cells were used.

11.2.3 Transient Transfection

Plasmid DNA was introduced into cultured mammalian cells using Lipofectamine 2000 (Invitrogen). Transfection was carried out for 16–18 h, and then cells were washed twice with phosphate-buffered saline and incubated in fresh medium for 24–48 h before harvesting. pGL-control, which contains a luciferase gene driven by the SV40 early region promoter/enhancer, and empty pGL-Basic vectors were used as positive and negative controls, respectively. To standardize the transfection efficiency, 0.1 μg of pRL-CMV vector (pRL Renilla Luciferase control reporter vector, Promega) was cotransfected in all experiments. Cells were harvested 48 h after transfection and lysed in 200 μl of reporter lysis buffer (Promega). A luciferase assay was performed using a dual-luciferase assay kit (Promega), and activity was measured with an Optocomp 1 luminometer (MGM Instruments, Inc., Hamden, CT). Promoter activity (mean ± SD of four samples in relative light units) of each construct is represented by relative light output normalized to pRL-CMV control. Graphs represent typical results of four separate experiments. The concentration of protein in the cell extracts was determined using the Bradford method (Bio-Rad, Hercules, CA).

11.2.4 Measurement of Taurine Transport

Taurine transport studies were performed on confluent monolayers 3 days after seeding cells. Briefly, cells were washed with Earle's Balanced Salt Solution (EBSS) at 37°C. Uptake was initiated by the addition of uptake buffer (2 mM KCl, 1 mM $MgCl_2$, 96 mM NaCl, 1.8 mM $CaCl_2$, 5 mM Hepes, pH 7.6) to which 50 μM unlabelled taurine and 0.5 μCi/ml ^{14}C-taurine (Perkin Elmer, Boston, MA) were added. After incubation for 30 min at room temperature, uptake was terminated by the removal of uptake buffer followed by three rapid washes with cold EBSS. Cells were solubilized in 1% SDS in 0.2 N NaOH and radioactivity counted in a Packard 2000-CA Liquid Scintillation Analyzer.

11.2.5 Western Blot Analysis

Cells were lysed in 50 μl M-PER mammalian protein extraction reagent (Pierce, Inc., Rockford, IL) supplemented with a protease inhibitors cocktail for use with mammalian cell and tissues extracts (Sigma, St. Louis, MO). The lysates were cleared by centrifugation at 14,000 ×*g* for 2 min, and the supernatants transferred to clean tubes. Equal amounts of protein (50 μg) were separated by electrophoresis on a 12% SDS-polyacrylamide gel and transferred to a nitrocellulose membrane (Millipore, Bedford, MA) using a semi-dry electrophoretic transfer system (Bio-Rad, Hercules, CA). Membranes were incubated in 5% nonfat dry milk in Tris base/sodium chloride (TBS) buffer with 0.2% Tween 20 (TBST) at 4°C overnight. The membranes were incubated with primary antibody for 1 h at room temperature, washed with TBST and incubated with horseradish peroxidase-linked secondary antibody (Sigma) for 1 h, and then the proteins of interest were detected using a chemiluminescent detection kit (Pierce, Inc.).

11.2.6 Apoptosis Assays

Apoptosis was measured by using a TACS Annexin V-FITC Apoptosis Detection kits (R&D Systems, Inc, Minneapolis, MN) following the manufacturer's instructions. Briefly, cells were detached by adding 5 ml trypsin to the cell cultures and collected by centrifugation at 500 ×*g* for 5 min at room temperature. Cells were washed by resuspending in 1 ml cold 1 × PBS and then pelleted by centrifugation. Then, cells were gently resuspended in the Annexin V Incubation Reagent at a concentration of 1 × 10^6cells/100 μl and incubated in the dark for 15 minutes at room temperature. Finally, 400 μl of 1 × binding buffer were added to each sample and apoptosis was analyzed by flow cytometry.

11.2.7 Materials

[^{14}C]-taurine (92.1 mCi/mmol) was purchased from Du Pont, Perkin-Elmer. LLC-PK1 cells were purchased from ATCC and cisplatin was purchased from Sigma.

11.2.8 Statistics

All experiments using tissue cultures were performed in triplicate. Luciferase assay results were expressed in units of relative light output. Statistical comparisons were made using one-way ANOVA and Student's *t* test to determine significant differences in the means between experimental groups.

11.3 Results

11.3.1 Cisplatin Impairs Taurine Transporter in LLC-PK1 Proximal Tubule Renal Cells

To determine whether cisplatin impairs the function of the taurine transporter, LLC-PK1 cells were treated with or without cisplatin (0–100 μM) for 24 hours and taurine transport activity by the cells was measured as described in the Methods. As shown in Fig. 11.1, cisplatin significantly decreased the taurine transport activity of LLC-PK1 cells in a dose-dependent manner.

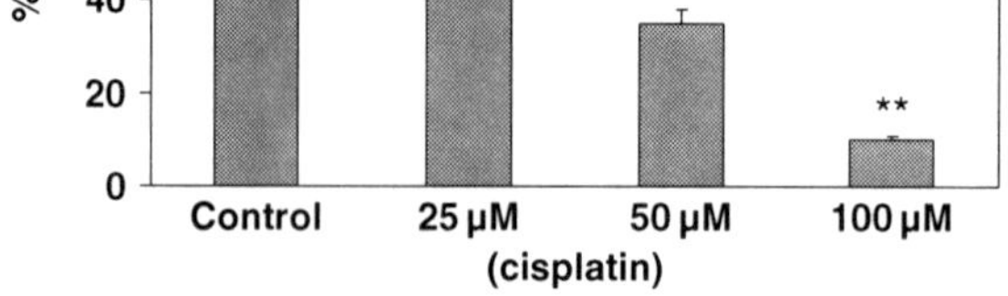

Fig. 11.1 Effect of cisplatin on taurine uptake. LLC-PK1 cells were cultured in DMEM/F12 medium containing cisplatin as indicated for 24 h, and then taurine uptake was measured. $^{*}p < 0.05$ vs. control, $^{**}p < 0.01$ vs. control

11.3.2 Cisplatin Down-Regulates Expression of TauT in LLC-PK1 Renal Cells

To determine whether cisplatin represses expression of the *TauT* gene, Western blot analysis was performed in cells treated with cisplatin (25–100 μM) for 24 hours using a specific *TauT* protein antibody produced by our laboratory. As shown in Fig. 11.2, expression of *TauT* was down-regulated by cisplatin in a dose-dependent manner.

11.3.3 Transcriptional Repression of TauT by Cisplatin

To determine if down-regulation of the *TauT* gene by cisplatin occurs at the transcriptional level, the reporter construct pGL-963 was transiently transfected into LLC-PK1 cells, and regulation of *TauT* promoter function by cisplatin was examined. As shown in Fig. 11.3, cisplatin decreased *TauT* promoter activity in LLC-PK1 cells in a manner similar to that observed in Figs. 11.1 and 11.2, suggesting that cisplatin represses *TauT* expression at the transcriptional level.

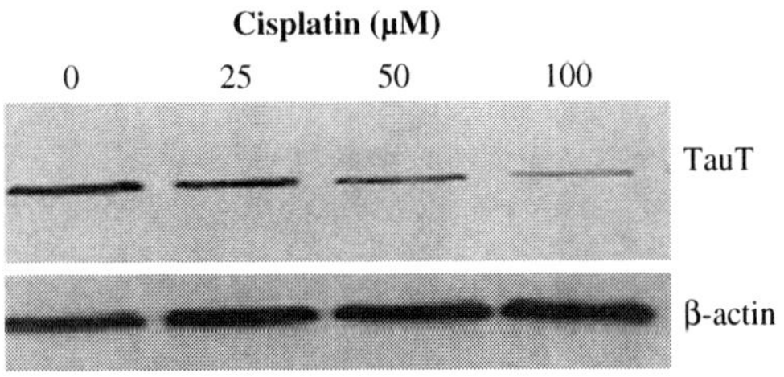

Fig. 11.2 Dose-dependent regulation of *TauT* by cisplatin. LLC-PK1 cells were cultured in medium containing cisplatin for 24 h, and then Western blot analysis of *TauT* was carried out

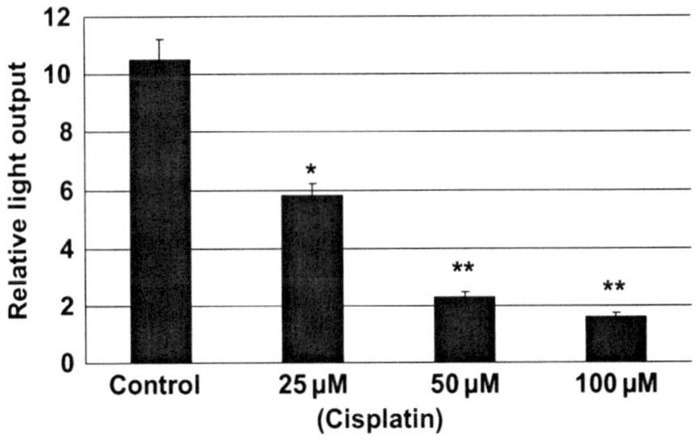

Fig. 11.3 Down-regulation of *TauT* promoter activity by cisplatin in renal cells. pGL-963 was transiently transfected into LLC-PK1 cells treated with cisplatin for 24 h . The luciferase assay was performed using cell lysates. $^{*}p < 0.05$ vs. control, $^{**}p < 0.01$ vs. control

11.3.4 Cisplatin Induces p53 Expression in LLC-PK1 Cells

To examine whether cisplatin would induce expression of endogenous p53, Western blot analysis was carried out in cells treated with cisplatin for 24 hours. We found that expression of p53 was increased by cisplatin in LLC-PK1 cells in a dose-dependent manner (Fig. 11.4). This result suggests that cisplatin down-regulates *TauT* expression, at least in part, through a p53-dependent pathway.

11.3.5 Effect of Forced Overexpression of TauT on Taurine Uptake

Studies have shown that cisplatin induces dose- and duration-dependent apoptosis in LLC-PK1 cells (Park et al. 2002). To test if *TauT* functions as an anti-apoptotic gene in LLC-PK1 cells, *TauT* cDNA was stably transfected into the LLC-PK1 cells selected by G418. To test whether forced overexpression of *TauT* blocks the effect of cisplatin on taurine uptake by LLC-PK1 cells, wild-type and *TauT*-overexpressing LLC-PK1 cells were treated with cisplatin (50 μM) for 24 hours, and taurine uptake was measured. The untreated cells and cells stably transfected with the vector alone were used as controls. As shown in Fig. 11.5, taurine uptake by wild-type LLC-PK1

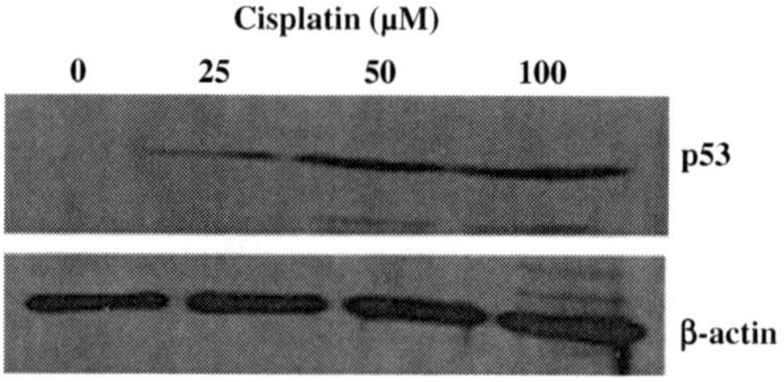

Fig. 11.4 Cisplatin-induced p53 expression in LLC-PK1 cells. Cells were cultured in medium containing cisplatin for 24 h, and then Western blot analysis of p53 expression was carried out

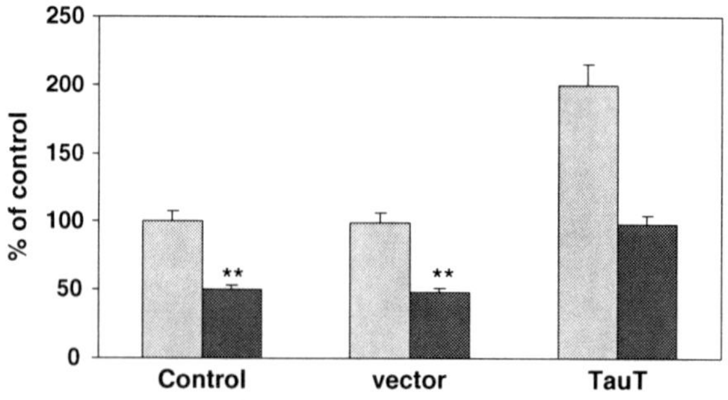

Fig. 11.5 Effect of cisplatin on taurine uptake. *TauT* cDNA was stably transfected into LLC-PK1 cells. Cells transfected with vector alone were used as a negative control. Cells were cultured in medium containing cisplatin (50 μM) for 24 h, and then taurine uptake was measured

cells and cells transfected with vector alone was decreased by more than 50% after treatment with cisplatin, as compared to the amount of taurine taken up by control cells. Overexpression of *TauT* doubled the taurine uptake and attenuated the effect of cisplatin on taurine uptake in *TauT*-overexpressing cells, as compared to control cells.

11.3.6 Effect of Forced Overexpression of TauT on Cisplatin-Induced Apoptosis

To determine if forced overexpression of *TauT* attenuates cisplatin-induced apoptosis, wild-type and *TauT*-overexpressing LLC-PK1 cells were treated with cisplatin (50 μM) for 24 h, then apoptosis was measured by flow cytometry as described in the Methods. As shown in Fig. 11.6, the number of apoptotic cells was significantly higher in cisplatin-treated wild-type cells (7.82%) than in *TauT*-overexpressing cells (1.34%). This result indicates that the cisplatin-induced apoptosis can be attenuated by forced overexpression of *TauT*.

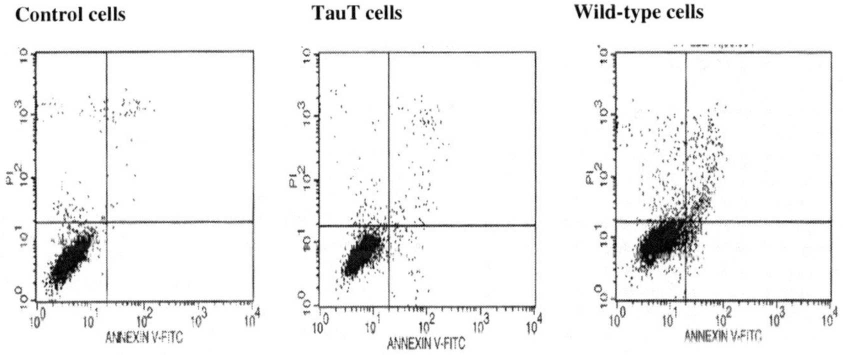

Fig. 11.6 Effect of forced overexpression of *TauT* on cisplatin-induced LLC-PK1 cell apoptosis. Wild-type and *TauT*-overexpressing LLC-PK1 cells was cultured in medium containing cisplatin for 24 h, and then apoptosis was measured in control, *TauT*-overexpression, and wild-type LLC-PK1 cells

11.4 Discussion

Cisplatin is one of the most widely used chemotherapeutic agents for the treatment of several human malignancies. Nephrotoxicity after cisplatin treatment is common and may manifest after a single dose with acute renal failure or may present with a chronic syndrome of renal electrolyte wasting.

Studies have shown that taurine can prevent cell apoptosis through several mechanisms, including inhibition of the generation of reactive oxygen species

(ROS), nitric oxide (NO), tumor necrosis factor alpha (TNF-alpha), and regulation of intracellular calcium flux (Seabra et al. 1998; Gurujeyalakshmi et al. 2000; Huang et al. 2003). However, such protective actions of taurine require the normal function of the taurine transporter located on the cell membrane (Maar et al. 1998). These findings led us to postulate that the normal function of the taurine transporter is impaired in cisplatin-induced renal injury, and cisplatin-induced nephrotoxicity may be prevented by increased expression of the *TauT* gene.

In this study we demonstrated that *TauT* is down-regulated by cisplatin in renal cells. Cisplatin represses the *TauT* gene at the transcriptional level through a p53-dependent manner, at least in part. Studies have shown that *TauT* is down-regulated by the p53 tumor suppressor gene in renal cells (Han et al. 2002), and overexpression of p53 results in progressive renal failure in p53 transgenic mice, which is similar to observations made in taurine-deficient kittens (Han et al. 2000; Godley et al. 1996). The present study confirms that forced overexpression of *TauT* can attenuate cisplatin-induced apoptosis of renal proximal tubule LLC-PK1 cells, suggesting that *TauT* plays an important role in protecting against drug-induced renal injury.

Acknowledgments The authors wish to thank Andrea Patters for her insightful comments and suggestions. This work was supported by grants from the National Kidney Foundation and the Le Bonheur Chair of Excellence in Pediatrics.

References

Godley LA, Eckhaus M, Paglino JJ, Owens J, Varmus HE (1996) Wild-type p53 transgenic mice exhibit altered differentiation of the ureteric bud and possess small kidneys. Genes Dev 10: 836–850

Gurujeyalakshmi G, Wang Y, Giri SN (2000) Suppression of bleomycin-induced nitric oxide production in mice by taurine and niacin. Nitric Oxide 4:399–411

Han X, Budreau AM, Chesney RW (2000) The taurine transporter gene and its role in renal development. Amino Acids 19:499–507

Han X, Budreau AM, Chesney RW (2002) Transcriptional repression of taurine transporter gene (TauT) by p53 in renal cells. J Biol Chem 277:39266–39273

Han X, Chesney RW (2006) Regulation of TauT by cisplatin in LLC-PK1 renal cells. Pediatr Nephrol 20:1067–1072

Huang XM, Zhu WH, Kang ML (2003) Study on the effect of doxorubicin on expressions of genes encoding myocardial sarcoplasmic reticulum Ca^{2+} transport proteins and the effect of taurine on myocardial protection in rabbits. J Zhejiang Univ Sci 4:114–120

Jiang M, Yi X, Hsu S, Wang CY, Dong Z (2004) Role of p53 in cisplatin-induced tubular cell apoptosis: dependence on p53 transcriptional activity. Am J Physiol Renal Physiol 287:F1140–F1147

Maar TE, Lund TM, Gegelashvili G, Hartmann-Petersen R, Moran J, Pasantes-Morales H, Berezin V, Bock E, Schousboe A (1998) Effects of taurine depletion on cell migration and NCAM expression in cultures of dissociated mouse cerebellum and N2A cells. Amino Acids 15:77–88

Miyaji T, Kato A, Yasuda H, Fujigaki Y, Hishida A (2001) Role of the increase in p21 in cisplatin-induced acute renal failure in rats. J Am Soc Nephrol 12:900–908

Park MS, De Leon M, Devarajan P (2002) Cisplatin induces apoptosis in LLC-PK1 cells via activation of mitochondrial pathways. J Am Soc Nephrol 13(4):858–65

Seabra V, Stachlewita RF, Thurman RG (1998) Taurine blunts LPS-induced increases in intracellular calcium and TNF-alpha production by Kupffer cells. J Leukoc Biol 64:615–621

Chapter 12
TauT Protects Against Cisplatin-Induced Acute Kidney Injury (AKI) Established in a TauT Transgenic Mice Model

Xiaobin Han and Russell W. Chesney

Abstract Cisplatin is a commonly used chemotherapeutic agent that has a major limitation because of its nephrotoxicity. Taurine is an important osmolyte that has been found to protect against cisplatin-induced apoptosis in renal cells *in vitro*. To determine whether over-expression of hTauT protects against cisplatin-induced acute kidney injury (AKI) *in vivo*, animals (wt and transgenic) were injected with cisplatin and the levels of BUN and serum creatinine were measured. Saline-injected mice were used as a control. The results showed that the levels of BUN and creatinine were significantly increased in the cisplatin-injected wild-type mice (110 ± 12 mg/dl and 0.98 ± 0.05 mg/dl) as compared to the saline-injected wild type mice (20 ± 2.5 mg/dl and 0.52 ± 0.06 mg/dl). However, over-expression of hTauT effectively prevented the progression of cisplatin-induced AKI in hTauT transgenic mice, as measured by the levels of BUN and serum creatininelevels (23 ± 3.5 mg/dl and 0.6 ± 0.05 mg/dl).

Abbreviations *AKI,* acute kidney injury; *hTauT,* human taurine transporter gene; *BUN,* blood urine nitrogen

12.1 Introduction

Acute renal failure due to ischemic or toxic renal injury is a common disorder with a mortality of about 50% (Kelly and Molitoris 2000 and Thadhani et al. 1996). A vast majority of research in the field has focused on the determination of events and factors that cause renal proximal tubular cell (RPTC) injury and death and lead to the development of ARF. Cisplatin-induced ARF is currently a topic of intense study. As a highly effective chemotherapeutic agent, cisplatin has been used to treat a wide variety of solid tumors (Lebwohl and Canetta 1998). However, 25–35% of patients experience a significant decline in renal function after the administration

X. Han (✉)
Department of Pediatrics, University of Tennessee Health Science Center and the Children's Foundation Research Center at Le Bonheur Children's Medical Center, Memphis, TN, USA

J. Azuma et al. (eds.), *Taurine 7*, Advances in Experimental Medicine and Biology 643,
DOI 10.1007/978-0-387-75681-3_12, © Springer Science+Business Media, LLC 2009

of a single dose of cisplatin (Ries and Klastersky 1986). Several mechanisms, including oxidation, inflammation, genotoxic damage and cell cycle arrest, have been implicated in cisplatin nephrotoxicity (Ramesh and Reeves 2002; Ramesh and Reeves 2003; Baliga et al. 1998; Megyesi et al. 1998; Price et al. 2004; Ramesh and Reeves 2004).

Elevated levels of the tumor suppressor gene p53 have been found in the kidneys of animal models of acute renal failure induced by cisplatin administration (Miyaji et al. 2001). Jiang et al. have demonstrated that p53 activation is an early signal in cisplatin-induced apoptosis in renal tubular cells (Jiang et al. 2004). Varmus' group has found that transgenic mice overexpressing p53 undergo progressive renal failure through a novel mechanism by which p53 appears to alter cellular differentiation, rather than by growth arrest or the direct induction of apoptosis (Godley et al. 1996). These findings suggest that altered expression of certain p53 target gene(s) involved in renal development may be responsible for p53-induced progressive renal failure in p53 transgenic mice.

Our studies have shown that *TauT* is negatively regulated by p53 in renal cells (Han et al. 2002). Interestingly, the progressive renal failure seen in p53 transgenic mice is similar to that previously observed in the offspring of taurine-deficient cats, which showed ongoing kidney damage and abnormal renal and retinal development (Sturman et al. 1985), suggesting that the taurine transporter gene is an important target of p53 during kidney development and renal injury. It is worth noting that cisplatin accumulates in cells from all nephron segments, but is preferentially taken up by the highly susceptible proximal tubule cells within the S3 segment, which is the site for renal adaptive regulation of *TauT* (Leibbrandt et al. 1995; Matsell et al. 1997). A recent study showed that taurine was able to attenuate cisplatin-induced nephrotoxicity and protect renal tubular cells from tubular atrophy and apoptosis (Saad and Al-Rikabi 2002). Therefore, down-regulation of *TauT* by p53 may play an important role in cisplatin-induced nephrotoxicity.

12.2 Materials and Methods

12.2.1 Animals

FVB mice were maintained at the Animal Research Facility at the University of Tennessee Health Science Center and all procedures were in accordance with the Guidelines for Care and Use of Experimental Animals.

12.2.2 Generation of hTauT Transgenic Mice

For generation of *TauT* transgenic mice, we used a pCAGGS expression vector (a generous gift from Dr. Jun-ichi Miyazaki at Osaka University Medical School) (Maruyama, et al. 2002) which has been widely used for creating transgenic mice (Gawlik and Quaggin 2004). Human *TauT* cDNA was fused to the EcoR I site of the pCAGGS vector tailed with a rabbit ß-globin poly A, which is driven by a chicken

β-actin promoter. The transgene was purified from vector sequences and 2 ng/ml of the DNA was injected into fertilized FVB/N mouse eggs to establish lines of transgenic animals using standard methodologies (Hogan et al. 1994) at the UTHSC transgenic facility.

12.2.3 Genotyping

The transgenic nature of these animals was tested by PCR using flanking primers, as shown in Fig 12.1, top Primer 1 (a specific human TauT forward primer 5′-AACCCCATCTTTGGCAGGCA-3′ residues 3691–3710 of GenBank Z18956), primer 2 (a specific rabbit β-globin reverse primer 5′-AGCCAGAA-GTCAGATGCT CAA-3′ residues 1486-1466 of GenBank V00882), primer 3 (a specific human and mouse TauT forward primer 5′-GGCCTGCCTGTGTTTTTCTT-3′ residues 501–520 of GenBank L03292) and primer 4 (a specific mouse TauT reverse primer 5′-GGTGAAGTTGGCAGTGCTAAGG-3′ residues 807–785 of GenBankL03292) were used for PCR. Amplification of heterozygous transgenic DNA resulted in two bands, wild-type DNA yielded only band I (1.3 kb) and homozygous transgenic

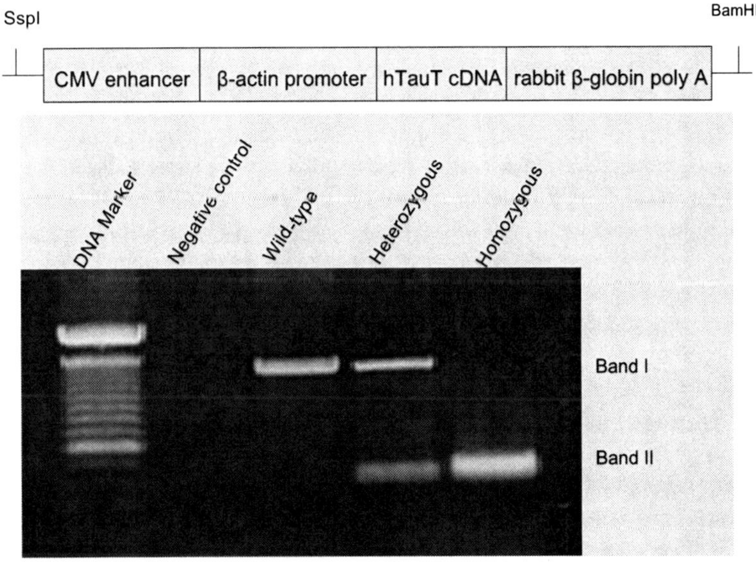

Fig. 12.1 Genotyping of *TauT* transgenic mice. Top: Transgene construct for generation of *TauT* transgenic mice; Bottom: PCR analysis of wild-type, *TauT* transgene heterozygotic and *TauT* transgene homozygotic mice shows the transgene and flanking mouse DNA (bottom band); corresponding to the wild-type mouse locus (top band). Primers 3 and 4 amplify band I only in samples from wild-type or heterozygous transgenic mice, whereas primers 1 and 2 amplify the shorter band II in heterozygotes and homozygotic transgenic animals. Thus, homozygotes produce only band II and heterozygotes produce both bands. All four primers were used in all samples. Lane 1, DNA standard; lane 2, non-template negative control; lane 3, wild-type mouse; lane 4, heterozygote; lane 5, homozygote

DNA yielded only band II (406 bp) (Fig. 12.1, bottom) Band I contains an approximate 1.0 kb intron of mouse *TauT*.

12.2.4 In Vivo Model of Cisplatin-Induced AKI

Male mice (wild-type and *TauT* transgenic), 10–12 weeks old and weighing 28 to 30 g, were assigned to treatment groups ($n = 8/\text{group}$). For the experiment, eight *TauT* transgenic mice and eight wild-type mice received a single dose of cisplatin (15 mg/kg body weight) by intraperitoneal injection. Eight saline-injected mice were used as controls. To determine cisplatin-induced nephrotoxicity, mice were sacrificed 3 days after cisplatin injection. Blood samples were collected. The levels of BUN and serum creatinine were measured by an enzymatic colorimetric assay kit (Sigma, St. Louis, MO).

12.2.5 Western Blot Analysis

Kidney tissues were lysed in 50μ lM – PER mammalian protein extraction reagent (Pierce, Inc., Rockford, IL) supplemented with a protease inhibitors cocktail for use with mammalian cell and tissues extracts (Sigma, St. Louis, MO). The lysates were cleared by centrifugation at $14,000 \times g$ for 2 min and the supernatants transferred to clean tubes. Equal amounts of protein (50 μ g) were separated by electrophoresis on a 12% SDS-polyacrylamide gel and transferred to a nitrocellulose membrane (Millipore, Bedford, MA) using a semi-dry electrophoretic transfer system (Bio-Rad, Hercules, CA). Membranes were incubated in 5% nonfat dry milk in Tris base/sodium chloride (TBS) buffer with 0.2% Tween 20 (TBST) at 4°C overnight. The membranes were incubated with primary antibody for 1 h at room temperature, washed with TBST and incubated with horseradish peroxidase-linked secondary antibody (Sigma) for 1 h and then the proteins of interest were detected using a chemiluminescent detection kit (Pierce, Inc.).

12.2.6 Histomorphometric Analysis of Kidney Injury

Five-micron transverse kidney sections were prepared and stained with hematoxylin and eosin for histomorphometric analysis. Tubular injury was rated on a scale of 0 to 4, where 0 = normal, 1 = <10%, 2 = 10 – 25%, 3 = 26 – 75% and 4 = >75% injury.

12.2.7 Statistics

All experiments were performed in triplicate. The data represent the mean ± SEM of 3 or 4 experiments. Statistical comparisons were made using one-way ANOVA and Student's *t*-test to determine significant differences in the means.

12.3 Results

12.3.1 TauT Transgenic Mice

hTauT overexpressing mice were created and used in this study. The transgenic nature of these animals was tested by PCR using flanking primers, as shown in Fig. 12.1. Primer 1 (a specific human *TauT* forward primer 5′-AACCCCATCT TTGGCAGGCA-3′ residues 3691–3710 of GenBank Z18956), primer 2 (a specific rabbit β-globin reverse primer 5′-AGCCAGAA-GTCAGATGCTCAA-3′ residues 1486–1466 of GenBank V00882), primer 3 (a specific human and mouse TauT forward primer 5′-GGCCTGCCTGTGTTTTTCTT-3′ residues 501–520 of GenBank L03292) and primer 4 (a specific mouse TauT reverse primer 5′- GGTGAAGTTGG-CAGTGCTAAGG-3′ residues 807–785 of GenBankL03292) were used for PCR. Amplification of heterozygous transgenic DNA resulted in two bands, wild-type DNA yielded only band I (1.3 kb) and homozygous transgenic DNA yielded only band II (406 bp) (Fig. 12.1, bottom) Band I contains an approximate 1.0 kb intron of mouse *TauT*. There were no visible differences between the *hTauT* transgene mice (heterozygotes and homozygotes) and wild-type animals regarding body hair, birth weight, organ weight, or growth curve (data not shown). Expression of *hTauT* was analyzed by RT-PCR using RNA extracted from organs, including brain, lung, heart, liver, spleen and kidney (Fig. 12.2a). Western blot analysis showed that expression

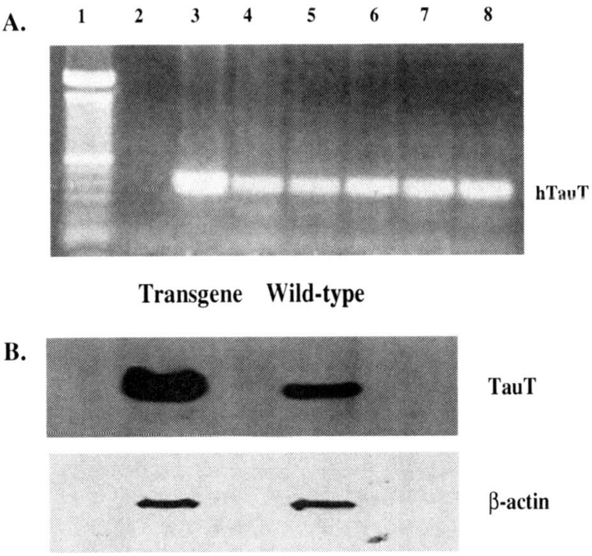

Fig. 12.2 Tissue expression of h*TauT* in *TauT* transgenic mice. (**a**): RT-PCR analysis of *hTauT* expression in various tissues of *hTauT* transgenic mice. Lane 1, DNA std; lane 2, negative control; lane 3, brain; lane 4, lung; lane 5, heart; lane 6, liver; lane 7, spleen; and lane 8, kidney; (**b**): Western blot analysis of *TauT* expression in kidney. Total kidney protein (50μg) was analyzed by Western blot using an antibody recognizing both human and mouse taurine transporters, as described in the Methods section. β-actin was used as an internal control

of *TauT* was elevated by 2.5-fold in the kidneys of *hTauT* transgenic mice as compared to wild-type control mice (Fig. 12.2, bottom). The expression pattern of the transgene was consistently steady after several generations of breeding (data not shown). Thus, this line of *TauT* transgenic mice was used for the following studies.

12.3.2 hTauT Transgenic Mice are Resistant to Cisplatin-Induced ARF

To determine whether overexpression of *hTauT* pretects against cisplatin-induced ARF, 12-week-old male FVB/N mice or *hTauT* transgenic mice (heterozygous) weighing 28 to 30 g were injected with a single intraperitoneal dose of cisplatin (15 mg/kg/body weight). Wild-type mice injected with saline were used as negative controls. Body weight changes and animal deaths were recorded. Cisplatin administration resulted in a significant weight loss in wild-type mice, but not in the *hTauT* transgenic mice, during the first 4 days after injection (Fig. 12.3a). Significant weight loss was first observed in *hTauT* transgenic mice at day 5 after cisplatin injection (from 28.5 ± 0.5 g to 25.8 ± 0.8 g), while the weight loss reached

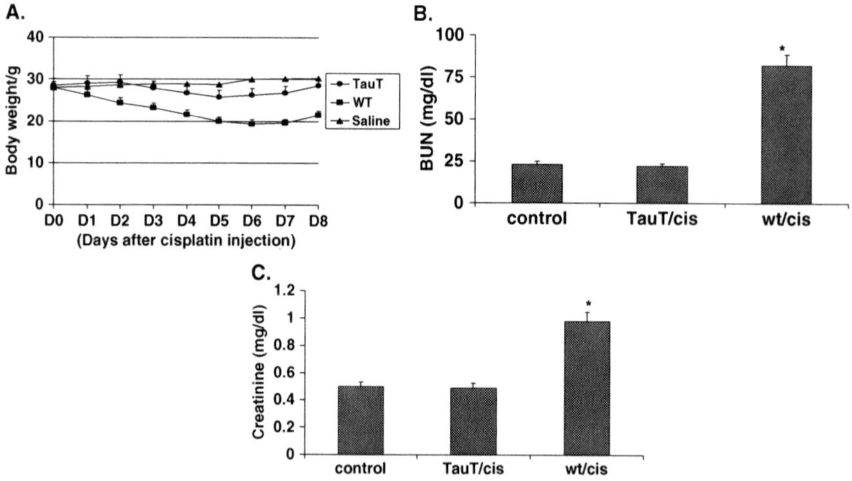

Fig. 12.3 Effect of overexpression of *TauT* on cisplatin-induced nephrotoxicity. **(a)** Eight (8) wild-type male 12-week-old (WT/cis) or *hTauT* transgenic mice (heterozygotes) (TauT/cis), weighing 28 to 30 g, were injected with a single intraperitoneal dose of cisplatin (15 mg/kg/body weight). Wild-type mice injected with saline were used as a negative control (WT/saline). Cisplatin-induced body weight changes and mortality rates were recorded in both wild-type and *hTauT* transgenic mice **(b)** Levels of BUN in wild-type saline-injected mice, *hTauT* transgenic mice injected with cisplatin and wild-type cisplatin-injected mice **(c)** Levels of serum creatinine in mice injected with saline or cisplatin. Data are mean ± SE; $n = 8$ mice in each experimental group. $*p < 0.05$ vs. WT/saline mice

maximum in cisplatin-injected wild-type mice (from 28.4±0.6 g to 20.0±0.9 g). All *hTauT* transgenic mice survived cisplatin injection, began to regain weight on day 6 and had recovered to their initial weight by day 8. In contrast, the first cisplatin-induced death was observed on day 4 in wild-type mice and the mortality rate was 75% during the 8-day experimental period. The cisplatin-injected wild-type mice who survived started to regain weight after day 7. By day 8, the average weight of cisplatin- injected wild-type mice was still 23.5% less than their initial weight (Fig. 12.3a).

Kidney function was evaluated in animals 3 days after cisplatin injection by determining blood urea nitrogen (BUN) and serum creatinine levels. Injection of cisplatin (15 mg/kg/body weight) resulted in a significant increase of both BUN and serum creatinine in the wild-type mice, but not in the *hTauT* transgenic mice, as compared to the wild-type mice injected with saline (Figs.12.3b,c).

12.4 Discussion

Recently, the mechanisms of renal cell repair and regeneration have garnered much attention (Price et al. 2004). Unfortunately, the development of therapeutic strategies that are efficacious in humans with AKI has proven problematic. This suggests that the development of more successful therapies requires approaching the problem from a different vantage point (Kelly and Molitoris 2000). The regenerative capacity of the kidney is well documented (Price et al. 2004) and the responses of surviving RPTC are thought to be crucial to the restoration of renal function following ARF. Consequently, identifying genes that are involved in RPTC protection, repair and regeneration may uncover new therapeutic targets that promote renal recovery and decrease the severity of ARF.

Unlike most amino acids, taurine is not metabolized or incorporated into protein but remains free in the intracellular compartment, where it plays an important role in cell volume regulation (Falktoft and Lambert 2004; Huang et al. 2006; Molchanova et al. 2004). Studies have demonstrated that taurine and the taurine transporter play an important role in kidney development (Han et al. 2000a; Heller-Stilb et al. 2002) In the F1 generation of inbred taurine-deficient cats, taurine deficiency results in renal malformation, with significantly diminished renal size and progressive kidney damage. Gross and histological examinations of these kidneys show ureteral dilatation, enlarged glomeruli, proximal tubular flattening and abnormal differentiation of renal tubular epithelium, especially in the distal tubule, where taurine has been found to be in highest concentration in immunohistochemical studies (Han et al. 2000a; Heller-Stilb et al. 2002). In mammals, the taurine total body pool is controlled at the site of the S3 segment of the renal proximal tubule through renal adaptive regulation of the taurine transporter gene (*TauT*) (Matsell et al. 1997). Studies have shown that *TauT* is regulated by many systems, such as transcriptional factors (p53, WT1 and Sp1), osmolarity and dietary taurine (Han et al. 2002; Han et al. 2000b; Han and Chesney 2003). Regulation of *TauT* by WT1 and Sp1 mainly occurs during the process of renal development, whereas p53 represses *TauT* expression in injured renal cells. Normally, *TauT* is specifically adaptively regulated by dietary taurine availability through a cis-element (**t**aurine **re**sponse **e**lement - TREE) located in the promoter region (−574 to −964) of the *TauT* gene (Han et al. 2000b). Heller-Stilb demonstrated that knockout of *TauT* results in severe and progressive retinal degeneration, a small brain and shrunken kidneys in a *TauT* -/- mouse model. These findings demonstrated that *TauT* is required for normal retinal, brain and kidney development.

Studies have shown that *TauT* is down-regulated by the p53 tumor suppressor gene in renal cells (Han et al. 2002) and overexpression of p53 results in progressive renal failure in p53 transgenic mice, which is similar to observations made in taurine-deficient kittens (Sturman et al. 1985). Furthermore, our current study using *TauT* transgenic mice has shown that elevated expression of *TauT* is capable of protecting against cisplatin-induced AKI *in vivo,* as measured by the levels of blood urea nitrogen (BUN) and serum creatinine in cisplatin-injected wild type and *TauT* transgenic animals. Since cisplatin displays a high rate of tissue uptake in the kidneys and exhibits a dose-related and cumulative nephrotoxicity that causes animal death as the result of acute renal failure, our results suggest that *TauT* functions as an anti-apoptotic gene and plays an important role in protecting against cisplatin-induced renal injury.

12.5 Conclusion

The present study using *TauT* transgenic mice has shown that elevated expression of *TauT* is capable of protecting against cisplatin-induced ARF *in vivo* as measured by the levels of blood urea nitrogen (BUN) and serum creatinine in cisplatin-injected wild type and *TauT* transgenic animals. Since Cisplatin mainly displays high tissue uptake in the kidneys and exhibits dose-related and cumulative nephrotoxicity which further causes animal death as the result of acute renal failure, suggesting that *TauT* functions as an anti-apoptotic gene and plays an important role in protecting against cisplatin-induced renal injury.

Acknowledgments The authors wish to thank Andrea Patters for insightful comments. This work was supported by grants from the National Kidney Foundation, Le Bonheur Children's Medical Center and the Le Bonheur Chair of Excellence in Pediatrics.

References

Baliga R, Zhang Z, Baliga M, Ueda N, Shah SV (1998) In vitro and in vivo evidence suggesting a role for iron in cisplatin-induced nephrotoxicity. Kidney Int 53:394–401

Falktoft B, Lambert IH (2004) Ca2+-mediated potentiation of the swelling-induced taurine efflux from HeLa cells: on the role of calmodulin and novel protein kinase C isoforms. J Membr Biol 201:59–75

Godley LA, Kopp JB, Eckhaus M, Paglino JJ, Owens J, Varmus HE (1996) Wild-type p53 transgenic mice exhibit altered differentiation of the ureteric bud and possess small kidneys. Genes Dev 10:836–850
Gawlik A, Quaggin SE (2004) Deciphering the renal code: advances in conditional gene targeting. Physiology (Bethesda) 19:245–252
Han X, Budreau AM, Chesney RW (2000a) The taurine transporter gene and its role in renal development. Amino Acids 19(3–4):499–507
Han X, Budreau AM, Chesney RW (2000b) Cloning and characterization of the promoter region of the rat taurine transporter (TauT) gene. Adv Exp Med Biol 483:97–108
Han X, Budreau AM, Chesney RW (2000c) Identification of promoter elements involved in adaptive regulation of the taurine transporter gene: role of cytosolic Ca^{2+} signaling. Adv Exp Med Biol 483:535–544
Han X, Chesney RW (2003) Regulation of taurine transporter gene (TauT) by WT1. FEBS Lett 540:71–76
Han X, Patters AB, Chesney RW (2002) Transcriptional repression of taurine transporter gene (TauT) by p53 in renal cells. J Biol Chem 277:39266–39273
Heller-Stilb B, van Roeyen C, Rascher K, Hartwig HG, Huth A, Seeliger MW (2002) Disruption of the taurine transporter gene (taut) leads to retinal degeneration in mice. Faseb J 16: 231–233
Hogan B, Beddington R, Constantini F, Lacy E (eds) (1994) Manipulating the mouse embryo. A laboratory manual. 2nd ed. Cold Spring Harbor Laboratory Press, New York
Huang Z, Hou Q, Cheung N.S, Li QT (2006) Neuronal cell death caused by inhibition of intracellular cholesterol trafficking is caspase dependent and associated with activation of the mitochondrial apoptosis pathway. J Neurochem 97:280–291.
Jiang M, Yi X, Hsu S, Wang CY, Dong Z (2004) Role of p53 in cisplatin-induced tubular cell apoptosis: dependence on p53 transcriptional activity. Am J Physiol Renal Physiol 287:F1140–F1147
Kelly KJ, Molitoris BA (2000) Acute renal failure in the new millennium: time to consider combination therapy. Semin Nephrol 20:4–19
Lebwohl D, Canetta R (1998) Clinical development of platinum complexes in cancer therapy: an historical perspective and an update. Eur J Cancer 34:1522–1534
Leibbrandt ME, Wolfgang GH, Metz AL, Ozobia AA, Haskins JR (1995) Critical subcellular targets of cisplatin and related platinum analogs in rat renal proximal tubule cells. Kidney Int 48:761–770
Maruyama H, Higuchi N, Nishikawa Y, Hirahara H, Iino N, Kameda S, et al. (2002) Kidney-targeted naked DNA transfer by retrograde renal vein injection in rats. Hum Gene Ther 13: 455–468
Matsell DG, Bennett T, Han X, Budreau AM, Chesney RW (1997) Regulation of the taurine transporter gene in the S3 segment of the proximal tubule. Kidney Int 52:748–754
Megyesi J, Safirstein RL, Price PM (1998) Induction of p21WAF1/CIP1/ SDI1 in kidney tubule cells affects the course of cisplatin-induced acute renal failure. J Clin Invest 101:777–782
Miyaji T, Kato A, Yasuda H, Fujigaki Y, Hishida A (2001) Role of the increase in p21 in cisplatin-induced acute renal failure in rats. J Am Soc Nephrol 12(5):900–908
Molchanova S, Oja SS, Saransaari P (2004) Characteristics of basal taurine release in the rat striatum measured by microdialysis. Amino Acids 27:261–268.
Price PM, Megyesi J, Saf Irstein RL (2004) Cell cycle regulation: repair and regeneration in acute renal failure. Kidney Int 66:509–514
Ramesh G, Reeves WB (2002) TNF-alpha mediates chemokine and cytokine expression and renal injury in cisplatin nephrotoxicity. J Clin Invest 110:835–842
Ramesh G, Reeves WB (2003) TNFR2-mediated apoptosis and necrosis in cisplatin-induced acute renal failure. Am J Physiol Renal Physiol 285:F610–F618
Ramesh G, Reeves WB (2004) Inflammatory cytokines in acute renal failure. Kidney Int S56–S61
Ries F, Klastersky J (1986) Nephrotoxicity induced by cancer chemotherapy with special emphasis on cisplatin toxicity. Am J Kidney Dis 8:368–379

Saad SY, Al-Rikabi AC (2002) Protection effects of Taurine supplementation against cisplatin-induced nephrotoxicity in rats. Chemotherapy 48:42–48

Sturman JA, Moretz RC, French JH, Wisniewski HM (1985) Taurine deficiency in the developing cat: persistence of the cerebellar external granule cell layer. Prog Clin Biol Res 179: 43–52

Thadhani R, Pascual M, Bonventrem JV (1996) Acute renal failure. N Engl J Med 334:1448–1460

Chapter 13
Perinatal Taurine Depletion Increases Susceptibility to Adult Sugar-Induced Hypertension in Rats

Sanya Roysommuti, Atchariya Suwanich, Dusit Jirakulsomchok, and J. Michael Wyss

Abstract This study tests the hypothesis that perinatal taurine depletion produces autonomic nervous system dysregulation and increases arterial pressure in young male rats maintained on a high sugar diet. Sprague-Dawley dams were either taurine depleted (beta-alanine 3% in water) or left untreated from conception to weaning. Their male offspring were fed normal rat chow with or without 5% glucose. At 7–8 weeks of age, the male offspring were either tested in a conscious, unrestrained state or after anesthetia. Body weight was slightly lower in the taurine-depleted rats although their heart or kidneys to body weight ratios were similar. Plasma potassium, blood urea nitrogen, plasma creatinine, hematocrit, fasting blood glucose concentrations and glucose tolerance were all similar. In the taurine-depleted, high glucose group, mean arterial pressure and sympathetic nervous system activity were increased while baroreflex function was impaired. These findings suggest that in this model perinatal taurine depletion causes autonomic nervous system dysfunction that may contribute to dietary high sugar-induced hypertension.

Abbreviations *SHR*, spontaneously hypertensive rat; *CG*, control with glucose; *CW*, control without glucose; *TDW*, taurine depletion without glucose; *TDG* taurine depletion with glucose

13.1 Introduction

The perinatal environment can greatly influence adult function and disease development (Barker et al. 2002; Langley-Evans 2006). For instance, perinatal diets that are very low in nutritional value (e.g., proteins) can result in low birth weights in infants and subsequently insulin resistance, diabetes mellitus or arterial hypertension in adults (Barker et al. 2007; Eriksson et al. 2007). Mechanisms potentially underlying

S. Roysommuti (✉)
Department of Physiology, Faculty of Medicine, Khon Kaen University, Khon Kaen 40002, Thailand

J. Azuma et al. (eds.), *Taurine 7*, Advances in Experimental Medicine and Biology 643,
DOI 10.1007/978-0-387-75681-3_13,

these changes include perinatal imbalances in circulating glucocorticoids, the renin-angiotensin system, oxidative stress and nephrogenesis. Substances that can inhibit or stimulate these factors either directly or indirectly (including perinatal angiotensin converting enzyme inhibitor treatment or taurine supplementation) can reduce these deleterious symptoms (Aerts and Van Assche 2002; Racasan et al. 2004; Wyss et al. 1994).

Taurine is a sulphur-containing beta-amino acid that plays many essential roles in prenatal and adult life (Sturman 1993), including intracellular volume regulation, cell membrane stabilization, neuromodulation, antioxidative stress, learning and memory, renal growth and differentiation, vasodilation and modulation of cardiac performance. Taurine is an essential amino acid during fetal life, a period in which there is limited endogenous biosynthesis (Aerts and Van Assche 2002). Maternal taurine supply to the fetus and the newborn can be severely limited due to abnormal maternal protein intake or an imbalance between protein and carbohydrate consumption (Barker et al. 2007; Forrester 2004; Langley-Evans 2006; Mendez et al. 2004). Though long-term effects of excess perinatal taurine exposure is not well established, perinatal taurine supplementation appears to be beneficial to the newborn (Chesney et al. 1998).

Previously, we have demonstrated that both prenatal and postnatal taurine depletion decrease renal blood flow and increase renal vascular resistance in adult, male rats (Roysommuti et al. 2004). In addition, both prenatal and postnatal taurine depletion significantly increase arterial pressure in adult, male offspring; however, they do not significantly alter natriuretic and diuretic responses to an acute intravenous saline load. Further, perinatal taurine depletion does not affect resting arterial pressure and heart rate. Many lines of evidence suggest that perinatal taurine depletion may increase an animal's sensitivity to risk factors of adult hypertension (Dawson et al. 1996; Schaffer et al. 2003).

High dietary carbohydrate intake is implicated in the pathogenesis of hypertension, and diets high in sugar increase arterial pressure in many animal models, including spontaneously hypertensive rats (SHR) and normotensive rats (Melancon et al. 2006; Reaven 1990; Shimamoto and Ura 2006). The underlying mechanism may involve the renin-angiotensin system, insulin resistance, sympathetic nerve activation, and renal damage (Roysommuti et al. 2002). The present study tests the hypothesis that perinatal taurine depletion leads to impairment of autonomic nervous system control of arterial pressure in normotensive, young adult, male rats. This impairment is exacerbated by high dietary sugar.

13.2 Materials and Methods

SD rats were bred in the animal unit of the Faculty of Medicine, Khon Kaen University and maintained at constant humidity (60±5%), temperature (24±1°C), and light cycle (0600–1800 h). Female SD dams were either taurine depleted (beta-alanine 3% in tap water, TD) or untreated (control, C) from conception to weaning. The male offspring were fed normal rat chow with (TDG, CG) or without (TDW, CW) 5% glucose in their tap water throughout the experiment. All experimental procedures were preapproved by the Universities Animal Care and Use Committee and were conducted in accordance with the National Institutes of Health guidelines.

At 7–8 weeks of age, all male rats were placed under thiopental anesthesia and then implanted with femoral arterial and venous catheters. Two days later and after an overnight fast, arterial blood samples were obtained in the conscious animal for determination of Na, K, blood urea nitrogen (BUN), creatinine, hematocrit, and fasting blood sugar. Thereafter, glucose tolerance testing was initiated by an intravenous injection of glucose (2 g/kg in saline), and blood glucose levels were measured at 0, 30, 60, and 120 minute intervals. Twenty-four hours later, non-fasting blood samples were collected and then arterial pressure pulses were continuously recorded (Biopac system, CA) in the conscious animal before and during infusion of phenylephrine (increased arterial pressure) or sodium nitroprusside (decreased arterial pressure).

Renal nerve function was tested one day later in animals subjected to thiopental anesthesia. Renal sympathetic nerve function was assessed using stainless steel electrodes (12 MW, 0.01 Taper, A-M System, FL) connected to DAM80 amplifier (WPI, Sarasota, FL) and Biopac (Goleta, CA) system. Single unit recordings of renal nerve activity were conducted only on nerve units that responded to changes in arterial pressure following nitroprusside or phenylephrine infusion. Baroreflex sensitivity was measured as changes in heart rate and/or renal nerve activity per changes in mean arterial pressure.

Mean arterial pressure, heart rate, baroreflex sensitivity (BS) following phenylephrine or sodium nitroprusside and power spectrum densities of arterial pressure pulse were determined offline by using Acknowledge software 3.8.1 (Biopac, Goleta, CA). Plasma Na, plasma K, BUN, and plasma creatinine concentrations were measured by an automatic analyzer, hematocrit by a standard technique and blood sugar by standard glucostrips and a glucometer (Accu-chek®, Germany). The autonomic nervous system control of arterial pressure was estimated from low frequency (0.3–0.5 Hz; sympathetic nerve activity) and high frequency (0.5–4.0 Hz; parasympathetic nerve activity) components of the power spectrum densities of baseline arterial pressure pulse using the Fourier analysis (Cerutti et al. 1991; Stauss and Kregel 1996).

All data were expressed as means ± SEM. Statistical comparisons among groups (p<0.05) were done by using one-way ANOVA and Duncan' Multi-Range (StatMost 3.6, Dataxiom Software, Los Angeles, CA).

13.3 Results

By 7–8 weeks of age, taurine depletion lowered body weight by about 10%; however, kidney and heart weights (as absolute weights or as ratios to body weight) were not significantly affected (Table 13.1). Further, fasting and non-fasting plasma potassium levels were not significantly different between the groups, while

Table 13.1 Body (BW), heart (HW), and kidney (KW) weights of male offspring

Treatment	BW (g)	KW (g)	HW (g)	KW/BW (%)	HW/BW (%)
CW ($n = 6$)	233 ± 4	1.12 ± 0.02	0.90 ± 0.02	0.48 ± 0.01	0.39 ± 0.01
CG ($n = 5$)	234 ± 7	1.13 ± 0.03	0.91 ± 0.02	0.48 ± 0.02	0.39 ± 0.02
TDW ($n = 5$)	210 ± 5*	1.07 ± 0.04	0.85 ± 0.02	0.51 ± 0.03	0.41 ± 0.01
TDG ($n = 5$)	214 ± 6	1.02 ± 0.06	0.89 ± 0.03	0.48 ± 0.04	0.42 ± 0.02

Data represent means±SEM. *$P < 0.05$ when compared to CW. See text for abbreviations.

Table 13.2 Fasting and non-fasting plasma sodium and potassium of male offspring

	Plasma sodium (mEq/L)		Plasma potassium (mEq/L)	
Treatment	Fasting	Non-fasting	Fasting	Non-fasting
CW ($n = 6$)	139.8 ± 0.31	139.6 ± 1.15	3.72 ± 0.19	3.75 ± 0.19
CG ($n = 5$)	137.0 ± 2.09	138.6 ± 1.50	3.80 ± 0.22	3.82 ± 0.04
TDW($n = 5$)	135.4 ± 1.81	139.4 ± 0.50	3.88 ± 0.07	3.86 ± 0.02
TDG ($n = 5$)	131.4 ± 2.38*	134.6 ± 2.11	3.68 ± 0.10	3.74 ± 0.18

Data represent means±SEM. *$P< 0.05$ when compared to CW. See text for abbreviations.

non-fasting plasma sodium levels were not significantly different between the groups. In contrast, fasting plasma sodium concentrations were slightly, but significantly lower in TDG compared to all other groups (Table 13.2). Blood urea nitrogen (BUN) and plasma creatinine were unaffected by perinatal taurine depletion or dietary sugar supplementation (Table 13.3). Hematocrit, fasting blood glucose (Table 13.4), and glucose tolerance (Fig. 13.1) were also not significantly different between the groups. In contrast, all glucose treated animals exhibited elevated non-fasting blood sugar levels.

Neither dietary glucose supplementation nor perinatal taurine depletion of control animals altered mean arterial pressure or heart rate. However, perinatal taurine depletion followed by high dietary glucose significantly increased mean arterial pressure (Fig. 13.2). Power spectrum analyses of arterial pressure indicated that taurine depletion blunted both sympathetic and parasympathetic components. In control rats glucose supplementation did not affect either component, but in the taurine-depleted rats, glucose excess increased the sympathetic nerve component (Fig. 13.3). Further, in the taurine-depleted rats compared to all other groups, glucose decreased

Table 13.3 Fasting and non-fasting blood urea nitrogen and plasma creatinine

	Blood urea nitrogen (mg/dl)		Plasma creatinine (mg/dl)	
Treatment	Fasting	Non-fasting	Fasting	Non-fasting
CW ($n = 6$)	17.38 ± 0.75	18.17 ± 0.59	0.45 ± 0.02	0.43 ± 0.02
CG ($n = 5$)	16.86 ± 0.83	20.08 ± 0.96	0.46 ± 0.02	0.46 ± 0.02
TDW ($n = 5$)	16.72 ± 0.62	18.84 ± 1.02	0.46 ± 0.02	0.42 ± 0.02
TDG ($n = 5$)	18.02 ± 0.69	17.96 ± 0.91	0.44 ± 0.02	0.46 ± 0.02

Data represent means±SEM. No significant differences were observed between the groups. See text for abbreviations.

Table 13.4 Fasting (FBG) and non-fasting (NFBG) blood glucose and hematocrit of male offspring

	Blood glucose (mg/dl)		Hematocrit (%)	
Treatment	NFBG	FBG	Fasting	Non-fasting
CW ($n = 6$)	83.5 ± 2.9	80.3 ± 3.7	42.7 ± 0.7	42.2 ± 0.8
CG ($n = 5$)	105.6 ± 5.3*	80.8 ± 1.6	43.2 ± 0.6	42.2 ± 1.0
TDW ($n = 5$)	89.8 ± 5.6	81.2 ± 4.1	42.8 ± 0.7	41.4 ± 0.6
TDG ($n = 5$)	107.0 ± 4.9*	84.6 ± 3.4	43.4 ± 0.6	42.2 ± 0.8

Data represent means±SEM. * $P < 0.05$ when compared to CW. See text for abbreviations.

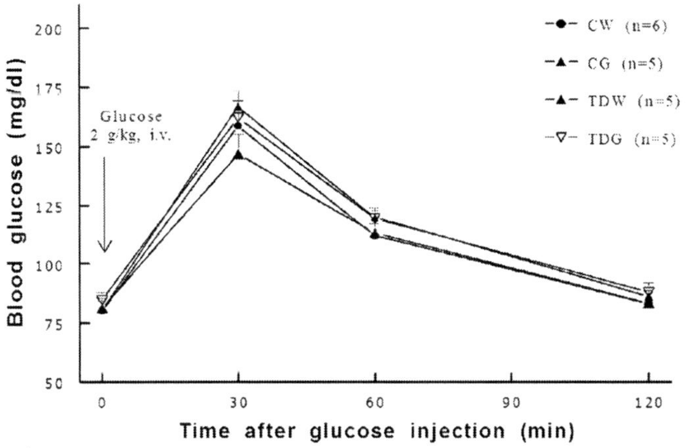

Fig. 13.1 All experimental groups displayed similar glucose tolerance (*$P < 0.05$ to CW). See text for abbreviations

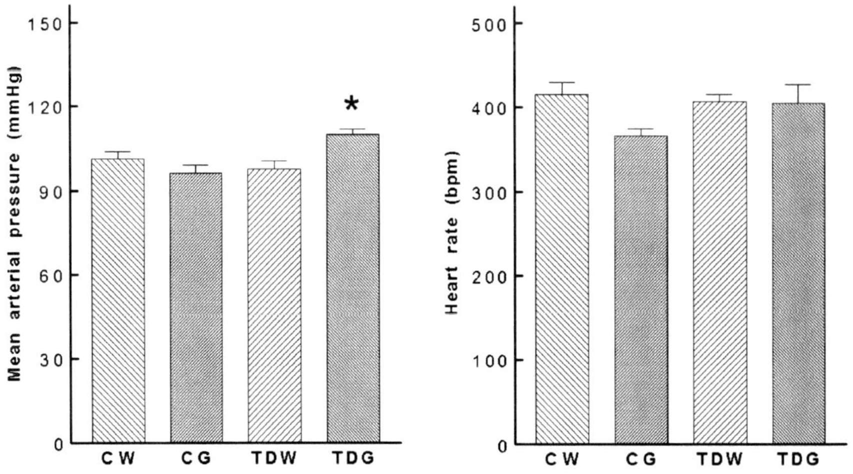

Fig. 13.2 High sugar intake increases mean arterial pressure in TDG while heart rate was not significantly different between the groups (*$P < 0.05$ to CW). See text for abbreviations

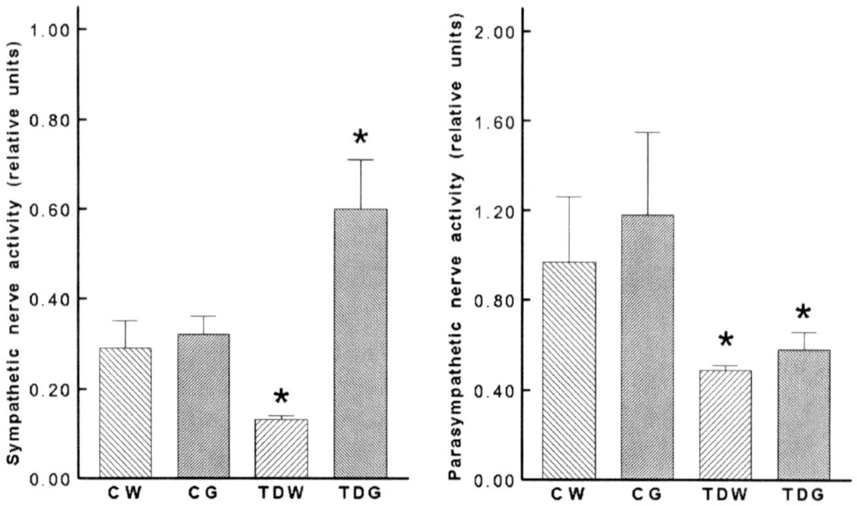

Fig. 13.3 Perinatal taurine depletion decreases both sympathetic and parasympathetic nerve activities in adult, male rats. High sugar intake heightened the sympathetic nerve activity only in the TDG (*$P<0.05$ to CW). See text for abbreviations

the parasympathetic nerve spectral component of arterial pressure. Thus, the ratio of sympathetic to parasympathetic activity was increased by taurine depletion.

Baroreflex control of heart rate was significantly decreased in both taurine-depleted groups (Fig. 13.4), but the decrease was significantly greater in taurine depleted rats maintained on a high glucose diet; a high glucose diet did not affect baroreflex regulation of control rats. Baroreflex control of renal nerve activity (an

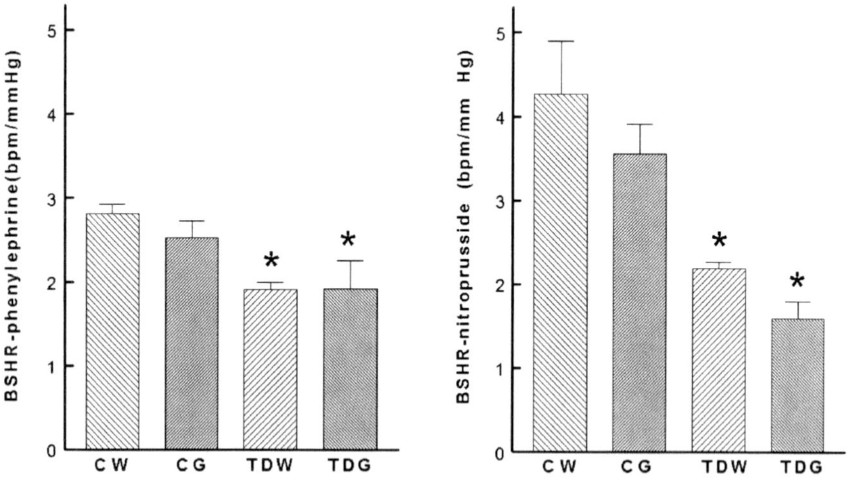

Fig. 13.4 Perinatal taurine depletion decreases the control of baroreflex sensitivity of heart rate in adult, male rats (*$P<0.05$ to CW). See text for abbreviations

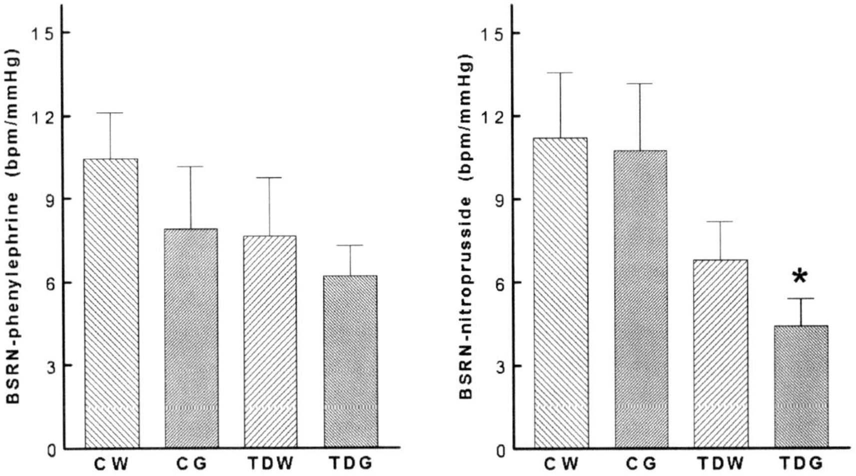

Fig. 13.5 High sugar intake aggravates baroreflex sensitivity control of renal nerve activity in perinatal taurine depleted rats ($^*P < 0.05$ to CW). See text for abbreviations

indicator of autonomic nervous system control of renal vascular resistance) was also blunted by perinatal taurine depletion and was further blunted by high dietary glucose in the taurine depleted rats (Fig. 13.5). Further, baroreflex control of arterial pressure (within the normal range of arterial pressure, i.e., 80–120 mm Hg) was consistently blunted by perinatal taurine depletion.

13.4 Discussion

During perinatal life, large alterations in nutritional intake can lead to obesity, insulin resistance, hypertension, and other cardiovascular symptoms. These effects may be initiated during pre- or postnatal development (Barker et al. 2002; Harding 2001; Langley-Evans 2006). The present study indicates that perinatal taurine depletion causes a dysregulation of the autonomic nervous system that is exacerbated by a high sugar diet. Further, these changes in autonomic nervous system regulation may contribute to increased arterial pressure in perinatal taurine-depleted rats on a high glucose diet. This provides another example of how the perinatal environment can alter function in adults and potentially contribute to disease (Barker et al. 2002; Harding 2001; Langley-Evans 2006).

Several lines of evidence indicate that imbalances in maternal protein to carbohydrate intake can result in low birth weight of offspring (Law et al. 2001; Shiell et al. 2001), which subsequently increases the risk of developing insulin resistance and cardiovascular disease. Imbalances in glucocorticoid levels and in the renin-angiotensin system, as well as abnormal nephrogenesis during early life, can modify the adaptive ability of the offspring in later life (Eriksson et al. 2007; Hanson et al. 2004; Mendez et al. 2004). The present study shows that perinatal taurine depletion in rats slightly decreases body weight, thus supporting the growth-promoting ability of taurine in early life (Lourenco and Camilo 2002). While perinatal taurine depletion could have caused taurine concentrations to remain low throughout life, thereby producing chronic taurine deficiency, this seems unlikely. After weaning, all animals were supplied with normal rat chow that appears to "normalize" taurine content within 4–5 weeks (Pacioretty et al. 2001). However, we did not independently measure taurine content in these rats. In addition, the animals in the present study displayed no signs of insulin resistance or diabetes mellitus. This suggests that insulin resistance may not be the key factor predisposing these animals to subsequent cardiovascular impairment (Reaven 1991). We have previously reported that in this model, high sugar diet can induce renal dysfunction without insulin resistance or hypertension (Roysommuti et al. 2002).

Taurine supplementation during perinatal or adult life reduces hypertension in animal models, including SHR, cyclosporine A-induced hypertension and sugar-induced hypertension (Militante and Lombardini 2002). The present findings indicate that perinatal taurine depletion predisposes animals to dietary sugar-induced hypertension. Lack of taurine in early life may underlie the effect of perinatal protein malnutrition on adult hypertension (Aerts and Van Assche 2002). Interestingly, taurine, which is a sulphur-containing beta-amino acid that is found mainly in animal meat, is not present in most plant proteins and is found in very low levels in cow's milk (Aerts and Van Assche 2002). In contrast, it is in high concentration in human and rat milk. In humans, taurine content is higher in non-vegetarians than vegetarians.

Dietary taurine supplementation appears to be directly related to taurine's ability to decrease sympathetic nerve activity, likely at the level of the central nervous system (Mizushima et al. 1996; Sato et al. 1987). Taurine supplementation also reduces oxidative stress (Aerts and Van Assche 2002; Racasan et al. 2004). Interestingly, all perinatal taurine depleted rats on a normal glucose diet display resting autonomic nerve hypoactivity but sympathetic nerve activity is only selectively increased in rats fed the high sugar diet, suggesting that perinatal taurine depletion may not retard growth and development of the autonomic nervous system, but may dysregulate it. The selective action of high sugar intake on the regulation of the sympathetic nervous system suggests that taurine deficiency in early life may alter the adult central nervous system. Taurine injection into the brain has been shown to decrease sympathetic outflow in animals (Inoue et al. 1985); however, this effect does not appear to be direct, but rather acts through some other mechanism, e.g., adenosine or glutamate receptor systems (Albrecht and Schousboe 2005; Kohlenbach and Schlicker 1990). The hypothalamic sympathetic pathways play a very important role in pathogeneses of many models of hypertension (Carlson et al. 2001). Taurine depletion during development may modify the function of this brain area.

Baroreceptor reflex plays a crucial role in minute-to-minute regulation of arterial pressure, and recent evidence suggests that baroreceptor dysfunction may contribute to hypertension in some animal models, e.g., SHR (Carlson et al. 2001). In sustained hypertension, baroreflex sensitivity is usually blunted, either by arterial vascular wall stiffness or central nervous system adjustments, i.e., baroreflex resetting. In the present study, baroreflex control of both heart rate and renal nerve activity was blunted in perinatal taurine-depleted rats, even though mean arterial pressure and heart rate were not significantly different in the taurine-depleted rats compared to the control rats. In contrast, glucose supplementation led to both baroreflex blunting and increased arterial pressure in the taurine-depleted but not in the control rats. Together, this suggests that postprandial hyperglycemia and/or hyperinsulinemia may lead to sustained sympathetic overactivity and hypertension due to insufficient baroreceptor reflex adjustments. Renal nerve activity reflects sympathetic nervous system activity but not parasympathetic nervous system activity. Thus, the blunted baroreflex control of renal nerve activity by a high sugar diet in the perinatal taurine-depleted rats indicates that the changes in cardiovascular control in these rats are due primarily to the actions of sympathetic (as opposed to parasympathetic) pathways. However, blunted renal excretory function is likely not to be the primary cause of hypertension in these animals, since natriuretic and diuretic function is not different between the groups (Roysommuti et al. 2004).

The renin-angiotensin system plays an important role in the pathogenesis of arterial hypertension, including sugar-induced hypertension. Our previous study indicates that high sugar intake impairs renal function before the development of insulin resistance and hypertension (Roysommuti et al. 2002). This effect is abolished by treatment with angiotensin converting enzyme inhibitor, captopril. The present experiment also further indicates that high dietary glucose does not alter autonomic nervous system function in control rats. It is possible that a glucose-induced rise in angiotensin II may suppress the central baroreflex pathway (DiBona and Jones 2003; McMullan et al. 2007) and activate sympathetic nerve activity (Gao et al. 2005; Johns 2005) in the perinatal taurine depleted rats, which are more susceptibility to pressor agents than the control. Overexpression of angiotensin II receptors has also been reported in perinatal protein restricted offspring (Pladys et al. 2004; Riviere et al. 2005), but not in adult, perinatal taurine depleted animals.

In summary, while perinatal taurine-depletion blunts baroreceptor reflexes and suppresses autonomic nervous system activity, excess dietary glucose increases sympathetic nerve activity. The baroreceptor reflex was also further blunted by high sugar intake. These data further support an important role for maternal dietary taurine during the perinatal period.

Acknowledgments This work was supported by grants from the Faculty of Medicine, Khon Kaen University and the National Institutes of Health grants (AT 00477 (JMW) from the National Center for Complementary and Alternative Medicine and the Office of Dietary Supplements.

References

Aerts L, Van Assche FA (2002) Taurine and taurine-deficiency in the perinatal period. J Perinat Med 30:281-286

Albrecht J, Schousboe A (2005) Taurine interaction with neurotransmitter receptors in the CNS: an update. Neurochem Res 30:1615–1621

Barker DJ, Eriksson JG, Forsen T, Osmond C (2002) Fetal origins of adult disease: strength of effects and biological basis. Int J Epidemiol 31: 1235–1239
Barker DJ, Osmond C, Forsen TJ, Kajantie E, Eriksson JG (2007) Maternal and social origins of hypertension. Hypertension 50:565–571
Carlson SH, Roysomutti S, Peng N, Wyss JM (2001) The role of the central nervous system in NaCl-sensitive hypertension in spontaneously hypertensive rats. Am J Hypertens 14: 155S–162S
Cerutti C, Gustin MP, Paultre CZ, Lo M, Julien C, Vincent M, Sassard J (1991) Autonomic nervous system and cardiovascular variability in rats: a spectral analysis approach. Am J Physiol 261:H1292–H1299
Chesney RW, Helms RA, Christensen M, Budreau AM, Han X, Sturman JA (1998) The role of taurine in infant nutrition. Adv Exp Med Biol 442:463–476
Dawson R, Jr., Eppler B, Patterson TA, Shih D, Liu S (1996) The effects of taurine in a rodent model of aging. Adv Exp Med Biol 403:37–50
DiBona GF, Jones SY (2003) Endogenous angiotensin affects responses to stimulation of baroreceptor afferent nerves. J Hypertens 21:1539–1546
Eriksson JG, Forsen TJ, Kajantie E, Osmond C, Barker DJ (2007) Childhood growth and hypertension in later life. Hypertension 49:1415–1421
Forrester T (2004) Historic and early life origins of hypertension in Africans. J Nutr 134:211–216
Gao L, Wang W, Li YL, Schultz HD, Liu D, Cornish KG, Zucker IH (2005) Sympathoexcitation by central ANG II: roles for AT1 receptor upregulation and NAD(P)H oxidase in RVLM. Am J Physiol Heart Circ Physiol 288:H2271–H2279
Hanson M, Gluckman P, Bier D, Challis J, Fleming T, Forrester T, Godfrey K, Nestel P, Yajnik C (2004) Report on the 2nd World Congress on Fetal Origins of Adult Disease, Brighton, U.K., June 7-10, 2003. Pediatr Res 55:894–897
Harding JE (2001) The nutritional basis of the fetal origins of adult disease. Int J Epidemiol 30: 15–23
Inoue A, Takahashi H, Lee LC, Iyoda I, Sasaki S, Okajima H, Takeda K, Yoshimura M, Nakagawa M, Ijichi H (1985) Centrally induced vasodepressor and sympathetic nerve responses to taurine. Jpn Circ J 49:1180–1184
Johns EJ (2005) Angiotensin II in the brain and the autonomic control of the kidney. Exp Physiol 90:163–168
Kohlenbach A, Schlicker E (1990) GABAB receptor-mediated inhibition of the neurogenic vasopressor response in the pithed rat. Br J Pharmacol 100:365–369
Langley-Evans SC (2006) Developmental programming of health and disease. Proc Nutr Soc 65:97–105
Law CM, Egger P, Dada O, Delgado H, Kylberg E, Lavin P, Tang GH, von Hertzen H, Shiell AW, Barker DJ (2001) Body size at birth and blood pressure among children in developing countries. Int J Epidemiol 30:52–57
Lourenco R, Camilo ME (2002) Taurine: a conditionally essential amino acid in humans? An overview in health and disease. Nutr Hosp 17:262–270
McMullan S, Goodchild AK, Pilowsky PM (2007) Circulating angiotensin II attenuates the sympathetic baroreflex by reducing the barosensitivity of medullary cardiovascular neurones in the rat. J Physiol 582:711–722
Melancon S, Bachelard H, Badeau M, Bourgoin F, Pitre M, Lariviere R, Nadeau A (2006) Effects of high-sucrose feeding on insulin resistance and hemodynamic responses to insulin in spontaneously hypertensive rats. Am J Physiol Heart Circ Physiol 290:H2571–H2581
Mendez MA, Wynter S, Wilks R, Forrester T (2004) Under- and overreporting of energy is related to obesity, lifestyle factors and food group intakes in Jamaican adults. Public Health Nutr 7: 9–19
Militante JD, Lombardini JB (2002) Treatment of hypertension with oral taurine: experimental and clinical studies. Amino Acids 23:381–393
Mizushima S, Nara Y, Sawamura M, Yamori Y (1996) Effects of oral taurine supplementation on lipids and sympathetic nerve tone. Adv Exp Med Biol 403:615–622

Pacioretty L, Hickman MA, Morris JG, Rogers QR (2001) Kinetics of taurine depletion and repletion in plasma, serum, whole blood and skeletal muscle in cats. Amino Acids 21:417–427

Pladys P, Lahaie I, Cambonie G, Thibault G, Le NL, Abran D, Nuyt AM (2004) Role of brain and peripheral angiotensin II in hypertension and altered arterial baroreflex programmed during fetal life in rat. Pediatr Res 55:1042–1049

Racasan S, Braam B, van der Giezen DM, Goldschmeding R, Boer P, Koomans HA, Joles JA (2004) Perinatal L-arginine and antioxidant supplements reduce adult blood pressure in spontaneously hypertensive rats. Hypertension 44:83–88

Reaven GM (1990) Insulin and hypertension. Clin Exp Hypertens A 12:803–816

Reaven GM (1991) Insulin resistance, hyperinsulinemia, and hypertriglyceridemia in the etiology and clinical course of hypertension. Am J Med 90:7S–12S

Riviere G, Michaud A, Breton C, VanCamp G, Laborie C, Enache M, Lesage J, Deloof S, Corvol P, Vieau D (2005) Angiotensin-converting enzyme 2 (ACE2) and ACE activities display tissue-specific sensitivity to undernutrition-programmed hypertension in the adult rat. Hypertension 46:1169–1174

Roysommuti S, Khongnakha T, Jirakulsomchok D, Wyss JM (2002) Excess dietary glucose alters renal function before increasing arterial pressure and inducing insulin resistance. Am J Hypertens 15:773–779

Roysommuti S, Malila P, Jirakulsomchok D, Jirakulsomchok S, Wyss JM (2004) Perinatal taurine status influences renal hemodynamics in adult conscious rats. FASEB J 18 (4 Part I):A292–A293

Sato Y, Ando K, Fujita T (1987) Role of sympathetic nervous system in hypotensive action of taurine in DOCA-salt rats. Hypertension 9:81–87

Schaffer S, Azuma J, Takahashi K, Mozaffari M (2003) Why is taurine cytoprotective? Adv Exp Med Biol 526:307–321

Shiell AW, Campbell-Brown M, Haselden S, Robinson S, Godfrey KM, Barker DJ (2001) High-meat, low-carbohydrate diet in pregnancy: relation to adult blood pressure in the offspring. Hypertension 38:1282–1288

Shimamoto K, Ura N (2006) Mechanisms of insulin resistance in hypertensive rats. Clin Exp Hypertens 28:543–552

Stauss HM, Kregel KC (1996) Frequency response characteristic of sympathetic-mediated vasomotor waves in conscious rats. Am J Physiol 271:H1416–H1422

Sturman JA (1993) Taurine in development. Physiol Rev 73:119–147

Wyss JM, Roysommuti S, King K, Kadisha I, Regan CP, Berecek KH (1994) Salt-induced hypertension in normotensive spontaneously hypertensive rats. Hypertension 23:791–796

Chapter 14
Sex Dependent Effects of Perinatal Taurine Exposure on the Arterial Pressure Control in Adult Offspring

Sanya Roysommuti, Atchariya Suwanich, Wichaporn Lerdweeraphon, Atcharaporn Thaeomor, Dusit Jirakulsomchok, and J. Michael Wyss

Abstract The present study tests the sex-dependent effect of perinatal taurine exposure on arterial pressure control in adults. Female Sprague-Dawley rats were fed normal rat chow with 3% beta-alanine (taurine depletion,TD), 3% taurine (taurine supplementation,TS) or water alone (C) from conception to weaning. Their male and female offspring were then fed normal rat chow and tap water with 5% glucose (C with glucose, CG; TD with glucose, TDG; TS with glucose, TSG) or water alone (CW, TDW or TSW). At 7–8 weeks of age, they were studied in a conscious condition. Body weights were lower in male and female TDG and male TDW rats. Kidney to body weights increased in female TSW but not TSG. Plasma sodium and potassium were not significantly different among males. Among females, plasma sodium levels were lower in all glucose treated groups while plasma potassium levels were lower only in TDG. Hematocrit, fasting blood glucose, and glucose tolerance were not significantly different between the sexes. Mean arterial pressure increased in male TDG, TSW, and TSG while in the females, mean arterial pressure increased in TabstractDW, TDG, and TSG. Heart rates were not significantly different between the sexes. The present data indicate that perinatal taurine exposure alters arterial pressure control of adult rats and this effect is gender specific.

Abbreviations *CW*, control with water alone; *CG*, control with glucose; *TDW*, taurine depletion with water; *TDG*, taurine depletion with glucose; *TSW*, taurine supplement with water; *TSG*, taurine supplement with glucose

---

S. Roysommuti (✉)
Department of Physiology, Faculty of Medicine, Khon Kaen University, Khon Kaen 40002, Thailand

J. Azuma et al. (eds.), *Taurine 7*, Advances in Experimental Medicine and Biology 643, DOI 10.1007/978-0-387-75681-3_14, © Springer Science+Business Media, LLC 2009

14.1 Introduction

In both human and animal models, the fetal environment in utero has a significant impact on adult health and disease (Harding 2001; Langley-Evans et al. 2003). Undernutrition or imbalanced food consumption (for instance low protein-high carbohydrate diet) in the prenatal period results in low birth weight and subsequently induces several cardiovascular disorders in adults, including coronary vascular diseases, hypertension, insulin resistance, diabetes mellitus, and ultimately renal damage (Barker et al. 2002; Forrester 2004). Hypertension and diabetes mellitus also appear to be related to obesity developed in later life (Mendez et al. 2004). Epidemiologic studies indicate that African-American women have a higher prevalence of low birth weight and adult obesity (Ventura et al. 2002). Although the mechanism(s) of these effects is still unclear, abnormalities in the renin-angiotensin and sympathetic nervous systems have been characterized in humans and animal models (Eriksson et al. 2007; Lackland et al. 2002). Perinatal programming of adult function and diseases has been recognized for a decade (Barker 2007). Low birth weight has been associated with many changes, including taurine deficiency in the perinatal period and later life (Aerts and Van Assche 2002). In adult animals, taurine supplementation decreases hypertension, presumably by increasing renal Na excretion, inhibiting the renin-angiotensin system, and decreasing sympathetic nerve activity (Militante and Lombardini 2002).

Taurine, 2-aminoethane sulfonic acid, is a phylogenetically ancient compound that is present in high concentration in many organs, including brain, heart, kidneys, and reproductive organs. Its content is highest in these organs during fetal life and gradually decreases after birth (Aerts and Van Assche 2002). During lactation it appears to be an essential amino acid, since taurine synthesis is minimal in the organism with maternal milk as its main source. Several lines of evidence indicate that perinatal taurine status programs cells for adult function, especially organs related to the cardiovascular system. Perinatal taurine supplementation prevents hypertension in spontaneously hypertensive rats (SHR), partly through its antioxidant activity (Racasan et al. 2004). Our previous experiments indicated that either taurine depletion or supplementation in early life alters renal function (Roysommuti et al. 2004) and autonomic nervous control of arterial pressure (Roysommuti et al. 2007; Suvanich et al. 2006) in adult, male rats. Perinatal taurine depletion increases arterial pressure but not heart rate in adult, female offspring. Their renal hemodynamics and excretory function are also modified by perinatal taurine exposure (Lerdweerapol et al. 2007). In addition, the autonomic nervous system and renal function responses to high sugar intake in young adult animals appear to be altered by perinatal taurine exposure. The protection of arterial pressure in females may be due to the putative antihypertensive effects of estrogen in female animal models, including the spontaneously hypertensive and the salt-induced hypertensive, ovariectomized rats (Clark et al. 2004; Peng et al. 2003). This study compares the long-term effect of perinatal taurine exposure on arterial pressure control in male and female adult offspring fed a high sugar diet.

14.2 Materials and Methods

Sprague-Dawley (SD) rats were bred from the animal unit of the Faculty of Medicine, Khon Kaen University and maintained at constant humidity (60 ± 5%), temperature (24 ± 1°C), and light cycle (0600–1800 h). Female SD rats were fed with normal rat chow and given free access to tap water alone (control, C), tap water with 3% beta-alanine (taurine depletion, TD) or tap water with 3% taurine (taurine supplementation, TS) from conception until weaning. Then, their male and female offspring were fed normal rat chow supplemented with either 5% glucose in tap water (TD with glucose, TDG; TS with glucose, TSG; C with glucose, CG) or tap water alone (TDW, TSW, and CW) throughout the experiment.

At 7–8 weeks of age, under thiopental anesthesia, all male and female offspring were implanted with femoral arterial and venous catheters. Two or three days later and after an overnight fast, arterial blood samples were obtained from conscious animals for Na, K, hematocrit, and fasting blood sugar determination. Glucose tolerance tests were started by intravenous injection of glucose (2 g/kg in saline) and blood glucose levels were subsequently measured 30, 60, and 120 minutes later. Twenty-four hours later, non-fasting blood samples were collected and then arterial pressure pulses were continuously recorded (Biopac system, CA) in the conscious animals. At the end of experiment, all animals were sacrificed and kidney and heart weights were obtained.

Experimental protocols and animal care were approved by the university animal committee. Plasma sodium and potassium concentrations were determined by flame photometry, hematocrit by a standard technique, fresh blood sugar by a glucometer and glucostrips (Accu-chek®, Germany), and mean arterial pressure and heart rate by Acknowledge software version 3.8.1 (Biopac system, CA).

All data were expressed as means ± SEM. Statistical comparisons among groups ($p < 0.05$) were done by using one-way ANOVA and Duncan'Multi-Range.

14.3 Results

Perinatal taurine depletion caused significant growth retardation in male offspring which could be partially prevented by high sugar supplementation after weaning (Table 14.1). In contrast, growth in female rats was not retarded by perinatal taurine depletion alone. However, when combined with a high sugar diet after weaning, it did retard growth. Perinatal taurine supplementation had no effect on body and heart weight but slightly increased kidney weight when compared to CG and CW female rats. High sugar treatment alone had no effect on growth. While plasma Na levels were not significantly different among male groups, they were significantly lower in all glucose-treated female offspring (Table 14.2). Both male and female offspring from all of the groups displayed similar plasma potassium concentrations, hematocrit, and fasting blood sugar. Among males, non-fasting blood sugar levels were slightly increased in CG rats, while they were significantly increased in all glucose-treated female offspring when compared to their corresponding controls (CW, TDW, and TSW). Both male and female offspring displayed similar glucose tolerance (Fig. 14.1).

Table 14.1 Body (BW), kidney (KW), and heart (HW) weights in male (M) and female (F) offspring

Treatment		BW (g)	KW (g)	HW (g)	KW/BW (%)	HW/BW (%)
CW	M($n = 6$)	233 ± 4	1.12 ± 0.02	0.90 ± 0.02	0.48 ± 0.01	0.39 ± 0.01
	F($n = 7$)	194 ± 5	1.66 ± 0.04	0.6 ± 0.02	0.85 ± 0.03	0.36 ± 0.01
CG	M($n = 5$)	234 ± 7	1.13 ± 0.03	0.91 ± 0.02	0.48 ± 0.02	0.39 ± 0.02
	F($n = 8$)	196 ± 4	1.61 ± 0.04	0.74 ± 0.02	0.82 ± 0.01	0.37 ± 0.00
TDW	M($n = 5$)	210 ± 5$^{*,\alpha}$	1.07 ± 0.04	0.85 ± 0.02	0.51 ± 0.03	0.41 ± 0.01
	F($n = 6$)	196 ± 2	1.72 ± 0.08	0.72 ± 0.04	0.88 ± 0.04	0.37 ± 0.02
TDG	M($n = 5$)	214 ± 6	1.02 ± 0.06	0.89 ± 0.03	0.48 ± 0.04	0.42 ± 0.02
	F($n = 10$)	179 ± 5$^{*,\alpha}$	1.59 ± 0.02	0.67 ± 0.03	0.89 ± 0.02	0.37 ± 0.01
TSW	M($n = 6$)	238 ± 6	1.08 ± 0.04	0.90 ± 0.02	0.45 ± 0.01	0.38 ± 0.01
	F($n = 6$)	197 ± 4	1.88 ± 0.0$^{*,\alpha}$	0.75 ± 0.01	0.95 ± 0.02$^{*,\alpha}$	0.38 ± 0.01
TSG	M($n = 6$)	228 ± 7	1.11 ± 0.05	0.93 ± 0.02	0.49 ± 0.02	0.41 ± 0.01
	F($n = 8$)	191 ± 5	1.67 ± 0.04	0.73 ± 0.02	0.88 ± 0.04	0.38 ± 0.01

Data represent means ±SEM.$^{*,\alpha}$ denotes significant difference ($P < 0.05$) when compared to CW or CG. See text for abbreviations.

Table 14.2 Plasma sodium, plasma potassium, hematocrit, non-fasting blood sugar (NFBS), and fasting blood sugar (FBS) in male (M) and female (F) offspring

Treatment		Na (mEq/L)	K (mEq/L)	Hematocrit (%)	NFBS (mg/dl)	FBS (mg/dl)
CW	M($n = 6$)	139.6 ± 1.2	3.75 ± 0.19	42.2 ± 0.83	83.5 ± 2.91	80.3 ± 3.73
	F($n = 7$)	132.5 ± 3.3	4.44 ± 2.23	41.9 ± 1.17	115.0 ± 5.17	86.4 ± 3.60
CG	M($n = 5$)	138.6 ± 1.5	3.82 ± 0.04	42.2 ± 1.02	105.6 ± 5.28*	80.0 ± 1.65
	F($n = 8$)	118.3 ± 3.1*	4.51 ± 0.33	40.8 ± 0.77	129.4 ± 3.02*	91.3 ± 2.74
TDW	M($n = 5$)	139.4 ± 0.5	3.86 ± 0.02	41.4 ± 0.60	89.8 ± 5.59	81.2 ± 4.13
	F($n = 6$)	122.7 ± 2.4	4.49 ± 0.08	40.5 ± 0.82	125.7 ± 2.86	91.2 ± 2.81
TDG	M($n = 5$)	134.6 ± 2.1	3.74 ± 0.18	42.2 ± 0.80	107.0 ± 4.94*	84.6 ± 3.37
	F($n = 10$)	117.3 ± 2.4*	3.45 ± 0.18	40.5 ± 0.93$^{*,\alpha}$	120.4 ± 5.35	86.6 ± 2.83
TSW	M($n = 6$)	139.0 ± 1.3	3.88 ± 0.05	42.8 ± 0.65	88.0 ± 5.95	82.8 ± 3.13
	F($n = 6$)	124.0 ± 2.4	3.97 ± 0.17	39.2 ± 1.09	126.7 ± 3.56	94.0 ± 3.44
TSG	M($n = 6$)	134.8 ± 1.8	3.80 ± 0.15	41.2 ± 0.79	102.5 ± 4.40*	85.7 ± 2.20
	F($n = 8$)	114.3 ± 6.0*	4.06 ± 0.12	41.3 ± 1.10	121.5 ± 3.92	90.3 ± 1.45

Data represent means±SEM.$^{*,\alpha}$ denotes significant difference ($P < 0.05$) when compared to CW or CG. See text for abbreviations.

Perinatal taurine depletion alone increased mean arterial pressure in female but not male TDW while taurine supplementation alone increased them in male but not female TSW (Fig. 14.2). These perinatal taurine effects were not altered by the high sugar diet. The high sugar diet significantly increased mean arterial pressure in perinatal taurine depleted males (CW, 101.2 ± 2.5 mm Hg; CG, 96.0 ± 3.0 mm Hg; TDW, 97.6 ± 2.7 mm Hg; TDG, 109.6 ± 2.1 mm Hg; P < 0.05) and perinatal

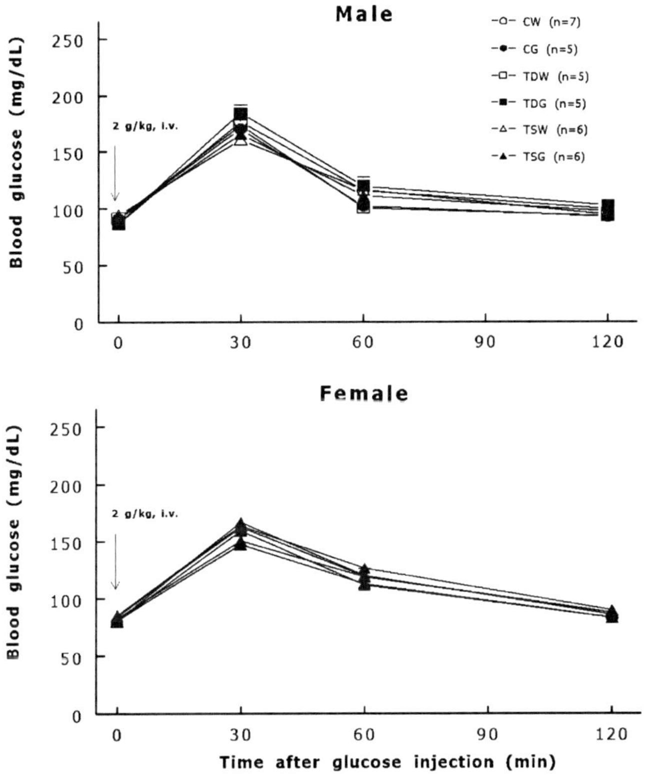

Fig. 14.1 Both male and female offspring displayed well glucose tolerance

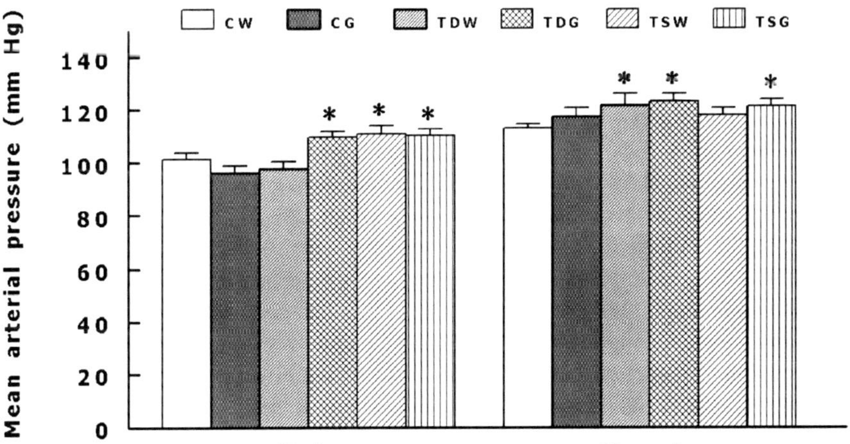

Fig. 14.2 Mean arterial pressures were significantly higher in male TDG, male TSW, male TSG, female TDW, female TDG, and female TSG (* denotes significant difference ($P < 0.05$) compared to CW). See text for abbreviations

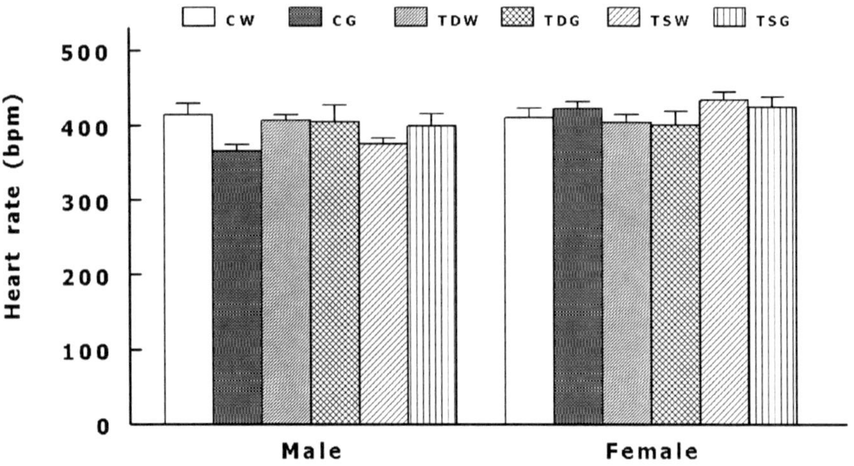

Fig. 14.3 Heart rates were not significantly different among male or female offspring. See text for abbreviations

taurine-supplemented female (CW, 112.8 ± 1.8 mm Hg; CG, 117.16 ± 3.39 mm Hg; TSW, 118.0 ± 2.5 mm Hg; TSG, 121.2 ± 3.0 mm Hg; $P < 0.05$) when compared to CW or CG rats (Fig. 14.2). Heart rates were not significantly different between male and female rats (Fig. 14.3). Although the females displayed lower body weight than the males (Table 14.1), their mean arterial pressures were significantly elevated relative to the males (Fig. 14.2).

14.4 Discussion

Perinatal treatment can significantly alter adult organ function and health, including cardiovascular function. Female rats treated with testosterone during the first 4 days of life develop a male pattern of gonadotropin secretion with abnormal female sexual behavior in mature life (Barraclough 1961; Clark et al. 2004; Peng et al. 2003). Handling or exposure to various stressors as a neonate results in permanent changes in hypothalamic structure and abnormal stress responses in the adult (Cella et al. 1990). Perinatal administration of angiotensin converting enzyme inhibitors attenuates hypertension in the adult SHR but does not prevent salt-induced hypertension (Wyss et al. 1994). Perinatal taurine treatment also attenuates hypertension in the adult SHR, likely via its antioxidant properties (Racasan et al. 2004). Our previous experiments indicate that pre- or postnatal (lactation) taurine supplementation increases mean arterial pressure in adult male offspring (Roysommuti et al. 2004). The present data further demonstrates that perinatal taurine supplementation can increase mean arterial pressure in adult male rats but not female rats. Further, perinatal taurine depletion can increase arterial pressure in adult female but not male rats. This study thus demonstrates gender specific responses to perinatal taurine exposure.

It is well-known that sugar consumption is a significant risk factor for the development of hypertension. Sugar-induced hypertension is associated with hyperinsulinemia, insulin resistance, renin-angiotensin system overactivity, sympathetic nervous system overactivity, and renal dysfunction (Johnson et al. 2007). However, previous experiments indicate that glucose supplementation induces renal dysfunction before insulin resistance and hypertension, and that these effects can be abolished by treatment with an angiotensin converting enzyme inhibitor (captopril) (Roysommuti et al. 2002). Taurine inhibits the renin-angiotensin system (Azuma et al. 2000; Schaffer et al. 2000) and prevents fructose-induced hypertension in rats (Harada et al. 2004). Moreover, in many forms of hypertension taurine supplementation in young or adult life reduces arterial pressure (Militante and Lombardini 2002), improves renal function and inhibits the sympathetic nervous system. The present study reports the interaction between perinatal taurine exposure and the subsequent effect of high sugar intake on arterial pressure in both sexes. Perinatal taurine depletion coupled with high sugar consumption induces a pressor effect in male but not female rats while elevation of blood pressure caused by perinatal taurine supplementation does not depend on sugar consumption in the male but does in the female. These opposite effects may be due to gender differences in body taurine content, sex hormones, autonomic nervous system, the renin-angiotensin system or renal function of the adult. Permanent changes and programming at an early age is hypothesized to be the primary factor.

Obesity, insulin resistance, hyperinsulinemia and electrolyte disturbances are associated with the development of hypertension; perinatal nutritional imbalances have been reported to be a predisposing factor to these dysfunctions (Barker et al. 2002; Harding 2001; Langleyevans 2006). Exposure to taurine in fetuses and neonates is primarily from diets through the placenta or maternal milk (Aerts and Van Assche 2002). Thus, taurine deficiency is observed in rat neonates of pregnant mothers that are fed low protein diets (Cherif et al. 1998);. Though body taurine content can return to normal levels within 5–6 weeks of terminating supplementation or depletion (Pacioretty et al. 2001), permanent changes appear to continue into adult life, as shown by the present data. In the present study, the pressor effect of perinatal taurine irrespective of sugar supplementation did not relate to adult body weight, insulin resistance or Na-K imbalance. Periodic fluctuations in blood glucose levels and hyperinsulinemia in all sugar-treated rats might play a role in sugar-induced hypertension in older rats. Also, sugar treatment of male Sprague-Daley rats has been reported to alter renal function without insulin resistance, glucose intolerance, and hypertension (Roysommuti et al. 2002).

Although taurine supplementation may improve or prevent hypertension in humans and animal models (Militante and Lombardini 2002), its mechanism of action is complicated. It prevents fructose-induced hypertension but exacerbates hyperinsulinemia and hypertriglyceridemia in rats (Anuradha and Balakrishnan 1999; Harada et al. 2004). This antihypertensive action is likely mediated by kinins and renal fluid excretion (Gentile et al. 1994; Nandhini et al. 2004; Nandhini and

Anuradha 2004). In contrast, hypertension, insulin resistance, and renal damage in adults that result from a perinatal nutritional imbalance may be prevented or attenuated by taurine treatment at an early age (Hoet et al. 2000; Militante and Lombardini 2002). In addition, taurine may inhibit sympathetic activity and the renin-angiotensin system in many forms of hypertension. Thus, the pressor effect of perinatal taurine exposure and its interaction with high sugar consumption, events that occur later in life, require further clarification.

Gender differences in the pathogeneses of cardiovascular diseases have been shown in many experimental models and in humans (D'Amore and Mora 2006; Meyer et al. 2006). Estrogen, rather than testosterone, protect against these diseases. By acting on blood vessels and on cardiovascular centers in the brain, estrogen protects against increases in arterial pressure (Ashraf and Vongpatanasin 2006; Maturana et al. 2007; Peng et al. 2003; Wyss and Carlson 2003). Recently, it has been reported that estrogen treatment improves or prevents hypertension in female, growth-restricted offspring (Ojeda et al. 2007). Taurine depletion has also been observed in these animal models (Aerts and Van Assche 2002). In addition, prenatal testosterone treatment can induce cardiovascular diseases in adult offspring, an effect similar to prenatal malnutrition (Dumesic et al. 2007; King et al. 2007). Thus, perinatal taurine exposure likely alters sex hormonal status at an early age and therefore programs subsequent organ function and affects adult diseases.

In summary, the present study indicates a gender disparity in the long-term effects of perinatal taurine depletion and supplementation on arterial pressure control. An imbalance of taurine exposure at an early age will thus predispose or program the pressor effect of high sugar consumption in later life.

Acknowledgments This work was supported by grants from the Faculty of Medicine, Khon Kaen University and the National Institutes of Health (AT 00477 (JMW) from the National Center for Complementary and Alternative Medicine and the Office of Dietary Supplements. Its contents are solely the responsibility of the authors and do not necessarily represent the official views of the National Center for Complementary and Alternative Medicine, the Office of Dietary Supplements, or the National Institutes of Health.

References

Aerts L, Van Assche FA (2002) Taurine and taurine-deficiency in the perinatal period. J Perinat Med 30:281–286

Anuradha CV, Balakrishnan SD (1999) Taurine attenuates hypertension and improves insulin sensitivity in the fructose-fed rat, an animal model of insulin resistance. Can J Physiol Pharmacol 77:749–754

Ashraf MS, Vongpatanasin W (2006) Estrogen and hypertension. Curr Hypertens Rep 8:368–376

Azuma M, Takahashi K, Fukuda T, Ohyabu Y, Yamamoto I, Kim S, Iwao H, Schaffer SW, Azuma J (2000) Taurine attenuates hypertrophy induced by angiotensin II in cultured neonatal rat cardiac myocytes. Eur J Pharmacol 403:181–188

Barker DJ(2007) The origins of the developmental origins theory. J Intern Med 261:412–417

Barker DJ, Eriksson JG, Forsen T, Osmond C (2002) Fetal origins of adult disease: strength of effects and biological basis. Int J Epidemiol 31:1235–1239

Barraclough CA (1961).Production of anovulatory, sterile rats by single injections of testosterone propionate. Endocrinology 68:62–67
Cella SG, Locatelli V, Mennini T, Zanini A, Bendotti C, Forloni GL, Fumagalli G, Arce VM, de GC, V, Wehrenberg WB, et al. (1990) Deprivation of growth hormone-releasing hormone early in the rat's neonatal life permanently affects somatotropic function. Endocrinology 127: 1625–1634
Cherif H, Reusens B, Ahn MT, Hoet JJ, Remacle C (1998) Effects of taurine on the insulin secretion of rat fetal islets from dams fed a low-protein diet. J Endocrinol 159:341–348
Clark JT, Chakraborty-Chatterjee M, Hamblin M, Wyss JM, Fentie IH (2004) Estrogen depletion differentially affects blood pressure depending on age in Long-Evans rats. Endocrine 25: 173–186
D'Amore S, Mora S (2006) Gender-specific prediction of cardiac disease: importance of risk factors and exercise variables. Cardiol Rev 14:281–285
Dumesic DA, Abbott DH, Padmanabhan V (2007) Polycystic ovary syndrome and its developmental origins. Rev Endocr Metab Disord 8:127–141
Eriksson JG, Forsen TJ, Kajantie E, Osmond C, Barker DJ (2007) Childhood growth and hypertension in later life. Hypertension 49:1415–1421
Forrester T (2004) Historic and early life origins of hypertension in Africans. J Nutr 134:211–216
Gentile S, Bologna E, Terracina D, Angelico M (1994) Taurine-induced diuresis and natriuresis in cirrhotic patients with ascites. Life Sci 54:1585–1593
Harada H, Tsujino T, Watari Y, Nonaka H, Emoto N, Yokoyama M (2004) Oral taurine supplementation prevents fructose-induced hypertension in rats. Heart Vessels 19:132–136
Harding JE (2001) The nutritional basis of the fetal origins of adult disease. Int J Epidemiol 30: 15–23
Hoet JJ, Ozanne S, Reusens B (2000) Influences of pre- and postnatal nutritional exposures on vascular/endocrine systems in animals. Environ Health Perspect 108 Suppl 3:563–568
Johnson RJ, Segal MS, Sautin Y, Nakagawa T, Feig DI, Kang DH, Gersch MS, Benner S, Sanchez-Lozada LG (2007) Potential role of sugar (fructose) in the epidemic of hypertension, obesity and the metabolic syndrome, diabetes, kidney disease, and cardiovascular disease. Am J Clin Nutr 86:899–906
King AJ, Olivier NB, Mohankumar PS, Lee JS, Padmanabhan V, Fink GD (2007). Hypertension caused by prenatal testosterone excess in female sheep. Am J Physiol Endocrinol Metab 292:E1837–E1841
Lackland DT, Egan BM, Syddall HE, Barker DJ (2002) Associations between birth weight and antihypertensive medication in black and white medicaid recipients. Hypertension 39:179–183
Langley-Evans SC(2006) Developmental programming of health and disease. Proc Nutr Soc 65:97–105
Langley-Evans SC, Langley-Evans AJ, Marchand MC (2003) Nutritional programming of blood pressure and renal morphology. Arch Physiol Biochem 111:8–16
Lerdweerapol W, Jirakulsomchok D, Wyss JM, Roysommuti S (2007) Perinatal taurine supplementation influences renal hemodynamics and excretory function in adult, female rats. FASEB J 21 (5 Part I):A501–502
Maturana MA, Irigoyen MC, Spritzer PM (2007) Menopause, estrogens, and endothelial dysfunction: current concepts. Clinics 62:77–86
Mendez MA, Wynter S, Wilks R, Forrester T (2004) Under- and overreporting of energy is related to obesity, lifestyle factors and food group intakes in Jamaican adults. Public Health Nutr 7: 9–19
Meyer MR, Haas E, Barton M (2006) Gender differences of cardiovascular disease: new perspectives for estrogen receptor signaling. Hypertension 47:1019–1026
Militante JD, Lombardini JB (2002) Treatment of hypertension with oral taurine: experimental and clinical studies. Amino Acids 23:381–393
Nandhini AT, Anuradha CV (2004) Hoe 140 abolishes the blood pressure lowering effect of taurine in high fructose-fed rats. Amino Acids 26:299–303

Nandhini AT, Thirunavukkarasu V, Anuradha CV (2004) Potential role of kinins in the effects of taurine in high-fructose-fed rats. Can J Physiol Pharmacol 82:1–8
Ojeda NB, Grigore D, Robertson EB, Alexander BT (2007) Estrogen protects against increased blood pressure in postpubertal female growth restricted offspring. Hypertension 50:679–685
Pacioretty L, Hickman MA, Morris JG, Rogers QR (2001) Kinetics of taurine depletion and repletion in plasma, serum, whole blood and skeletal muscle in cats. Amino Acids 21:417–427
Peng N, Clark JT, Wei CC, Wyss JM (2003) Estrogen depletion increases blood pressure and hypothalamic norepinephrine in middle-aged spontaneously hypertensive rats. Hypertension 41:1164–1167
Racasan S, Braam B, van der Giezen DM, Goldschmeding R, Boer P, Koomans HA, Joles JA (2004) Perinatal L-arginine and antioxidant supplements reduce adult blood pressure in spontaneously hypertensive rats. Hypertension 44:83–88
Roysommuti S, Khongnakha T, Jirakulsomchok D, Wyss JM (2002) Excess dietary glucose alters renal function before increasing arterial pressure and inducing insulin resistance. Am J Hypertens 15:773–779
Roysommuti S, Malila P, Jirakulsomchok D, Jirakulsomchok S, Wyss JM (2004) Perinatal taurine status influences renal hemodynamics in adult conscious rats. FASEB J 18 (4 Part I): A292–A293
Roysommuti S, Suvanich A, Jirakulsomchok D, Wyss JM (2007) Perinatal taurine depletion causes autonomic dysregulation in rats on a high glucose diet. FASEB J 21 (6 Part II):A887
Schaffer SW, Lombardini JB, Azuma J (2000) Interaction between the actions of taurine and angiotensin II. Amino Acids 18:305–318
Suvanich A, Jirakulsomchok D, Muchimapura S, Wyss JM, Roysommuti S (2006) Perinatal taurine depletion impairs autonomic control in conscious rats. FASEB J 20 (5 Part II):A1406–A1407
Ventura SJ, Martin JA, Curtin SC, Mathews TJ, Park MM, Hyattsville, MD (eds) (2002) Department of Health and Human Services. National Vital Statistics Report, pp 1–100
Wyss JM, Carlson SH (2003) Effects of hormone replacement therapy on the sympathetic nervous system and blood pressure. Curr Hypertens Rep 5:241–246
Wyss JM, Roysommuti S, King K, Kadisha I, Regan CP, Berecek KH (1994) Salt-induced hypertension in normotensive spontaneously hypertensive rats. Hypertension 23:791–796

Chapter 15
Perinatal Taurine Alters Arterial Pressure Control and Renal Function in Adult Offspring

Sanya Roysommuti, Wichaporn Lerdweeraphon, Pisamai Malila, Dusit Jirakulsomchok, and J. Michael Wyss

Abstract The present study investigates the effect of perinatal taurine exposure on renal function in adult, female rats on a high sugar diet. Perinatal taurine depleted (TD), supplemented (TS) or untreated control (C) female offspring were fed normal rat chow and tap water (CW,TDW or TSW) or tap water with 5% glucose (CG, TDG or TSG) after weaning. At 7–8 weeks of age, renal function was studied in the conscious, restrained rats. Mean arterial pressure was significantly higher in TDW, TDG, and TSG rats. Plasma sodium concentration was significantly lower in all glucose treated animals, but the greatest decrease was in TDW rats. Basal renal blood flow was lowest in TSW and TSG, and the responses to a saline load were also lowest in those two groups. These changes were consistent with increased renal vascular resistance. The basal glomerular filtration rate was lowest in TSW, but the responses to a saline load were similar in all of the groups. Water excretion was lower in TSG and TSW, consistent with increased renal tubular water reabsorption. These data suggest that perinatal taurine exposure alters normal renal function and renal responses to dietary sugar in adult female offspring.

Abbreviations *KW*, kidney weight; *ERBF*, effective renal blood flow; *GFR*, glomerular filtration rate; *RVR*, renal vascular resistance

15.1 Introduction

The kidney is a major regulator of long-term arterial pressure; as such, renal dysfunction or damage can lead to sustained hypertension (Cusi and Bianchi 1991; Rettig et al. 1990). Renal dysfunction can be genetically inherited, as shown by renal transplantation studies in spontaneously hypertensive and normotensive rats (Grisk et al. 2002; Rettig and Grisk 2005). The normotensive recipients of spontaneously hypertensive rat kidneys develop hypertension as adults while the hypertensive donors receiving kidneys from normotensive rats remain normotensive.

S. Roysommuti (✉)
Department of Physiology, Faculty of Medicine, Khon Kaen University, Khon Kaen 40002, Thailand

J. Azuma et al. (eds.), *Taurine 7*, Advances in Experimental Medicine and Biology 643,
DOI 10.1007/978-0-387-75681-3_15,

Prenatal food imbalances can result in low birth weight in neonates and in adults, and can produce a high risk of cardiovascular disease, insulin resistance, obesity, and hypertension (Barker et al. 2002; Harding 2001; Langley-Evans et al. 2003). Decreases in nephron number and excretory function also result from early malnutrition (Langley-Evans et al. 2003; Sturman 1993) and may underlie hypertensive development in adult offspring. Perinatal taurine depletion appears to induce low birth weight, multiple organ damage (Sturman 1993) and low tissue taurine content in many organs and plasma (Aerts and Van Assche 2002). While the mechanisms underlying the adverse effects of perinatal taurine exposure remain ambiguous, a decrease in taurine content likely leads to adult diseases.

Taurine is a free sulphur-containing beta-amino acid found in various organs including brain, the heart, and the kidney (Aerts and Van Assche 2002). Taurine content is highest during the perinatal period and then subsequently declines with advancing age (Sturman 1993). Renal dysfunction with aging, diabetes mellitus, hypertension, and obesity are inversely correlated to body taurine content (Cruz et al. 2000; Dawson et al. 1999; Eppler and Dawson 2001; Militante and Lombardini 2002). Thus, taurine supplementation could prevent age-related renal damage (Dawson et al. 1999), sugar-induced hypertension (Anuradha and Balakrishnan 1999; Harada et al. 2004), ethanol-induced hypertension (Harada et al. 2000), and drug-induced diabetes (Franconi et al. 2006; Mozaffari et al. 2003; Tenner et al. 2003). In addition, perinatal taurine supplementation could prevent cardiovascular diseases in the adult offspring following maternal malnutrition (Militante and Lombardini 2002; Zelikovic et al. 1990) and in spontaneously hypertensive rats (Racasan et al. 2004). Our previous experiments indicate that changes in either prenatal or postnatal taurine exposure alter renal hemodynamics in the adult, male offspring. However, the renal diuretic and natriuretic responses to an acute saline load remain within the normal control range (Roysommuti et al. 2004). Further, perinatal taurine depletion by addition of beta-alanine to the drinking water increases the pressor effect of a high sugar diet in these animals (Suvanich et al. 2006). Sympathetic nerve overactivity and baroreceptor reflex dysfunction underlie these alterations. Moreover, dietary sugar supplementation may induce renal dysfunction before hypertension and glucose intolerance in male Sprague-Dawley rats (Roysommuti et al. 2002), suggesting that renal dysfunction is not due to kidney damage from chronic hypertension. The present study tests the hypothesis that perinatal taurine exposure alters renal function in the adult, female offspring, and this change predisposes them to sugar-induced renal dysfunction.

15.2 Materials and Methods

Sprague-Dawley (SD) rats were bred in the animal unit of the Faculty of Medicine, Khon Kaen University and maintained at constant humidity (60 ± 5%), temperature (24 ± 1°C), and light cycle (0600–1800 h). Female Sprague-Dawley rats were fed normal rat chow with 3% beta-alanine (taurine depletion, TD), 3% taurine (taurine supplementation, TS) or water alone (Control, C) from conception to weaning.

Female offspring were fed normal rat chow with either 5% glucose in tap water (TD with glucose, TDG; TS with glucose, TSG; C with glucose, CG) or tap water alone (TDW, TSW, and CW) throughout the experiment. Experimental protocols and animal care were approved and monitored by the university animal committee and were in conformity with the *Guide for the Care and Use of Animals* and the US National Institutes of Health guidelines.

At 7–8 weeks of age, all female rats were placed under thiopental anesthesia and then implanted with femoral arterial, femoral venous and urinary bladder catheters. Forty-eight hours later, arterial pressure, heart rate, basal renal function and renal function after an acute intravenous saline load (a mixture of 0.5% inulin and 0.5% p-aminohippuric acid (PAH) in isotonic saline, 5% of body weight, 0.5 ml/min) were studied in conscious, restrained rats, that had been acclimated to the rat restrainer three hours per day for a week prior to the experiment (Roysommuti et al. 2002). Two days later, blood glucose and glucose tolerance were tested after an overnight fast. At the end of the experiment, all rats were sacrificed and kidney (KW) and heart (HW) weights were determined.

Arterial pressure pulse was continuously recorded during the experiment using the Biopac system, and mean arterial pressure and heart rate were analyzed offline using Acknowledge software (version 3.5.5, Biopac system, CA). Urine volumes were measured gravitationally, urine and plasma potassium and sodium were determined using flame photometry, hematocrit by a standard method, urine and plasma inulin and PAH by colorimetric methods and blood glucose by glucostrips and a glucometer (Accu-chek®, Germany). The glomerular filtration rate (GFR) was estimated by inulin clearance, effective renal blood flow (ERBF) by PAH clearance and hematocrit, and renal vascular resistance (RVR) was calculated by MAP/ERBF. The filtration fraction was calculated from the ratio of GFR to effective renal plasma flow (PAH clearance). Fractional Na excretion was calculated from the ratio of Na excretion per filtered Na load (GFR x plasma Na) while water excretion was the ratio of urinary flow per GFR. All renal excretory parameters were expressed as grams per KW.

All data were expressed as means ± SEM and were statistically analyzed using one-way ANOVA and appropriate post hoc tests (Duncan's Multi-Range) with significant criteria of $p < 0.05$.

15.3 Results

Compared to the control, body weights significantly decreased ($P<0.05$) only in perinatal taurine depleted rats on glucose supplementation since weaning (CW 194±5 g, CG 197±5 g, TDW 196±2 g, TDG±179 5 g, TSW 197±4 g, TSG 191± 5 g). Kidney weights significantly increased ($P<0.05$) in offspring of perinatal taurine supplemented rats (CW 1.66±0.04 g, CG 1.61±0.04 g, TDW 1.72±0.08 g, TDG 1.59±0.02 g, TSW 1.88±0.07, 1.64±0.04 g). Heart weights were not significantly different between the groups. All experimental groups displayed similar fasting blood sugar and glucose tolerance. Basal plasma sodium concentrations were significantly decreased in all glucose treated animals, but in response to a glucose challenge, plasma sodium significantly decreased only in the TDW rats (Fig. 15.1). Hematocrit was not different between the groups throughout the experiment.

Basal mean arterial pressures were significantly higher in TDW (121.7± 4.7 mm Hg), TDG (123.0±3.2 mm Hg), and TSG (121.2±3.0 mm Hg), than in CW (112.8±1.8 mm Hg) rats while arterial pressure in TSW (118.0±2.5 mm Hg) was not significantly different from that of the CW rat. The arterial pressures responses to an acute intravenous saline load were similar in CG, TDG, TSW, TSG and CW rats (Fig. 15.2). Perinatal taurine depletion or supplementation did not increase the pressor effects of a high sugar diet at rest or following an acute saline load (TDW versus TDG or TSW versus TSG was not significantly different). Basal heart rate was not significantly different between the groups and remained relatively constant after an acute saline load.

Basal ERBF was significantly lower in TSW (2.2±0.8 mm/min/g KW) and TSG (3.0±1.0 mm/min/g KW) than in the CW (5.7±0.3 mm/min/g KW) or CG groups (6.3±1.6 mm/min/g KW). Moreover, blunted renal blood flow responses to an intravenous saline load were observed in TDW, TDG, TSW, and TSG compared to the responses in the control group (Fig. 15.3). These changes were consistent with increased renal vascular resistance. A significant increase in basal renal vascular resistance was present in TSW rats but not in TDW rats (Fig. 15.4). Compared to CW (0.8±0.2 mm/min/g KW), the basal glomerular filtration rate was lower in TSW

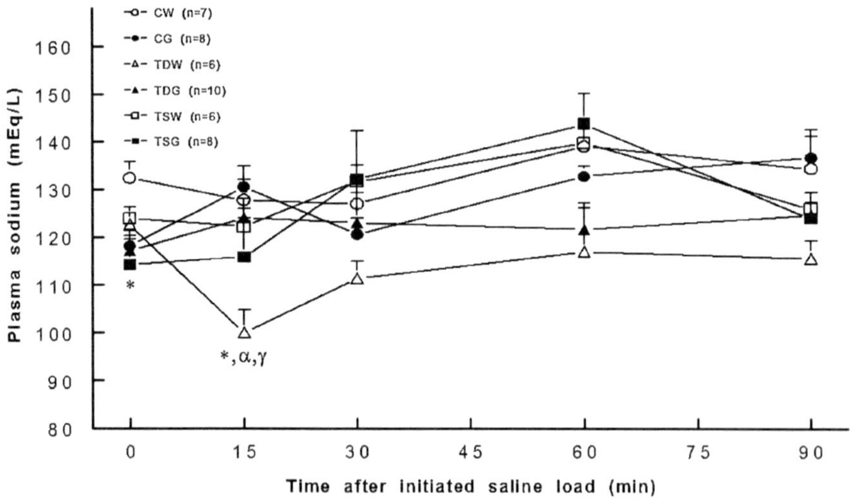

Fig. 15.1 Plasma sodium concentration was significantly lower in controls in CG, TDG and TSG and at 15 min in TDW rats. ∗, $P<0.05$ compared to CW; α, compared to CG; γ, compared to TSW; see text for abbreviations

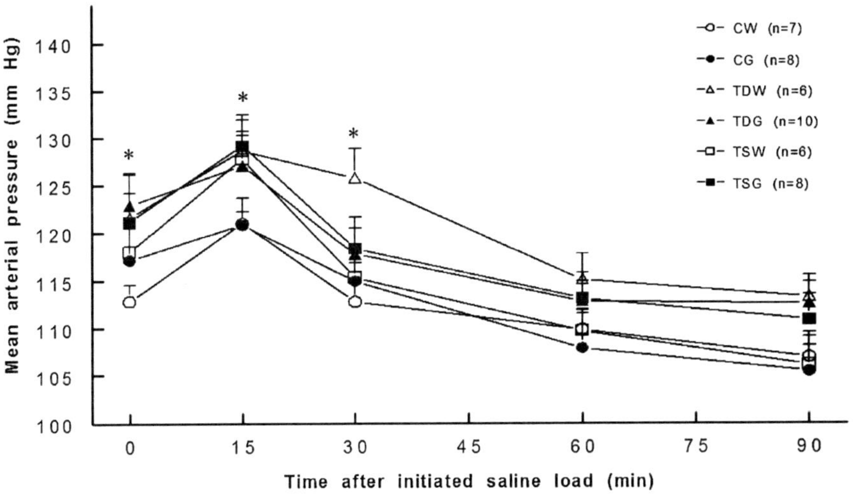

Fig. 15.2 Mean arterial pressure was significantly elevated in TDW (0,30 min), TDG (0 min), TSW (15 min) and TSG (0,15 min) rats compared to controls. *, $P<0.05$ compared to CW; see text for abbreviations

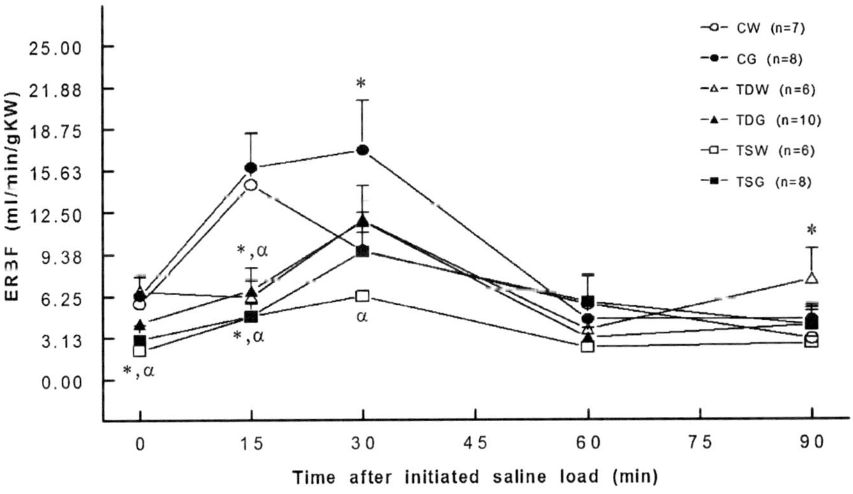

Fig. 15.3 Effective renal blood flow was lower in TDW (15 min), TDG (15 min), TSW (0,15 min), and TSG (0,15 min) rats than in the controls. *, $P<0.05$ compared to CW; α, compared to CG; see text for abbreviations

(0.3±0.0 mm/min/g KW), but the responses to a saline load were not significantly different between the groups (Fig. 15.5).

Renal water excretion at rest was significantly reduced in both TSW (17.4±3.3 μl/min/g KW, P<0.05) and TSG (11.9±2.2 μl/min/g KW; $P<0.05$) when compared to CW (28.2±4.7 μl/min/g KW) or CG (23.0±2.3 μl/min/g KW) rats, and they

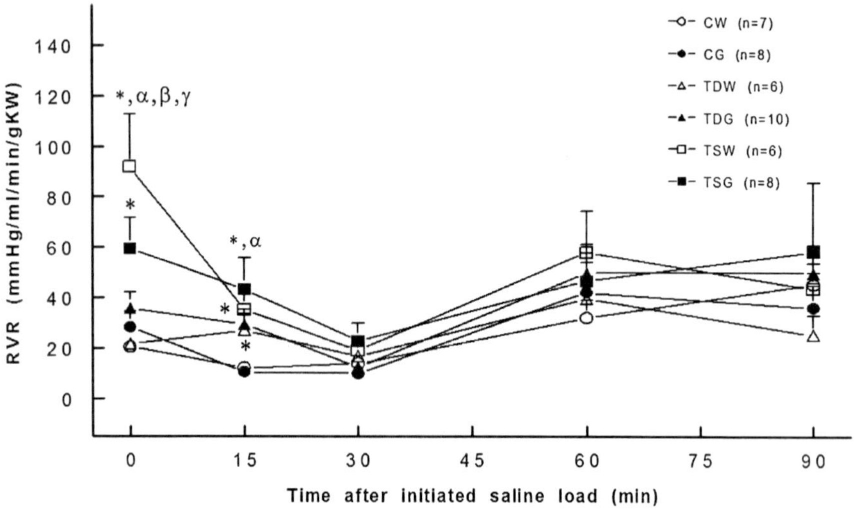

Fig. 15.4 Renal vascular resistance was significantly higher in TDW (15 min), TSW (0,15 min), TSG (0,15 min) rats than in the controls. ∗, P<0.05 compared to CW; α, compared to CG; β, compared to TDW; γ, compared to TSW; see text for abbreviations

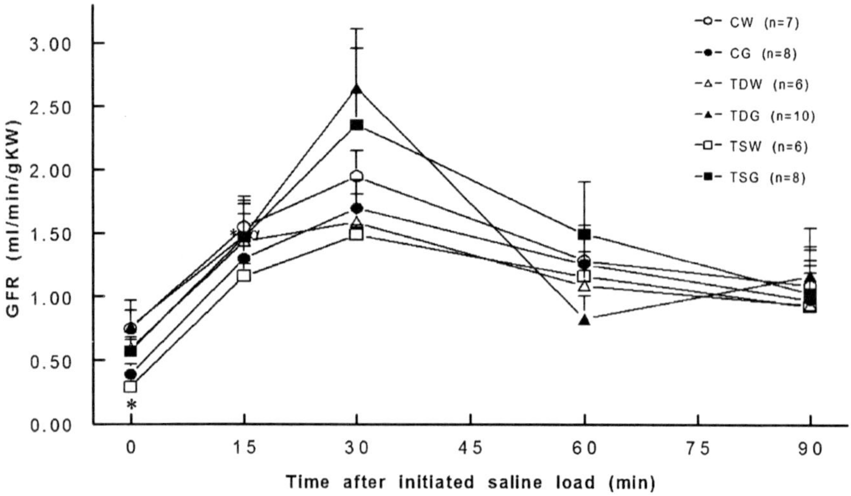

Fig. 15.5 Glomerular filtration rates were similar in all groups. ∗, $P<0.05$ compared to CW at rest; see text for abbreviations

tended to be lower than the CW or CG rats throughout the experiment (Fig. 15.6). The reduction in water excretion of the TSG rat was consistent with an increase in renal tubular water reabsorption, as indicated by a significant decrease in fractional water excretion (Fig. 15.7). In addition, fractional water excretion of both TDG and

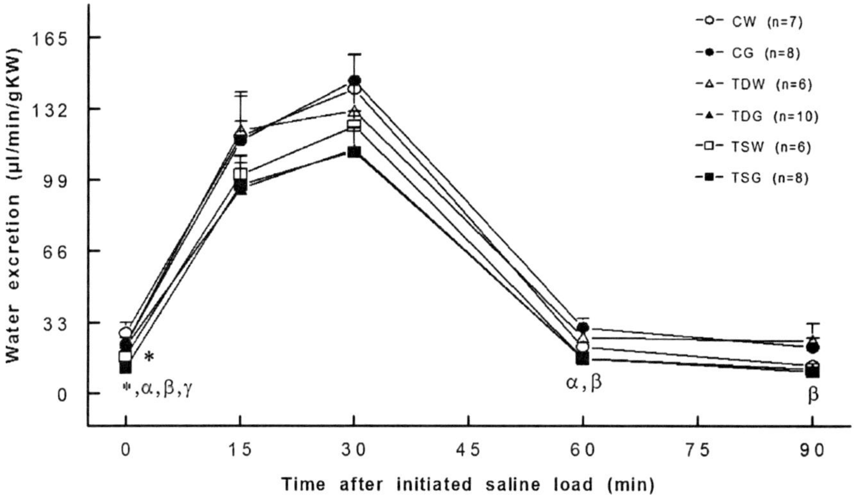

Fig. 15.6 Basal water excretion was lower in TSW and TSG rats. ∗, $P<0.05$ compared to CW; α, compared to CG; β, compared to TDG; γ, compared to TSW; see text for abbreviations

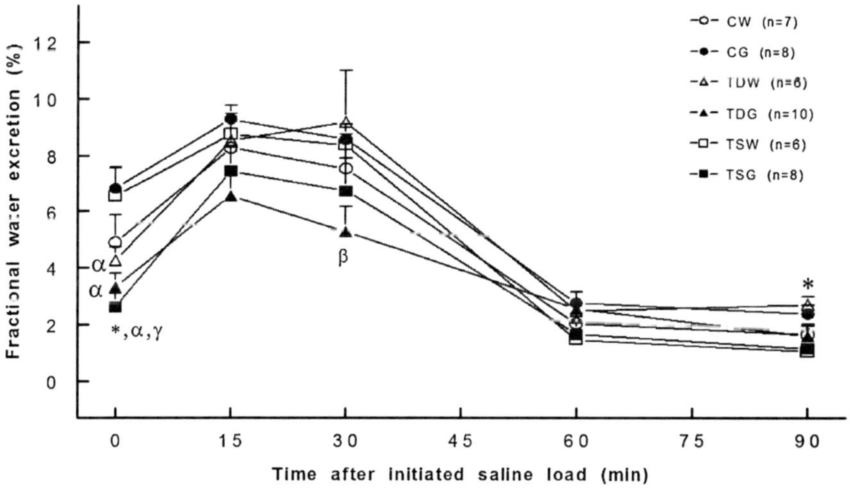

Fig. 15.7 Fractional water excretion was lower in TDW (0 min), TDG (0 min), TSG (0,30 min), TSG and TSW rats. ∗, $P<0.05$ compared to CW; α, compared to CG; β, compared to TDW; γ, compared to TSW; see text for abbreviations

TSG rats but not of the TDW and TSW rats was significantly lower than that of the CG rats.

Sodium excretion at rest was not significantly different between the groups, but the natriuretic response to an acute saline load was higher at 30 minutes for the TSW rat and at 60 minutes for the CG rat compared to the CW rat (Fig. 15.8). Moreover, compared to the CG rat, TDG and TSG rats displayed slightly blunted natriuretic

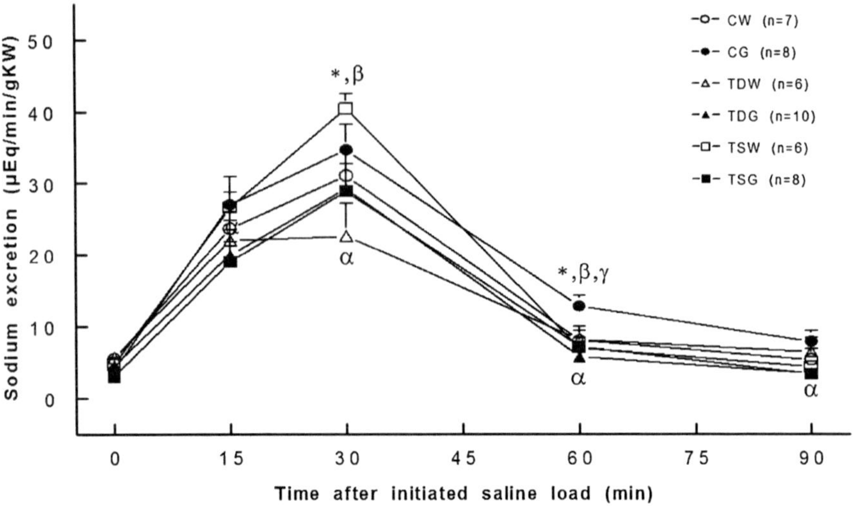

Fig. 15.8 Sodium excretion was higher in CG (60 min) and TSW (30 min) and decreased in TDW (30 min), TDG (90 min), and TSG (90 min) rats. *, $P<0.05$ compared to CW; α, compared to CG; β, compared to TDW; γ, compared to TSW; see text for abbreviations

responses to an acute saline load. Fractional Na excretion at rest and following an acute saline load was higher only in the TSW rat compared to the CW or TSG groups (Fig. 15.9). Though the fractional Na excretions of TDG and TSG rats were not significantly different from the other groups, they tended to be lower than those of the CG rats.

15.4 Discussion

The perinatal origin of adult function and disease has been widely accepted. Perinatal nutritional imbalances play a key role in adult diseases of the offspring, as supported by many lines of evidence in animal models and human epidemiologic studies (Barker et al. 2002; Harding 2001; Lackland et al. 2002; Langley-Evans et al. 2003). Hypertension in adult life has been reported to be programmed during perinatal exposure to protein over- or under-nutrition, high sugar intake, high salt intake, intrauterine hypoxia and other stressors. Animals and humans subjected in utero to such environments display decreased nephrogenesis that causes a permanent reduction in nephron number throughout adult life (Gopalakrishnan et al. 2005; Langley-Evans et al. 2003). This decrease in nephron number results in altered renal weight and glomerular filtration rate and blunted pressure-natriuretic and diuretic responses, anomalies that are commonly observed in many forms of arterial hypertension (Hall et al. 1992). Although the present study supports the perinatal origin of adult renal function, perinatal taurine depletion or supplementation did not decrease kidney weights in female offspring, as previously reported in male offspring (Roysommuti et al. 2007). In addition, perinatal taurine supplementation increased renal weight and despite a decrease in the basal glomerular filtration rate, the kidneys responded to an acute intravenous saline load similarly in all of the groups. Moreover, these changes in kidney weight and glomerular filtration were restored to normal by maintenance of a postnatal high sugar diet. Thus, these data suggest that perinatal nutritional imbalances have long-term effects on renal growth and function in the adult offspring. Moreover, both excess sugar consumption during postnatal life and perinatal taurine alterations influence the final renal functional phenotype. Similar effects are observed in response to low birth weight and adult obesity and hypertension. Low birth weight also triggers obesity and hypertension in rats on a basal diet or humans from a poor socioeconomic environment, i.e., malnourished (Barker 2007; Militante and Lombardini 2002). In addition, in the present study perinatal taurine depletion coupled with sugar supplementation since weaning led to a decrease in body weight but perinatal taurine depletion alone did not have any long-term effect on body, kidney, and heart weight.

Pressure-diuresis/natriuresis is the main mechanism by which the kidney regulates arterial pressure (Hall et al. 1996). Though mean arterial pressure of the perinatal taurine depleted rat is significantly elevated and was higher than any of the other groups at rest and following an acute saline load, their diuretic and natriuretic responses were similar. This indicates a blunted pressure-diuretic/natriuretic function in perinatal taurine depleted rats but not offspring of the perinatal taurine supplemented rats. This renal dysregulation could not be restored or heightened by maintenance on a high sugar diet. However, the natriuretic responses to a saline load are different in perinatal taurine depleted rats depending on their dietary sugar

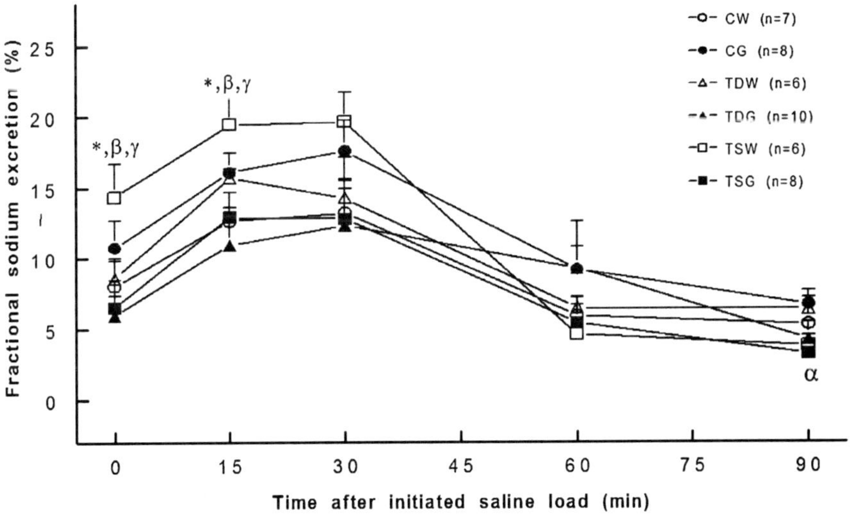

Fig. 15.9 Fractional sodium excretion was higher in TSW (0, 15 min) and TSG (90 min) rats. ∗, $P<0.05$ compared to; α, compared to CG; β, compared to TDW; γ, compared to TSW; see text for abbreviations

intake. Since their glomerular filtration rates were similar but plasma sodium is sharply decreased only in rats on a basal sugar diet, the maintenance of pressure-natriuresis likely resulted from decreased renal tubular sodium reabsorption; the slight increase in fractional sodium excretion observed during the initial phase of the responses to the saline load in the present study supports this interpretation. In contrast, increases in mean arterial pressure may underlie higher natriuretic but not diuretic responses in perinatal taurine supplemented rats, indicating that the offspring exhibit a normal pressure-natriuretic relationship. Thus, the high sugar diet since weaning did not restore mean arterial pressure of the TSG rats but natriuretic responses to an acute saline load were not perturbed, indicating that the high sugar diet induced a blunted pressure-natriuretic response in the perinatal taurine supplemented offspring. Though pressure-diuretic responses were similar in perinatal taurine supplemented and depleted rats, the taurine supplemented rats appear to depend on renal tubular reabsorption rather than changes in glomerular filtration to maintain homeostasis. In the perinatal taurine supplement group, the basal glomerular filtration rate was lower while fractional sodium excretion was higher.

Perinatal nutritional imbalances have been reported to induce hypertension in offspring. Exposure to diets with high protein and low carbohydrate or low protein and high carbohydrate in the perinatal period can produce hypertension in the adult offspring, likely via sympathetic nerve overactivity, insulin resistance, renin-angiotensin system overactivity, and renal dysfunction (Hall 2003; Langley-Evans et al. 2003). Significantly, altered perinatal taurine exposure has also been reported to contribute to these abnormalities. Thus, taurine supplementation in prenatal or later life attenuates hypertension in the offspring (Militante and Lombardini 2002). However, this action of taurine may not be related to the improvement in insulin resistance or hyperinsulinemia (Hultman et al. 2007; Nandhini and Anuradha 2004). The present data support the hypothesis that an imbalance in perinatal nutrition alters adult offspring function. In addition, the present study supports our previous experiment regarding male offspring, namely, that perinatal taurine can induce hypertension independent of insulin resistance and diabetes mellitus (Roysommuti et al. 2007). Moreover, high sugar intake since weaning did not exacerbate hypertension in the perinatal taurine depletion or supplementation groups or induce hypertension in untreated control rats as previously reported in the male offspring (Roysommuti et al. 2007). Nevertheless, high sugar intake blunted renal pressure-natriuretic and diuretic function in these rats, suggesting that the perinatal taurine status interacts with dietary sugar supplementation to impair renal function before the appearance of either insulin resistance or significant hypertension.

Previous experiments indicate that inappropriate taurine exposure during prenatal or postnatal periods affects renal hemodynamics in male offspring (Roysommuti et al. 2004). The present data further suggest that perinatal taurine status also affects these parameters in female offspring. Perinatal taurine supplementation decreased renal blood flow at rest and following an acute saline load, consistent with a rise in renal vascular resistance. These changes could not be restored to control levels, suggesting a permanent deficit in these animals. Though perinatal taurine depletion did not alter resting renal blood flow, its response to an acute saline load was attenuated (slight increase in renal vascular resistance). Mean arterial pressure is unlikely to underlie these abnormalities. In contrast, a rise in mean arterial pressure in the perinatal taurine supplemented groups may contribute to increased renal vascular resistance.

In summary, the present data indicate that a modification in perinatal taurine exposure alters renal function and blood pressure in adult female offspring and these effects enable high sugar diets to impair kidney function in later life. The complexity of this phenomenon is dependent on a change in perinatal programming and the subsequent development and adaptation in later life. Thus, perinatal taurine status should be considered carefully in pregnant animals and humans.

Acknowledgments This work was supported by grants from the Faculty of Medicine of Khon Kaen University, the National Institutes of Health (AT 00477) (JMW), the National Center for Complementary and Alternative Medicine and the Office of Dietary Supplements. Its contents are solely the responsibility of the authors and do not necessarily represent the official views of the National Center for Complementary and Alternative Medicine, the Office of Dietary Supplements, or the National Institutes of Health.

References

Aerts L, Van Assche FA (2002) Taurine and taurine-deficiency in the perinatal period. J Perinat Med 30:281–286

Anuradha CV, Balakrishnan SD (1999) Taurine attenuates hypertension and improves insulin sensitivity in the fructose-fed rat, an animal model of insulin resistance. Can J Physiol Pharmacol 77:749–754

Barker DJ (2007) Obesity and early life. Obes Rev 8 Suppl 1:45–49

Barker DJ, Eriksson JG, Forsen T, Osmond C (2002) Fetal origins of adult disease: strength of effects and biological basis. Int J Epidemiol 31:1235–1239

Cruz CI, Ruiz-Torres P, del Moral RG, Rodriguez-Puyol M, Rodriguez-Puyol D (2000) Age-related progressive renal fibrosis in rats and its prevention with ACE inhibitors and taurine. Am J Physiol Renal Physiol 278:F122–F129

Cusi D, Bianchi G (1991) The kidney in the pathogenesis of hypertension. Semin Nephrol 11: 523–527

Dawson R Jr, Liu S, Eppler B, Patterson T (1999) Effects of dietary taurine supplementation or deprivation in aged male Fischer 344 rats. Mech Ageing Dev 107:73–91

Eppler B, Dawson R, Jr. (2001) Dietary taurine manipulations in aged male Fischer 344 rat tissue: taurine concentration, taurine biosynthesis, and oxidative markers. Biochem Pharmacol 62: 29–39

Franconi F, Loizzo A, Ghirlanda G, Seghieri G (2006) Taurine supplementation and diabetes mellitus. Curr Opin Clin Nutr Metab Care 9:32–36

Gopalakrishnan GS, Gardner DS, Dandrea J, Langley-Evans SC, Pearce S, Kurlak LO, Walker RM, Seetho IW, Keisler DH, Ramsay MM, Stephenson T, Symonds ME (2005) Influence of maternal pre-pregnancy body composition and diet during early-mid pregnancy on cardiovascular function and nephron number in juvenile sheep. Br J Nutr 94:938–947

Grisk O, Kloting I, Exner J, Spiess S, Schmidt R, Junghans D, Lorenz G, Rettig R (2002) Long-term arterial pressure in spontaneously hypertensive rats is set by the kidney. J Hypertens 20:131–138

Hall JE (2003) The kidney, hypertension, and obesity. Hypertension 41:625–633

Hall JE, Guyton AC, Brands MW (1996) Pressure-volume regulation in hypertension. Kidney Int Suppl 55:S35–S41

Hall JE, Mizelle HL, Brands MW, Hildebrandt DA (1992) Pressure natriuresis and angiotensin II in reduced kidney mass, salt-induced hypertension. Am J Physiol 262:R61–R71
Harada H, Kitazaki K, Tsujino T, Watari Y, Iwata S, Nonaka H, Hayashi T, Takeshita T, Morimoto K, Yokoyama M (2000) Oral taurine supplementation prevents the development of ethanol-induced hypertension in rats. Hypertens Res 23:277–284
Harada H, Tsujino T, Watari Y, Nonaka H, Emoto N, Yokoyama M (2004) Oral taurine supplementation prevents fructose-induced hypertension in rats. Heart Vessels 19:132–136
Harding JE (2001) The nutritional basis of the fetal origins of adult disease. Int J Epidemiol 30: 15–23
Hultman K, Alexanderson C, Manneras L, Sandberg M, Holmang A, Jansson T (2007) Maternal taurine supplementation in the late pregnant rat stimulates postnatal growth and induces obesity and insulin resistance in adult offspring. J Physiol 579:823–833
Lackland DT, Egan BM, Syddall HE, Barker DJ (2002) Associations between birth weight and antihypertensive medication in black and white medicaid recipients. Hypertension 39:179–183
Langley-Evans SC, Langley-Evans AJ, Marchand MC (2003) Nutritional programming of blood pressure and renal morphology. Arch Physiol Biochem 111:8–16
Militante JD, Lombardini JB (2002) Treatment of hypertension with oral taurine: experimental and clinical studies. Amino Acids 23:381–393
Mozaffari MS, Miyata N, Schaffer SW (2003) Effects of taurine and enalapril on kidney function of the hypertensive glucose-intolerant rat. Am J Hypertens 16:673–680
Nandhini AT, Anuradha CV (2004) Hoe 140 abolishes the blood pressure lowering effect of taurine in high fructose-fed rats. Amino Acids 26:299–303
Racasan S, Braam B, van der Giezen DM, Goldschmeding R, Boer P, Koomans HA, Joles JA (2004) Perinatal L-arginine and antioxidant supplements reduce adult blood pressure in spontaneously hypertensive rats. Hypertension 44:83–88
Rettig R, Folberth C, Kopf D, Stauss H, Unger T (1990) Role of the kidney in the pathogenesis of primary hypertension. Clin Exp Hypertens A 12:957–1002
Rettig R, Grisk O (2005) The kidney as a determinant of genetic hypertension: evidence from renal transplantation studies. Hypertension 46:463–468
Roysommuti S, Khongnakha T, Jirakulsomchok D, Wyss JM (2002) Excess dietary glucose alters renal function before increasing arterial pressure and inducing insulin resistance. Am J Hypertens 15:773–779
Roysommuti S, Malila P, Jirakulsomchok D, Jirakulsomchok S, Wyss JM (2004) Perinatal taurine status influences renal hemodynamics in adult conscious rats. FASEB J 18 (4 Part I): A292–A293
Roysommuti S, Suvanich A, Jirakulsomchok D, Wyss JM (2007) Perinatal taurine depletion causes autonomic dysregulation in rats on a high glucose diet. FASEB J 21 (6 Part II):A887
Sturman JA (1993) Taurine in development. Physiol Rev 73:119–147
Suvanich A, Jirakulsomchok D, Muchimapura S, Wyss JM, Roysommuti S (2006) Perinatal taurine depletion impairs autonomic control in conscious rats. FASEB J 20 (5 Part II):A1406–A1407
Tenner TE Jr, Zhang XJ, Lombardini JB (2003) Hypoglycemic effects of taurine in the alloxan-treated rabbit, a model for type 1 diabetes. Adv Exp Med Biol 526:97–104
Zelikovic I, Chesney RW, Friedman AL, Ahlfors CE (1990) Taurine depletion in very low birth weight infants receiving prolonged total parenteral nutrition: role of renal immaturity. J Pediatr 116:301–306

Part II
Effect of Taurine on Brain and Retina

Chapter 16
Taurine Protects Immature Cerebellar Granullar Neurons against Acute Alcohol Administration

Andrey G. Taranukhin, Elena Y. Taranukhina, Irina M. Djatchkova, Pirjo Saransaari, Markku Pelto-Huikko, and Simo S. Oja

Abstract Acute ethanol administration causes extensive apoptosis throughout the nervous system. We studied the protective effect of taurine on alcohol-induced apoptosis in the cerebellum of developing mice. Taurine rescued a part of immature neurons by markedly reducing caspase-3 immunoreactivity and the number of TUNEL-positive cells in most cerebellar lobules.

Abbreviations *caspase-3 IR*, caspase-3-immunoreactive; *GL*, grey level

16.1 Introduction

The central nervous system is extremely sensitive to ethanol during its development and the periods of vulnerability are temporally well defined. Exposure to ethanol during the last trimester of human gestation can produce a broad spectrum of neuropathological consequences (Clarren et al. 1978 Famy et al. 1998 Spadon et al. 2007). The approximate equivalent of this period in rodents is the first postnatal week (Dobbin and Sands 1979; Rice and Barone 2000). Acute alcohol administration to mice during this period causes extensive apoptosis throughout the central nervous system (Olney et al. 2002a). Taurine has been shown to interact with the effects of ethanol (Olive et al. 2002). For instance, it modulates ethanol-stimulated locomotion (Aragón et al. 1992) and prolongs ethanol-induced sedation when given intracerebroventricularly to mice (Ferko 1987; Ferko 1988). Furthermore, ethanol administration has been shown to elicit an increase of extracellular taurine in the rat cerebral cortex and hippocampus (Dachhour and De Witte 2000). In the present work we studied the possible neuroprotective effects of taurine in ethanol-induced apoptosis in the developing mouse cerebellum. The experiments

A.G. Taranukhin (✉)
Brain Research Center, University of Tampere Medical School, Finland, Laboratory of Comparative Somnology and Neuroendocrinology, Sechenov Institute of Evolutionary Physiology and Biochemistry, Russia

J. Azuma et al. (eds.), *Taurine 7*, Advances in Experimental Medicine and Biology 643, DOI 10.1007/978-0-387-75681-3_16,

were performed on 7-day-old male mice. This age was chosen as the most sensitive to ethanol-induced apoptotic neurodegeneration. Moreover, spontaneous apoptosis is a common prominent phenomenon at this age (Wood et al. 1993).

16.2 Materials and Methods

16.2.1 Animals and Treatments

Seven-day-old infant male NMRI mice were used in all experiments. The animals were divided into three groups: ethanol-treated, ethanol+taurine-treated and controls. Ethanol was mixed in sterile saline to a 20% solution and administered subcutaneously at a total dose of 5 g/kg (2.5 g/kg at time zero and 2.5 g/kg again at 2 h) to the ethanol and ethanol+taurine groups. The ethanol+taurine group also received two injections of taurine (1 g/kg diluted with saline). The first taurine injection was given one hour before the first ethanol injection and the second taurine injection one hour after the second ethanol injection. The control animals were given saline subcutaneously. Eight hours after the first ethanol injection all animals were killed by decapitation. Their cerebella were rapidly excised, fixed in 4% paraformaldehyde, embedded in paraffin and cut with a microtome into 5-μm thick mid-sagittal sections containing lobules II-X of the cerebellum.

16.2.2 Immunohistochemistry

The sections were deparaffinized with xylene and hydrated in a graded ethanol series to distilled water. After antigen retrieval by microwave (20 min at 1000 W in 0.01 M citrate buffer (pH 6.0), washing in phosphate buffered saline (PBS) and blocking with 0.5% hydrogen peroxide in PBS for 20 min, specimens were preincubated for 30 min in serum-blocking solution (1% bovine serum albumin and 0.3% Triton X-100 in PBS). Thereafter the specimens were incubated with polyclonal activated caspase-3 antibody [cleaved caspase-3 (Asp 175) antibody] Cell Signaling Technology Inc., diluted 1:200 in serum-blocking solution] in moist chambers overnight at 4°C. After incubation with the primary antibody, the sections were incubated with biotinylated secondary antibody (goat anti-rabbit 1:500 in blocking solution) and ABC complex (Vectastain Elite ABC Kit, Vector Laboratories, Inc.) each for 30 min. Diaminobenzidine (DAB) was used as a chromogen to visualize the sites expressing activated caspase-3 immunoreactivity. The control sections were incubated without the primary antibodies to rule out nonspecific staining. Finally, the sections (without additional counter-staining) were dehydrated and mounted.

16.2.3 Semi-Quantitative Analysis of Caspase-3

The sections were processed under standardized conditions in every experiment, which allowed semi-quantitative analysis of the protein amount in the histological slices (Smolen 1990). An image analysis system comprising IBM PC, Nikon

Microphot-FXA microscope, SensiCam digital camera (PCO Computer Optics GmbH), Image-Pro Plus (Media Cybernetics) was used for a semi-quantitative analysis of caspase-3 expression in the histological sections of the cerebellum. Five sections cut at the same level of the cerebellum vermis from every animal were analyzed. The sections were reviewed at 250-fold magnification under a light microscope. The optical density was evaluated by two parameters reflecting the expression level of this protein in the lobules of the cerebellum. As the first parameter, the number of caspase-3-immunoreactive (IR) cells was calculated in every slice in each lobule and the average number of activated caspase-3-IR cells per slice counted. As the second parameter, the relative optical density of DAB precipitates in the perikaryons of individual cells was estimated in every section and the average optical density with its SEM calculated. Optical density was analyzed by the software as a "grey level" (GL). The optical density reflecting the content of the proteins studied in neurons was calculated as the GL of an IR field of the cell by subtracting the background GL. The optical density of the background was estimated in the same slice in the field of non-immunoreactive cerebellar tissue.

16.2.4 Detection of Cell Death In Situ

DNA fragmentation is one of the most precise markers by which to recognize apoptotic cells in tissue. In order to detect DNA fragmentation in cell nuclei, terminal deoxynucleotidyl transferase-mediated dUTP nick end-labeling (TUNEL) reaction was applied to the paraffin sections using the In Situ Cell Death Detection Kit, POD (Roche Applied Science, Germany). After deparaffinization, the sections were irradiated with microwaves in 0.01 M citric acid buffer (pH 6) for 10 min at 750 W. No inhibition of endogenous peroxidase was performed because H_2O_2 weakens terminal deoxynucleotidyl transferase activity (Migheli et al. 1995) and induces DNA breaks (Wijsman et al. 1993). Sections were incubated with TUNEL reaction mixture for 60 min at 37°C. Further incubation with peroxidase conjugated antibody was performed for 30 min at 37°C. The sections were stained with DAB for 10 min at room temperature and then counterstained with hematoxylin-eosin.

16.2.5 Statistical Analysis

Statistical significance was determined by Student's *t* test. Each value was expressed as mean ± SEM. Differences were considered significant when the calculated p value was <0.05.

16.3 Results

16.3.1 Effects of Taurine on Caspase-3 Activation

There were randomly activated caspase-3-IR cells visible in the control group, reflecting the rate at which spontaneous (physiological) cell death occurs at this

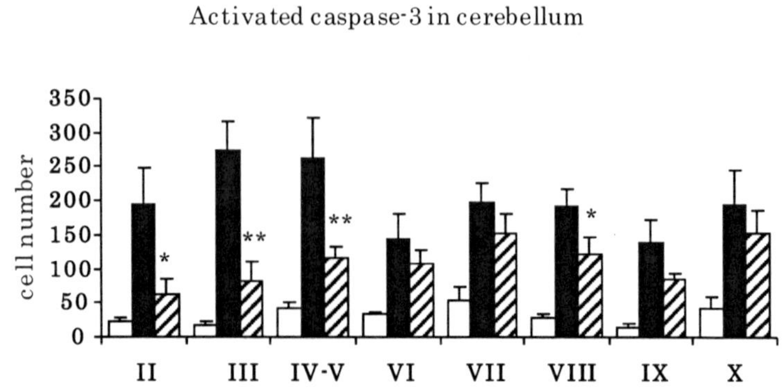

Fig. 16.1 The number of activated caspase-3-IR cells in the cerebellar lobules in control (*open bars*), ethanol-treated (*filled bars*) and ethanol+taurine-treated (*hatched bars*) mice. The results are given per mm^2 with SEMs. The number of animals in each group is 5. The significance of differences between ethanol and ethanol+taurine groups: $^{*}P < 0.05$, $^{**}P < 0.01$

age (Fig. 16.1). Following ethanol administration, there was a marked increase in activated caspase-3-IR granular neurons in all cerebellar lobules, indicating that ethanol triggered apoptotic neurodegeneration. Taurine treatment tended to reduce the number of activated caspase-3-IR cells in all cerebellar lobules when compared to the ethanol-treated group, the effect being significant in lobules II, III, IV-V and VIII, but it did not abolish it totally in any lobule. The content of activated caspase-3 in individual immunoreactive neurons was approximately the same in all experimental groups in all cerebellar lobules (Fig. 16.2).

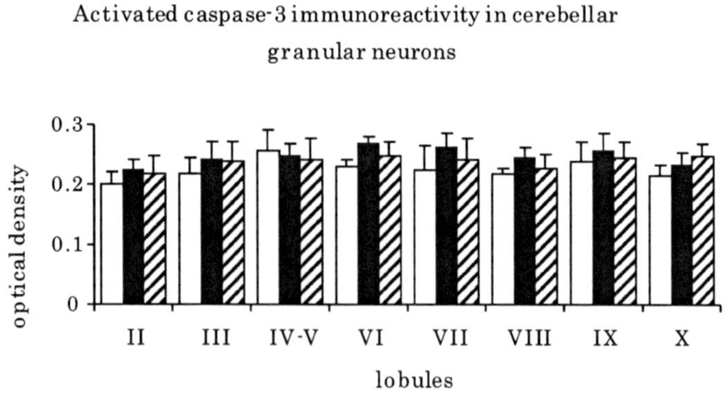

Fig. 16.2 The content of activated caspase-3 in individual cells in the cerebellar lobules in control (*open bars*), ethanol-treated (*filled bars*) and ethanol+taurine-treated (*hatched bars*) mice. The results show the optical density with SEMs. The number of animals in each group is 5

16.3.2 Effects of Taurine on Ethanol-Induced Apoptosis

TUNEL-positive cells were counted in each lobule of the cerebellar vermis (Fig. 16.3). Occasional TUNEL-positive cells were observed in the cerebellar lobules of control mice. Eight hours after ethanol administration the number of cells with fragmented DNA labeled by TUNEL assay was significantly increased in all lobules. In mice treated with taurine, the tendency towards a decrease in TUNEL-positive neurons was discernible in all cerebellar lobules except lobule VI, being statistically significant in lobules III, IV-V, VIII and IX.

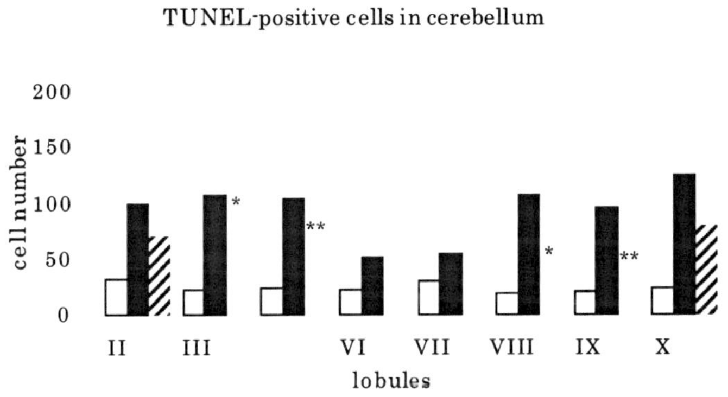

Fig. 16.3 The number of TUNEL-positive cells in the cerebellar lobules in control (*open bars*), ethanol-treated (*filled bars*) and ethanol+taurine-treated (*hatched bars*) mice. The results are given per mm^2 with SEMs. The number of animals in each group is 5. The significance of differences between ethanol and ethanol+taurine groups: $^*P < 0.05$, $^{**}P < 0.01$

16.4 Discussion

As already stated in the Introduction, acute exposure of rodents to ethanol during the period of developmental synaptogenesis causes extensive apoptotic neurodegeneration throughout the brain (Olney et al. 2000, 2002a). The degenerating neurons exhibit biochemical and ultrastructural features of apoptosis such as activation of caspase-3 (Olney et al. 2002a, 2002b) internucleosomal DNA fragmentation (Ikonomidou et al. 2000; Kumral et al. 2005) clumping of nuclear chromatin, formation of spherical chromatin masses and nuclear membrane fragmentation (Ikonomidou et al. 2000; Dikranian et al. 2001).

Two major apoptotic pathways have been established, one known as "extrinsic" and the other as "intrinsic" (Boatright and Salvesen 2003). In both pathways, the caspases play an important role in initiation, signal transduction and execution of apoptosis. The extrinsic pathway is triggered by activation of death receptors localized on the cell membrane surface and induces caspase-8 processing. The activated caspase-8 can directly or indirectly activate effector caspases such as caspases-3, -6 and -7. In the intrinsic pathway, many factors such as NO, oxidants and proapoptotic proteins, e.g. Bax, increase the mitochondrial membrane permeability and release cytochrome C into the cytoplasm. Cytochrome C binds to Apaf-1 and procaspase-9, forming an apoptosome and leading to caspase-9 activation (Purring-Koch and McLendon 2000). The active caspase-9 cleaves and activates effector caspases, including caspase-3. The activated effector caspases cleave many structural and functional proteins, activate DNase, which destroys chromosomes and leads to cell death (Budihardjo et al. 1999; Earnshaw et al. 1999). Recent studies have shown that caspases-6, -7 and -8 are not involved in ethanol-induced apoptosis (Olney et al. 2002a; Young et al. 2003). Apoptosis caused by ethanol in infant rodents is Bax-dependent and manifests itself mainly through the intrinsic mitochondrial pathway (Young et al. 2003; Nowoslawski et al. 2005).

Taurine is a sulfur-containing amino acid with multiple functions, including neuroprotection (Oja and Saransaari 2007). Earlier investigations on the interactions of taurine and ethanol in the brain have indicated that responses depend largely on the experimental set-up and the doses of ethanol and taurine administered (Oja and Saransaari 2007). The protective effects of taurine are not limited to ethanol. For instance, it has been shown in studies in vitro that taurine also protects cardiomyocytes (Takatani et al. 2004) and hypothalamic neurons (Taranukhin et al. 2007) from apoptosis induced by ischemia. Taurine suppresses the formation of the Apaf-1/caspase-9 apoptosome in cardiomyocytes and thereby prevents caspase-9 activation and apoptosis (Takatani et al. 2004). In our previous experiments we have shown that taurine reduces caspase-9 expression in hypothalamic neurons under ischemic conditions (Taranukhin et al. 2007). These findings and the knowledge that ethanol-induced apoptosis emerges preferentially via the mitochondrial pathway were the impetus to use taurine as a possible neuroprotector against ethanol-induced apoptosis. The present experimental set-up has been used as the model of ethanol-induced neurodegeneration (Ikonomidou et al. 2000; Olney et al. 2002a; 2002b; Young et al. 2003, 2005; Kumral et al. 2005; Nowoslawski et al. 2005), the time-points of which are well-defined. Because caspase-3 activation is considered an important step in the execution of apoptotic neuronal death and activated caspase-3 is widely expressed after acute ethanol administration (Olney et al. 2002a, b), the immunocytochemical detection of activated caspase-3 was adopted as a marker of apoptosis.

In the cerebellar lobules of control mice we saw only rarely activated caspase-3-IR neurons undergoing physiological cell death. As could be expected, eight hours after ethanol administration an increase in the number of activated caspase-3-IR cells was detected throughout all lobules. Taurine application reduced the number of caspase-3-IR neurons in the cerebellar lobules, but did not abolish them completely. The tendency towards a decrease in caspase-3-IR neurons was observed in all cerebellar lobules but taurine treatment did not alter the content of activated caspase-3 in individual immunoreactive neurons. The absence of graded effects suggests that taurine completely abolishes the activation of caspase-3 by ethanol in only one group of neurons, but has no effect in the others.

In the light of recent publications (Oomman et al. 2004; Rosado et al. 2006) caspase-3 activation is not always related to apoptosis. Furthermore, caspase-3 activation is a prominent feature but it is not an essential step in developmental ethanol-induced neuroapoptosis (Young et al. 2005). We therefore used an additional marker for apoptotic cells, DNA fragmentation using TUNEL staining. Only few cells with fragmented DNA labeled by TUNEL assay were found in the cerebellar lobules in control mice. Eight hours after ethanol administration an increase in the number of TUNEL-positive cells was observed throughout all vermis lobules. The picture of TUNEL staining in the cerebellum under ethanol effects was very similar to that of caspase-3-IR, confirming caspase-3 involvement in ethanol-induced developmental neuronal apoptosis. In the cerebella of ethanol-exposed taurine-treated mice, the number of apoptotic cells decreased significantly when compared to the ethanol-exposed mice. The number of caspase-3-positive neurons was greater than that of TUNEL-positive neurons, because caspase-3 activation preceeds eventual cell death.

We here show that taurine can rescue immature neurons from apoptosis induced by acute ethanol administration by suppressing activation of caspase-3. Which are the possible mechanisms of this? It might be supposed that taurine interferes with the Bax-dependent mitochondrial pathways of ethanol-induced apoptosis (Young et al. 2003), prevents apoptosome formation and caspase-9 activation (Takatani et al. 2004) and in this manner blocks caspase-3 activation and cell death. In this case however, the question remains why taurine blocks the apoptosis and caspase-3 activation only in some neurons, but does not completely abolish cell death. One possible answer is that ethanol induces apoptosis via several different mechanisms in different populations of neurons. Then, in those neurons in which ethanol induces apoptosis via caspase-9-dependent pathways taurine can have a neuroprotective effect and blocks caspase-9-dependent caspase-3 activation. On the other hand in other neurons in which caspase-3 activation is independent from caspase-9, taurine does not have an antiapoptotic effect. This assumption requires further experiments using specific blockers of the apoptotic proteins involving caspase-3 activation.

Acknowledgments The authors are deeply grateful to Mrs. Ulla M. Jukarainen for excellent technical assistance. This study was supported by the Finnish Center for International Mobility (grants EH-04-2615 and MK-05-3550) and the competitive research funding of the Pirkanmaa Hospital District (grants 9F070, 9G051, 9G203).

References

Aragón CM, Trudeau LE, Amit Z (1992) Effect ofTtaurine on Ethanol-Induced changes in open-field locomotor activity. Psychopharmacology 107:337–340

Boatright KM, Salvesen GS (2003) Mechanisms of caspase activation. Curr Opin Cell Biol 15:725–731

Budihardjo I, Oliver H, Lutter M, Luo X, Wang X (1999) Biochemical pathways of caspase activation during apoptosis. Annu Rev Cell Dev Biol 15:269–290

Clarren SK, Alvord AC, Sumi SM, Streissguth AP, Smith DW (1978) Brain malformations related to prenatal exposure to ethanol. J Pediat 92:64–67

Dachhour A, De Witte P (2000) Ethanol and amino acids in the central nervous system, Assessment of the pharmacological actions of acamprosate. Prog Neurobiol 60:343–362

Dikranian K, Ishimaru MJ, Tenkova T, Labruyere J, Qin YQ, Ikonomidou C, Olney JW (2001) Apoptosis in the in vivo mammalian forebrain. Neurobiol Dis 8:359–379

Dobbin J, Sands J (1979) The brain growth spurt in various mammalian species Early Human Dev 3:79–84

Earnshaw WC, Martins LM, Kaufmann SH (1999) Mammalian caspases: structure activation substrates and functions during apoptosis. Annu Rev Biochem 68:383–424

Famy C, Streissguth AP, Unis AS (1998) Mental illness in adults with fetal alcohol syndrome or fetal alcohol effects. Am J Psychiat 155:552–554

Ferko AP, Babyock E (1988) Effect of taurine on ethanol-induced sleep in mice genetically bred for differences in ethanol sensitivity. Pharmacol Biochem Behav 31:667–673

Ferko AP (1987) Ethanol-induced sleep time: interaction with taurine and a taurine antagonist. Pharmacol Biochem Behav 27:235–238

Ikonomidou C, Bittigau P, Ishimaru MJ, Wozniak DF, Koch C, Genz K, Price MT Stefovska V, Hörster F, Tenkova T, Dikranian K, Olney JW (2000) Ethanol-induced apoptotic neurodegeneration and fetal alcohol syndrome. Science 287:1056–1060

Kumral A, Tugyan K, Gonenc S, Genc K, Genc S, Sonmez U, Yilmaz O, Duman N, Uysal N, Ozkan H (2005) Protective effects of erythropoietin against ethanol-induced apoptotic neurodegeneration and oxidative stress in the developing C57BL/6 mouse brain. Brain Res Dev Brain Res 160:146–156

Migheli A, Atanasio A, Schiffer D (1995) Ultrastructural detection of DNA strand breaks in apoptotic neural cells by in situ endlabelling techniques. J Pathol 176:27–35

Nowoslawski L, Klocke BJ, Roth KA (2005) Molecular regulation of acute ethanol-induced neuron apoptosis. J Neuropathol Exp Neurol 64:490–497

Oja SS, Saransaari P (2007) Taurine. In: SS Oja A Schousboe, P Saransaari (eds) Handbook of Neurochemistry and Molecular Neurobiology vol. 6 Amino Acids and Peptides in the Nervous System Springer, New York, pp 156–206

Olive MF, Nannini MA, Ou CJ, Koenig HN, Hodge CW (2002) Effects of acute acamprosate and homotaurine on ethanol intake and ethanol-stimulated mesolimbic dopamine release. Eur J Pharmacol 437:55–61

Olney JW, Tenkova T, Dikranian K, Muglia LJ, Jermakowicz WJ, D'Sa C, Roth KA (2002a) Ethanol-induced caspase-3 activation in the in vivo developing mouse brain. Neurobiol Dis 9:205–219

Olney JW, Tenkova T, Dikranian K, Qin YQ, Labruyere J, Ikonomidou C (2002b) Ethanol-induced apoptotic neurodegeneration in the developing C57BL/6 mouse brain. Brain Res Dev Brain Res 133:115–126

Olney JW, Ishimaru MJ, Bittigau P, Ikonomidou C (2000) Ethanol-induced apoptotic neurodegeneration in the developing brain. Apoptosis 5:515–521

Oomman S, Finckbone V, Dertien J, Attridge J, Henne W, Medina M, Mansouri B, Singh H, Strahlendorf H, Strahlendorf J (2004) Active caspase-3 expression during postnatal development of rat cerebellum is not systematically or consistently associated with apoptosis. J Comp Neurol 476:154–173

Purring-Koch C, McLendon G (2000) Cytochrome c binding to Apaf-1: the effects of dATP and ionic strength Proc Natl Acad Sci USA 97:11928–11931

Rice D, Barone S Jr (2000) Critical periods of vulnerability for the developing nervous system: evidence from humans and animal models. Environ Health Perspect 108 Suppl 3:511–533

Rosado JA, Lopez JJ, Gomez-Arteta E, Redondo PC, Salido GM, Pariente JA (2006) Early caspase-3 activation independent of apoptosis is required for cellular function. J Cell Physiol 209: 142–152

Smolen AJ (1990) Image analytic techniques for quantification of immunohistochemical staining in the nervous system Methods Neurosci 3:208–229

Spadoni AD, McGee CL, Fryer SL, Riley EP (2007) Neuroimaging and fetal alcohol spectrum disorders Neurosci Biobehav Rev 31:239–245

Takatani T, Takahashi K, Uozumi Y, Shikata E, Yamamoto Y, Ito T, Matsuda T, Schaffer S W Fujio Y, Azuma J (2004) Taurine inhibits apoptosis by preventing formation of the Apaf-1/caspase-9 apoptosome Am J Physiol Cell Physiol 287:C949–C953

Taranukhin AG, Taranukhina EY, Saransaari P, Djatchkova IM, Pelto-Huikko M, Oja SS (2007) Taurine reduces caspase-8 and caspase-9 expression induced by ischemia in the mouse hypothalamic nuclei. Amino Acids DOI 101007/s00726-006-0405-z

Wijsman JH, Jonker RR, Keijzer R, van de Velde CJH, Cornelisse CJ, van Dierendonck JH (1993) A new method to detect apoptosis in paraffin sections: ISEL of fragmented DNA. J Histochem Cytochem 41:7–12

Wood KA, Dipasquale B, Youle RJ (1993) In situ labeling of granule cells for apoptosis-associated DNA fragmentation reveals different mechanisms of cell loss in developing cerebellum. Neuron 11:621–632

Young C, Klocke BJ, Tenkova T, Choi J, Labruyere J, Qin YQ, Holtzman DM, Roth KA, Olney JW (2003) Ethanol-induced neuronal apoptosis in vivo requires BAX in the develsoping mouse brain. Cell Death Differ 10:1148–1155

Young C, Roth KA, Klocke BJ, West T, Holtzman DM, Labruyere J, Qin YQ, Dikranian K, Olney JW (2005) Role of caspase-3 in ethanol-induced developmental neurodegeneration. Neurobiol Dis 20:608–614

Chapter 17
Mechanism of Neuroprotective Function of Taurine

Jang-Yen Wu, Heng Wu, Ying Jin, Jianning Wei, Di Sha, Howard Prentice, Hsin-Hsueh Lee, Chun-Hua Lin, Yi-Hsuan Lee, and Ling-Ling Yang

Abstract Taurine has potent protective function against glutamate-induced neuronal injury presumably through its function in regulation of intracellular free calcium level, $[Ca^{2+}]_i$. In this communication, we report that taurine exerts its protective function through one or more of the following mechanisms: 1. Inhibition of glutamate-induced calcium influx through L-, N- and P/Q-type voltage-gated calcium channels and NMDA receptor calcium channel; 2. Attenuation of glutamate-induced membrane depolarization; 3. Prevention of glutamate-induced apoptosis via preventing glutamate-mediated down-regulation of Bcl-2; 4. Prevention of cleavage of Bcl-2 by calpain. This action of taurine is due to its inhibition on glutamate induced calpain activation. Based on these observations, we propose that taurine protects neurons against glutamate-induced neurotoxicity in part, by preventing glutamate-induced membrane depolarization, elevation of $[Ca^{2+}]_i$, activation of calpain, reduction of Bcl-2 and apoptosis.

Abbreviations *VGCC*, voltage-gated calcium channel; *BME*, basal medium eagle; *div*, days in vitro; *EBSS*, Earle's balanced salt solution; *Glu*, Glutamate

17.1 Introduction

The physiological role of taurine (2-amino-ethanesulfonic acid), one of the most abundant amino acids in mammals, has received considerable attention since the reports that cats fed with a taurine deficient diet developed central retinal degeneration (Hayes et al. 1975) and cardiomyopathy (Pion et al. 1987). Taurine has been shown to be involved in many important physiological functions (Bianchi et al. 2006) e.g., as a trophic factor in the development of the CNS (Sturman 1993; Young 2004). It also serves in maintaining the structural integrity of the membrane (Moran et al. 1988), regulating calcium homeostasis (El Idrissi, 2006), as an osmolyte (Wade et al. 1988; Schaffer et al. 2000), as a neurotransmitter (Okamoto et al. 1983;

J.-Y. Wu (✉)
Department of Biomedical Science, Florida Atlantic University, Boca Raton, FL, USA

J. Azuma et al. (eds.), *Taurine 7*, Advances in Experimental Medicine and Biology 643, DOI 10.1007/978-0-387-75681-3_17, © Springer Science+Business Media, LLC 2009

Lin et al. 1985) and as a neuroprotector against Glu-induced neurotoxicity (Tang et al. 1996; Ward et al. 2006).

One important function of taurine is neuroprotection. Taurine has been shown to effectively prevent Glu-induced neuronal injury in cultured neurons (El Idrissi 2006; Chen et al. 2001; Tang et al. 1996). It is generally believed that the neuroprotective functions of taurine are due to its role in reducing the intracellular free Ca^{2+}, $[Ca^{2+}]_i$, concentration, and its anti-oxidative stress capacity (Chen et al. 2001; Schaffer et al. 2003). It was reported that taurine reduced Glu-induced elevation of $[Ca^{2+}]_i$ was through inhibition of Ca^{2+} influx via the reverse mode of the Na^+/Ca^{2+} exchanger (Chen et al. 2001). In addition to the Na^+/Ca^{2+} exchanger, we found taurine also inhibits various voltage-gated calcium channels (VGCC) including L-, N- and P/Q-type (Wu et al. 2005). We also found that taurine can protect neurons against Glu-induced apoptosis by preventing the activation of calpain, a calcium-dependent protease, after Glu stimulation.

17.2 Methods

17.2.1 Cell Culture and Experiments

Whole-brain primary neuronal cell cultures were prepared from fetal rat brains. Briefly, brains were dissected from 17-day fetal rats, kept in GME (BME supplemented with 20% fetal bovine serum). The brains were mechanically dissociated and cell suspension was centrifuged at 200 g for 3 minutes. The pellet obtained was resuspended in GME and plated in 24-well plates pre-coated with poly-d-lysine. For confocal microscopic study, 20 mm circular glass coverslips held by 35 mm dishes were used. Cells were allowed to adhere for hour in an incubator (37°C, 5% CO_2). The incubation medium was then replaced with serum-free BME.

Cultured neurons at 14 day in vitro (DIV) were used. For the Glu treatment group, cultured neurons were treated with 0.25 mM Glu for 10 minutes or 100 μM for 24 hours. For the taurine plus Glu group, cultured neurons were pre-incubated with 25 mM taurine for 10 minutes prior to the Glu stimulation.

17.2.2 Detection of Intracellular Calcium and ^{45}Ca Influx Study

Cultured neurons plated on 20 mm circular glass coverslips were used. After 14 DIV, the original media was replaced with Earle's balanced salt solution (EBSS) (116.4 mM NaCl, 5.4 mM KCl, 0.8 mM $MgSO_4$, 1.0 m M NaH_2CO_3, 1.8 m M $CaCl_2$, 5.6 m M glucose, pH 7.2). Cultures were equilibrated for 2 hours. Fluo-3 was added and further incubated for 30 minutes. After rinsing 3 times with EBSS to eliminate excessive dye, the coverslip was mounted on a customized perfusion chamber. The neurons were then treated with Glu alone, or in the presence of taurine and other compounds as described. The neurons were washed after each treatment.

The fluorescent dye was excited at 488 nm, and the changes in fluorescence were monitored using a computer connected to a confocal microscope.

The Ca^{2+} influx was assayed using ^{45}Ca. After 14 DIV, the cultures were washed twice with fresh EBSS, the media were replaced with fresh EBSS and incubated for 1 hour in the incubator. The media were changed to EBSS containing 55.5 KBq/ml $^{45}CaCl_2$, and the cultures were equilibrated for 10 minutes prior to a 10-minute exposure to 50 μM Glu. The uptake of ^{45}Ca was terminated by the removal of media, followed by 3 quick rinses with 1ml ice-cold 4 mM EGTA/0.9% NaCl. The cells were solubilized in 700 μl 0.3 M NaOH, and the radioactivity was determined using a liquid scintillation counter (the efficiency for ^{45}Ca is 95%).

17.2.3 Hoechst Staining for Nuclear Morphology

Nuclear morphology was detected by Hoechst 33342 staining 24 hours after initial treatment. Before dye application, cultures were washed twice with PBS and fixed for 10 minutes with 4% paraformaldehyde. After fixation, cultures were washed twice with PBS and exposed to Hoechst 33342 (2 mg/l) for 15 minutes. Cells with bright blue fragmented nuclei showing condensation of chromatin were identified as apoptotic cells.

17.2.4 Immunoblotting and DNA Electrophoresis

For immunoblotting, equal amount of protein samples were first separated on an SDS-PAGE and then blotted onto nitrocellulose membranes followed by a 2-hour blocking with 3% non-fat milk at room temperature. After a 2-hour incubation with the primary antibody and a 2-hour incubation with the secondary antibody at room temperature, the protein immuno-complex was visualized using ECL^{TM} detection reagents.

For DNA electrophoresis, 7 μg of genomic DNA was loaded onto 1.8% agarose gel containing ethidium bromide and run at 80 volts until the dye reached half of the gel. The DNA was then visualized using UV light.

17.3 Results

17.3.1 Effect of Taurine on Glu-Induced Neurotoxicity and Intracellular Calcium, $[Ca^{2+}]_i$, Elevation

Primary neuronal cell cultures of 14 DIV were treated with 250μM Glu in the presence or absence of 25 mM taurine. The neuronal damage was evaluated by LDH releasing assay. Glu stimulation caused cell damage in primary neuronal culture, as indicated by the increased LDH release, and this increased LDH release was largely inhibited by taurine (Fig. 17.1a).

Intracellular calcium level was determined by confocal microscopy. As shown in Fig. 17.1b, compared with the control (panel A) which shows the basal level of intracellular free calcium (blue color), the application of 50 μM Glu markedly increased intracellular calcium level (panel B, bright red color). In the presence of 25 mM taurine, Glu failed to induce an increase in $[Ca^{2+}]_i$ (compare panel H with B), indicating that taurine blocks the Glu-induced $[Ca^{2+}]_i$ accumulation. The cultured neurons used remained viable as indicated by a sharp increase in $[Ca^{2+}]_i$ when they were treated with Glu again at the end of the experiment (panel J).

Furthermore, extracellular calcium influx into neurons was measured using ^{45}Ca. As shown in Fig. 17.1c, 50 μM Glu stimulation induced ^{45}Ca accumulation inside the neurons, and taurine, at concentrations of 25 mM and 5 mM, inhibited Glu-induced intracellular calcium accumulation.

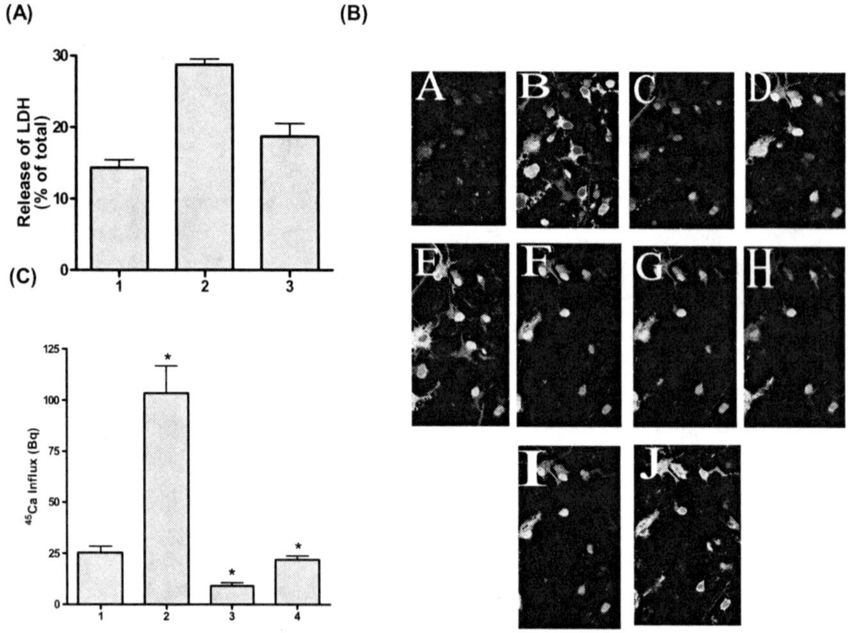

Fig. 17.1 (**a**) Effect of taurine on Glu-induced LDH release. 1. Control; 2. Glu; 3. Glu plus taurine. (**b**) Effect of taurine on Glu induced $[Ca^{2+}]_i$ accumulation (confocal). A. Baseline; B & J. Glu; C, F & I. After washing; D. Nifedipine; E. Nifedipine plus Glu; G. Taurine; H. Taurine plus Glu. Color coding indicates $[Ca^{2+}]_i$, red being the highest and blue the lowest C. Effect of taurine on Glu induced [*Ca*2+]i accumulation ($^{45}Ca^{2+}$ influx). 1. Control; 2. Glu; 3. Glu plus 25mM taurine; 4. Glu plus 5mM taurine

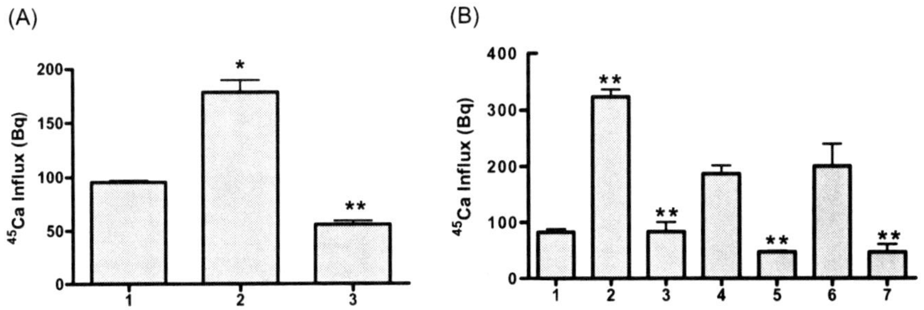

Fig. 17.2 (**a**) Effect of taurine on L-type VGCC-mediated calcium influx. 1. Control; 2. Bay K 8644; 3. Bay K 8644 plus taurine. (**b**) Effect of taurine on P/Q-, N-type VGCC-mediated calcium influx. 1. Control; 2. Glu; 3. Glu plus taurine; 4. P/Q-type VGCC (Glu plus channel blockers mixture); 5. Same as 4 plus taurine; 6. P/Q-type VGCC (Glu plus channel blockers mixture); 7. Same as 6 plus taurine

17.3.2 Effect of Taurine on Calcium Influx Through Various Voltage-Gated Calcium Channels (VGCCs)

Since extracellular calcium influx was confirmed, the effect of taurine on major types (including L-, P/Q- and N-type) of VGCCs was determined. In confocal microscopic study, nifedipine, a specific L-type VGCC blocker, inhibited Glu-induced intracellular calcium accumulation (Fig. 17.1b, compare panel E and B). As shown in Fig. 17.2a, Bay K 8644, a specific L-type VGCC activator, increased the $[Ca^{2+}]_i$ by 90%. However, when neurons were pretreated with taurine, the Bay K 8644-induced increase in $[Ca^{2+}]_i$ was completely inhibited suggesting that taurine blocks the calcium influx through L-type VGCC.

To study the effect of taurine on P/Q- and N-type VGCCs, the Glu-induced ^{45}Ca influx was conducted in the presence of a mixture of calcium channel blockers to inhibit all the major calcium channels except for the P/Q- or N-type VGCC. For example, P/Q-type VGCC was studied in the presence of a combination of nifedipine, N-conotoxin GVI A, 2,4-dichlorobenzamil, MK801 which blocks L-, N-type VGCCs, reverse mode of Na^+/Ca^{2+} exchanger and NMDA receptor, respectively. As shown in Fig. 17.2b, P/Q-type (column 4) and N-type (column 6) VGCCs contributed, to a similar extent, to Glu-induced calcium influx. Taurine blocked the calcium influx through both P/Q-type (column 5) and N-type (column 7) VGCCs.

17.3.3 Effect of Taurine on Calcium Influx Through NMDA Receptor Calcium Channel

Besides VGCCs, the NMDA receptor calcium channel represents another major calcium influx pathway. When the experiments were performed with regular EBSS (Mg^{2+} is present), calcium influx through the NMDA receptor calcium channel was completely blocked by taurine (Fig. 17.3 a column 3). When neurons were treated in

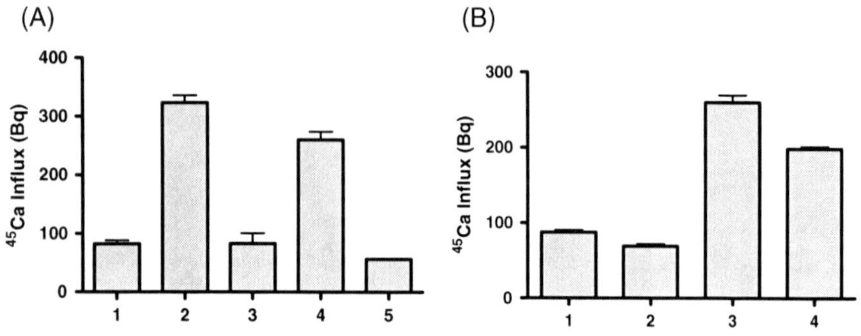

Fig. 17.3 (**a**) Effect of taurine on NMDA-mediated calcium influx in EBSS. 1. Control; 2. Glu; 3. Glu plus taurine; 4. NMDA calcium channel (Glu plus channel blockers mixture); 5. Same as 4 plus taurine. (**b**) Effect of taurine on NMDA-mediated calcium influx in Mg^{2+}-freeEBSS. 1. Control; 2. Taurine; 3. NMDA; 4. NMDA plus taurine

$Mg^{2+} - free$ EBSS media containing CNQX (a specific antagonist for AMPA/KA receptors), NMDA induced calcium influx, however, taurine failed to block the calcium influx through NMDA receptors (Fig. 17.3b, column 4). These results indicate that taurine does not act directly on NMDA receptors.

17.3.4 Effect of Taurine, Calpastatin on the Glu-Induced Nucleus Condensation

To study the effect of taurine on Glu-induced apoptosis, neuronal cell cultures at 12–14 DIV were treated with 50 μM Glu for 24 hours. Hoechst 33342 was used to show nuclear morphology. As shown in Fig. 17.4, compared with the control group (panel A), a 20-hour treatment of 50μM Glu induced nuclear condensation

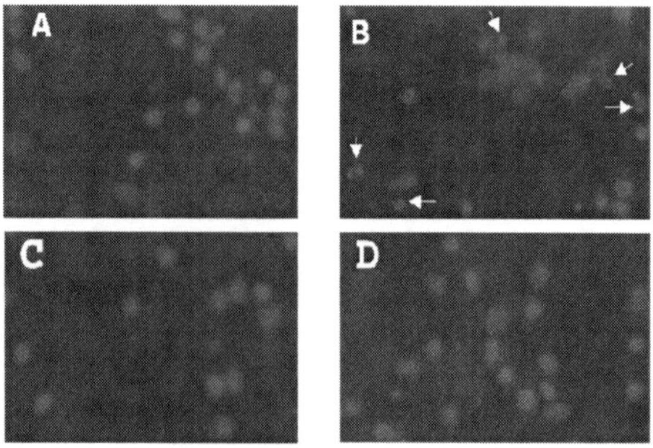

Fig. 17.4 Effect of taurine on the Glu-induced apoptosis: Hoechst 33342 staining. (**A**) Control; (**B**) Glu; (**C**) Glu plus taurine. (**D**) Glu plus calpastatin. Arrows indicate the condensed nuclei

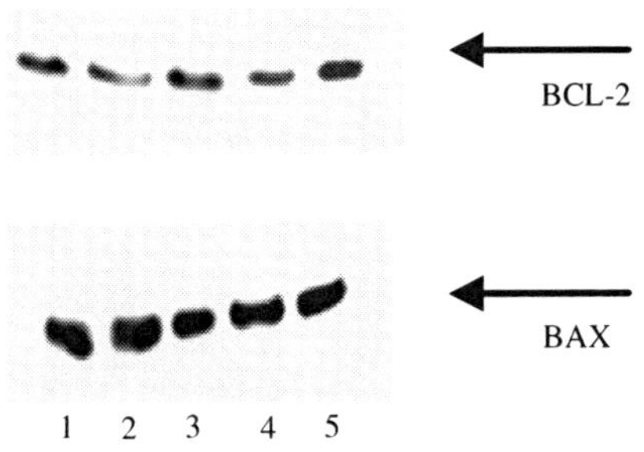

Fig. 17.5 Effect of taurine on the Glu-induced change in BCL-2 and BAX. 1. Control; 2. Glu; 3. Glu plus taurine; 4. Glu; 5.Glu plus calpastatin

(panel B), indicating that Glu stimulation induced apoptosis in primary neuronal cultures. The Glu-induced nuclear condensation was inhibited by calpastatin (panel D), a specific calpain inhibitor, suggesting the involvement of calpain, a calcium-dependent cysteine protease, in the Glu-induced apoptosis. The Glu-induced nuclear condensation was also prevented by the presence of taurine (panel C), suggesting that taurine can protect neurons against the Glu-induced apoptosis.

17.3.5 Effect of Taurine and Calpastatin on the Glu-Induced Change of BCL-2 and BAX

The BCL-2 family is a family of proteins involved in regulating apoptosis. Among these, BCL-2 antagonizes apoptosis, while BAX promotes the occurrence of apoptosis, and the balance between BCL-2 and BAX regulates apoptosis. Based on the immuno-blotting results (Fig. 17.5), the level BCL-2 was decreased by the Glu treatment (upper panel, lane 2 and 4). Interestingly, in the presence of either taurine or calpastatin, Glu-induced decrease of BCL-2 level was not observed (upper panel lane 3 and 5) indicating that the Glu-induced down-regulation of BCL-2 was prevented by taurine or calpastatin.

Contrary to BCL-2, the level of BAX was not affected by treatment of all 3 groups: Glu alone, Glu plus taurine, and Glu plus calpastatin (Fig. 17.5, lower panel).

17.3.6 Effect of Taurine and Calpastatin on the Glu-Induced Calpain Activation

It was reported that BCL-2 is a substrate of calpain, and we showed that calpastatin inhibited Glu-induced nuclear condensation in the primary neuronal cell culture. The effect of Glu on calpain activity was examined by measuring the cleavage of spectrin, a specific endogenous calpain substrate. The ratio of the cleaved to the total

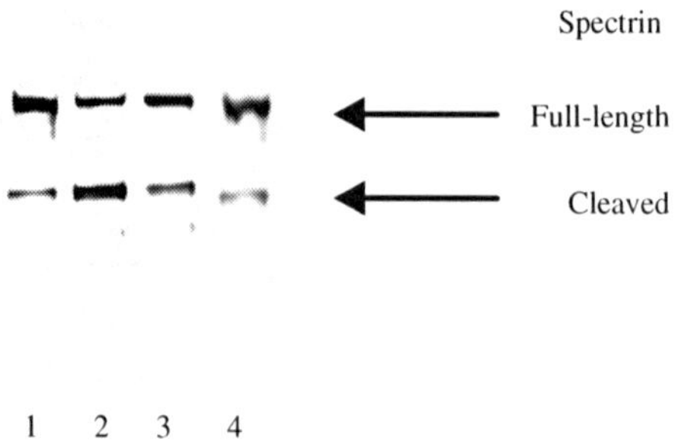

Fig. 17.6 Effect of taurine on the Glu-induced cleavage of spectrin. 1. Control; 2. Glu; 3. Glu plus taurine; 4. Glu plus calpastatin

spectrin increased by 2.5 folds (Fig. 17.6, lane 2) after Glu treatment compared to the control group, indicating the activation of calpain by Glu stimulation. Furthermore, the Glu-induced activation of calpain was largely inhibited by taurine (lane 3) and calpastatin (lane 4).

17.4 Discussion

Glu is the major excitatory amino acid neurotransmitter in the mammalian central nervous system. Glu plays an important role in neuronal differentiation, migration and survival in the developing brain (Ikonomidou et al. 1999; Behar et al. 1999; Hirai et al. 1999; Wu et al. 1996; Komuro and Rakic 1993) as well as in synaptic maintenance and plasticity (Wu et al. 1996). However, excessive extracellular Glu can cause cell damage, even cell death, by increasing intracellular free calcium. It has been well established that taurine protects neurons against Glu-induced neuronal damage in the primary neuronal culture. The neuroprotective effect of taurine is attributed to its functions in maintaining intracellular calcium homeostasis, membrane integrity (Pasantes-Morales and Cruz 1984; 1985) and as an antioxidant.

In our studies, neurons survived only in the presence of a high concentration of taurine, but not with a low concentration of taurine that failed to prevent the Glu-induced calcium influx (data not shown), suggesting an important role of taurine-mediated calcium homeostasis. We further found that taurine blocked calcium influx through L-, P/Q-, and N-type VGCCs, whose opening depends on the membrane depolarization. One interesting finding is that taurine blocked calcium influx through the NMDA receptor calcium channel only when regular EBSS with Mg^{2+} was used, but not with the use of Mg^{2+}-free EBSS. It is known that NMDA receptor is ligand- and membrane-potential dependent when $Mg^{2+}$is trapped in its channel pore and is only ligand-dependent when Mg^{2+} is absent. This finding implies that taurine does not act directly on the NMDA receptor, but via its effect on membrane potential. Interestingly, a report by Chen et al. (2001) that taurine inhibits calcium influx through the reverse mode of Na^+/Ca^{2+} exchanger (whose activity is membrane potential-dependent) suggests that the neuroprotection of taurine is mediated by its function of preventing Glu-induced membrane depolarization as we also demonstrated previously (Wu et al. 2005). Here, we propose that taurine protects neurons against Glu excitotoxicity by preventing Glu-induced membrane depolarization, probably through its effect in opening of chloride channels and, therefore, preventing the Glu-induced increase in calcium influx and other downstream events.

The exposure to excessive Glu causes neuronal damage including both necrosis and apoptosis depending on the recovery of the mitochondrial function (Ankarcrona et al. 1995). It is reported that the Glu-induced apoptosis accounts for the neuronal degeneration in some neurological diseases such as Alzheimer's disease (Masliah et al. 1998) and Huntington's disease (Portera-Cailliau et al. 1995). In our study, we also found that taurine prevented Glu-induced DNA fragmentation and nuclear condensation, indicating that taurine can prevent the Glu-induced apoptosis. We also found that BCL-2 protein (an anti-apoptotic protein) level was down-regulated by Glu treatment, and this down-regulation was prevented in the presence of taurine. On the other hand, Glu stimulation caused no change in BAX protein (a pro-apoptotic protein) level. BAX can form homodimers, which attack mitochondria and lead to the release of cytochrome C, which in turn initiates the caspase cascade and apoptosis (Henshall et al. 2002). Since BCL-2 forms heterodimers with BAX to prevent mitochondria from attack by homodimers of BAX, the restoration of the balance of BCL-2/BAX by taurine protects neurons from apoptosis. Furthermore, we found that taurine treatment prevented Glu-induced cleavage of spectrin, a specific

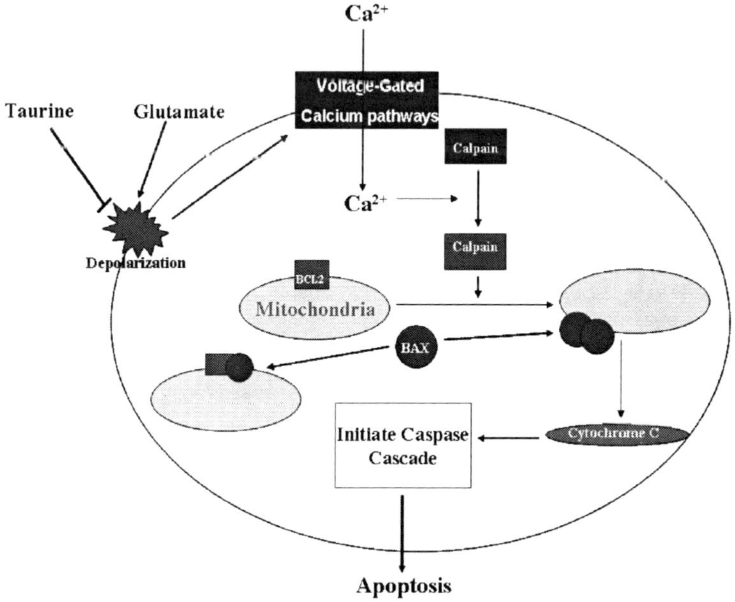

Fig. 17.7 A model depicting the pathway that taurine exerts its function against Glu-induced apoptosis (see discussion)

endogenous substrate of calpain, suggesting that taurine inhibits Glu-induced activation of calpain, a calcium-dependent cysteine protease. It was reported that BCL-2 is a substrate of calpain, therefore, taurine restores the balance of BCL-2/BAX by preventing Glu-induced activation of calpain. This was confirmed by the experiment with calpastatin, a specific calpain inhibitor, which also prevented the Glu-induced nuclear condensation and down-regulation of BCL-2. Here we propose the sequence of events leading from Glu stimulation to apoptosis and the mode of action of taurine in preventing Glu-induced apoptosis are as follows (Fig. 17.7): (1) Glu stimulation induces $[Ca^{2+}]_i$ elevation which in turn activates calpain; (2) The activated calpain cleaves BCL-2; (3) Glu stimulation also induces a conformational change of BAX resulting in formation and activation of homodimers of BAX; (4) With decreased protection from BCL-2, the homodimers of BAX target the mitochondria and lead to the release of cytochrome C; (5) Released cytochrome C activates Apaf-1 which in turn activates a downstream caspase cascade leading to apoptosis; (6) The anti-apoptotic function of taurine is due to its capability of preventing Glu-induced $[Ca^{2+}]_i$ elevation by preventing Glu-induced depolarization resulting in the blockade of all the downstream events.

References

Ankarcrona M, Dypbukt JM, Bonfoco E, Zhivotovsky B, Orrenius S, Lipton SA, Nicotera P (1995) Glutamate-induced neuronal death: a succession of necrosis or apoptosis depending on mitochondrial function. Neuron 15:961–973

Behar T N, Scott CA, Greene CL, Wen X, Smith SV, Maric D, Liu QY, Colton CA, Barker JL (1999) Glutamate acting at NMDA receptors stimulates embryonic cortical neuronal migration. J. Neurosci 19:4449–4461

Bianchi L, Colivicchi MA, Ballini C, Fattori M, Venturi C, Giovannini MG, Healy J, Tipton KF, Della Corte L (2006) Taurine, taurine analogues, and taurine functions: overview. Adv Exp Med Biol 583:443–448

Chen WQ, Nguyen M, Carr J, Lee YJ, Jin H, Foos T, Hsu CC, Davis KM, Schloss JV, Wu J-Y (2001) Role of taurine in regulation of intracellular calcium level and neuroprotective function in cultured neurons. J Neurosci Res 66:612–619

El Idrissi A (2006) Taurine increases mitochondrial buffering of calcium: role in neuroprotection. Amino Acids [Epub ahead of print]

Hayes KC, Carey RE, Schmidt SY (1975) Retinal degeneration associated with taurine deficiency in the cat. Science 188:949–951

Henshall DC, Araki T, Schindler CK, Lan JQ, Tiekoter KL, Taki W, Simon RP (2002), Activation of Bcl-2-associated death protein and counter-response of Akt within cell populations during seizure-induced neuronal death. J Neurosci 22:8458–8465

Hirai K, Yoshioka H, Kihara M, Hasegawa K, Sakamoto T, Sawada T, Fushiki S (1999) Inhibiting neuronal migration by blocking NMDA receptors in the embryonic rat cerebral cortex: a tissue culture study. Brain Res Dev Brain Res 114:63–67

Ikonomidou C, Bosch F, Miksa M, Bittigau P, Vockler P, Dikranian K, Tenkova TI, Stefovska V, Turski L, Olney JW (1999) Blockade of NMDA receptors and apoptotic neurodegeneration in the developing brain. Science 283:70–74

Komuro H, Rakic P (1993) Modulation of neuronal migration by NMDA receptors. Science 260:95–97

Lin CT, Su YYT, Song GX, Wu JY (1985) Is taurine a neurotransmitter in rabbit retina? Brain Res 337:293–298
Masliah E, Mallory M, Alford M, Tanaka S, Hansen LA (1998) Caspase dependent DNA fragmentation might be associated with excitotoxicity in Alzheimer disease. J Neuropathol Exp Neurol 57:1041–1052
Moran J, Salazar P, Pasantes-Morales H (1988) Effect of tocopherol and taurine on membrane fluidity of retinal rod outer segments. Exp Eye Res 45:769–776
Okamoto K, Kimura H, Sakai Y (1983) Evidence for taurine as an inhibitory neurotransmitter in cerebellar stellate interneurons: Selective antagonism by TAG (6-aminomethyl-3-methyl-4H, 1,2,4-benzothiadiazine-1,1-dioxide). Brain Res 265:163–168
Pasantes-Morales H, Cruz C (1984) Protective effect of taurine and zinc on peroxidation-induced damage in photoreceptor outer segments. J Neurosci Res; 11(3):303–11
Pasantes-Morales H, Cruz C (1985) Taurine and hypotaurine inhibit light-induced lipid peroxidation and protect rod outer segment structure. Brain Res 330:154–157
Pion PD, Kittleson MD, Rogers QR, Morris JG (1987) Myocardial failure in cats associated with low plasma taurine: A reversible cardiomyopathy. Science 237:764–768
Portera-Cailliau C, Hedreen JC, Price DL, Koliatsos VE (1995) Evidence for apoptotic cell death in Huntington disease and excitotoxic animal models. J Neurosci 15:3775–3787
Schaffer S, Takahashi K, Azuma J, (2000) Role of osmoregulation in the actions of taurine. Amino Acids 19:527–546
Schaffer S, Azuma J, Takahashi K, Mozaffari M (2003) Why is taurine cytoprotective? In: Lombardini JB, Schaffer S W and Azuma J (eds) Taurine 5, Kluwer Academic/Plenum Publishers, pp 307–321
Sturman JA (1993) Taurine in development. Physiol Rev 73:119-147
Tang XW, Deupree DL, Sun Y, Wu JY (1996) Biphasic effect of taurine on excitatory amino acid-induced neurotoxicity. In: Huxtable RJ, Azuma J, Nakagawa M, Kuriyama K, Bala A (eds) Taurine: Basic and Clinical Aspects, Plenum Publishing Co, pp 499–506
Wade JV, Olson JP, Samson FE, Nelson SR, Pazdernik TL (1988) A possible role for taurine in osmoregulation within the brain. J Neurochem 51:740–745
Ward R, Cirkovic-Vellichovia T, Ledeque F, Tirizitis G, Dubars G, Datla K, Dexter D, Heushling P, Crichton R (2006) Neuroprotection by taurine and taurine analogues. Adv Exp Med Biol 583:299–306
Wu G, Malinow R, Cline HT (1996) Maturation of central glutamatergic synapse. Sciece 274: 972–976
Wu H, Jin Y, Wei J, Jin H, Sha D, Wu JY (2005) Mode of action of taurine as a neuroprotector. Brain Res 1038:123–131
Young TL, Cepko CL (2004) A role for ligand-gated ion channels in rod photoreceptor development. Neuron 41:867–879

Chapter 18
Taurine and Guanidinoethanesulfonic Acid (GES) Differentially Affects the Expression and Phosphorylation of Cellular Proteins in C6 Glial Cells

Hemanta K. Sarkar, Thanh T. Tran, and Rao Papineni

Abstract Effects of taurine and guanidinoethanesulfonic acid (GES), a taurine transport inhibitor, on the expression and phosphorylation of cellular proteins in C6 glial cells were examined using two-dimensional (2-D) gel electrophoresis and 2-D immunoblots. 2-D gels stained with Coomassie Blue or SYPRO Ruby showed differential distribution patterns of cellular proteins in the control, taurine-supplemented and GES-supplemented cells. 2-D immunoblot analysis using the anti-phosphotyrosine antibody recognized only few immuno-reactive proteins in all three samples, and their distribution patterns were different. On the other hand, 2-D immunoblot analysis using the anti-phosphothreonine antibody recognized many immuno-reactive proteins with distinctly different distribution patterns in the control, taurine-supplemented and GES-supplemented cells. In GES-supplemented cells, the relative number of the anti-phosphotyrosine immuno-reactive protein spots increased modestly, whereas the relative number of the anti-phosphothreonine immuno-reactive protein spots decreased markedly than those in the control and taurine-supplemented cells.

Abbreviations *GES*, guanidinoethanesulfonic acid; *2-D*, two-dimensional; *IPG*, immobilized pH gradient; *IEF*, isoelectric focussing

18.1 Introduction

The beta-amino acid taurine is highly abundant in its free form in the heart, brain, retina, skeletal muscle, kidney, and leukocytes. In mammals, the plasma taurine concentration is 20–50 μM, whereas the intracellular taurine concentration can reach 10–50 mM depending on the tissue/cell (for reviews, see Huxtable 1992;

H.K. Sarkar (✉)
BBTech Inc., 11 Back River Drive, Dartmouth, MA, USA

J. Azuma et al. (eds.), *Taurine 7*, Advances in Experimental Medicine and Biology 643,
DOI 10.1007/978-0-387-75681-3_18, © Springer Science+Business Media, LLC 2009

Chesney et al. 1998). A membrane bound taurine transporter is primarily responsible for maintaining the intracellular taurine concentration (Huxtable 1992).

Taurine is considered as conditionally essential amino acid in preterm infants and neonates, where it is thought to play a vital role in growth and development of brain (for reviews, see Huxtable 1992; Chesney et al. 1998; Redmond et al. 1998; Lima et al. 2001). Some of the other proposed roles of taurine include membrane stabilization, bile salt formation, osmoregulation, calcium homeostasis maintenance, antioxidation, glycolysis and glycogenesis stimulation, and in vision (for reviews, see Huxtable 1992; Chesney et al. 1998; Redmond et al. 1998;).

The physiological role of taurine in adults, and consequently its significance in pathologic disease states, is however emerging. Several recent studies suggested that taurine might have potential immunoregulatory properties(Redmond et al. 1998). and therapeutic applications (Schulle-Levis and Park 2003; Birdsall 1998). Thus, taurine supplementation was shown to ameliorate TPN-associated cholestasis (Muhlfeld et al. 2003), protect against ischemia reperfusion injury (Kingston et al. 2004), inhibit cell apoptosis (Maher et al. 2005), attenuate bleomycin-induced pulmonary fibrosis (Gurujeyalakshmi et al. 2000), attenuate hypertrophy induced by angiotensin II in cultured neonatal rat cardiac myocytes (Azuma et al. 2000), enhance cholesterol degradation in rats (Yokogoshi and Oda 2002), protect against early and late skeletal muscle dysfunction secondary to ischaemia reperfusion injury (McLaughlin et al. 2000), reverse endothelial cell dysfunction(Moloney et al. 2003), and improve memory in mice treated with a variety of amnestic agents (Vohra and Hui 2001). More importantly, supplemental taurine was shown to act as a potent neuroprotectant against glutamate excitotoxicity, cerebellar ischaemia, oxidative-stress and build-up of several toxins (Hilgier et al. 2003; Rivas-Arancibia et al. 2001). Taurine depletion studies, where intracellular taurine was depleted using the taurine transport inhibitor GES or beta-alanine, further added taurine's importance in cellular physiology. Thus, taurine depletion affected neuronal migration (Maar et al. 1995), caused degenerative changes in the rat retina (Lake and Malik 1987), and altered vascular reactivity in rats (Abebe and Mozaffari 2003). Consistent with these studies, disruption of the taurine transporter gene, product of which is primarily responsible for maintaining the intracellular taurine concentration, led to the retinal degeneration (Heller-Stilb et al. 2002), and also resulted in severe skeletal muscle impairment (Warskulat et al. 2004) in mice.

Some of taurine's action is thought to be mediated by altering the calcium homeostasis of the cell (El Idrissi and Trenkner 2003). However, taurine-associated numerous physiologic functions and pharmacologic actions cannot be explained by a calcium-based mechanism alone, and thus, warrants exploring the involvement of other cellular pathways, such as alteration in protein expression and/or phosphorylation. In this communication, we examined how taurine and GES affected the expression and phosphorylation patterns of cellular proteins in rat C6 glial cells. Here, using a combination of two-dimensional (2-D) gel electrophoresis and immunoblot assays, we demonstrated that taurine and GES differentially modulated the expression and phosphorylation patterns of cellular proteins in the C6 cells.

18.2 Methods

18.2.1 Cell Culture

C6 glial cells were grown in DMEM supplemented with 4 mM L-glutamine, 4.5 g/L glucose and 10% fetal serum albumin at 37°C in growth chamber in the presence of 5 mM of taurine (taurine-supplemented) or 5 mM GES (GES-supplemented) for 48 h.

18.2.2 Preparation of Total Cellular Proteins for 2-D Gel Electrophoresis

Cells were lysed in 350 μl of lysis buffer (8M urea, 2M thiourea, 4% CHAPS, 0.2% BioLyte 3/10, 65 mM DTT, 40 mM Tris) freshly supplemented with a protease inhibitor mixture containing 2.5 mM sodium pyrophosphate, 1 mM sodium vanadate, 1% protease inhibitor cocktail (Boehringer-Mannheim, Indianapolis, IN) with gentle shaking for 1 h at room temperature.

18.2.3 Two-Dimensional (2-D) Gel Electrophoresis

For 2-D gel electrophoresis, protein samples were first subjected to isoelectric focusing (IEF) and then the electro-focused proteins were separated on a gradient SDS-PAGE. A BioRad 2-D starter kit (BioRad, Ventura, CA) was used for the 2-D gel electrophoresis. IEF was carried out using the PROTEAN IEF Cells (BioRad, CA) following the manufacturer's protocol. Samples ($100 - -150\mu g$ of protein) were loaded onto immobilized pH gradient (IPG) strips (7 cm Ready Strip, pH 3–10, BioRad, CA) by active-rehydration protocol [rehydration buffer:8M urea, 0.5% CHAPS, 10 mM DTT, 0.2% (w/v) Bio-Lytes 3/10, 0.001% bromophenol blue]. Focusing was carried out using a ramping protocol in which the voltage was gradually increased up to 5000 V. After the IEF was completed, and prior to running the second dimension, strips were first equilibrated for 15-min in equilibration buffer I (6M urea, 2% SDS, 0.375 M Tri-HCl/pH 8.8, 20% glycerol, 135 mM DTT) and then blocked for another 15-min in equilibration buffer II (6 M urea, 2% SDS, 0.375 M Tris-HCl/pH 8.8, 20% glycerol, 135 mM iodoacetamide) to reduce the -S-S- bonds. IPG strips were then placed on the top of pre-cast gradient (4–20% polyacrylamide) SDS-PAGE gels (BioRad, CA) and the second dimension electrophoresis was performed at 80 mA for 7 h using Mini-PROTEAN 3 (BioRad, CA).

18.2.4 Gel Staining

Gels were stained with either Coomassie Brilliant Blue G-250 (Sigma Chemicals, St. Louise, MO) or SYPRO Ruby protein stain (BioRad, Hercules, CA). Coomassie

Blue stained gels were visualized on a white light box and the Ruby stained gels were visualized on a blue light transilluminator. Gels were photographed using the Alpha Imager 2000 (Alpha Innetech Co., San Leandro, CA).

18.2.5 Immunoblot Assay

Immunoblot assays were performed essentially as described (Tran et al. 2000) using the commercially available polyclonal anti-phosphotyrosine (Cell Signaling Technology, MA), anti-phosphoserine (Zymed Laboratory, CA), and anti-phosphothreonine (Zymed Laboratory) antibodies. Total cellular proteins were separated on polyacrylamide gels (1-D or 2-D) and then electrophoretically transferred onto the Hybond-P polyvinylidene difluoride (PDVF) membranes. The membranes were blocked with 3% BSA and 2% non-fat dry milk, and then incubated with 1:500 to 1:5,000 dilution of one of the polyclonal anti-phosphoamino acid antibodies for 1–3 h at room temperature. Next, the blots were washed several times with TBST (137 mM NaCl, 2.6 mM KCl, 25 mM Tris-HCl, pH 7.4, and 0.1% Tween 20) and then incubated for 1-2 h at room temperature with a 1:10,000 dilution of the horseradish peroxidase-conjugated goat anti-rabbit IgG (Pierce, Rockford, IL). The bound antibody-antigen complex was subsequently detected using the ECL Plus Chemiluminescent Detection kit (Amersham Pharmacia Biotech, Arlington Height's, IL).

18.3 Results

18.3.1 Taurine and GES Differentially Affect Expression of Cellular Proteins

Visual inspections of stained 2-D gels of cell extracts prepared from the control (cells grown in normal growth medium), taurine-supplemented and GES-supplemented cells revealed modest but easily detectable differences in protein expression patterns. Figure 18.1 shows the representative results of 2-D gel electrophoresis from two experimental sets of cells, one set was stained with Coomassie Brilliant Blue G-250 and the other set was stained with SYPRO Ruby. Both staining procedures revealed that the distribution pattern of cellular proteins on the 2-D gels of samples prepared from the control cells differed from those of the taurine-supplemented and GES-supplemented cells. For example, the protein spot, indicated by an arrow in Fig. 18.1 (Fig. 18.1, Right panel, CB), seems to be expressed under all three conditions, whereas expression of several other proteins were decreased in taurine- and GES-supplemented cells (compare the boxed rectangle, circle, hexagon and diamond areas of the three top panels in Fig. 18.1, CB). Expressions of a number of proteins were highest in GES-supplemented cells, modest in taurine-supplemented cells and lowest in control cells (compare areas boxed with hexagon

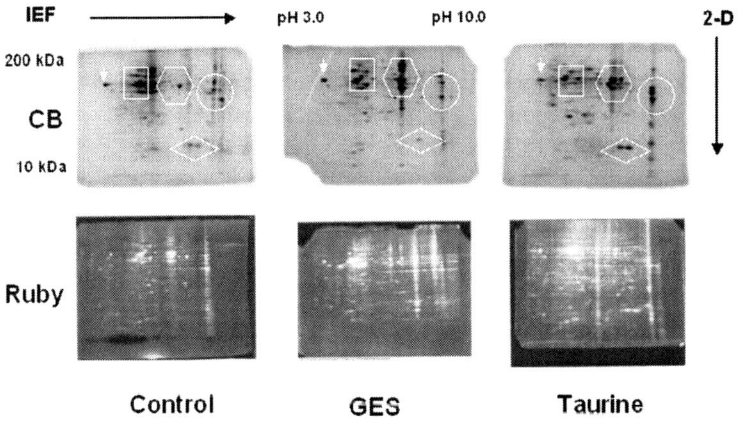

Fig. 18.1 Total proteins from taurine-adapted, GES-treated and control cells were analyzed by 2-D gel electrophoresis as described in Methods. Arrows indicate directions of IEF (using Biolyte 3/10) and 2-D electrophoresis (4–20% precast polyacrylamide gels). Subsequently, gels were stained with Coomasie Blue (CB) and SYPRO Ruby (Ruby). Photographs of stained gels were taken using Alpha Imager 2000. Approximate positions of the 200 kDa and 10 kDa MW protein markers are also indicated

in top three panels in Fig. 18.1, CB). Additionally, expression of two proteins in taurine-supplemented cells decreased modestly in GES-supplemented cells, which were absent in control cells (compare areas boxed with diamond in Fig. 18.1, CB, bottom right).

Visual inspection of the SYPRO Ruby-stained gels (Fig. 18.1, bottom panels) also showed differences in protein expression patterns among the control, taurine-supplemented and GES-supplemented cells. For example, two protein spots found in the Coomasie Blue stained 2-D gels of taurine-supplemented cells (Fig. 18.1, top right panel, boxed with diamond) also appeared in the SYPRO Ruby stained 2-D gels of taurine-supplemented cell sample (Fig. 18.1, Ruby, near bottom right). Apparent decrease/increase in the relative intensity of a number of other protein spots was also observed by Ruby-staining procedure (compare three bottom panels). Taken together, these results suggested that taurine- and GES-supplementation differentially affected the expression of cellular proteins in C6 cells. Similar changes in the distribution patterns of proteins spots were also found between the control and taurine-supplemented neuroblastoma NG-108 cells, which are widely used as a model for neurons (results not shown).

18.3.2 *Taurine and GES Altered Phosphorylation of Cellular Proteins*

Next, we determined the effect of taurine- and GES-supplementation on phosphorylation of cellular proteins. Immunoblots of total proteins extracts prepared from the control, taurine-supplemented, and GES-supplemented cells and separated on a 10% SDS-PAGE revealed modest and minor differences in immuno-reactive protein bands when assayed with the anti-phosphothreonine and anti-phosphotyrosine antibodies, respectively, and no apparent difference when assayed with the anti-phosphoserine antibody (results not shown). Therefore, we did not use the anti-phosphoserine antibody for the 2-D immunoblot analysis described below.

As shown in Fig. 18.2, both anti-threonine and anti-tyrosine antibodies produced distinctly different 2-D immunoblots. Only a few anti-phosphotyrosine immuno-reactive protein spots were visible on the 2-D immunoblots of both control and taurine-supplemented C6 cells (Fig. 18.2, top panels). However, the number of anti-phosphotyrosine immuno-reactive protein spots increased in GES-supplemented cells (Fig. 18.2, top panels). Figure 18.2 also shows the immunoblots using the anti-phosphothreonine antibody (bottom panels).

As shown, relatively large number of anti-phosphothreonine immuno-reactive protein spots appeared in the control and taurine-supplemented cells, although there were apparent differences in their distribution patterns (compare the bottom left and middle panels). Visible inspection of the 2-D immunoblots of the control and taurine-supplemented cells clearly identified several anti-phosphothreonine immuno-reactive spots whose intensities increased significantly upon taurine-supplementation. In stark contrast, GES-supplementation led to a dramatic decrease in the number of the anti-phosphothreonine immuno-reactive proteins spots (compare the bottom right panel with the bottom left and middle panels). Such a dramatic decrease in the relative numbers and intensities of immuno-reactive protein spots were likely due to the dephosphorylation of the threonine residues in GES-supplemented cells, as we have not observed similar decrease in protein expression

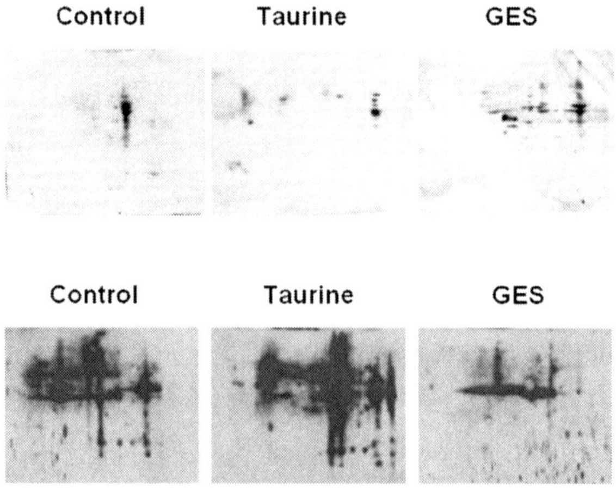

Fig. 18.2 Two-dimensional immunoblots showing differential expression of phosphorylated proteins in control, taurine-supplemented and GES-supplemented C6 cells. Experiments were carried out following the procedure described in Methods using the anti-phospho-tyrosine antibody (*top panels*) or anti-phospho-threonine antibody (*bottom panels*)

in these cells. Taken together these result suggested that taurine and GES differentially modulate the phosphrylation status of a number of yet to be identified C6 cellular proteins. Furthermore, GES-supplementation, that depletes intracellular taurine content, led to the dephosphorylation of the threonine residues in a number of proteins in C6 cells.

18.4 Discussion

Even though taurine supplementation seems to have some potential therapeutic, proinflammatory and immunoregulatory properties (Chesney et al. 1998; Redmond et al. 1998; Lima et al. 2001; Schulle Levis and Park 2003; Birdsall 1998), an understanding of the underlying mechanism of taurine's action at the cellular level is lacking. Results described above strongly suggest that taurine modulates the expression and as well as phosphorylation of many cellular proteins.

Taurine-supplementation was previously shown to down-regulate its own transporter (Bitoun and Tappaz 2000), a protein that is primarily responsible for maintaining the intracellular taurine concentration. Taurine-supplementation also down-regulated the multidrug resistance protein 2 (Mrp2) and bile salt export pump (Bsep) (Muhlfeld et al. 2003). In addition, taurine-supplementation induced expression of the mRNAs of other transporters in the brain and heart (Labudova et al. 1999). A recent study using cDNA microarray revealed that taurine modulated the expression of 477 genes in HepG2 cells, out of which 44 genes belonged to the MAPK and protein kinase C signaling pathways, thus suggesting that taurine is a modulator of the MAPK and protein kinase C signaling pathways in these cells (Park et al. 2006). Studies on taurine-mediated modulation of cellular protein expression and phosphorylation are however very limited. Notably, Lombardini and his group showed that taurine-depletion by GES increased the phosphorylation of a 20 kDa and a 44 kDa protein in the rat heart and retina, and subsequently identified these proteins as histone H2B and pyruvate kinase, respectively (Lombardini 1998).

GES, a taurine transport inhibitor, has been widely used to deplete intracellular taurine content both in animal and cell culture models (Moran and Pasantes-Morales 1991; Maar et al. 1995; Lake and Malik 1987; Abebe and Mozaffari 2003). In cultured mouse cerebellar granule neurons and astrocytes, progressive and rapid decrease in intracellular taurine concentration was found upon exposure to 2 mM GES; a reduction of 20% of intracellular taurine was found as early as 1 h after exposure to GES, the loss continued until the taurine pool was reduced by about 90% and no further decrease was observed even after 3 weeks of exposure to GES (Moran and Pasantes-Morales 1991).

Although we did not measure the intracellular content of taurine in GES-supplemented C6 cells, it is likely that a decrease in the intracellular taurine concentration in these cells is responsible for the observed changes in the cellular protein phosphorylation pattern. The observed changes in the phosphorylation patterns in taurine- and GES-supplemented cells could be explained, at least to some extent, by the taurine-mediated regulation (up- and/or down-regulation) of MAPK and protein kinase C signal pathway-related genes(Park et al. 2006).

Even though the results presented here are of qualitative nature, several noteworthy conclusions can be drawn from these studies. First, as taurine is commonly known to play an osmoregulatory role in several tissues, including brain, heart and kidney (for reviews, see (Huxtable 1992; Lima et al. 2001; Schulle-Levis and Park 2003), some of the changes observed using the C6 cells are likely to occur in other cell types. Second, it is clear that taurine-supplementation alters both expression and phosphorylation patterns in C6 cells; however, based on these results, we cannot conclude whether the associated changes are due to an alteration in the extracellular or intracellular taurine concentration. Third, GES-supplementation resulted in modest changes in the expression pattern, but caused profound changes in the phosphorylation pattern of cellular proteins. Since, GES is known to deplete intracellular taurine in animal models as well as in cultured cells (Moran and Pasantes-Morales 1991; Maar et al. 1995; Lake and Malik 1987), the observed changes are likely associated with the altered intracellular taurine concentration, suggesting that the intracellular taurine might be influencing, directly or indirectly, one or more signaling pathway(s). At present we do not know of any intracellular taurine receptor through which taurine might be exerting its effect. However, intracellular taurine could also influence its effect by binding to a taurine-responsive promoter element or by modulating the expression of proteins involved in the signaling pathways. Whether there exists a taurine-specific new signaling pathway still remains to be established. Fifth, our results strongly indicate that whereas GES-supplementation led to the dephosphorylation of the phospho-threonine residues in a number of proteins, it induced phosphorylation of tyrosine residues in several proteins.

As a dietary supplement, taurine has been shown to provide protection against glutamate-induced toxicity, cerebral ischaemia, oxidative stress, toxin build-up and injury-induced brain damage and toxin build-up (Hilgier et al. 2003; Rivas-Arancibia et al. 2001; Khan et al. 2000). Dietary supplementation of taurine is also increasingly used these days to attenuate pathophysiological conditions and improve living conditions in patients with cardiovascular diseases (Birdsall 1998). Although the identity of the taurine-mediated signaling pathway(s) remains elusive, the significant effects of taurine on cellular protein expression and phosphorylation patterns in glial cells, nevertheless, may have far reaching implications in glial cell physiology, and also in determining taurine's role in neuroprotection.

Finally, even though our results seem to suggest that the depletion of intracellular taurine in GES-supplemented cells might influence a signaling pathway, we interpret the significance of these results with a cautionary note. This is primarily because we have previously shown that the GES, at the concentration used in our experiments, could also significantly inhibit the creatine uptake of the cloned human creatine transporter (Dai et al. 1999). Although inhibition of creatine transport may not be of physiological importance in glial cells, it may significantly affect the cellular physiology in high-energy consuming cells/tissues, where a reduction in the intracellular creatine may affect the cellular ATP concentration. Obviously, further in-depth and quantitative studies are warranted to identify proteins whose expression and/or phosphorylation are modulated by taurine. Such studies will not only help identifying the signaling pathway(s) influenced by taurine, but will also lead to a better understanding on the cellular role of taurine.

Acknowledgments Authors thank Dr. Susan Hamilton for the use of the BioRad PROTEAN IEF Cells Alpha Imager 2000, and Cynthia Edwards for the technical help. This work was supported with a grant from USDA/Agriculture Research Services under Cooperative Agreement 58-6255-6001 (to HKS). This work was carried out at Baylor College of Medicine, Houston, Texas.

References

Abebe W, Mozaffari MS (2003) Effect of taurine deficiency on adenosine receptor-mediated relaxation of the rat aorta. Vascul Pharmacol 40(4):219-28

Azuma M, Takahashi K, Fukuda T, Ohyabu Y, Yamamoto I, Kim S, Iwao H, Schaffer SW, Azuma J (2000) Taurine attenuates hypertrophy induced by angiotensin II in cultured neonatal rat cardiac myocytes. Eur J Pharmacol 403:181–188

Birdsall TC (1998) Therapeutic applications of taurine. Altern Med Rev 3:128–136

Bitoun M, Tappaz M (2000) Taurine down-regulates basal and osmolarity-induced expression of its transporter but not the gene expression of its iosynthetic enzymes in astrocytes primary cultures. J Neurochem 75:919–924

Chesney RW, Holms RA, Christensen M, Burdreau AM, Han X, Sturman JA (1998) The role of taurine in infant nutrition. Adv Exp Med Biol 442:463–476

Dai W, Vinnakota S, Qian X, Kunze D, Sarkar HK (1999) Molecular characterization of the human CRT-1 creatine trnsporter expressed in Xenopus oocytes. Arch Biochem Biophys 361:75–84

El Idrissi A, Trenkner E (2003) Taurine regulates mitochondrial calcium homeostasis. In: Lombardini JB, Schaffer SW, Azuma J (eds) Taurine 5 Beginning the 21st Century, Adv Exp Med Biol, Vol 526, Kluwer Press, NewYork, pp 527-536

Gurujeyalakshmi G, Wang Y, Giri SN (2000) Suppression of bleomycin-induced nitric oxide by taurine and niacin. Nitric Oxide 4:399–411

Heller-Stilb B, van Roeyen C, Rascher K, Hartwig HG, Huth A, Seeliger MW, Warskulat U, Haussinger D (2002) Disruption of taurine transporter gene (taut) leads to retinal degeneration in mice. FASEB J 16:231–233

Hilgier W,,erzhanova E, Oja SS, Saransaari P, Albrecht J (2003) Taurine reduces ammonia- and N-methyl-D-aspartate-induced accumulation of cGMP and hydroxyl radicals in microdialysates of the rat striatum. Eur J Pharmacol 468:21–35

Huxtable RJ (1992) Physiological actions of taurine. Physiol Rev 72:101–163

Khan SH, Banigesh A, Baziani A, Todd KG, Miyashita H, Eweida M, Shuaib A (2000) The role of taurine in neuronal protection following transient global forebrain ischemia. Neurochem Res 25:217–223

Kingston R, Kelly CJ, Murray P (2004) The therapeutic role of taurine in ischaemia-reperfusion injury. Curr Pharm Des 10:2401–2410

Labudova O, Yeghiazarjan C, Hoer H, Lubec G (1999) Taurine modulates expression of transporters in rat brain and heart. Amino Acids 17:301–313

Lake N, Malik N (1987) Retinal morphology in rats treated with a taurine transport antagonist. Exp Eye Res 44(3):331-46

Lima L, Obregon F, Cubillos S, Fazzino F, Jaimes I (2001) Taurine as a micronutrient in development and regeneration of central nervous system. Nutr Neurosci 4:439–443

Lombardini JB (1998) Increased phosphorylation of specific rat cardiac and retinal proteins in taurine-depleted animals:isolation and identification of the phosphoproteins. Adv Exp Med Biol 442:441–447

Maar T, Moran J, Schousboe A, Pasantes-Morales H (1995) Taurine deficiency in dissociated mouse cerebellar cultures affects neuronal migration. Int J Dev Neurosci 13:491–502

Maher SG, Condron CE, Bouchier-Hayes DJ, Toomey DM (2005) Taurine attenuates CD3/interleukin-2-induced T cell apoptosis in an in vitro model of activation-induced cell death (AICD). Clin Exp Immunol 139:279–286

McLaughlin R, Bowler D, Kelly CJ, Kay E, Bouchier-Hayes D (2000) Taurine protects against early and late skeletal muscle dysfunction secondary to ischaemia reperfusion injury. Eur J Surg 166:375–379

Moloney MA, Kelly JJ, Condron CM, Roche-Nagle G, O'Donnell DH, Fennessy FM, Bouchier-Hayes DJ (2003) Taurine supplementation reverses endothelial cell dysfunction and promotes mobilization of endothelial progenitor cells. J Surg Res 114:298–299

Moran J, Pasantes-Morales H (1991) Taurine-deficient cultured cerebellar astrocytes and granule neurons obtained by treatment with guanidinoethane sulfonate. J Neurosci Res 29:533–537

Muhlfeld A, Kuitz R, Dransfeld O, Haussinger D, Wettstein M (2003) Taurine supplementation induces multidrug resistance protein 2 and bile salt export pump expression in rats and prevents endotoxin-induced cholestasis. Arch Biochem Biophys 413:32–40

Park SH, Lee H, Park KK, Kim HW, Park T (2006) Taurine-responsive genes related to signal transduction as identified by cDNA microarray analyses of HepG2 cells. J Med Food 9:33–34

Redmond HP, Stapleton PP, Neary P, Bouchier-Hayes DJ (1998) Immunonutrition:The role of taurine. Nutrition 14:599–604

Rivas-Arancibia S, Rodriguez AI, Zigova T, Willing AE, Brown WD, Cahill DW, Sanberg PR (2001) Taurine increases rat survival and reduces striatal damage caused by 3-nitropropionic acid. Int J Neurosci 108:55–67

Schulle-Levis GB, Park E (2003) Taurine: new implications for an old amino acid. FEMS Microbiol Lett 226:195–202

Tran TT, Dai W, Sarkar HK (2000) Cyclosporin A inhibits creatine uptake by altering the surface expression of the creatine transporter. J Biol Chem 275:35708–35714

Vohra BP, Hui X (2001) Taurine protects against carbon tetrachloride toxicity in the cultured neurons and in vivo. Arch Physiol Biochem 109:90–94

Warskulat U, Flogel U, Jacoby C, Hartwig HG, Theswissen M, Merx MW, Molojavyi A, Heller-Stilb B, Schrader J, Haussinger D (2004) Taurine transporter knockout depletes muscle taurine levels and results in severe skeletal muscle impairment but leaves cardiac function uncompromised. FASEB J 18:577–579

Yokogoshi H, Oda H (2002) Dietary taurine enhances cholesterol degradation and reduces serum and liver cholesterol concentrations in rats fed a high-cholesterol diet. Amino Acids 23: 433–439

Chapter 19
Taurine Improves Congestive Functions in a Mouse Model of Fragile X Syndrome

Abdeslem El Idrissi, Latifa Boukarrou, Carl Dokin, and W. Ted Brown

Abstract Increased seizure susceptibility is a feature of the mouse model for fragile X that has parallels in the hyperarousal and prevalence of seizures in the fragile X syndrome. Our investigation of the basis for the increased seizure susceptibility of the fragile X mouse indicated a reduction in $GABA_A$ receptor expression and increased expression of glutamic acid decarboxylase (GAD), the enzyme responsible for GABA synthesis. Taurine-fed mice also show these GABAergic alterations. However, unlike fragile X mice, taurine-fed mice show a significant increase in memory acquisition and retention. This discordance implies that there may be divergent events downstream of the biochemical changes in the GABAergic system in these two mouse models. To investigate the divergence of these two models we fed taurine to fragile X mice. Our preliminary data shows that taurine supplementation to fragile X mice resulted in a significant improvement in acquisition of a passive avoidance task. Since taurine is an agonist for $GABA_A$ receptor, we suggest that chronic activation of $GABA_A$ receptors and the ensuing alterations in the GABAergic system may have beneficial effects in ameliorating the learning deficits characteristic of the fragile X syndrome.

Abbreviations *Tau*, taurine; *GAD*, glutamic acid decarboxylase; *WT*, wild type; *KO*, fragile X knockout; *IRB*, infrared beam

19.1 Introduction

The fragile X syndrome includes hyperarousal, hypersensitivity to sensory stimuli and an increased prevalence of seizures (Hagerman 2002; Wisniewski et al. 1991). The mouse model for this disorder (Bakker et al. 1994) has increased seizure susceptibility (Musumeci et al. 2000; Chen and Toth 2001; Yan et al. 2004) and this may be a direct parallel to elements of the syndrome that suggest reduced inhibition/increased excitability. Our investigations of the molecular basis of increased

A. El Idrissi (✉)
Biology/Neuroscience, College of Staten Island, City University of New York Graduate School, New york

J. Azuma et al. (eds.), *Taurine 7*, Advances in Experimental Medicine and Biology 643, DOI 10.1007/978-0-387-75681-3_19,

seizure susceptibility in the fragile X mouse indicated a reduction in $GABA_A$ receptor expression (El Idrissi et al. 2005). Since these receptors play a major role in inhibition, their reduction helps explain the increased seizure susceptibility of this mouse model for fragile X and suggest that the GABAergic system may be affected in the fragile X syndrome.

We also found increased expression of the enzyme responsible for the synthesis of GABA, the neurotransmitter agonist for $GABA_A$ receptors. This increase is likely to be a response of the brain to reduced inhibition – a response that has been observed in other models of elevated excitability (Riback et al. 1993). The excitability of neuronal circuits is kept within a normal range through feed-forward and -backward inhibition, mediated by inhibitory interneurons. These neurons continuously adjust their inhibitory output to match the level of excitatory input. Thus, when there is reduced inhibition of postsynaptic neurons, feedback from these neurons causes the presynaptic neurons to increase their inhibitory output. In the example of fragile X mouse brain, reduced $GABA_A$ receptor expression on postsynaptic membranes would induce an increase in GAD expression, thus increasing the bioavailability of GABA in presynaptic terminals. Therefore, increased GAD may represent a secondary response to the direct effects of Fmrp depletion.

In our previous studies (El Idrissi et al. 2003; El Idrissi and Trenkner 2004), we have shown that mice chronically supplemented with taurine in their drinking water showed biochemical changes in the GABAergic system similar to those observed in fragile X mouse, including reduced $GABA_A$ receptor and increased GAD expression as well as a lower threshold for seizure induction. However, unlike fragile X mice, taurine-fed mice showed a significant improvement in learning (acquisition and retention). The discrepancies in learning abilities signal dissimilarities between the two models. Therefore, we used a comparative approach, between the fragile X and taurine-fed mice, and examined divergent events downstream of the biochemical changes in the GABAergic system. Furthermore, we looked for neuronal markers that are differentially expressed in fragile X and taurine-fed mice that might explain the phenotypic discrepancies between these two mouse models (mainly learning deficit). This neuronal marker should show a correlation with at least some fragile X-specific features. Our preliminary data indicate that this neuronal marker could be somatostatin.

19.2 Methods

19.2.1 Passive Avoidance Test

We conducted the repetitive training passive avoidance test when the animals were 8 weeks of age. The test was carried out during the light phase (13:00–17:00 hrs), and each animal was housed individually during the test. The apparatus has a bright and a dark compartment with a computer-controlled door between them. The delivery of electric shocks (0.5 mA for 2 sec) and the raising and lowering of the door and the latencies at which the animals stepped into the dark from the bright compartment were controlled by the computer. Each animal was gently placed in the light compartment for 10 sec, after which the guillotine door was raised and the time the animal waited before crossing to the dark (shock) compartment was recorded as the latency. The trial ended when an animal waited more than 180 sec to cross to the dark side, or if it received an electrical shock in the dark side after crossing. Once the animal crossed with all four paws to the dark compartment, the door was closed and a 0.5 mA foot shock was delivered for 2 sec. This shuttle box apparatus has 8 infrared sensors on each side that allow to measure locomotor activity by measuring the number of infrared beams breaks.

19.2.2 Statistic Analysis

Multifactorial analysis of variance was used to identify overall condition effects. Significant changes were determined by post hoc comparisons of means using the Tukey HSD test. Significance was set at a confidence level of 95 %. Data are presented as mean ± SEM.

19.3 Results

19.3.1 Behavioral Consequences of Taurine on Fragile X Mice

To test if there were changes induced by taurine on the cognitive function of fragile X mice, we measured learning ability using a repetitive training passive avoidance test. In this paradigm, we tested the ability of mice to learn to avoid a mild electrical foot shock (0.5 mA). We gave the mice six repetitive trials and measured their learning ability. Retention of the learned task was determined 24 hrs later. The apparatus has a bright and a dark compartment with a computer-controlled door between them. The delivery of electric shocks and the raising and lowering of the door and the latencies at which the animals stepped into the dark from the bright compartment were controlled by the computer. Each animal was gently placed in the light compartment for 10 sec, after which the guillotine door was raised and the time the animal waited before crossing to the dark (shock) compartment was recorded as the latency. The trial ended when an animal waited more than 180 sec to cross to the dark side, or if it received an electrical shock in the dark side after crossing. Once the animal crossed with all four paws to the dark compartment, the door was closed and a 0.5 mA foot shock was delivered for 2 sec. This shuttle box apparatus has 8 infrared sensors on each side that allow measure of locomotor activity. On the first trial, all mice had the same level of activity as measured by the number of infrared beams breaks. By the third trial, KO-Tau mice were relatively more active than WT mice (Fig. 19.1). The Baseline activity of KO-Tau mice was similar to all other groups, if not these mice were slightly hypoactive. The hyperactivity emerges

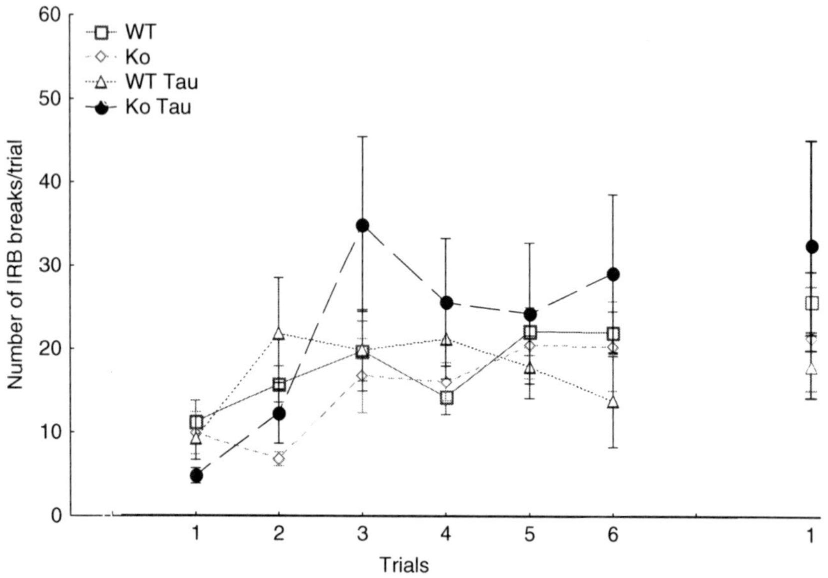

Fig. 19.1 Stress-Induced activity in KO-Tau mice. Activity is measured as the number of infrared beam (IRB) breaks. The graph shows activity over six trials and 24hrs later. KO-Tau mice have very low baseline activity and become very active following the foot shock. All mice were 2 months old. Taurine (0.05%) was supplemented in the drinking water for four weeks prior to the test. WT, $n = 20$; KO, $n = 12$; WT-Tau, $n = 12$; KO-Tau, $n = 12$

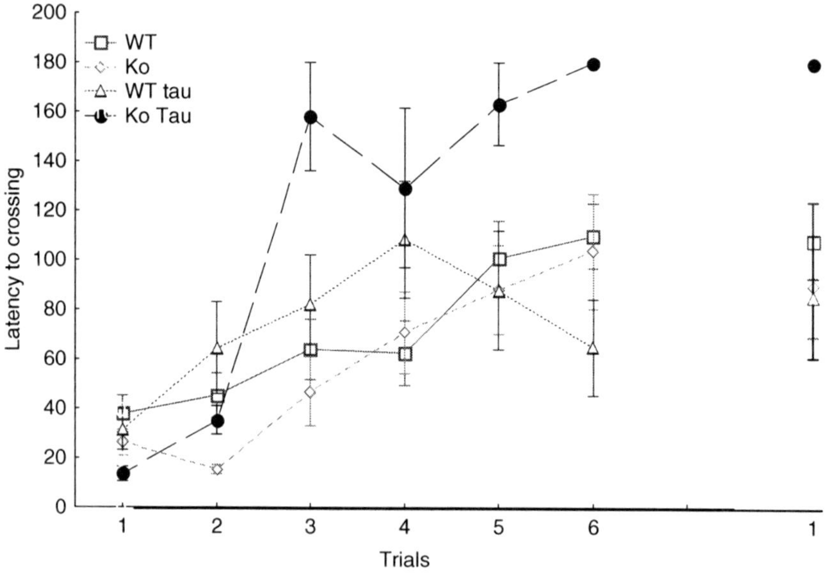

Fig. 19.2 Taurine enhances learning in KO-Tau mice. Learning is expressed by increases in the latency to cross to the dark side of the box. The graph shows latencies over six trials and 24 hrs later. KO-Tau mice were the only group to completely learn the task and they retained the same learning level 24 hrs later

after these mice were exposed to stress (foot shock). This increase in activity is probably induced by stress and should be further investigated in the context of stress and psychostimulant-induce locomotor and behavioral sensitization. If these mice are hypersensitive to stress then this sensitivity might be mediated through the dopaminergic and adrinergic systems and could help to make these mice more vigilant and hence improve their cognitive function as demonstrated in Fig. 19.2 (i.e. improvement in passive avoidance test through the increase in latency to choice).

Despite stress-induced hyperactivity, it took only two trials for the KO-Tau group to learn the task and by the end of the sixth trial, all mice in this group had a perfect score. There were no significant differences between the other groups throughout the 6 trials (WT, KO and WT-Tau). Interestingly, 24 hrs later all mice retained the same level of learning, with an impressive perfect score for the KO-Tau group. Thus, the appearance of cognitive improvement correlates well with the appearance of hyperactivity in the KO-Tau mice. The learning of the passive avoidance task was associated with fewer errors. Figure 19.3 shows, as expected, that all groups made 100% errors on the first trial and overall on the first day. By the end of the sixth trial, there were no significant differences between groups except the KO-Tau group that did exceptionally well making no errors at all by the sixth trial. Similarly this graph shows the rapid stress-induced learning in the KO-Tau group. After 24hrs

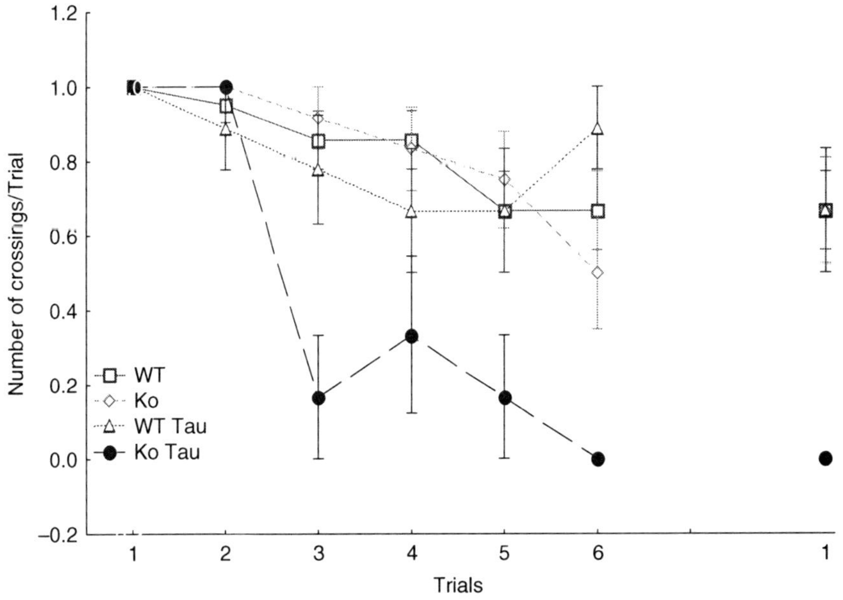

Fig. 19.3 Taurine enhances learning in KO mice. Learning was associated with decreased errors. Errors are defined by the number of attempts to cross to the dark side before the end of the trial. The graph shows errors in the six trials and 24 hrs later. All mice made the same number of errors in the first trial of the first day. KO-Tau mice were the only group to completely learn the task and they retained the same learning level 24 hrs later

later, there was clear separation between the KO-Tau mice and the other groups (WT, KO, WT-Tau). The KO-Tau were the only mice that made no errors by the end of the sixth trial and retained this memory when tested 24 hrs later. Thus, supplementation of taurine in drinking water to KO mice which increases motoric activity also seem to have a positive effect on learning/memory mechanisms studied by a passive avoidance behavior model. Interestingly, in this test the effects of taurine are more conspicuous only in the KO mice that have deficits in the inhibitory system.

19.4 Discussion

The excitability of neuronal circuits is controlled by inhibitory GABAergic interneurons. In this study, we used two models of hyper-excitability the fragile X mouse, where hyper excitability is genetically induced, and the taurine-fed mice where hyper-excitability can be induced by supplementing taurine in drinking water (0.05%) for 4 continuous weeks then we tested the effects of increased neuronal excitability on GABAergic plasticity. We found that like fragile X KO mice, taurine-fed mice showed increased susceptibility to KA-induced seizures (El Idrissi et al. 2003). Associated with this increased state of brain excitability, we found some biochemical changes in the GABAergic system, mainly, increase in the expression and activity of the enzyme responsible for GABA synthesis, glutamic acid decarboxylase (GAD). We also found a reduced expression of the β subunit of $GABA_A$ receptors. These biochemical and functional changes were similar in both mouse models. In addition, supplementing taurine in the drinking water resulted in an increase in somatostatin expression in both the WT and KO mice.

The neonatal brain contains high levels of taurine (Huxtable 1989; Sturman 1993). As the brain matures its taurine content declines and reaches stable adult concentrations that are second to those of glutamate, the principal excitatory neurotransmitter in the brain. Taurine levels in the brain significantly increase under stressful conditions (Wu et al. 1998), suggesting that taurine may play a vital role in neuroprotection. A possible mechanism of taurine's neuroprotection lies in its calcium modulatory effects (El Idrissi and Trenkner 1999; El Idrissi and Trenkner 2003; El Idrissi and Trenkner 2004) and agonistic role on GABAA receptors (El Idrissi et al. 2003; El Idrissi and Trenkner 2004).

Taurine has been shown to play a role in neurotransmission, but taurine does not satisfy the criteria of a classical neurotransmitter. However, there is increasing evidence supporting a functional interaction between GABA, glycine and taurine (Kuriyama and Hashimoto 1998; El Idrissi and Trenkner 2004). Taurine has been shown to increase plasma membrane chloride conductances by affecting bicuculine-sensitive chloride channels (del Olmo et al. 2000; Mellor et al. 2000). Taurine has also been shown to act as a partial agonist of $GABA_A$ receptors in synaptic membranes (Quinn and Harris 1995). In addition to modulating neuronal transmission, the observed effects of taurine on the up-regulation of somatostatin expression are not well understood and could be mediated at the transcription level. The levels of somatostatin and somatostatin-positive neurons were increased in the brain of taurine-fed mice (Levinskaya et al. 2006). Numerous experimental and clinical studies have demonstrated that somatostatin neurotransmission plays an important role in the modulation of several brain functions, including learning and memory processes. It is possible that the increase in somatostatin levels observed after taurine supplementation may be responsible for the increased learning observed in fragile X mice. Taurine supplementation therefore could be beneficial as a naturally occurring pharmaco-therapeutic agent for the improvement of learning in fragile X mice and perhaps patients.

19.5 Conclusion

In summary, this study shows that taurine supplementation to fragile X mice induced a significant increase in acquisition and retention of a hippocampal-dependent memory task, interpreted here as improvement in cognitive functions. The taurine enhancing effects are mediated though interaction with, and modification of the GABAergic system.

Acknowledgments We are greatly thankful to Ekaterina Zavyalova, Khrystyna Splavnyk, George Malliaros, for helping with the behavioral testing of mice. This work was supported by FRAXA research foundation, PSC-CUNY and NYS-IBR.

References

Bakker CE, Verheij C, Willemsen R, van der Helm R, Oerlemans F, Vermeij M, Bygrave A, Hoogeveen AT, Oostra BA, Reyniers E, De Boulle K, D'Hooge R, Cras P, van Velzen D, Nagels G, Marti JJ, De Deyn P, Darby JK, Willems PJ (1994) Fmr1 knockout mice: a model to study fragile X mental retardation. Cell 78:23–33

Chen L, Toth M (2001) Fragile X mice develop sensory hyperreactivity to auditory stimuli Neurosci 103:1043–1050

del Olmo N, Bustamante, J, del Rio RM, Soli J (2000) Taurine activates GABA(A) but not GABA(B) receptors in rat hippocampal CA1 area. Brain Res 864 298–307

El Idrissi A, Ding X-H, Scalia J, Trenkner E, Brown WT, Dobkin C (2005) Decreased GABAA receptor expression in the seizure-prone fragile X mouse. Neurosci Lett 377:141–146

El Idrissi A, Trenkner E (2003) Taurine regulates mitochondrial calcium homeostasis. In: Lombardini JB, Schaffer SW, Azuma J (eds) Taurine 5 Beginning the 21st Century, Adv Exp Med Biol, Vol 526, Kluwer Press, NewYork, pp 527-536

El Idrissi A, Trenkner E (2004) Taurine as a modulator of excitatory and inhibitory neurotransmission. Neurochem Res 29:189–197

El Idrissi A, Messing J, Scalia J, Trenkner E (2003) Prevention of Epileptic Seizures through taurine. Adv Exp Med Biol 526: 515–525

El Idrissi A, Trenkner E (1999) Growth factors and taurine protect against excitotoxicity by stabilizing calcium homeostasis and energy metabolism. J Neurosci 19:9459–9468

Hagerman RJ (2002) Physical and behavioral phenotype. In: Hagerman RJ, Hagerman PJ (eds) Fragile X syndrome Diagnosis, treatment and research, Johns Hopkins University Press, Baltimore, pp 3–109

Huxtable RJ (1989) Taurine in the central nervous system and the mammalian action actions of taurine. Prog Neurobiol 32:471–533

Levinskaya N, Trenkner E, El Idrissi A (2006) A Increased GAD-positive neurons in the cortex of taurine-fed mice. Adv Exp Med Biol 583 411–417

Kuriyama K, Hashimoto T (1998) Interrelationship between taurine and GABA. Adv Exp Med Biol 442:329–337

Mellor JR, Gunthorpe MJ, Randall AD (2000) The taurine uptake inhibitor guanidinoethyl sulphonate is an agonist at gamma-aminobutyric acid(A) receptors in cultured murine cerebellar granule cells. Neurosci Lett 286:25–28

Musumeci SA, Bosco P, Calabrese G, Bakker C, De Sarro GB, Elia M, Ferri R, Oostra BA (2000) Audiogenic seizures susceptibility in transgenic mice with fragile X syndrome. Epilepsia 41:19–23

Quinn MR, Harris CL (1995) Tautine allosterically inhibits binding of [35S]-t-butylbicyclophosphorothionate (TBPS) to rat brain synaptic membranes. Neuropharmacol 34:1607–1613

Riback CE, Lauterborn JC, Navetta MS, Gall CM (1993) The inferior colliculus of GEPRs contains greater numbers of cells that express glutamate decarboxylase (GAD67) mRNA. Epilepsy Res 14:105–113

Sturman JA (1993) Taurine in development. Physiol Rev 73:119–147

Wisniewski KE, Segan SM, Miezejeski CM, Sersen EA, Rudelli RD (1991) The Fra(X) syndrome: neurological, electrophysiological, and neuropathological abnormalities. Am J Med Genet 38:476–480

Wu JY, Tang XW, Schloss JV, Faiman MD (1998) Regulation of taurine biosynthesis and its physiological significance in the brain. Adv Exp Med Biol 442:339–345

Yan QJ, Asafo-adjei PK, Arnold HM, Brown RE, Bauchwitz RP (2004) A phenotypic and molecular characterization of the fmr1-tm 1 Cgr fragile X mouse. Genes Brain Behav 3:337–359

Chapter 20
Functional Implication of Taurine in Aging

Abdeslem El Idrissi, Latifa Boukarrou, Khrystyna Splavnyk, Ekaterina Zavyalova, Edward F. Meehan, and William L'Amoreaux

Abstract Age-related impairment of central functions is though to result from alterations of neurochemical indices of synaptic function. These neurochemical modifications involve structural proteins, neurotransmitters, neuropeptides and related receptors. Several studies demonstrated that GABA receptors, glutamic acid decarboxylase (GAD65&67), and different subpopulations of GABAergic neurons are markedly decreased in experimental animal brains during aging. Thus, the age-related decline in cognitive functions could be attributable, at least in part, to decrements in the function of the GABAergic inhibitory neurotransmitter system. In this study we show that chronic supplementation of taurine to aged mice significantly ameliorated the age-dependent decline in memory acquisition and retention, and caused alterations in the GABAergic system. These changes include increased levels of the neurotransmitters GABA and glutamate, increased expression of glutamic acid decarboxylase and the neuropeptide somatostatin and increased in the number of somatostatin-positive neurons. These specific alterations of the inhibitory system caused by taurine treatment oppose those naturally-occurring during aging, and suggest a protective role of taurine in this process.

Increased understanding of age-related neurochemical changes in the GABAergic system will be important in elucidating the underpinnings of the functional changes of aging. Taurine might help forestall the age-related decline in cognitive functions through interaction with the GABAergic system.

Abbreviations *Tau*, taurine; *GAD*, glutamic acid decarboxylase; *IRB*, infrared beam

20.1 Introduction

Aging of the brain is characterized by several neurochemical modifications involving structural proteins, neurotransmitters, neuropeptides and related receptors (Marczynski 1998). Alterations of neurochemical indices of synaptic function have

A. El Idrissi (✉)
Department of Biology, City University of New York Graduate School, NewYork

J. Azuma et al. (eds.), *Taurine 7*, Advances in Experimental Medicine and Biology 643,
DOI 10.1007/978-0-387-75681-3_20, © Springer Science+Business Media, LLC 2009

been considered as indicators of age-related impairment of central functions, such as locomotion, memory and sensory performances.

Gamma-aminobutyric acid (GABA) is the primary inhibitory neurotransmitter in the central nervous system and is present in neurons in all brain regions. A number of GABAergic parameters have been reported to undergo changes during senescence (Araki et al. 1996). Several studies demonstrated that GABA receptors are markedly decreased in experimental animal brains during aging (Govoni et al. 1980; Hunter et al. 1989; Milbrandt et al. 1996). Significant age-related decreases in glutamic acid decarboxylase (GAD65&67), the enzyme responsible for GABA synthesis, were observed in the cortex and hippocampus of aged rats relative to their young adult cohorts, suggesting an age-dependent down-regulation of normal adult inhibitory GABA neurotransmission (Marczynski 1998). Consistent with this, functional studies in primate visual and auditory cortices demonstrated sensory coding changes suggestive of altered inhibitory processing in aged animals. Such age-related loss of normal adult GABA neurotransmission in the auditory cortex would likely alter temporal coding properties and could contribute to the loss in speech understanding observed in the elderly. Thus the age-related central sensory processing deficits could be attributable, at least in part, to decrements in GABA inhibitory neurotransmission (Caspary et al. 1990, 2002). Indeed, the auditory midbrain shows significant age-related changes related to GABA neurotransmission (Banay-Schwartz et al. 1989; Caspary et al. 1990, 1995; Gutiérrez et al. 1994, Milbrandt et al. 1996; Raza et al. 1994). Furthermore, different subpopulations of GABAergic neurons such as somatostatin- and parvalbumin-containing neurons are reduced in aged rats (Kuwahara et al. 2004). These observations seem to indicate that age-related changes in GABAergic function may be an important determinant of cognitive function. In the present study, therefore, we focused on GABA, the major inhibitory neurotransmitter system.

The neonatal brain contains high levels of taurine (Huxtable 1989; Huxtable 1992; Sturman 1993; Kuriyama and Hashimoto 1998). As the brain matures its taurine content declines and reaches stable adult concentrations that are second to those of glutamate, the principal excitatory neurotransmitter in the brain. Taurine levels in the brain significantly increase under stressful conditions (Wu et al. 1998), suggesting that taurine may play a vital role in neuroprotection. A possible mechanism of taurine's neuroprotection lies in its calcium modulatory effects. We have shown that taurine modulates both cytoplasmic and intra-mitochondrial calcium homeostatsis (El Idrissi et al. 1999; El Idrissi et al. 2003; El Idrissi and Trenkner 2004). Furthermore, taurine acts as an agonist of $GABA_A$ receptors (Quinn and Harris 1995; Wang et al. 1998; del Olmo et al. 2000; Mellor et al. 2000; El Idrissi et al. 2003; El Idrissi and Trenkner 2004). Thus, we hypothesize that the age-dependent deterioration in GABAergic function and the resulting decline in cognitive function could be ameliorated by supplementing exogenous taurine.

We found an age-dependent decline in learning and memory as measured by acquisition and retention of a passive avoidance task between young and old mice. Young mice (2 months) learned at a significantly quicker rate and much greater amplitude than aged mice (16 months). Furthermore, the young group of mice performed significantly better on a recall test than aged mice. Interestingly, when aged mice were supplemented with taurine in drinking water for four weeks, they showed a significant increase in acquisition and retention of a passive avoidance task as compared to age-matched controls. On the other hand, young mice supplemented with taurine learned to the same extent as their age-matched controls. Furthermore, we found several biochemical changes that accompanied the increased performance in memory tasks of taurine-fed mice. The brains of these mice have elevated levels of both the excitatory and inhibitory neurotransmitters (glutamate and GABA, respectively) and the GABA synthesizing enzyme, glutamic acid decarboxylase (GAD). The levels of somatostatin and somatostatin-positive neurons were increased in the brain of taurine-fed mice. These chances seem to be opposing those naturally-induced by aging. Interestingly, electrophysiological recordings from hippocampal slices prepared from the brain of taurine-fed mice showed an increased in the amplitude and duration of population spikes recorded from CA1 in response to Schaefer collaterals stimulation. Such increased excitability of hippocampal slices of taurine-fed mice is consistent with lower threshold for LTP induction, which would explain the increased learning in these mice.

Using this paradigm of taurine treatment we will gain significant understanding of the mechanisms by which taurine influences the inhibitory GABAergic systems in the brain and explore the potential of taurine in reversing the age-dependent alteration in the inhibitory system. Most importantly, the identification of specific age-dependent alterations in the GABAergic system will enhance our understanding of the basis for the long-lasting altered cellular and synaptic properties that contribute to the decline in cognitive function, characteristic of senescence.

20.2 Methods

20.2.1 Passive Avoidance Test

We conducted the repetitive training passive avoidance test when the animals were 2 months or 16 months old. The test was carried out during the light phase (13:00–17:00 hrs), and each animal was housed individually during the test. The apparatus has a bright and a dark compartment with a computer-controlled door between them. The delivery of electric shocks (0.5 mA for 2 sec) and the raising and lowering of the door and the latencies at which the animals stepped into the dark from the bright compartment were con-trolled by the computer. Each animal was gently placed in the light compartment for 10 sec, after which the guillotine door was raised and the time the animal waited before crossing to the dark (shock) compartment was re-corded as the latency. The trial ended when an animal waited more than 180 sec to cross to the dark side, or if it received an electrical shock in the dark side after crossing. Once the animal crossed with all four paws to the dark compartment, the door was closed and a 0.5 mA foot shock was de-livered for 2 sec.

This shuttle box apparatus has 8 infrared sensors on each side that allow to measure locomotor activity by measuring the number of infrared beams breaks.

20.2.2 Statistic Analysis

Multifactorial analyses of variance was used to identify overall condition effects. Significant changes were determined by post hoc comparisons of means using the Tukey HSD test. Significance was set at a confidence level of 95%. Data are presented as mean ± SEM.

20.3 Results

20.3.1 Taurine Improves Acquisition and Retention in Aged Mice

Taurine has been shown to act as an agonist of GABAA receptors (del Olmo et al. 2000; El Idrissi and Trenkner 2004; Wang et al. 1998). Since senescence is characterized by a decline in the GABAergic neurotransmission, we supplemented taurine in drinking water to determine if chronic taurine intake alleviates the age-dependent decline in cognitive function. Using the passive avoidance paradigm, we tested the acquisition and retention in both young (2 months-old FVB/NJ males) and aged (16 months-old FVB/NJ males) mice supplemented with taurine (0.05%) in drinking water. Mice were given six trials a day for five days and after repetitive training, the level of learning was measured on the fifth day. Figure 20.1 shows that the performances of young and old mice during the last day of training. The apparatus has a bright and a dark compartment with a computer-controlled door between them. The delivery of electric shocks, the raising and lowering of the door and the latencies at which the animals stepped into the dark from the initial bright compartment were controlled and measured by the computer. Each animal was gently placed in the light compartment for 10 sec, after which the guillotine door was raised. The amount of time the animal waited before crossing to the dark (shock) compartment was recorded as the latency. The trial ended when an animal waited more than 180 sec to cross to the other side, or if it received an electrical shock in the dark side after crossing. Once the animal crossed with all four paws to the next compartment, the door was closed and a 1.5 mA foot shock was delivered for 5 sec. Young mice were supplemented with taurine for four weeks and old mice were fed taurine for 8 months prior to, and during testing. Figure 20.1 shows that taurine had no significant effects on young mice. However, aged mice supplemented with taurine showed a significant increase in learning when compared to aged-matched controls (Fig. 20.1). Furthermore, aged mice approached learning levels observed in the young group after repetitive training. When we tested retention two weeks later, the performance of every experimental group correlated with learning. However, the aged control group showed the highest decline in the retention of previously learned

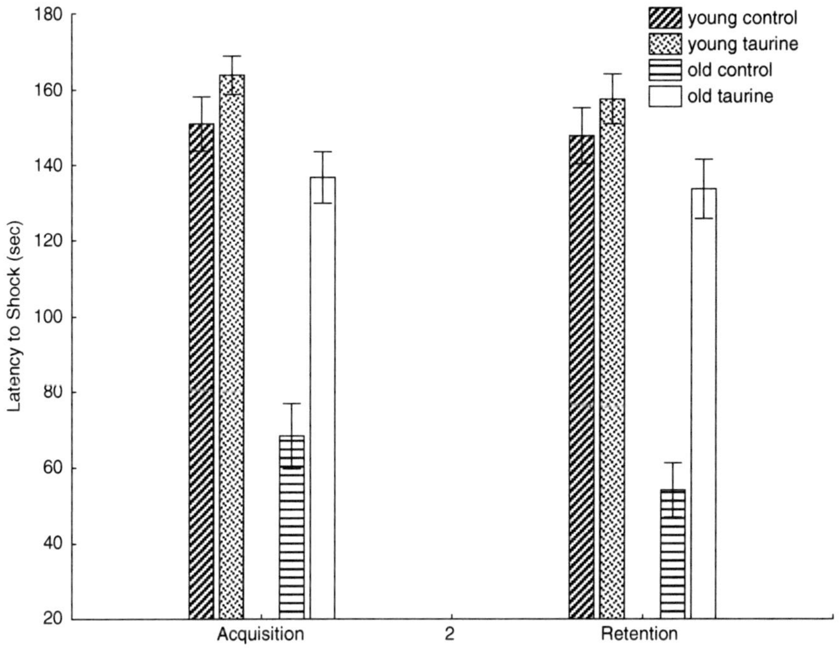

Fig. 20.1 Acquisition and retention of a passive avoidance task in young and old mice. Each animal was familiarized with the behavioral apparatus for 2–3 min the day before the training session and 10 sec before the test. All training and testing was carried out between 08:00 and 12:00 h. Mice (FVB/NJ males) were individually housed throughout the testing. Young group (2 months-old), aged group (16 months-old). Data represent the mean ± SEM of latencies to the end of the trial on the fifth day of training (acquisition) and two weeks after the last day of training (retention). Each group consisted of 10 mice. $^{*}p < 0.01$

task (extinction) when compared against all groups. Though, this decrease did not reach significance.

20.4 Discussion

We found an age-dependent decline in learning and memory as measured by acquisition and retention of a passive avoidance task between young and old mice. Taurine supplementation in drinking water for eight months significantly increased the performances of aged mice as compared to untreated controls. We have previously shown that chronic taurine supplementation in drinking water resulted in several biochemical changes in the inhibitory system. These taurine-induced alterations oppose those observed during aging. The brains of taurine-fed mice have elevated levels of both the excitatory and inhibitory neurotransmitters (glutamate and GABA, respectively) and the GABA synthesizing enzyme, GAD (El Idrissi and Trenkner 2004). The levels of somatostatin and somatostatin-positive neurons were increased in the brain of taurine-fed mice (Levinskaya et al. 2006). Numerous experimental and clinical studies have demonstrated that somatostatin neurotransmission plays an important role in the modulation of several brain functions, including learning and memory processes. Expression of somatostatin in the brain declines during aging (Dournaud et al. 1996). A prominent decrease in this neuropeptide also represents a pathological characteristic of Alzheimer disease (Saito et al. 2005). Furthermore, one of the most consistent neurochemical abnormalities in Alzheimer's disease is a reduction in cortical somatostatin (Davies et al. 1980; Rossor et al. 1980). This may be attributed to the loss of intrinsic cortical neurons or could be caused by a decrease in synthesis or an increase in degradation of the peptide. In our previous study, we reported that taurine supplementation resulted in a significant increase in the number of somatostatin-positive neurons in the cortex and hippocampus (Levinskaya et al. 2006). Therefore, within the GABAergic population of interneurons, somatostatin-positive neurons seem to be more vulnerable in aging. Taurine supplementation selectively enhances the survival of this population of GABAergic neurons (somatostatin-positive). Although the mechanisms mediating this observation are not currently understood, it is possible that taurine may have a trophic effect on this subpopulation of GABAergic neurons.

These biochemical changes resulting from taurine supplementation are opposing those naturally-induced by aging and suggest that taurine improves learning and memory in aged mice through amelioration of the age-dependent decline in GABAergic function. In the central nervous system, the effects of taurine are not limited to interactions with the inhibitory GABAergic system. Taurine also activates glycine receptors (Häusser et al. 1992), acts as anti-oxidant (Aruoma et al. 1988) and regulates intracellular calcium homeostasis (El Idrissi and Trenkner 1999). Therefore, these neuroprotective effects of taurine could also contribute to the improvement of cognitive functions observed after chronic supplementation with taurine. Young mice on the other hand showed no improvement in learning and retention above controls. This could possibly be due to the limited sensitivity of the behavioral test used in this study that could not detect subtle differences between the two groups of mice. However, taurine induced several biochemical changes to the inhibitory GABAergic system at early ages that could be beneficial in aging.

20.5 Conclusion

In summary, this study shows that supplementation of taurine in drinking water to senescent mice significantly improved their ability to learn and retain memory tasks. Taurine induces biochemical changes in the inhibitory system opposing those naturally occurring during aging.

Acknowledgments We thank George Malliaros, Lorenz Neuwirth for helping with the behavioral testing of mice. This work was supported by PSC-CUNY and CSI.

References

Araki T, Kato H, Fujiwara T, Itoyama Y (1996) Regional age-related alterations in cholinergic and GABAergic receptors in the rat brain. Mech Ageing Dev 88:49–60

Aruoma O I, Halliwell B, Hoey BM, Butler J (1988) The antioxidant action of taurine, hypotaurine and their metabolic precursors. Biochem J 256:251–255

Banay-Schwartz M, Lajtha A, Palkovits M (1989) Changes with aging in the levels of amino acids in rat CNS structural elements II Taurine and small neutral amino acids. Neurochem Res 14:563–570

Banay-Schwartz M, Lajtha A, Palkovits M (1989) Changes with aging in the levels of amino acids in rat CNS structural elements I Glutamate and related amino acids. Neurochem Res 14:555–562

Caspary DM, Raza A, Lawhorn Armour BA, Pippin J, Arnerić SP (1990) Immunocytochemical and neurochemical evidence for age-related loss of GABA in the inferior colliculus: implications for neural presbycusis. J Neurosci 10(7):2363-72.

Caspary DM, Milbrandt JC, Helfert RH (1995) Central auditory aging: GABA changes in the inferior colliculus. Exp Gerontol 30:349–360

Caspary DM, Palombi PS, Hughes LF (2002) GABAergic inputs shape responses to amplitude modulated stimuli in the inferior colliculus. Hear Res 168:163–173

Davies P, Katzman R, Terry RD (1980) Reduced somatostatin-like immunoreactivity in cerebral cortex from cases of Alzheimer's disease and Alzheimer senile dementia Nature 288:279–280

del Olmo N, Bustamante J, del Rio RM, Soli J (2000) Taurine activates GABA(A) but not GABA(B) receptors in rat hippocampal CA1 area. Brain Res 864:298–307

Dournaud P, Jazat-Poindessous F, Slama A, Lamour Y, Epelbaum J (1996) Correlations between water maze performance and cortical somatostatin mRNA and high-affinity binding sites during ageing in rats Eur J Neurosci 8:476–485

El Idrissi A, Messing J, Scalia J, Trenkner E (2003) Prevention of Epileptic Seizures through taurine. In: Lombardini JB, Schaffer SW, Azuma J (eds) Taurine 5 Beginning the 21st Century, Adv Exp Med Biol, Vol 526, Kluwer Press, NewYork, pp 515–525

El Idrissi A, Trenkner E (2004) Taurine as a modulator of excitatory and inhibitory neurotransmission. Neurochem Res 29:189–197

El Idrissi A, Trenkner E (1999) Growth factors and taurine protect against excitotoxicity by stabilizing calcium homeostasis and energy metabolism. J Neurosci 19:9459–9468

Govoni S, Memo M, Saiani L, Spano PF, Trabucchi M (1980) Impairment of brain neurotransmitter receptors in aged rats. Mech Ageing Dev 12:39–46

Gutiérrez A, Khan ZU, Morris SJ, De Blas AL (1994) Age-related decrease of GABAA receptor subunits and glutamic acid decarboxylase in the rat inferior colliculus. J Neurosci 14: 7469–7477

Häusser MA, Yung WH, Lacey MG (1992) Taurine and glycine activate the same Cl- conductance in substantia nigra dopamine neurons. Brain Res 571:103–108

Hunter C, Chung E, Van Woert MH (1989) Age-dependent changes in brain glycine concentration and strychnine-induced seizures in the rat. Brain Res 482:247–251

Huxtable RJ (1989) Taurine in the central nervous system and the mammalian action actions of taurine. Prog Neurobiol 32:471–533

Huxtable RJ (1992) The physiological actions of taurine. Physiol Rev 72:101–163

Kuriyama K, and Hashimoto T (1998) Interrelationship between taurine and GABA. Adv Exp Med Biol 442:329–337

Kuwahara S, Kesuma Sari D, Tsukamoto Y, Tanaka S, Sasaki F (2004) Age-related changes in growth hormone (GH)-releasing hormone and somatostatin neurons in the hypothalamus and in GH cells in the anterior pituitary of female mice. Brain Res 1025:113–122

Levinskaya N, Trenkner E, El Idrissi A (2006) Increased GAD-positive neurons in the cortex of taurine-fed mice. Adv Exp Med Biol 583:411–417

Marczynski TJ (1998) GABAergic deafferentation hypothesis of brain aging and Alzheimer's disease revisited. Brain Res Bull 45:341–379

Mellor JR, Gunthorpe MJ, and Randall AD (2000) The taurine uptake inhibitor guanidinoethyl sulphonate is an agonist at gamma-aminobutyric acid(A) receptors in cultured murine cerebellar granule cells. Neurosci Lett 286:25–28

Milbrandt JC, Albin RL, Turgeon SM, Caspary DM (1996) GABAA receptor binding in the aging rat inferior colliculus. Neuroscience 73:449–458

Quinn MR, and Harris CL (1995) Tautine allosterically inhibits binding of [35S]-t-butylbicyclophosphorothionate (TBPS) to rat brain synaptic membranes. Neuropharmacol 34:1607–1613

Raza A, Milbrandt JC, Arneric SP, Caspary DM (1994) Age-related changes in brainstem auditory neurotransmitters: measures of GABA and acetylcholine function. Hear Res 77:221–230

Rossor MN, Emson PC, Mountjoy CQ, Roth M, Iversen LL (1980) Reduced amounts of immunoreactive somatostatin in the temporal cortex in senile dementia of Alzheimer type. Nenrosci Lett 20:373–377

Saito T, Iwata N, Tsubuki S, Takaki Y, Takano J, Huang SM, Suemoto T, Higuchi M, Saido TC (2005) Somatostatin regulates brain amyloid beta peptide, beta42 through modulation of proteolytic degradation. Nat Med 11:434–439

Sturman JA (1993) Taurine in development. Physiol Rev 73:119–147

Wang DS, Xu TL, Pang ZP, Li JS, Akaike N (1998) Taurine-activated chloride currents in the rat sacral dorsal commissural neurons. Brain Res 792:41–47

Wu JY, Tang XW, Schloss JV, Faiman MD (1998) Regulation of taurine biosynthesis and its physiological significance in the brain. Adv Exp Med Biol 442:339–345

Chapter 21
Effects of Taurine on Anxiety-Like and Locomotor Behavior of Mice

Abdeslem El Idrissi, Latifa Boukarrou, Wally Heany, George Malliaros, Chaichan Sangdee, and Lorenz Neuwirth

Abstract Taurine is one of the most abundant free amino acids especially in excitable tissues, with wide physiological actions. We have previously reported that chronic supplementation of taurine in drinking water to mice increases brain excitability, mainly through alterations in the inhibitory GABAergic system. In this study we investigated the effects of chronic versus acute taurine treatment on anxiety-like and locomotor behaviors using two behavioral tests: elevated plus-maze and open-field. These two test conditions generated different levels of anxiety, and both anxiolytic and anxiogenic effects of taurine could be assessed. We used two paradigms for taurine treatment: Acute injection versus chronic supplementation. In the open field test, taurine supplementation increased whereas taurine injection suppressed locomotor activity. We found that taurine supplementation induced an increase in the total distance traveled, the overall movement speed, the time the animals spent mobile, the number of line crossings, and the time the animals entered the center zone. In the elevated arm maze, taurine injection suppressed anxiety whereas taurine supplementation was anxiogenic. The major findings of this are two folds: First these results suggest that taurine might play a role in the modulation of anxiety and locomotor activity. Second, taurine when injected acutely had opposite effects than when administered chronically.

Abbreviation *Tau*, taurine; *KA*, kainic acid; *Inj*, injected

21.1 Introduction

Anxiety disorders are considered the most common psychiatric diagnoses, affecting between 10 and 30% of the general population. Excess anxiety can be debilitating and damage the quality of life. Benzodiazepines have been extensively used for the treatment of several forms of anxiety, although these compounds have well-known side-effects such as sedation, muscle relaxation, amnesia, and dependence (Rickels and Schweizer 1997). The development of new anxiolytic drugs has been an area of

A. El Idrissi (✉)
Department of Biology, City University of New York Graduate School, NewYork

J. Azuma et al. (eds.), *Taurine 7*, Advances in Experimental Medicine and Biology 643,
DOI 10.1007/978-0-387-75681-3_21,

interest. In a search for new anxiolytic compounds, various types of non-traditional medicines have been used in the world today (Rex et al. 2002).

The GABAergic system plays a very important role in the regulation of anxiety. Since taurine interacts with GABAA receptors and mimics the actions of GABA, we determined the effects of taurine on this type of behavior. Furthermore, we investigated the relationship between taurine and anxiety in to two treatment paradigms: acute and chronic. We examined the effects of taurine on the behavior of mice in the elevated plus-maze, a most commonly used animal models of anxiety and in an open-field test that provides a simple method for measuring the response of an animal to an unfamiliar environment.

Taurine, 2-aminoethane-sulphonic acid, is a sulfur-containing amino acids found in relatively high concentrations in the central nervous system of mammals. Taurine has been shown to be essential for the development, survival, growth of vertebrate neurons (Hayes et al. 1975). Taurine deficiency has been confirmed in many neuropathological conditions, such as epilepsy (Barbeau et al. 1975; Joseph and Emson 1976), mental depression(Perry 1976), and the alcohol withdrawal syndrome (Ikeda 1977). Furthermore, we have shown that acute taurine injections increased the threshold of pharmacologically-induced convulsions (El Idrissi et al. 2003). Taurine (43 mg/kg, s.c) significantly increased the latency and decreased the duration of convulsions induced by kainic acid. While is well established that taurine acts as an agonist for $GABA_A$ receptors, chronic supplementation of taurine to mice, induces biochemical alterations in the inhibitory system (El Idrissi and Trenkner 2004). These alterations tend to increase neuronal excitability, eliciting therefore the opposite effects than acute injections of taurine. Thus, we examined in this study the effects of chronic versus acute treatment with taurine.

21.2 Methods

21.2.1 Open-Field Apparatus

The open-field used is a 60 × 60 cm square arena with 15 cm high walls. It is subdivided into 3 square sectors designated as outer, middle and inner zone. A male laboratory mouse is placed in the centre, covered by a small dome which was pulled up by an operator when the learners' recording activity begins. A video camera is positioned at about 1.5 m above the arena, immediately inside the vertical projection of a wall, covering the entire view of the arena. Animals were monitored for 10 min. After each run the mouse was returned to its home cage and the maze was cleaned with a damp sponge to remove any trace of odor.

21.2.2 Elevated-Plus Maze Apparatus

The elevated plus-maze used in this study comprised two opposing open arms (30 × 5cm) and two closed arms (30 × 5 × 15cm), which joined at a square central area

(5×5cm) to form a plus sign. The maze floor was constructed of black Plexiglas and the side/end walls (15cm height) of the enclosed arms of clear Plexiglas. To reduce the likelihood of falling-over, a slight raised edge (0.25 cm) around the perimeter of the open arms provided additional grip for the animals. The entire apparatus was elevated to a height of 45 cm above the floor by a single central support and four 25-W red fluorescent lights arranged as a cross at 100 cm above the maze were used as the source of illumination (Chen et al. 2003). Testing commenced by placing a mouse on the central platform of the maze facing an open arm. Its behavior on the plus-maze was recorded for 5 min by a vertically mounted video camera linked to a monitor and video recorder in an adjacent laboratory. After each run the mouse was returned to its home cage and the maze was cleaned with a damp sponge to remove any trace of odor.

21.2.3 Data Analysis

Data were recorded as digital video clips using an analog-digital converter. The movies were analyzed using AnyMaze software. Tracking of the animal was based on contrast relative to background. Different zones were labeled and indicated on the monitor. Two tracking points were specified one the head and the other the center of gravity of the animal. An excel spreadsheet was generated containing all the parameters specified.

21.2.4 Statistic Analysis

Multifactorial analysis of variance was used to identify overall condition effects. Significant changes were determined by post hoc comparisons of means using the Tukey HSD test. Significance was set at a confidence level of 95%. Data are presented as mean $\pm$ SEM.

21.3 Results

21.3.1 Effects of Taurine on Locomotor Activity in an Open-Field

Open-field test provides a simple method for measuring the response of an animal to an unfamiliar environment and can be used to detect emotionality, anxiety and/or responses to stress in animals. Using this behavioral test, we found that taurine injection (43 mg/kg, s.c) decreased all parameters measured in the open-field test when compared to controls. On the other hand, taurine supplementation in the drinking water (0.05%) for four weeks induced the opposite effects than acute injection. Table 21.1 shows the different parameters measured. Injection of taurine (43 mg/kg, s.c) 15 min before the test significantly decreased the total distance

Table 21.1 Effects of taurine on locomotor activity

	Con	Tau-Fed	Tau-inj
Total distance traveled 9 m	69.83	72.81	30.36
Overall average speed (m/s)	0.12	0.12	0.05
Total time mobile (s)	473.24	481.53	285.44
Total time immobile (s)	126.76	118.47	314.56
Total mobile episodes	34.79	33.29	31.00
Total immobile episodes	34.21	33.00	30.36
Number of line crossing	208.71	250.00	101.00
Total distance traveled by the head	89.31	96.91	42.25

Con controls, *Tau-Fed* taurine-fed (0.05% for 4 weeks), *Tau-inj* taurine-injected (43 mg/kg, s.c). All mice were 2 months old. Taurine (0.05%) was supplemented in the drinking water for 4 weeks. control, n=6; Tau-fed, n=7, Tau-inj, n=7.

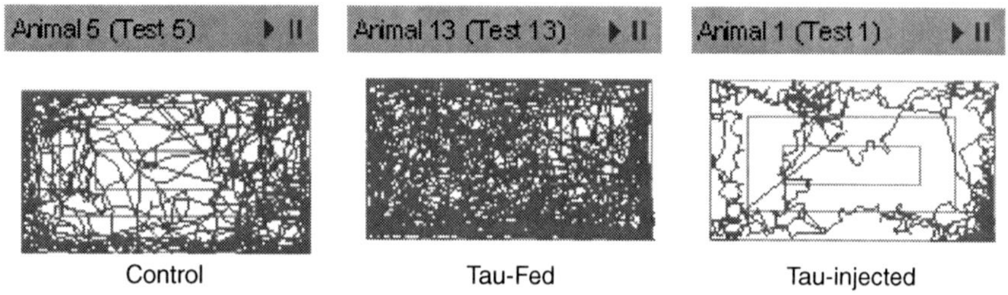

Fig. 21.1 Representative track plots depicting locomotor activity in an open-field. Mice were tracked for 10 min. Control mice spent most of the time around the periphery of the apparatus. On the other hand, taurine-fed mice moved significantly more. Injection of taurine (43 m/kg s.c) significantly suppressed locomotor activity and mice spent most of the time immobile in the corner of the platform

traveled, the average speed, the time the animal spent moving, and the number of line crossings. On the other, supplementation of taurine in the drinking water did not affect all parameters. However, the animals spent more time moving and therefore traveled longer distances. In doing so the taurine-fed mice crossed more lines than controls, although the speed of travel was the same as controls. Figure 21.1 shows a representative tracking plot of a control, a taurine-fed and a taurine-injected mice during the open-field test. Control mice (saline injected) spent more time around the periphery of the apparatus whereas taurine supplemented mice showed a significant increase in locomotor activity. On the other hand taurine injected mice showed a drastic reduction in the overall locomotor activity spending the bulk of the test time immobile around the corner of the apparatus.

21.3.2 Effects of Taurine on Anxiety-Like Behavior

To assess the effects of taurine on anxiety-like behavior, we used an elevated-plus maze. The elevated plus-maze is a well-established animal model for testing anxiolytic drugs (Dawson and Tricklebank 1995; Kulkarni and Reddy 1996). In the test,

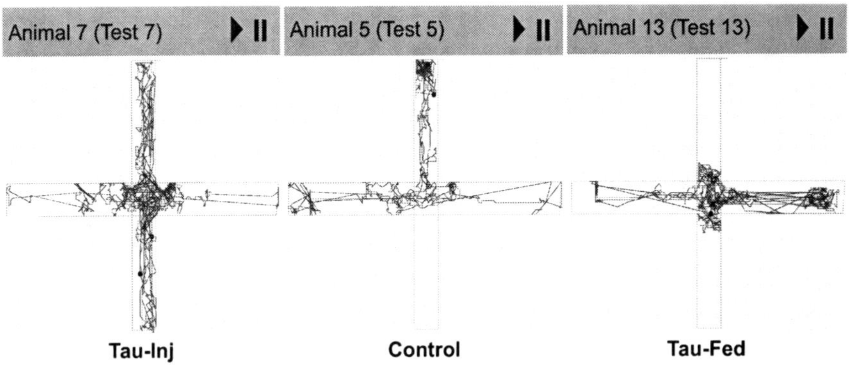

Fig. 21.2 Representative track plots showing motoric activity in an elevated plus maze. Mice were tracked for 5 min. Control mice spent approximately 20% of the test time exploring the open arm of the apparatus, shown here in the vertical position. On the other hand, taurine-injected (43 m/kg s.c) mice spent significantly more time exploring the open arm of the maze. Supplementation of taurine in drinking water (0.05% for 4 weeks) heightened the anxiety level and in some case the mice did not explore the open arm and spent all the time in the closed arm (shown here in the horizontal position)

the percentage of entries into the open arms and of the time spent in open arms have generally been used as indices of the anxiety.

Consistent with the open-field data, injection of taurine reduced the overall speed of locomotion (Table 21.1). However, the effects of taurine were more pronounced

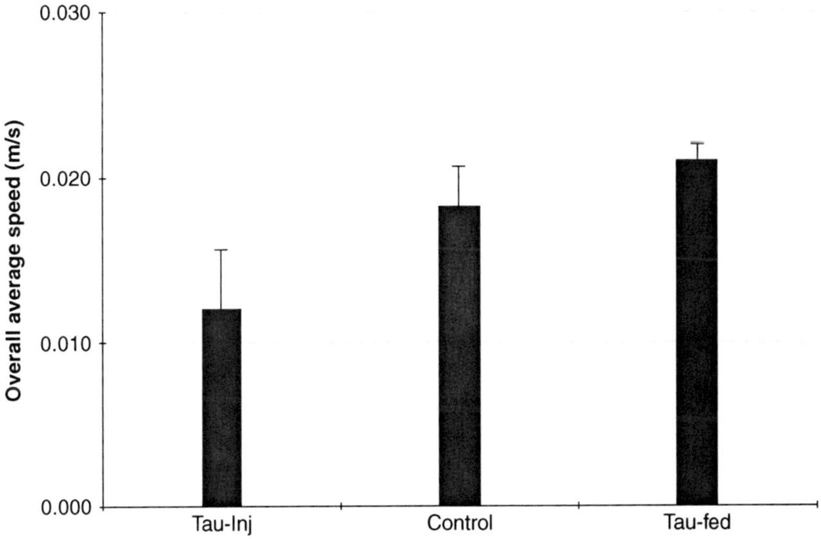

Fig. 21.3 Effects of taurine on speed of movement in the elevated plus maze. All mice were 2 months old. Taurine (0.05%) was supplemented in the drinking water for 4 weeks. Data represent mean ± SD. control, n=6; Tau-fed, n=7, Tau-inj, n=7

on anxiety–like behavior. In contrast to taurine-fed mice, taurine-injected mice spend more time in the open arm and less time in the closed arm (Fig. 21.2). In some cases, the anxiety level was so high in the taurine-fed mice that they did not explore the open arm at all. Thus, it seems that taurine, depending on the time and duration of treatment may have completely opposite effects on anxiety-like behavior. Acute injection of taurine has an axiolytic effect whereas chronic supplementation of taurine has an anxiogenic effect.

Consistent with the open-field data, injection of taurine reduced and supplementation of taurine to the drinking water increased the overall speed of locomotion in the elevated plus maze (Fig. 21.3). Therefore, the two behavioral test used in this study are consistent on the effects of taurine on locomotor activity. Acute taurine injection suppresses locomotor activity whereas chronic supplementation of taurine increases locomotor activity.

21.4 Discussion

Taurine is one of the major constituents of the free amino acid pool in the CNS (Shaw and Heine 1965; Guidotti et al. 1972). Taurine has a heterogeneous distribution in the brain with high levels found in cerebral cortical areas, hippocampus, caudate-putamen, and in cerebellum. Taurine enters the brain via a high affinity, saturable, sodium and chloride dependent carrier from blood to the endothelial cell (Benrabh et al. 1995). Once in the brain taurine may exert several biological effects. With respect to it neuromodulatory effects, taurine has been shown to interact with $GABA_A$ and glycine receptors. In recent years, the agonistic action of taurine on the inhibitory neurotransmission has been the focus of several studies. Taurine has been shown to be a low affinity agonist for $GABA_A$ receptors. Moreover, it has been found that taurine interacted with $GABA_A$ receptor-linked benzodiazepine receptor binding sites (Medina and DeRobertis 1984). Since the GABAergic system plays an important role in modulating anxiety-like behaviors and Diazepan/benzodiazepan, a typical anxiolytic drugs interact with $GABA_A$ receptor to the same binding site as taurine, we sought to determine the anxiety-modulatory role of taurine. Taurine binds to, and mimics the effects of GABA on the $GABA_A$ receptors. However, the functional consequence of taurine interaction with the inhibitory system is highly dependent on the duration of treatment. This is consistent with our finding of the effects of taurine on seizure threshold. Acute injection of taurine prior to seizure induction with kainic acid elevated seizure threshold exerting an anti-epileptic effect, whereas chronic supplementation of taurine has a pro-epileptic effect and lowered seizure threshold. We used the same treatment with taurine (acute and chronic) and evaluated the effects of taurine on anxiety and locomotor activity.

There are two major findings associated with the current work. Most importantly, taurine produced behavioral effects in both elevated plus maze and the open-field. The effect of taurine however was dependent on the treatment paradigm. Acute injection of taurine (43 mg/kg, s.c) has an anxiolytic effect whereas chronic supplementation in drinking water (0.05% for 4 weeks) had an anxiogenic effect.

Since acute injection of taurine activates $GABA_A$ receptors and chronic supplementation of taurine induces alteration in the inhibitory system, these data may suggest that taurine acting through the GABAergic system modulate anxiety-like behaviors. It has been shown that the strychnine-sensitive glycine receptors function to modulate anxiety-like behaviors *in vivo* (Danober and Pape 1998; McCool and Botting 2000). Since taurine interacts with these receptors as well, it is conceivable that the interaction of taurine with both inhibitory systems and the cooperative actions of these two systems is responsible for the observed effects of taurine on anxiety-like behaviors. Previous work has shown that $GABA_A$ receptor activation in the amygdala, either directly with agonist (Bueno et al. 2005) or indirectly with benzodiazepine agonist (Zangrossi and Graeff 1994), causes anxiolysis while inhibition with receptor antagonists cause anxiogenesis (Sanders and Shekhar 1995). Given that both glycine and $GABA_A$ receptors are ligand-gated chloride channels, we hypothesized that $GABA_A$ and/or glycine receptors activation by taurine would also produce anxiolytic effects under the conditions of our behavioral assays. On the other hand, chronic supplementation of taurine in drinking water, which increases neuronal excitability, produces anxiogenic effects.

Taurine elicited slightly distinct effects in the open-field and the elevated-plus maze where acute injection of taurine suppressed locomotor activity and chronic taurine supplementation increased locomotor activity in the open-field test. This is consistent with previous findings. In open-field test, injection of taurine significantly decreased ambulation levels, increased latency scores, and increased thigmotaxis(Sanberg and Ossenkopp 1977). While it may seem that acute taurine injection may be anxiogenic, since the number of visits to the center zone of taurine-injected mice was drastically reduced in the open-field test (Fig. 21.1), we think that this anxiogenic-like effect is a consequence of reduced locomotor activity. The assay-dependent taurine effects may also reflect distinct neurobiological contributions to anxiety-like behaviors expressed in each apparatus. The plus maze and open-field test share many characteristics. They are both perceived as unconditioned responses to naturally aversive environments; and, anxiety-like behaviors expressed in both assays are sensitive to many of the same pharmacological manipulations (Belzung and Griebel 2001; Bourin and Hascoet 2003; Rodgers and Dalvi 1997; Wall and Messier 2001). Despite these similarities, anxiety-like behavior is clearly multidimensional (Ramos and Mormede 1998); and, the unique environments in each assay are likely to recruit unique neurobiological processes. Thus, the test-specific effects of taurine may merely provide an additional demonstration of the complex and multidimensional character of anxiety.

21.5 Conclusion

In summary, this study shows that taurine regulates both locomotor and anxiety-like behavior in mice. Acute taurine injection reduced locomotor activity and anxiolytic. Chronic taurine supplementation induced a state of neuronal hyper-excitability characterized by increased ambulatory levels and heightened anxiety.

Acknowledgments We thank Ekaterina Zavyalova, Candice Cruz and Labentina Shala for helping with the behavioral testing. This work was supported by PSC-CUNY and CSI.

References

Barbeau A, Inoue N, Tsukada Y, Butterworth RF (1975) The neuropharmacology of taurine. Life Sci 17:669–678

Belzung C, Griebel G (2001) Measuring normal and pathological anxiety-like behaviour in mice:a review. Behav Brain Res 125:141–149

Benrabh H, Bourre JM, Lefauconnier JM (1995) Taurine transport at the blood-brain barrier: an in vivo brain perfusion study. Brain Res 692:57–65

Bourin M, Hascoet M (2003) The mouse light/dark box test. Eur J Pharmacol 463:55–65

Bueno CH, Zangrossi Jr. H, Viana MB (2005) The inactivation of the basolateral nucleus of the rat amygdala has an anxiolytic effect in the elevated T-maze and light/dark transition tests. Braz J Med Biol Res 38:1697–1701

Chen SW, Xin Q, Kong WX, Min L, Li JF (2003) Anxiolytic-like effect of succinic acid in mice. Life Sci 73:3257–3264

El Idrissi A, Trenkner E (2004) Taurine as a modulator of excitatory and inhibitory neurotransmission. Neurochem Res 29:189-197

Danober L, Pape HC (1998) Strychnine-sensitive glycine responses in neurons of the lateral amygdala: an electrophysiological and immunocytochemical characterization. Neuroscience 85:427–441

Dawson and Tricklebank MD (1995) Use of the elevated plus-maze in the search for novel anxiolytic agents. Trends in Pharm Sci 16:33–36

Guidotti A, Badiani G, Pepeu G (1972) Taurine distribution in cat brain. J Neurochem 19:431–435

Hayes KC, Carey SY, Schmidt SY (1975) Retinal degeneration associated with taurine deficiency in the cat. Science 188:949

Ikeda HC (1977) Effects of taurine on alcohol withdrawal. Lancet 2 (8036):509

Joseph and Emson (1976) Taurine and cobalt induced epilepsy in the rat: a biochemical and electrocorticographic study. J Neurochem 27:1495–1501

Kulkarni and Reddy DS (1996) Animal behavioral models for testing antianxiety agents. Methods and Findings in Experimental and Clinical Pharmacology 18:219–230

McCool BA, Botting SK (2000) Characterization of strychnine-sensitive glycine receptors in acutely isolated adult rat basolateral amygdala neurons. Brain Res 859:341–351

Medina JH, DeRobertis E (1984) Taurine modulation of the benzodiazepine gamma-aminobutyric acid receptor complex in brain membranes. J Neurochem 42:1212–1217

Perry TL (1976) Hereditary mental depression with taurine deficiency: futher studies, including a therapeutic trial of taurine administration. In: Huxtable R, Barbeau A (eds) Taurine, Raven Press, New York, pp 365–374

Ramos A, Mormede P (1998) Stress and emotionality: a multidimensional and genetic approach. Neurosci Biobehav Rev 22:33–57

Rex A, Morgenstern E, Fink H (2002) Anxiolytic-like effects of Kava-Kava in the elevated plus maze test – a comparison with diazepam. Prog Neuro-Psychopharmacol Biol Psychiatry 26:855–860

Rickels K, Schweizer E (1997) The clinical presentation of generalized anxiety in primary-care setting: practical concepts of classification and management. J Clin Psychiatry 58:4–9

Rodgers RJ, Dalvi A (1997) Anxiety, defence and the elevated plus-maze. Neurosci Biobehav Rev 21:801–810

Sanberg RP, Ossenkopp KP (1977) Dose-response effects of taurine on some open-field behaviors in the rat. Psychopharmacology 53:207–209

Sanders SK, Shekhar A (1995) Regulation of anxiety by GABAA receptors in the rat amygdale. Pharmacol Biochem Behav 52:701–706
Shaw RK, Heine JD (1965) Ninhydrin positive substances present in different areas of normal rat brain. J Neurochem 12:151–155
Wall PM, Messier C (2001) Methodological and conceptual issues in the use of the elevated plus-maze as a psychological measurement instrument of animal anxiety-like behavior. Neurosci Biobehav Rev 25:275–286
Zangrossi H, Graeff FG (1994) Behavioral effects of intra-amygdala injections of GABA and 5-HT acting drugs in the elevated plus-maze. Braz J Med Biol Res 27:2453–2456

Chapter 22
Taurine Transporter in Lymphocytes of Patients with Major Depression Treated with Venlafaxine Plus Psychotherapy

Fili Fazzino, Francisco Obregón, Margarita Morles, Andrés Rojas, Luis Arocha, Salvador Mata, and Lucimey Lima

Abstract The taurine transporter and taurine are present in lymphocytes, where taurine functions as an antioxidant and an anti-inflammatory agent. Taurine levels are elevated in lymphocytes of subjects with major depression, but returns to control levels after treatment with the antidepressant mirtazapine. Patients (40) were diagnosed using the Diagnostic and Statistical Manual IV of the American Psychiatric Association, and the severity of their condition was determined by the Hamilton Scale of Depression. One group of patients was treated with venlafaxine and the other with venlafaxine plus Neuro-Linguistic Programming. Lymphocytes were isolated from the peripheral blood by Ficoll/Hypaque. The coexistence of the taurine transporter with a subpopulation of CD4+ and CD8+ lymphocytes was measured by immunofluorescence. The levels of the pro-inflammatory, IL-2, and the anti-inflammatory, IL-4, cytokines were determined by ELISA while plasma amino acid levels were determined by HPLC. The percentage of CD4+ cells significantly decreased after both treatments, whereas the levels of CD8+ cells remained unchanged. The taurine transporter of CD4+ and CD8+ cells decreased after integrate treatment. No differences were found in the levels of IL-2 while IL-4 levels increased after integrate treatment. The observed effects of treatment on the taurine transporter and IL-4 content might modify lymphocyte activity during depression.

Abbreviations *APA*, American Psychiatric Association; *CNS*, central nervous system; *DSM-IV*, Diagnostic and Statistical Manual IV; *HAM-D*, Hamilton Scale of Depression; *IFN-γ*, interferon gamma; *Il*, interleukin; *NLP*, Neuro-Linguistic Programming; *Tau-Cl*, taurine cloramine; *TAUT*, taurine transporter; *VF*, venlafaxine

F. Fazzino (✉)
Laboratorio de Neuroquímica, Centro de Biofísica y Bioquímica, Instituto Venezolano de Investigaciones Científicas, Caracas, Venezuela

J. Azuma et al. (eds.), *Taurine 7*, Advances in Experimental Medicine and Biology 643,
DOI 10.1007/978-0-387-75681-3_22,

22.1 Introduction

Both taurine and the taurine transporter (TAUT) are present in lymphocytes (Vinton et al. 1987, Fazzino et al. 2006). Taurine plays an important role in several biological processes, such as development and regeneration of the central nervous system (CNS), as well as the modulation of calcium homeostasis, membrane stabilization, reproduction, and regulation of immune function (Huxtable 1992; Sturman 1993; Schuller-Levis et al. 1990). One of the functions of taurine in these cells might be related to its antioxidant activity (Learn et al. 1990), its regulation of pro-inflammatory cytokines, and its involvement in the formation of taurine cloramine (Tau-Cl) (Grimble 2006). Exogenous taurine was shown to possess significant anti-inflammatory activity in various *in vivo* and *in vitro* models of inflammation, and to protect against oxidative damage in inflammatory bowel disease, pancreatitis, and gastric mucosal injury (Kim et al. 1996; Marcinkiewicz et al. 1995). This anti-inflammatory action of taurine was shown to be a direct result of its antioxidant effects, including inhibition of lipid peroxidation and neutrophil activation (Kim et al. 1996). The ability of taurine to combat oxidative damage arises from its ability to scavenge hypochlorous acid (HOCl) forming in the process the relatively harmless Tau-Cl (Marquez and Dunford 1994). Tau-Cl inhibits the activation of NF-kappaB, a potent signal transducer of inflammatory cytokines (Schuller-Levis and Park 2003).

Major depression is a serious condition characterized by low mood and loss of interest or pleasure which lasts at least two weeks (American Psychiatric Association 1994). Considerable evidence supports the notion that this disorder causes an increase in the development of medical illnesses, including dysfunction of the immune response (Leonard 2006). An analysis of clinical data of immune abnormalities in depressed patients found that they exhibit an overall leukocytosis, manifested as relative neutrophilia and lymphopenia, an increase in the $CD4^+/CD8^+$ ratio, a reduction in natural killer-cell cytotoxicity, a reduction in lymphocyte proliferative responses to mitogens and an increase in circulating haptoglobin, prostaglandin E_2 and interleukin-6 (IL-6) (Zorrilla et al. 2001; González et al. 2007).

Previous studies have shown that taurine levels are elevated in lymphocytes of subjects with an episode of major depression, although, after treatment with the antidepressant mirtazapine, taurine returned to control levels (Lima et al. 2003). The purpose of this study was to evaluate whether taurine levels or the number of lymphocytes that express the taurine transporter change in patients with major depression before and after treatment with venlafaxine (VF) and with VF plus Neuro-Linguistic Programming (NLP).

22.2 Methods

22.2.1 Subjects

Twenty patients from 18 to 60 years of age were diagnosed using the Diagnostic and Statistical Manual IV of the American Psychiatric Association (DSM-IV)

(APA 1994), with the severity of depression diagnosed using the Hamilton Scale of Depression (HAM-D) (Hamilton 1960). None of the subjects had other medical or psychiatric illnesses of the Axis I, and were free of drugs for one month prior to starting the study. One group of patients was treated with VF and another group with VF plus weekly sessions of NLP for 6 weeks.

22.2.2 Preparation of Blood Peripheral Lymphocytes

Blood samples were taken between 7 and 9 a.m. by venipunture and placed in tubes containing 0.6 ml of heparin, 1000 U/ml. The blood was centrifuged at 1000 rpm with a vasculant rotor for 10 min at room temperature. The plasma was collected for determination of individual cytokine and amino acid levels. Moreover, the layer of white cells plus some red blood cells was taken and transferred to tubes with 10 ml of saline 0.1 M sodium phosphate buffer pH 7.4 (PBS). These suspensions were placed on 5 ml of Ficoll/Hypaque (1077 g/l). After centrifugation at 2000 rpm for 30 min the mononuclear cell layer was taken, washed twice with PBS and centrifuged at 1200 rpm for 10 min. To achieve an enriched lymphocyte preparation with minimal monocyte contamination, the resulting pellet was diluted in Roswell Park Memorial Institute Medium 1640 (RPMI) free of bovine serum albumin (BSA), and incubated in a plastic flask for 45 min at 37°C and 5% of CO_2. After the incubation, lymphocytes, which were non-adherent cells (>80% of cells), were dislodged from adherent monocytes, transferred to plastic tubes and washed twice. The integrity of the isolated lymphocytes was determined by the Trypan blue exclusion test, and was shown to be greater than 90%. For immunolabelling, aliquots of the cell suspension (1,000,000 cells/aliquot) were fixed in 2% para-formaldehyde.

22.2.3 Immunolabeling

For staining, the cell suspension was fixed with 2% para-formaldehyde, was washed 3 times in 1 ml of PBS and re-suspended in 100μl of PBS. The cell suspension was pre-incubated for 30 min with 100μl of blocking solution containing 10% BSA in PBS and 0.3% Triton X-100. It was then incubated with mouse monoclonal antibody directed against human CD4 and CD8 (Santa Cruz Biotechnology, Inc, California, U.S.A.) diluted 1:50 in PBS plus 1.5% BSA and 0.3% Triton X-100, overnight at 4°C. After two washing cycles with PBS, the cells were labeled with 100μl of bovine rodamine-conjugated antibody against mouse IgG diluted 1:100 with PBS and 1.5% BSA, and incubated for 1 hour at room temperature in a dark chamber. The cells were then washed 3 times with PBS. For the co-localization experiment, the above procedure was repeated, but using a rabbit anti-rat polyclonal antibody against TAUT, 1:50 from original preparation (Pow et al. 2002) as the primary antibody, and anti-rabbit IgG-fluorescein (1:100) as the second antibody. Finally, the cells were re-suspended in 100μl of 2% paraformaldehyde. Slides were examined using fluorescence microscopy with appropriate filters; 500 cells were evaluated per condition. The number of cells that were positive for either CD4 or CD8 with TAUT were counted.

22.2.4 *Measurement of Plasma Interleukin-2 and Interleukin-4*

Plasma, collected for IL-2 and IL-4 assays, was stored at −80°C. The levels of IL-2 and IL-4 were measured by the ELISA Endogen kits (Pierce Endogen, Cambridge, MA). Briefly, 50μl of sample and biotinylated anti-IL-2 (or IL-4) were added to 96 wells coated with rat IL-2 or IL-4 antibody and incubated for 2 hours at room temperature. The wells were then washed 4 times and 100μl of stabilized chromogen was added to each well, and incubated for 30 min at room temperature in the dark. The reaction was stopped with 100μl of stop solution and the absorbance was read at 450 nm. The levels of ILs are expressed as pg/ml.

22.2.5 *Determination of Amino Acids*

Amino acids were determined by HPLC using a modified fluorescent detection method as previously described (Lima et al. 1998). The HPLC system consisted of a Waters 2690 Separation System and a Shimadzu RF-551 fluorescent detector. A Sulpeco LC-18 column 4.6 × 100mm, 5μm was employed for amino acid separation. Platelet poor plasma, 300μl, was acidified with 50μl of 20% sulfosalicylic acid. Centrifugation was carried out at 17,000 rpm for 20 min, at 4°C, and the supernatant was kept at 80°C until chromatographic analysis. Immediately before injection, 50μl of the supernatant plus 150μl of potassium borate buffer pH 10.4 and 200μl of the mixture: 25 mg o-phtaldehyde, 500μl methanol, 25μl β-mercaptoethanol (1 g/ml), and 4.5 ml 0.4 M potassium borate buffer pH 10.4 was used for derivatization. Then, 15μl of the derivatized preparation was injected into the chromatographic system. The levels of amino acids were calculated (as nmol/ml) from the area under the curve using an external standard and the program Millenium.

22.2.6 *Analysis of Data*

Data are expressed as the arithmetic means ± standard error of the mean (SEM). Differences were statistically analyzed using the Student's t-test. Statistical significance was designated as $P < 0.05$.

22.3 Results

Figure 22.1A shows that the percentage of CD4+ cells before treatment was approximately 60% but significantly decreased to 52% after treatment for depression. However, CD8+ cells (Fig. 22.1B) remained unchanged by treatment. Immunolabeling revealed that VF treatment reduced the percentage of CD4+ and CD8+ cells containing TAUT to approximately 84% and 89%, respectively. By comparison, treatment with VF plus NLP reduced the TAUT containing cells to 77% for CD4+ and 80% for CD8+. The percentage of TAUT found in all T lymphocytes cells was approximately 60%, a significant reduction compared to the untreated group.

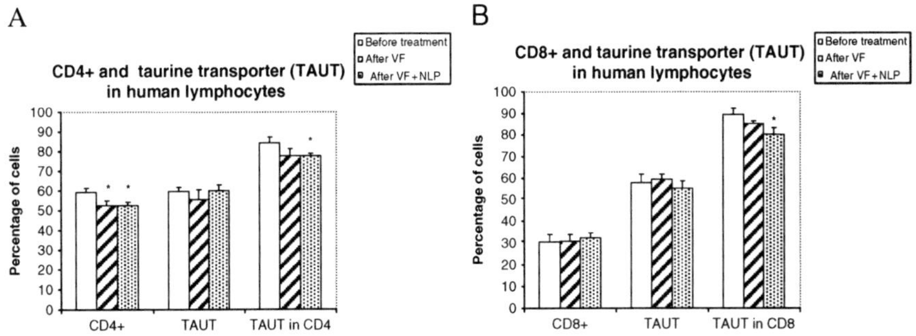

Fig. 22.1 Percentage of positive CD4+ cells and positive TAUT containing cells with respect to CD4+ subpopulations of T lymphocytes (A) and of positive CD8+ cells and positive TAUT with respect to CD8+ subpopulations of T lymphocytes (B), before and after treatment with venlafaxine (VF) and VF plus Neuro-Linguistic Programming (NLP). Each value represents the mean ± SEM, $P < 0,05$ with respect to before treatment group

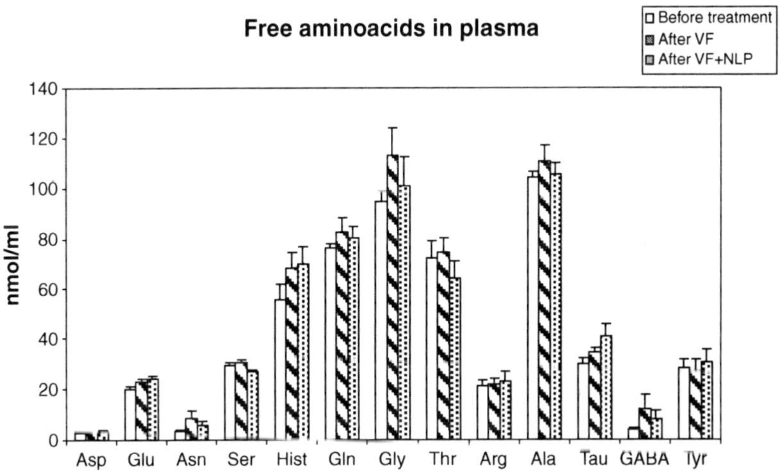

Fig. 22.2 Concentration of free plasma amino acids in nmol/ml before and after treatment with venlafaxine (VF) and VF plus Neuro-Linguistic Programming (NLP)

The concentration of the pro-inflammatory cytokine IL-2 remained unchanged by treatment, while, IL-4 significantly increased after treatment with VF plus NLP Table 22.1.

Table 22.1 Concentration of pro-inflammatory cytokine IL-2 and anti-inflammatory cytokine IL-4 before and after treatment with venlafaxine (VF) and VF plus Neuro-Linguistic Programming (NLP)

Cytokine	Before treatment (pg/ml)	After VF (pg/ml)	After VF + NLP (pg/ml)
IL-2	10.04 ± 0.84	11.07 ± 1.93	11.65 ± 1.23
IL-4	1.47 ± 0.13	1.72 ± 0.18	2.38 ± 0.17*

Each value represents the mean ± SEM, $^{*}P < 0,05$ with respect to before treatment.

Figure 22.2 shows the levels of amino acids in plasma, which were not significantly modified by treatment.

22.4 Discussion

Several studies have demonstrated that major depression is mediated by abnormalities in neurotransmitter function (deficient neurotransmission and decreased levels of serotonin), and alterations in endocrine and immune function (Syvalahti 1994; Connor and Leonard 1998; Anisman and Merali 2002; Capuron and Dantzer 2003; Schiepers et al. 2005). Characteristics of immune function in depression include increases in the number of circulating lymphocytes, systemic immune activation (Maes et al. 1995, González et al. 2007) and increased release of pro-inflammatory cytokines (Schiepers et al. 2005). Kubera et al. (1999) demonstrated that patients with treatment-resistant major depression had a significantly higher percentage of CD4+ than healthy controls. In the present study, the percentage of CD4+ cells significantly decreased after treatment, indicating that therapy against depression reduced some T lymphocytes subpopulations and affected the presence of TAUT in these cells. Thus, treatment might lead to functional modifications related to differential proliferation of cell types.

The significant decrease in TAUT in CD4+ and CD8+ cells after treatment with VF plus NLP could modify autocrine-linked regulation of lymphocytes. The fact that the total number of cells expressing TAUT did not change indicates the presence of TAUT in other populations of circulating immune cells. It is possible that the reduction in taurine levels after antidepressant therapy (Lima et al. 2003) could be explained by the observed decrease in cells with the transporter. Voss et al. (2004) demonstrated that transport activity, expression and nuclear localization of TAUT are reduced in a reversible manner following long-term exposure to high extracellular taurine content. The lower number of TAUT in cells can produce a temporal increase in circulating taurine, which could be involved in protection against oxidants and inhibition of pro-inflammatory cytokine-mediated damage through the formation of Tau-Cl. Several reports have demonstrated that Tau-Cl downregulates the production of superoxide anion, IL-6, and IL-8 by activated human polymorphonuclear leukocytes (Park et al. 2002), and suppresses indoleamine-2,3 dioxygenase activation via downregulation of interferon gamma (IFN-γ) production (Wirleitner et al. 2004). These effects of Tau-Cl on the actions of the pro-inflammatory cytokines and tryptophan degradation might be of clinical relevance in major depression.

It has been reported that the levels of the pro-inflammatory ILs are elevated in depressed patients (Dunn et al. 2005), but in this experimental group no difference was found between the concentration of IL-2 before and after treatment. Therefore, there does not appear to be a relationship between taurine and the pro-inflammatory cytokine, IL-2. The anti-inflammatory cytokine IL-4 significantly increased after treatment with VF plus NLP. IL-4 regulates the Th1-Th2 response such that an increase in production of IL-4 inhibits Th1 differentiation, leading to a reduction in IFN-γ production (Mosmann and Moore 1991). Therefore, the increase in the levels of the anti-inflammatory cytokine IL-4 might attenuate the increase in the pro-inflammatory cytokines and of tryptophan degradation of subjects treated for major depression.

Finally, the levels of amino acids in plasma were not significantly modified by treatment, which could be due to the large size of the plasma compartment and the relatively low number of patients.

In conclusion, in this particular group of patients there was no modification in plasma taurine, the TAUT and the levels of IL-2, although it has been reported that depression is associated with an increase in the levels of the pro-inflammatory ILs (Schiepers et al. 2005). Interestingly, treatment of depression led to a reduction in CD4+, CD4+/TAUT and CD8+/TAUT cells, with a resulting change in function. The increase in the levels of the anti-inflammatory cytokine IL-4 might counter the increase in the pro-inflammatory cytokines that accompanies the treatment of depressed subjects with VF plus NLP.

Acknowledgments Supported by Fondo Nacional de Ciencia, Tecnología e Innovación (FONACIT G-1387), Venezuela.

References

American Psychiatric Association (1994) American Psychiatric Association, Diagnostic and Statistical Manual of Mental Disorders DSM-IV (4th ed), American Psychiatric Press, Washington, DC, USA

Anisman H, Merali Z (2002) Cytokines, stress, and depressive illness. Brain Behav Immun 16: 513–524

Capuron L, Dantzer R (2003) Cytokines and depression: the need for a new paradigm. Brain Behav Immun Suppl 1:S119–S124

Connor T, Leonard B (1998 Depression, stress and immunological activation: the role of cytokines in depressive disorders. Life Sci 62:583–606

Dunn AJ, Swiergiel AH, de Beaurepaire R (2005) Cytokines as mediators of depression: what can we learn from animal studies? Neurosci Biobehav Rev 29:891–909

Fazzino F, Urbina M, Mata S, Lima L (2006) Taurine transport and transporter localization in peripheral blood lymphocytes of controls and major depression patients. Adv Exp Med Biol583:423–426

González A, Fazzino F, Castillo M, Mata S, Lima L (2007) Serotonin, 5-HT1A serotonin receptors and proliferation of lymphocytes in major depression patients. Neuroimmunomodulation 14: 8–15

Grimble RF (2006) The effects of sulfur amino acid intake on immune function in humans. J Nutr 136:1660S–1665S

Hamilton M (1960) A rating scale for depression. J Neurol Neurosurg Psychiatry 23:56-62

Huxtable R, (1992) Physiological actions of taurine. Physiol Rev 72:101–163

Kim S, Kim H, Yang W, Kim B (1996) Protective effect of taurine on indomethacin-induced gastric mucosal injury. In: RJ Huxtable, J Azuma, K Kuriyama, M Kakagawa, A Baba (eds) Taurine, Plenum Press, New York, pp 147–155

Kubera M, Van Bockstaele D, Maes M (1999) Leukocyte subsets in treatment-resistant major depression. Pol J Pharmacol 51:547–549

Learn DB, Fried VA, Thomas EL (1990) Taurine and hypotaurine content of human leukocytes. J Leukoc Biol 48:174–182

Leonard BE (2006) HPA and immune axes in stress: involvement of the serotonergic system. Neuroimmunomodulation13:268–276

Lima L, Obregon F, Urbina M, Carreira I, Baccichet E, Pena S (2003) Taurine concentration in human blood peripheral lymphocytes: major depression and treatment with the antidepressant mirtazapine. Adv Exp Med Biol 526:297–304

Lima L, Obregon F, Matus P (1998) Taurine, glutamate and GABA modulate the outgrowth from goldfish retinal explants and its concentrations are affected by the crush of the optic nerve. Amino Acids15:195–209

Maes M, Smith R, Scharpe S (1995) The monocyte-T-lymphocyte hypothesis of major depression. Psychoneuroendocrinology 20:111–116

Marcinkiewicz J, Grabowska A, Bereta J, Stelmaszynska T (1995) Taurine chloramine, a product of activated neutrophils, inhibits the generation of nitric oxide and other macrophage inflammatory mediators. J Leukocyte Biol 58:667–674

Marquez L, Dunford H (1994) Chlorination of taurine by myeloperoxidase; kinetic evidence for an enzyme-bound intermediate. J Biol Chem 269:7950–7956

Mosmann TR, Moore KW (1991) The role of IL-10 in crossregulation of TH1 and TH2 responses. Immunol Today 12:A49–A53

Park E, Jia J, Quinn M, Schuller-Levis G (2002) Taurine chloramine inhibits lymphocyte proliferation and decreases cytokine production in activated human leukocytes. Clin Immunol 102:179–184

Pow DV, Sullivan R, Reye P, Hermanussen S (2002) Localization of taurine transporters, taurine, and (3)H taurine accumulation in the rat retina, pituitary, and brain. Glia 37:153–168

Schiepers O, Wichers M, Maes M (2005) Cytokines and major depression. Progress in Neuro-Psychopharmacology and Biological Psychiatry 29:201–217

Schuller-Levis G, Park E (2003) Taurine: new implications for an old amino acid. FEMS Microbiol Lett 26:195–202

Schuller-Levis G, Mehta PD, Rudelli R, Sturman J (1990) Immunologic consequences of taurine deficiency in cats. J Leukoc Biol 47:321–331

Sturman JA (1993) Taurine in development. Physiol Rev 73:119–147

Syvalahti E (1994) Biological aspects of depression, Acta Psychiatr Scand Suppl 377:S11–S15

Vinton NE, Laidlaw SA, Ament ME, Kopple JD (1987) Taurine concentrations in plasma, blood cells, and urine of children undergoing long-term total parenteral nutrition. Pediatr Res 21: 399–403

Voss J, Pedersen S, Christensen S, Lambert I (2004) Regulation of the expression and subcellular localization of the taurine transporter TauT in mouse NIH3T3 fibroblasts. Eur J Biochem 271:4646–4658

Wirleitner B, Neurauter G, Nagl M, Fuchs D (2004) Down-regulatory effect of N-chlorotaurine on tryptophan degradation and neopterin production in human PBMC. Immunol Lett 93:143–149

Zorrilla E, Luborsky L, McKay J, Rosenthal R, Houldin A, Tax A, McCorkle R, Seligman D, Schmidt K (2001) The Relationship of Depression and Stressors to Immunological Assays: A Meta-Analytic Review. Brain Behav Immun 15:199

Chapter 23
Effect of Medium Osmolarity and Taurine on Neuritic Outgrowth from Goldfish Retinal Explants

Lisbeth Cubillán, Francisco Obregón, and Lucimey Lima

Abstract Taurine stimulates outgrowth of goldfish retinal explants in a concentration- and time-dependent manner, an effect related to calcium movement and protein phosphorylation. Since taurine is an osmoregulator in the central nervous system, and osmolality might influence regeneration, the purpose of this work was to evaluate the possible effect of hypo-osmolality on basal outgrowth and on the trophic action of the amino acid. Accordingly, goldfish retinal explants obtained after crushing the optic nerve were cultured in iso- and hypo-osmotic medium, the latter achieved by diluting the medium 10% 24 and 72 h after plating. The length and density of the neurites, measured after 5 days in culture, were significantly lower in the hypo- than in the iso-osmotic medium. Taurine stimulated the outgrowth under both conditions, but the percentage of increase was greater in iso-osmotic medium. Taurine concentration, determined by HPLC, did not significantly change in explants. Co-administration of β-alanine and taurine impaired the trophic effect of taurine to a greater extent in the iso- than in hypo-osmotic medium, indicating a possible differential interaction with the taurine transporter which could be altered by osmotic stress. The exact mechanism of outgrowth regulation by hypotonicity requires further clarification, taking into considering possible modification of the taurine transporter.

Abbreviations *PKC*, Protein kinase C; *NMDA*, N-methyl-D-aspartic acid; *GAP 43*, Growth associated protein 43; B50, HLA-B50 (B50), human leukocyte antigen, class I

23.1 Introduction

Taurine is involved in many cellular functions, including cell volume regulation (Bouckenooghe et al. 2006; Bridi and Choay 2003; Chesney 1985; Huxtable 1992; Warskulat et al. 2007; Yancey 2005). In cultured goldfish retina, taurine increases

L. Cubillán (✉)
Laboratorio de Neuroquímica, Centro de Biofísica y Bioquímica, Instituto Venezolano de Investigaciones Científicas, Venezuela

J. Azuma et al. (eds.), *Taurine 7*, Advances in Experimental Medicine and Biology 643,
DOI 10.1007/978-0-387-75681-3_23, © Springer Science+Business Media, LLC 2009

axonal outgrowth in a concentration-dependent manner, with optimal effects around 4 mM in the medium (Lima et al. 1988, 1989). The mechanisms underlying this trophic effect also include regulation of calcium flux (Lima et al. 1993), and protein phosphorylation catalyzed by protein kinase C (PKC) (Lima and Cubillos 1998). In addition, blockade of taurine transport into the cells results in attenuated stimulation of outgrowth (Nusetti et al. 2006. Efflux of taurine from frog retina takes place in hypo-osmotic medium (Petrosian et al. 2000). Moreover, subjecting hippocampal slices to hypo-osmotic shock produces a prominent efflux of taurine (Tranberg et al. 2007). Interestingly, among the organic osmolytes, taurine also possesses cytoprotective and antioxidant activity (Yancey, 2005).

Due to the above evidence the aims of this work were to evaluate the outgrowth from post-crush goldfish retinal explants in iso- and hypo-osmotic conditions in the presence and absence of taurine, to determine its levels in cultured tissue and medium, and to study the possible effect of the taurine transport inhibitor, β-alanine, on these conditions.

23.2 Materials and Methods

Retina of goldfish (*Carassius auratus*), 5–6 cm in length anesthetized in 0.05% tricaine, were dissected 10 days after a crush of the optic nerve with fine forceps. Squares of 500μm (5–6 per retina) were prepared by sectioning the retina with a McIlwain tissue chopper and placed, 10-14, on poly-L-lysine pre-coated tissue flasks (25 mm^2). The nutrient medium was Leibovitz, L-15, 3 ml per dish (Sigma) with 0.1 mg/ml of gentamicin and 20 mM of (*N*-[2-hydroethyl]piperazine-*N*'-[2-ethanesulfonic acid]) (HEPES). Osmolarity of the medium was changed after 24 and 72 h in culture by diluting the medium in 10%. Sucrose was added in some experiments at a concentration of 0.5 mM. Taurine, 4 mM, or β-alanine, 0.5 mM, were added to some cultured retinal explants. The density of neurites was evaluated 5 days after plating using a predetermined scale and length was measured by using the program SigmaScanPro (Jandel). Amino acids were determined by HPLC with fluorescence detection employing a modified method described previously (Lima et al. 1989). The HPLC system consisted of a Waters 2690 Separations Module and a Shimadzu RF-551 fluorescence detector. A Supelco LC-18 column 4.6 × 100 mm, 5μm was employed for amino acid separation. Immediately before injection, 100μl of the supernatant was derivatized with 100μl of a mixture of 25 mg *o*-phtalaldehyde, 500μl MeOH, 25μl β-mercaptoethanol (1 g/ml) and 4.5 ml 0.4 M potassium buffer pH 10.4. Aliquots of the derivatized preparation were injected into the chromatographic system. The column was eluted with increasing concentrations of acetonitrile in a 50 mM pH 6.4 sodium phosphate buffer containing 5% acetonitrile and 0.1% tetrahydrofurane. The main step gradient used was 0–10 min, 95% of buffer solution, 10 - 55 min 70%, 55–59 min 20%, 59–62 min 98%. Taurine was quantified using an external standard and concentrations were expressed in nmol/ml for medium and nmol/mg protein for tissue. Results are expressed as means ± standard error of the mean, analysis of variance was performed and Student's *t* test for comparing results. Statistical significance was defined as $P<0.05$.

23.3 Results

The reduction of osmolarity resulted in a significant decrease in the length of neurites as compared to results in iso-osmotic medium (Fig. 23.1). The density of neurites was significantly reduced in explants exposed to hypo-osmotic medium (P = 0.03), while the addition of taurine increased density of the explants exposed to iso-osmotic medium, although taurine had no effect when the explants were exposed to hypo-osmotic medium (data not shown). In another set of experiments the effect of hypo-osmolarity and that of taurine were evaluated. Hypo-osmolality significant ly decreased neurite length; taurine increased the outgrowth of the explants by 33% in iso-osmotic medium and by 25% in hypo-osmotic medium Fig. 23.2).

The addition of sucrose to the medium did not affect neuritic outgrowth, either length or density of neurites (data not shown). The presence of β-alanine in the medium reduced the stimulating effect of taurine by 57% and 18% in iso- and hypo-osmotic media, respectively (Fig. 23.3).

The levels of taurine in the explants increased after its addition to culture flasks. The increase was unaffected by the osmotic state of the medium (iso- vs hypo-osmotic) after 72 h and 5 days exposure to taurine containing medium (Figs. 23.4 and 23.5).

23.4 Discussion

The reduction in outgrowth caused by exposure to hypo-osmotic medium could be related to the relatively fast efflux of endogenous taurine, a process necessary for exerting its trophic effect. Hypo-osmotic stress leads to the release of organic osmolytes. In epithelial cells the release of organic osmolytes involves PKC and

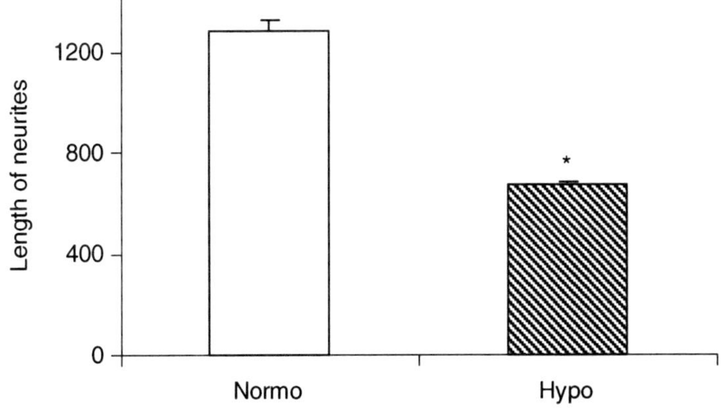

Fig. 23.1 Outgrowth of post-crush goldfish retinal explants in iso- (Normo) and hypo-osmotic (Hypo) media expressed in μ m after 5 days in culture. * $P < 0.05$ respect to Normo

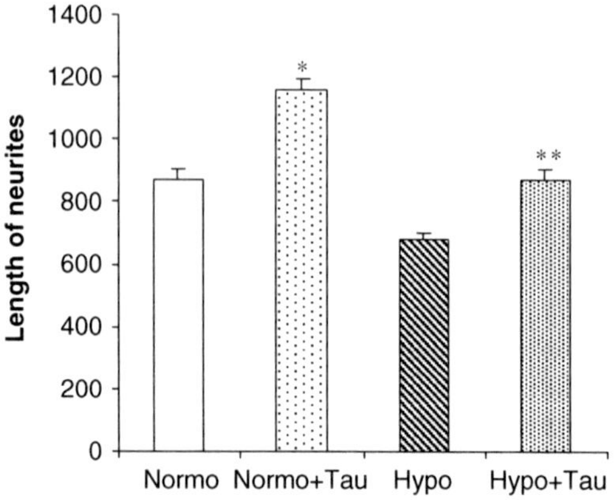

Fig. 23.2 Outgrowth of post-crush goldfish retinal explants in iso- (Normo) and hypo-osmotic (Hypo) media containing or lacking 4 mM taurine (Tau). The data are expressed in μm measured after 5 days in culture. $^{*}P < 0.05$ respect to Normo, and $^{**}P < 0.05$ respect to Hypo

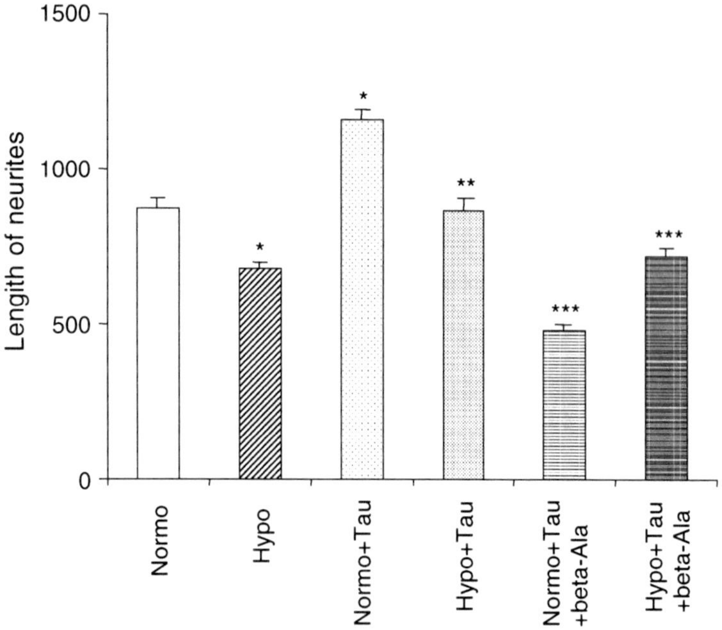

Fig. 23.3 Outgrowth of post-crush goldfish retinal explants in iso- (Normo) and hypo-osmotic (Hypo) media in the absence and in the presence of 4 mM taurine (Tau), or 0.5 mM β-alanine (beta-Ala). Values are expressed in μm after 5 days in culture. $^{*}P < 0.05$ respect to Normo, $^{**}P < 0.05$ respect to Normo+Tau, $^{***}P < 0.05$ respect to Normo+Tau or Hypo+Tau

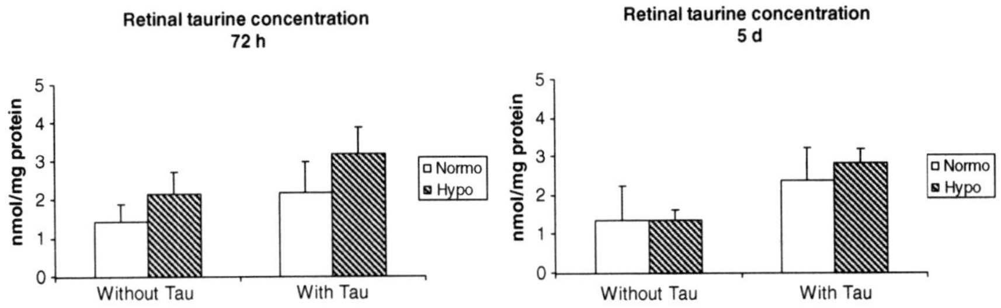

Fig. 23.4 Taurine (Tau) concentration of retinal explants in iso- (Normo) and hypo-osmotic (Hypo) conditions after 72 h or 5 d in culture

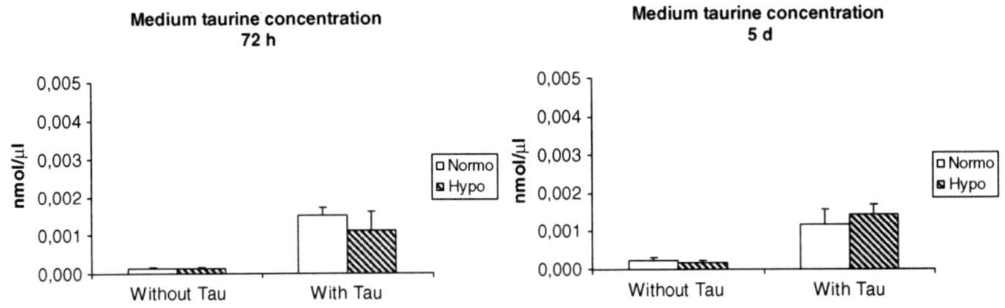

Fig. 23.5 Taurine (Tau) concentration of media in iso- (Normo) and hypo-osmotic (Hypo) conditions after 72 h or 5 d in culture

is independent of volume-regulated anion channels (Tomassen et al. 2004). This might be in agreement with the previously described role of PKC in goldfish retinal outgrowth (Lima and Cubillos 1998).

Exposure to hyper-osmotic stress increases the activity of the taurine transporter by increasing Vmax in retinal cells, such as pigment epithelium, ganglion and M`uller cells (El-Sherbeny et al. 2004). In those cells mRNA is osmotically regulated to protect retinal neuronal function in the rat (Morimura et al. 1997). In the osmotically stressed hippocampus, taurine is rapidly lost from the cells through membrane channels, but there are relevant reductions in the capacity of taurine transporter too (Olson and Martinho 2006). Neuromodulin, known as GAP43 and B50, a protein implicated in axonal growth and synaptic plasticity, is also an osmosensory protein which links phospho-inositide metabolism with changes in calcium flux in response to hypo-tonicity (Caprini et al. 2003). The impairment in retinal outgrowth seen in hypo-osmotic medium might be related to changes in calcium flux, considering that taurine partially exerts its trophic effect by modulation of intracellular calcium (Lima et al. 1993).

In addition, it is likely relevant that taurine is one of the amino acids that increases in rat brain 24 h after injury(Pascual et al. 2007). The lack of changes in taurine concentration in the tissue and medium upon a change in medium osmolality does not exclude the movement of taurine between the tissue and medium taking place early after dilution of the medium. Hypo-osmotic shock for 5 min produces efflux of amino acids from hippocampal slices, but taurine efflux is still prominent after the shock (Tranberg et al. 2007). Thus, a time course must be obtained to adequately assess changes in taurine levels in various compartments. Moreover, the relevance of NMDA glutamate receptors in osmoregulation seem to be associated with amino acid efflux after swelling (Morales et al. 2007; Tranberg et al. 2007).

β-Alanine reduced the effect of taurine in iso-osmotic medium, but was less effective in hypo-osmotic medium, a finding in accordance with potential changes in the activity of the taurine transporter. The reduction in taurine's effect on axonal outgrowth during hypo-osmotic stress might be related to the reduction in taurine transport.

These data indicate that goldfish retinal outgrowth following crushing of the optic nerve is partially influenced by the osmolarity of the medium. Taurine supplementation stimulates neurite length in an osmotic-dependent manner. The fact that β-alanine suppresses the trophic action of taurine to a greater extent in iso- than in hypo-osmotic medium might reflect the status of the retinal taurine transporter. Further studies should focus on the mechanism underlying these osmotic-linked changes.

Acknowledgments Fondo Nacional de Ciencia, Tecnología e Innovación, FONACIT S1-903.

References

Bouckenooghe T, Remacle C, Reusens B (2006) Is taurine a functional nutrient? Curr Opin Clin Nutr Metab Care 9:728–733

Bridi M, Choay P (2003) Taurine: a particular amionoacid with multiple functions. Ann Pharm Fr 61:385–391

Caprini M, Gomis A, Cabedo H, Planells-Cases R, Belmonte C, Viana F, Ferrer-Montiel A M, Gomis A, Cabedo H, Planells-Cases R, Belmonte C, Viana F, Ferrer-Montiel A (2003) GAP43 stimulates inositol trisphosphate-mediated calcium release in response to hypotonicity. EMBO J 22:3004–3014

Chesney RW (1985) Taurine: its biological role and clinical implications. Adv Pediatr 32:1–42

El-Sherbeny A, Naggar H, Miyauchi S, Ola MS, Maddox DM, Martin PM, Ganapathy V, Smith SB (2004) Osmoregulation of taurine transporter function and expression in retinal pigment epithelial, ganglion, and müller cells. Invest Ophthalmol Vis Sci 45:694–701

Huxtable RJ (1992) Physiological actions of taurine. Physiol Rev 72:101–63

Lima L, cubillos S (1998) Taurine might be acting as a trophic factor in the retina by modulating phosphorylation of cellular proteins. J Neurosci Res 53:377–384

Lima L, Matus P, Drujan B (1988) Taurine effect on neuritis growth from goldfish retinal explants. Int J Devel Neurosci 6:417–420

Lima L, Matus P, Drujan B (1989) The interaction of substrate and taurine modulates the ogrowth from regenerating goldfish retinal explants. Int J Develop Neurosci 7:375–382

Lima L, Matus P, Drujan B (1993) Taurine-induced regeneration of goldfish retina in culture may involved a calcium mediated mechanism. J Neurochem 60:2153–2157

Morales I, Dopico JG, Sabate M, Gonzalez-Hernandez T, Rodriguez M (2007) Substantia nigra osmoregulation: taurine and ATP involvement. Am J Physiol Cell Physiol 292:C1934–C1941

Morimura H, Shimada S, Otori Y, Saishin Y, Yamauchi A, Minami Y, Inoue K, Ishimoto I, Tano Y, Tohyama M (1997) The differential osmoregulation and localization of taurine transporter mRNA and Na^{+}/myo-inositol cotransporter mRNA in rat eyes. Brain Res Mol Brain Res 44:245–252

Nusetti S, Obregón F, Lim L (2006) Neuritic outgrowth from goldfish retinal explants, interaction of taurine and zinc. In: Oja SS, Saransaari P (eds) Taurine 6, Springer, New York, pp 435–40

Olson JE, Martinho E (2006) Taurine transporter regulation in hippocampal neurons. In: Oja SS, Saransaari P (eds) Taurine 6, Springer, New York, pp 307–14

Pascual JM, Solivera J, Prieto R, Barrios L, López-Larrubia P, Cerdán S, Roda JM (2007) Time course of early metabolic changes following diffuse traumatic brain injury in rats as detected by (1)H NMR spectroscopy. J Neurotrauma 24:944–959

Petrosian AM, Haroutounian JE, Fugelli K, Kanli H (2000) Effects of osmotic and light stimulation on 3H-taurine efflux from isolated rod outer segments and synthesis of tauret in the frog retina. Adv Exp Med Biol 483:441–451

Tomassen SF, Fekkes D, de Jonge HR, Tilly BC (2004) Osmotic swelling-provoked release of organic osmolytes in human intestinal epithelial cells. Am J Physiol Cell Physiol 286: C1417–C1422

Tranberg M, Abbas AK, Sandberg M (2007) In vitro studies on the putative function of N-acetylaspartate as an osmoregulator. Neurochem Res 32:1248–1255

Warskulat U, Heller-Stilb B, Oermann E, Zilles K, Haas H, Lang F, Häussinger D (2007) Phenotype of the taurine transporter knockout mouse. Methos Enzymol 428:439–458

Yancey PH (2005) Organic osmolytes as compatible, metabolic and counteracting cytoprotectants in high osmolarity and other stresses. J Exp Biol 208:2819–2830

Tomassen SF, Fekkes D, de Jonge HR, Tilly BC (2004) Osmotic swelling-provoked release of organic osmolytes in human intestinal epithelial cells. Am J Physiol Cell Physiol 286: C1417–C1422

Chapter 24
Localization of Taurine Transporter, Taurine, and Zinc in Goldfish Retina

Sonia Nusetti, Víctor Salazar, and Lucimey Lima

Abstract Taurine and zinc interact during structural and functional development of the rat retina and during the process of regeneration of retinal fragments from goldfish. These observations formed the basis for evaluating the regional correlation between the localization of the taurine transporter, taurine content and labile zinc content in goldfish retina. In the retina, the taurine transporter is expressed in photoreceptors, the outer plexiform layer, the inner nuclear layer and the ganglion cell layer. Taurine was detected in photoreceptors, the outer and inner nuclear layers, the outer plexiform layer and the ganglion cell layer. A large amount of labile zinc was detected in photoreceptors and to a lesser extent in ganglion cells. The taurine transporter, taurine and zinc coexist in photoreceptors and the ganglion cell layer. Their co-localization in photoreceptors may be related to the neuro-protective role of taurine and zinc in this layer. By contrast, their co-existence in ganglion cells may be related to their involvement in cell differentiation, development, and regeneration. This reveals the importance of taurine and zinc in maintaining normal cellular function in these particular layers of the retina

Abbreviations *CNS*, central nervous system; *FITC*, fluoresceineisotiosianato; *GC*, ganglion cells; *HBSS*, Hank's balanced salt solution; *HE*, hetoxiline-eosin; *INL*, inner nuclear layer; *NG*, Newport Green; *ONL*, outer nuclear layer; *OPL*, outer plexiform layer; *PAP*, peroxidase-antiperoxidase; *Phot*, photoreceptors layer; *TAUT*, taurine transporter; *TPEN*, N,N,N',N'-tetratikis-(2-pyridylmethyl)ethylendiamine

24.1 Introduction

Taurine (2-aminoethane sulfonic acid), a β-amino acid, is present in high levels in the retina of many vertebrates (Militante and Lombardini 2002). Taurine possesses neuroprotective and neurotrophic properties in the central nervous system

S. Nusetti (✉)
Laboratorio de Neuroquímica, Centro de Biofísica y Bioquímica, Instituto Venezolano de Investigaciones Científicas, Caracas, Venezuela

J. Azuma et al. (eds.), *Taurine 7*, Advances in Experimental Medicine and Biology 643,
DOI 10.1007/978-0-387-75681-3_24, © Springer Science+Business Media, LLC 2009

(CNS) during development and regeneration (Lima et al. 2004). Retinal function is dependent on high levels of retinal taurine, which are maintained mainly through transport from the blood (Lake and Verodone-Smith 1989). Taurine deficiency is associated with retinal degeneration, in particular ultrastructural changes in the photoreceptor outer segments. According to Lombardini (1991), high level of retinal taurine may be involved in: (1) protection of the photoreceptor outer segments; (2) regulation of Ca^{2+} transport; and (3) regulation of signal transduction. Some taurine functions are similar to those of zinc, an element involved in metabolic, genetic and neurotropic processes (Grahn et al. 2001; Redenti and Chappell 2004). Zinc is highly concentrated in the retina and its physiological function have been studied in retina, where it is believed to interact with taurine, to modify photoreceptor plasma membrane structure, to modulate synaptic transmission and to serve as an antioxidant (Grahn et al. 2001). The retina is one of the most prominent CNS areas that is dependent on taurine action, actually serving as a model for exploring the interaction of zinc and taurine during development and regeneration of the CNS. Both taurine and zinc form complexes that appear to insert into and stabilize the plasma membrane. Some studies have provided evidence for a role of their interaction during development, resulting in improved morphology and function of the rat retina (Gottschall-Pass et al. 1997). Taurine and zinc increase neurites outgrowth from goldfish retinal fragments and these effects are concentration dependent, exhibiting a bell shaped pattern. Also taurine requires an optimal concentration of intracellular zinc to mediate trophic effects on neuritic outgrowth (Nusetti et al. 2005). These observations have led us to evaluate the regional localization of the taurine transporter, taurine and labile zinc in the goldfish retina.

24.2 Methods

24.2.1 Immunolabeling of Goldfish Retinas for the Taurine Transporter

Goldfish (*Carassius auratus*) of 3–5 cm in length were adapted to darkness for 30 min, anesthetized with tricaine 0.05% and sacrificed by transcardial perfusion with 4% paraformaldehyde in 0.1 M phosphate buffer, pH 7.4. Eyes were dissected from the orbit and post-fixed for 4 h. Retinas were then removed and crioprotected overnight with 0.1 M phosphate buffer, pH 7.4 containing 30% sucrose. They were then transversely sectioned (8 μm) using a cryostat. The sections were incubated in an affinity-purified rabbit anti-rat TAUT-1 IgG (Abbot Laboratories) diluted 1:500 in 2% bovine serum albumin (BSA) and 0.3% Triton X-100 in 0.1 M saline phosphate buffer (PBS), pH 7.4 for 12 h at 4°C. Sections were then washed with PBS and incubated in fluoresceineisotiosanate (FITC) goat anti-rabbit IgG (Sigma) diluted 1:250 in PBS containing 2% BSA and 0.3% Triton X-100 for 1 h at room temperature (Pow et al. 2002).

24.2.2 *Immunohistochemistry for Taurine*

Goldfish were adapted to darkness, anesthetized and perfused with 4% paraformaldehyde and 2.5% glutaraldehyde in 0.1 M phosphate buffer, pH 7.4 containing 0.06 M $MgCl_2$. Eyes were removed and postfixed for 2 h (Omura and Inagaki 2000; Pow et al. 2002). Retinas were then removed and crioprotected overnight with 30% sucrose. The sections (8 μm) were incubated with the primary antibody against taurine (rabbit polyclonal, Chemicon) diluted 1:500 in PBS containing 2% BSA and 0.5% Triton X-100 for 48 h at 4°C. Inmunoreactivity was visualized using the peroxidase anti-peroxidase (PAP) method (Lake 1994).

24.2.3 *Localization of Intracellular Zinc Ions*

Goldfish were adapted to darkness, anesthetized and retinas were removed and bathed in Hank's balanced salt solution (HBSS) without Ca^{2+} and Mg^{2+}. For intracellular zinc fluorescence, retinas were preloaded with 20 μM of the membrane-permeable form of the Zn^{2+}-sensitive fluorescent dye, Newport Green diacetate (NG) (excitation: 485 nm; emission: 530 nm) and 0.01% dimethyl sulfoxide for 15 min at 25°C in the presence or absence of the intracellular zinc chelator *N, N, N', N'* tetrakis-(2-pyridylmethyl) ethylenediamine (TPEN) (Canzoniero et al. 1999; Li et al. 2001; López-García 2002; Redenti and Chappell 2005). Retinas were then washed with HBSS, fixed in cold, cut in 5 μm sections and viewed using fluorescence in a Nikon microscope.

To observe retinal structure in control sections, a nearby section was hematoxylin-eosin (HE) stained with Mayer's hemalum solution and 1% eosin Y.

24.3 Results

24.3.1 *Localization of Taurine Transporter, Taurine and Zinc*

The taurine transporter was evident in the photoreceptor (Phot), the outer plexiform layer (OPL), the inner nuclear layer (INL) and the ganglion cell layers (GC). Immunostaining for taurine was demonstrated in photoreceptors (Phot), the outer nuclear layer (ONL), the inner nuclear layer (INL), the outer plexiform layer (OPL) and the GC. NG-Zn fluorescence was evident in the photoreceptor (Phot) and the ganglion cell layers (GC) (Table 24.1). Figure 24.1 demonstrates the localization of the taurine transporter, taurine and zinc in the goldfish retina.

24.4 Discussion

24.4.1 *Localization of Taurine Transporter and Taurine*

Several reports have demonstrated the immunocytochemical localization of taurine in the retina of rat, cat, guinea pig (Lake and Verodone-Smith 1989), rabbit

Table 24.1 Localization of taurine transporter, taurine and zinc in the goldfish retina

Localization	TAUT	TAU	Zn^{2+}
Phot	++	++	+++
ONL	−−	++	−−
OPL	++	++	−−
INL	++	++	−−
IPL	−−	−−	−−
GC	++	+	+

TAUT taurine transporter, *TAU* taurine, Zn^{2+} zinc ionsl; + + +: Extremely strong; ++: Strong; +: Dectectable; −− Undectectable.

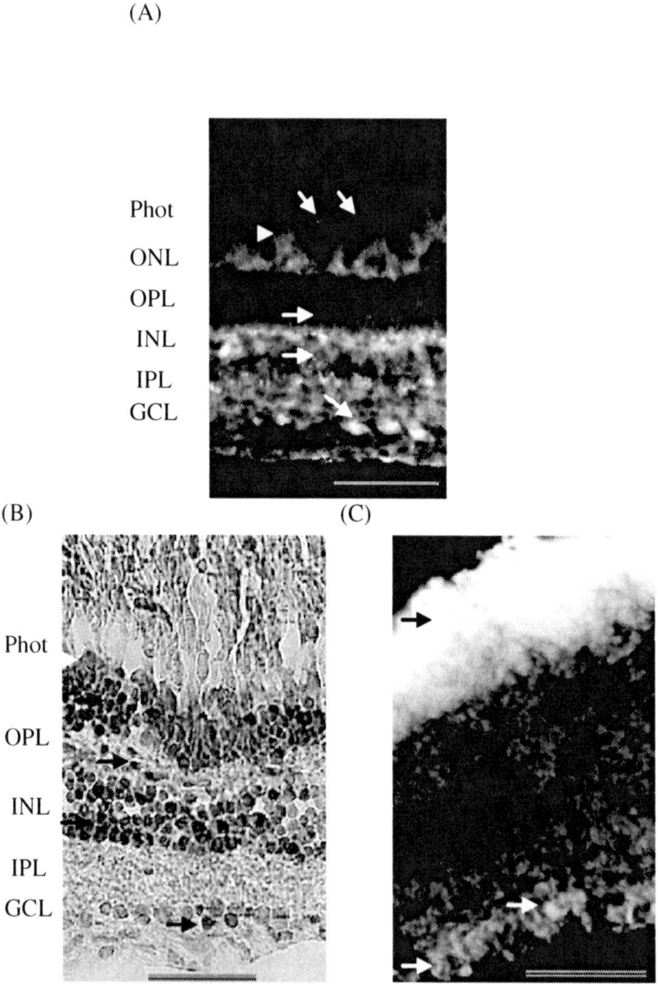

Fig. 24.1 Sections of goldfish retina. A: Illustrates immunolabeling of the taurine transporter. B: immunolabeling for taurine. C: NG fluorescence of intracellular free zinc. Scale Bar: 20 μm for all panels

(Pow et al. 1994), lungfish (Pow 1994), adult human and monkey (Nag et al. 1998), lefteye flounder (Omura and Yoshimura 1999) and goldfish (Omura and Inagaki 2000). All of these studies indicate that taurine is mainly concentrated in the photoreceptor layer, the outer nuclear layer and the synaptic terminals of the photoreceptor cells.

In this research we also obtained immunostaining in the photoreceptor, the external nuclear layer and the outer plexiform layer. It has been reported that taurine is released from photoreceptors in response to light, suggesting that it may be involved in retinal signal transduction, particularly based of its ability to inhibit protein phosphorylation. Among the other roles proposed for taurine in the retina are protection of the rod outer segments against damaging light and chemicals (Lombardini 1991), osmoregulation and modulation of cell membrane permeability, ion transport and water accumulation (Pasantes-Morales and Schousboe 1988; Pasantes-Morales et al. 2000).

The distribution of TAUT-1 corresponds to that of taurine immunoreactivity in the retinal layers, the exception being the outer nuclear layer. The presence of TAUT-1 in the photoreceptor and outer plexiform layer confirms the requirement of the regions for exogenous taurine. It is well established that taurine deficiency severely compromises retinal function, leading to profound degeneration of the photoreceptor in cat, rat, human, monkey and other species (Lake 1986; Lima 1999; Heller-Stilb et al. 2002). The outer nuclear layer did not show TAUT-1 immunoreactivity, suggesting that taurine in this layer may be mainly endogenous and synthesized by the enzyme cysteine-sulfinic acid decarboxylase. Lima and Obregón (1996) found that cysteine-sulfinic acid decarboxylase is present in the whole retina of goldfish and rat, however its distribution in the goldfish retina is still unknown.

Human taurine transporter plays a relevant role in the inner blood-retinal barrier (Masatoshi et al. 2007), in retinal pigment epithelium (Hillenkamp et al. 2004), and in ganglion and Müller cells (El-Sherbeny et al. 2004), probably related to taurine's osmoregulatory function.

Little is known about these particular distributions in fish retina, although there is intense taurine immunoreactivity in cone outer segments, rod inner segments, photoreceptor supranuclear region and outer plexiform layer *Anguila japonica* and *Carassius auratus* (Omura and Inagaki 2000), and changes occur in the developing retina of the fish *Paralichthys olivaceus* (Omura and Yoshimura 1999).

24.4.2 Localization of Zinc in the Goldfish Retina

The sensitive dye Newport Green diacetate was employed to localize free and weakly bound zinc in the goldfish retina. Fluorescence was found predominantly in the photoreceptors and to a lesser extent in the ganglion cell layers. Newport Green fluorescence was abolished by pretreatment with TPEN, demonstrating that the fluorescence reflected the distribution of zinc in goldfish retina.

Previous studies have obtained a similar distribution of zinc in other species. Wu et al. (1993) showed using Neo-Timm staining that photoreceptors in tiger salamander retina contain intracellular zinc associated with glutamatergic synaptic vesicles and may act as a diffusible molecular switch regulating neurotransmitter signaling. Hirayama (1990) detected using ditizone that zinc only localizes in rat photoreceptor outer segments and concluded that this element seems to be a component of metalloenzymes that play an important role in retinol metabolism.

Free or loosely bound zinc not only associates with the photoreceptors, but also with both plexiform layers, with perikarya in the vitreal part of the inner nuclear layer and with cells in the ganglion cell layer of rat (Ugarte and Osborn 1998; Akagi et al. 2001) and mouse retina (Kaneda et al. 2000; Wang et al. 2006). By electron microscopy, the presence of zinc in the plexiform layers was associated with neural processes, and it is believed to act as a modulator of synaptic transmission. In the outer nuclear layer zinc is associated with the nucleus, while in the inner nuclear layer and the ganglion cell layer it is associated with the Golgi apparatus, where it is assumed to catalyze metalloenzyme reactions (Akagi et al. 2001).

Zinc is essential for normal embryogenesis, development and regeneration of CNS (Huaqing and Amemiya 2001; Nusetti et al. 2005). The retinas of zinc deficient weanling rats possess vesicles, are disorganized, and eventually exhibit degeneration of the photoreceptor outer segments (Leure-duPree and Bridges 1982). Zinc deficiency imposed throughout pre- and postnatal retinal development depresses the electroretinogram, and zinc acts synergistically with taurine to modulate those effects. Nusetti et al. (2006) demonstrated that intracellular and extracellular zinc are necessary for the normal outgrowth of retinal fibers from cultured goldfish retinal explants, with a decrease in outgrowth produced by zinc chelators *in vivo* and *in vitro*. In the presence of taurine the zinc chelator was largely ineffective and there was an increase in outgrowth. Also it was shown that taurine requires an optimal concentration of intracellular zinc to exert its stimulatory effect.

In light adapted murine retina, the zinc transporter ZnT-3 and zinc appear to be localized in the outer plexiform, the inner nuclear, the inner plexiform and the ganglion cell layers (Redenti and Chappell 2004). In addition, by using the neo-Timm sulfate silver method, zinc was demonstrated to be mainly in the outer plexiform layer of carp retina (Du and Bao 1999).

In this study, the taurine transporter, taurine and zinc co-localized to the photoreceptor and the ganglion cells layers, where they may be necessary for the normal operation of these retinal layers. A functional relationship between the metal and taurine during regeneration of goldfish retina is possible (Nusetti et al. 2005).

24.4.3 Coexistence of Taurine Transporter, Taurine and Zinc in the Photoreceptors and Ganglion Cell Layers

The coexistence of the taurine transporter, taurine and zinc in goldfish retina photoreceptors are consistent with their important functions in the photoreceptor layer.

Sturman et al. (1981) reported that both taurine and zinc are localized to the periphery of tapetal rods and that they contribute to the stability of the cell membrane. Isolated frog rod outer segments exposed to ferrous sulfate display extensive disruption of their structure, characterized by acute swelling and disc membrane disorganization. The addition of zinc or taurine alone to the culture medium is ineffective in protecting against this outer segment damage. However, combined zinc and taurine treatment provide protection in a synergistic manner (Pasantes-Morales and Cruz 1984). They also reported that the protective effect of taurine and zinc against peroxidative damage may be related to the maintenance of the regulatory properties of the cell membrane relative to ion transport. In addition, taurine and zinc interaction are closely related to the development of retinal structure and function, as demonstrated by oscillatory potentials and light microscopy in rat (Gottschall-Pass et al. 1997).

In the present study, the taurine transporter, taurine and zinc were detected in the ganglion cell layers of goldfish retina. The taurine transporter may be involved in taurine uptake from the neuroretina, from which it is axonally transported to the optic tectum. Alternatively, the taurine transporter could deliver taurine from the optic tectum to the neuroretina (Guerra et al. 2000). In agreement with these observations, radiolabeled taurine injected into rat, rabbit or goldfish eyes is axonally transported along the optic nerve to the lateral geniculate nucleus and the superior colliculus (Politis and Ingoglia 1979; Sturman 1979; Guerra et al. 2000), where it is involved in the stabilization of electronic properties and the cytoskeleton of axons during synaptic arrangement during development and regeneration (Sturman 1979).

Ganglional cells of the goldfish retina have regenerative and proliferate capacity throughout life (Yung-Kang and Elam 2002), which could explain the localization of taurine in this layer. The distribution of the amino acid has been associated with cell proliferation, differentiation and maturation of rat, human fetus, monkey and goldfish retina (Lima et al. 1998; Lake 1994; Nag et al. 1998). In addition, Magnusson (1996) showed that the distribution of taurine seems to be spatio-temporally related in the hippocampus and cerebellum during development, and in the adult dentate gyrus of rat brain during plasticity. Therefore, taurine may play a role in dendritic outgrowth and synapse formation.

The localization of zinc in goldfish ganglion cells could be related to a function similar to that of taurine. This element is essential for brain and retinal development and function (Gottschall-Pass et al. 1997; Grahn et al. 2001). Taurine is also found in proliferate tissues of the brain, such as the dentate gyrus and olfactory bulb (Frederickson et al. 2000; Tarohda and Yamamoto 2004). Valente and Auladell (2002) observed an increase in zinc rich terminal fields during development of telencephalic circuits and concluded that zinc might be relevant for the establishment and the maturation of these circuits.

Taurine and zinc are essential for development and regeneration of the CNS. Their coexistence in goldfish retina may be related to the necessity of them interacting to achieve normal retinal function.

Acknowledgments Supported by Fondo Nacional de Ciencia, Tecnología e Innovación (FONACIT Grant S1-903), Venezuela.

References

Akagi T, Kaneda M, Ishii K, Hashikawa T (2001) Differential subcellular localization of zinc in the rat retina. J Histochem Cytochem 49:87–96

Canzoniero L, Turetsky D, Choi D (1999) Measurement of intracellular free zinc concentrations accompanying zinc-induced neuronal death J Neurosci 19:1–6

Du W Bao Y (1999) Light and electron microscopic observation of zinc distribution in the carp retina. Sheng Li Xue Bao 51:279–284

El-Sherbeny A, Naggar H, Miyauchi S, Shamsul M, Maddox D, Moore P, Ganapathy V, Smith S (2004) Osmoregulation of taurine transporter function and expression in retinal pigment epithelial ganglion and müller cells. Inves Ophthal Visual Sci 45:694–701

Frederickson C, Suh S, Silva D, Thompson R (2000) Importance of zinc in the central nervous system: the zinc containing neuron. J Nutr 13:1471S–1483S

Grahn B, Paterson P, Gottschall-Pass K, Zhang Z (2001) Zinc and the eye. J Am Coll Nutr 20: 106–118

Gottschall-Pass K, Grahn B, Gorecki D, Paterson P (1997) Oscillatory potentials and tight microscopic changes demonstrate an interaction between zinc and taurine in the developing rat retina. J Nutr 127:1206–1213

Guerra A, Urbina M, Lima L (2000) Modulation of taurine uptake in the goldfish retina and axonal transport to the tectum Effect of optic crush and axotomy. Amino Acids 19:1–17

Heller-Stilb B, van Roeyen C, Rascher K (2002) Disruption of the taurine transporter gene (TauT) leads to retinal degeneration in mice. FASEB J 16:231–233

Hillenkamp J, Hussain AA, Jackson TL, Cunningham JR, Marshall J (2004) Taurine uptake by human retinal pigment epithelium: implications for the transport of small solutes between the choroid and the outer retina. Invest Ophthalmol Vis Sci 45:4529–4534

Hirayama Y (1990) Histochemical localization of zinc and copper in rat ocular tissue. Acta Histochem 89:107–111

Huaqing G, Amemiya T (2001) Optic nerve changes in zinc deficient rats. Exp Eye Res 72:363–369

Kaneda M, , rásfalvy B, Kaneko A (2000) Modulation by Zn^{+2} of GABA responses in bipolar cells of the mouse retina. Vis Neurosci 17:273–281

Lake N (1986) Electroretinographic deficits in rats treated with guanidinoethyl sulfonate a depletor of taurine. Exp Eye Res 42:87–91

Lake N, Verodone-Smith C (1989) Immunocytochemical localization of taurine in the mammalian retina. Curr Eye Res 8:163–173

Lake N (1994) Taurine and GABA in the rat retina during postnatal development. Vis Neurosci 11:253–260

Leure-duPree A, Bridges C (1982) Changes in retinal morphology and vitamin A metabolism as a consequence of decreased zinc availability. Retina 2:294–302

Li Y, Houg C, Suh S, Sarvey J, Frederickson C (2001) Rapid translocation of $Zn(2^{+})$ from presynaptic terminals into postsynaptic hippocampal neurons after physiological stimulation. J Neurophysiol 86:2597–2604

Lima L, Obregón F (1996) Cystein sulfinate decarboxilase in retina of rat and goldfish Effect of goldfish optic nerve crush 1996 Annual Meeting of Society of Neuroscience Washington DC, USA

Lima L, Obregón F, Matus P, (1998) Taurine glutamate and GABA modulate the outgrowth from goldfish retinal explants and its concentrations are affected by the crush of the optic nerve. Amino Acids 15:195–209

Lima L (1999) Taurine and its trophic effects in the retina. Neurochem Res 24:1333–1338

Lima L, Obregón F, Roussó T, Quintal M, Benzo Z, Auladell C, (2004) Content and concentration of taurine hypotaurine and zinc in the retina the hippocampus and the dentate gyrus of the rat at various postnatal days. Neurochem Res 29:247–255
Lombardini J (1991) Taurine: Retinal function. Brain Res Rev 16:151–169
López-García C, Varea E, Palop J, Nacher J, Ramirez C, Ponsoda X, Molowny A (2002) Cytochemical techniques for zinc and heavy metals Localization in nerve cells. Microsc Res Tec 56:318–331
Magnusson KR (1996) Distributions of taurine, glutamate, and glutamate receptors during postnatal development and plasticity in the rat brain. Adv Exp Med Biol 403:435–444
Masatoshi T, Tomoyuki T, Tomoyuki I, Fuminobu E, Akihisa M, Michio K, Sumio O, Young-Sook K, Tetsuya T, Ken-ichi H (2007) Function and regulation of taurine transport at the inner blood–retinal barrier. Microvasc Res 73:100–106
Militante J, Lombardini J (2002) Taurine: evidence of physiological function in the retina. Nutr Neurosci 5:75–90
Nag T, Jotwani G, Wadhwa S (1998) Immunohistochemical localization of taurine in the retina of developing and adult human and adult monkey. Neurochemistry 33:195–200
Nusetti S, Obregon F, Quintal M, Benzo Z, Lima L (2005) Taurine and zinc modulate outgrowth from goldfish retinal explants. Neurochem Res 30:1483–1492
Nusetti S, Obregón F, Lima L (2006) Neuritic outgrowth from goldfish retinal explants interaction of taurine and zinc. Adv Exp Med Biol TAURINE 6:1300–1305
Omura Y, Yoshimura R (1999) Immunocytochernical localízation of taurine in the developing retina of the lefteye flounder Paralichfhys olivaceus. Arch Histol Cytol 62:441–446
Omura Y, Inagaki M (2000) Immonocytochemical localization of taurine in the fish retina under light and dark adaptations. Amino Acids 19:593–604
Pasantes-Morales H, Cruz C (1984) Protective effect of taurine and zinc on peroxidation induced damage in photoreceptor outer segments. J Neurosci Res 11:303–311
Pasantes-Morales H and Schousboe A (1988) Role of taurine in osmoregulation in brain cells: Mechanisms and functional implications. Amino Acids 12:281–292
Pasantes-Morales H, Cardin V, Tuz K (2000) Signaling events during swelling and regulatory volume decrease. Neurochem Res 25:1301–1314
Politis M, Ingoglia N (1979) Axonal transport of taurine along neonatal and young adult rat optic axons. Brain Res 166:221–231
Pow D (1994) Taurine amino acid transmitters and related molecules in the retina of the Australian lungfish Neoceratodus forstert a light-microscopic immunocytochemical and electron-microscopic study. Cell Tissue Res 278:311–326
Pow D, Crook D, Wong R (1994) Early appearance and transient expression of putative amino acid neurotransmitters and related molecules in the developing rabbit retina an immunocytochemical study. Vis Neurosci 11:1115–1134
Pow D, Sullivan R, Reye P, Hermanussen S (2002) Localization of taurine transporters taurine and ^{3}H taurine accumulation in the rat retina pituitary and brain. Glia 37:153–168
Redenti S, Chappell R (2004) Localization of zinc transporter-3 (ZnT-3) in mouse retina. Vis Res 44:3317–3321
Redenti S, Chappell R (2005) Neuroimaging of zinc released by depolarization of rat retinal cells. Vis Res 45:3520–3525
Sturman J (1979) Taurine in the developing rabbit visual system: changes in concentration and axonal transport incluiding a comparison with axonally transported proteins. J Neurobiol 10:221–237
Sturman J, Wen G, Wisniewski H, Hayes K (1981) Histochemical localization of zinc in the feline tapetum: Effect of taurine depletion. Histochem 72:341–350
Su YK, Elam JS (2002) Differential growth of goldfish retinal explants on regenerating and non-regenerating optic tract membranes. Brain Res Dev Brain Res 139:319–323
Tarohda T, Yamamoto M, Amamo R (2004) Regional distribution of manganese, iron, copper, and zinc in the rat brain during development. Anal Bioanal Chem 380(2):240–6

Ugarte M, Osborn N (1998) The localization of endogenous zinc and the in vitro effect of reactive oxygen species in the retina. Gen Pharmacol 30:297–303

Valente T, Auladell C (2002) Developmental expression of ZnT3 in mouse brain: correlation between the vesicular zinc transporter protein and chelatable vesicular zinc (CVZ) cells. Glial and neuronal CVZ cells interact. Mol Cell Neurosci 21(2):189–204

Wang X, Wang Z, Gao H, Danscher G (2006) Localization of ZnT7 and zinc ions in mouse retina immunohistochemistry and selenium autometallography. Brain Res Bull 71:91–96

Wu S, Qiao X, Noebels J, Yang X (1993) Localization and modulatory actions of zinc in vertebrate retina. Vision Res 33:2611–2616

Part III
Effect of Taurine on Skeletal Muscle

Chapter 25
Effects of Taurine Administration on Exercise

Yoshihisa Yatabe, Shumpei Miyakawa, Hajime Ohmori, Hajime Mishima, and Takako Adachi

Abstract Taurine concentration in rat skeletal muscles after endurance running, with and without taurine administration was studied. Taurine concentrations in skeletal muscles was significantly decreased in exercised groups without taurine administration. However, taurine administration reduced the decrease of taurine concentration in skeletal muscles in exercise. Oral administration of taurine has effect for maintaining taurine concentration in skeletal muscles in exercise. The duration of running time to exhaustion of rats, with and without taurine administration were studied. The duration of running time to exhaustion was significantly increased by taurine administration. Oral administration of taurine increases the ability of physical endurance. Rat urinary excretions of creatinine, creatine, 3-methylhistidine (3-MH) after treadmill running, with and without taurine administration were studied. Rat urinary excretions of creatinine, creatine, 3-MH after treadmill running was significantly decreased with taurine administration. Taurine administration was considered to reduce the exercise-induced muscle fatigue.

Abbreviations 3-*MH*, 3-methylhistidine; *EDL*, extensor digitorum longus muscle; *GC*, gastrocnemius muscle; *SOL*, soleus muscle

25.1 Introduction

Taurine has various physiological functions, ranging from osmoregulation to neurotransmission. In both excitable and nonexcitable tissues taurine modulates cell function through its effects on ion channel activity. In excitable tissues, taurine has a recognized modulatory role on different ion channels, thus controlling membrane excitability and consequently tissue function (Camerino et al. 2004).

Recently, the roles of taurine in relation to exercise and the effects of taurine with exercise have been reported. (Baum and Weiss 2001; Cuisinier et al. 2002) Further,

Y. Yatabe (✉)
Department of Orthopaedic Surgery, Moriya Daiichi Hospital, Japan

J. Azuma et al. (eds.), *Taurine 7*, Advances in Experimental Medicine and Biology 643,
DOI 10.1007/978-0-387-75681-3_25,

Matsuzaki et al. (2002) have reported that taurine concentration in rat skeletal muscles were decreased in relation to exercise duration.

We have continued some experimental studies to investigate the effects of taurine administration on exercise. In this paper, we want to summarize our studies.

25.2 Materials and Methods

25.2.1 Animals and Grouping

Male Sprague-Dawley rats (Japan SLC, Shizuoka, Japan) of about 200 g were cared for from the age of six weeks. The animals were maintained for 14 days at two or three per cage, with free access to non-taurine containing food (MF, Oriental Yeast, Tokyo, Japan), constant room temperature (20 – 22°C) and exposed to a light cycle of 12 hr/day (8:00 A.M.–8:00 P.M.) throughout the course of the experiments. Body weight of rat was measured in every morning. All animals received humane care in accordance with the guidelines of the University of Tsukuba for the care of laboratory animals.

Study (1) Forty male SD rats and they were divided into two groups; endurance exercise (n=20) and sedentary control (n=20) groups, each of them was further divided into two groups; one received distilled water (n=10) and the other received 500 mg/kg/day taurine solution water for a further two weeks (n=10). Thus, there were four groups; No taurine administration and no exercise (Control) group [NC, n=10], Taurine administration and no exercise (Control) group [TC, n=10], No taurine administration and Exercise group [NE, n=10], Taurine administration and Exercise group [TE, n=10]. Then, taurine concentration in rat skeletal muscles were studied.

Study (2) Ten male SD rats were divided into two groups; one (n=5) received 500 mg/kg/day taurine administration orally for two weeks and the other (n=5) received no taurine administration for two weeks, then both two groups were dared to treadmill running to exhaustion: Exhaustive exercise and Taurine administration group [ET] and Exhaustive exercise and No taurine administration group [EN], then the duration of running time to exhaustion of rats were studied.

Study (3) Twenty-seven male SD rats were divided into 3 groups: taurine administration at 100 mg/kg/day for 2 weeks [T100, n=9], taurine administration at 500 mg/kg/day for 2 weeks [T500, n=9] and controls with administration of distilled water only [CNT, n=9]. Then, rat urinary excretions of creatinine, creatine, 3-mthylhistidine (3-MH) after treadmill running were studied.

25.2.2 Exercise Program

In all Studies, the exercise program was planned treadmill running (using Nazme KN-73 treadmill device, Nazme, Tokyo, Japan) once only after their nursing period.

The slope of the treadmill was constantly fixed at 0 degree and the rats were loaded on continuous running for 60 minutes at 25 m/min. speed.

25.2.3 Taurine Administration

The dose of taurine administration were 100 mg/kg/day taurine or 500 mg/kg/day, dissolved in distilled water was administrated orally with a catheter to the rats in volumes of about 2 to 3 ml, once a day in the morning for all the taurine administration groups. The no taurine administration [: CNT] groups received distilled water by the same manner. For the rest of the day all rats had free access to taurine-free drinking water.

25.2.4 HPLC Analysis

HPLC analysis was used to measure plasma and skeletal muscle taurine concentration. Immediately after finishing the last program or exercise, all rats were sacrificed. Blood samples were taken by aspiration from the hearts prepared for measurement for Plasma lactic acid content. The soleus (SOL), extensor digitorum longus (EDL), gastrocnemius (GC) muscles were removed as quickly as possible, cleaned from adipose and connective tissue and weighted. Portions of each muscle were fully homogenized in ice-cold physiologic salt water for the next examinations. All muscular samples were centrifuged at 6200 g for 30 min. at 6°C. The supernatants were stored at −80°C until assay. Derivatization with-phthalaldehyde was performed and samples were processed for HPLC taurine determination.

25.2.5 Urinary Excretions of Creatinine, Creatine, 3-MH

Before the last 24hour, rats were individually placed in the metabolic cage to collect all their urine in Study 3. After treadmill running, rats were individually replaced in the metabolic cage again, to continue collecting urine for another 48 hours, and urinary excretions of creatinine, creatine, 3-MH were measured at every 24 hours.

25.2.6 Statistical Analysis

All data are reported as the mean±SD. Statistical analysis was performed by using a one-way analysis of variance (ANOVA) followed by Fisher's PLSD or unpaired student's *t* test to determine differences between groups. Differences of $p<0.05$ were considered to be significant.

25.3 Results

25.3.1 Study 1

Growth of rats in all groups was accompanied with an increase in their body weight till the end of this experiment (14 days). There was no significant difference in both body weight and muscle weight between each of the four groups. There were no significant differences in muscle weight of EDL, GC and SOL muscles between each of the taurine-treated groups and the corresponding non-treated groups also (no data shown).

The plasma lactic acid concentrations of the exercised groups (TE, NE) were higher than the non-exercised groups (TC, NC), but not significantly. Further, there were no significant differences between TE and NE (exercise groups), TC and NC (non-exercise groups) also. There were no significant differences in the plasma taurine concentration of the four groups (Table 25.1).

Taurine concentration in the skeletal muscle was significantly higher by 22.7% in the EDL muscle of taurine-treated groups (TC; 19.0 ± 1.5 nmol/mg) with respect to the untreated rats (NC; 15.5 ± 1.9 nmol/mg). There was a significant difference between the taurine-treated exercise groups (TE; 17.3 ± 3.0 nmol/mg) and non taurine treated exercise groups (NE; 12.8 ± 3.2 nmol/mg). Furthermore, taurine concentration in skeletal muscle of the NC group was significantly higher than the NE group. However, taurine concentration in the TE group was higher than the NC group despite their heavy exposure to running. The results of GC and SOL were the same as with EDL (Fig. 25.1).

25.3.2 Study 2

There were no significant differences between in plasma lactic acid concentrations of the taurine administration group (ET) and no taurine administration group (EN).

There were no significant differences between in plasma taurine concentration of the taurine administration group (ET) and no taurine administration group (EN) also (Table 25.2).

Table 25.1 Plasma lactic acid and taurine concentration in Study 1

Group	Lactic acid(mg/dl)	Taurine(nmol/ml)
TC	22.2 ± 7.6	282.9 ± 116.4
NC	21.6 ± 1.4	265.7 ± 83.7
TE	33.0 ± 7.0	270.1 ± 119.9
NE	32.5 ± 16.0	230.3 ± 95.3

Data are the mean±SD. There is no significant difference.

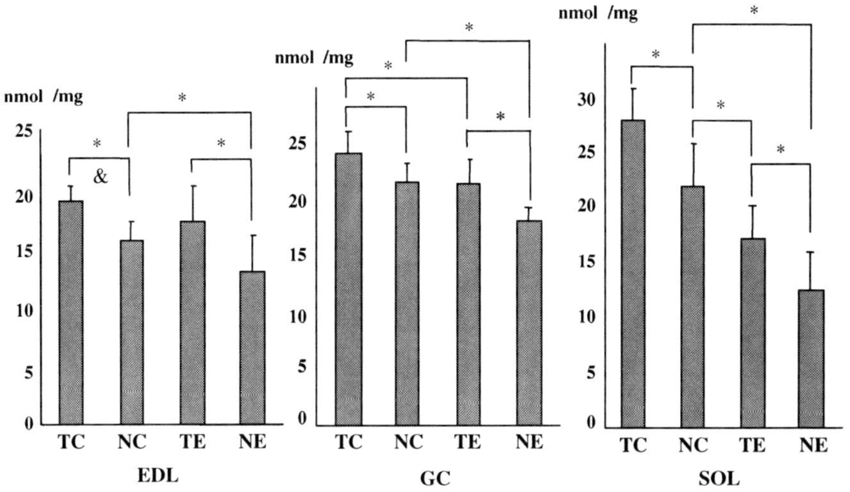

Fig. 25.1 Skeletal muscle taurine concentration in Study 1. Data are the mean ± SD. *: There is significant difference ($P < 0.05$)

Table 25.2 Plasma lactic acid and taurine concentration in Study 2

Group	Lactic acid(mg/dl)	Taurine(nmol/ml)
ET	64.6 ± 32.4	304.0 ± 44.0
EN	60.8 ± 26.0	247.8 ± 62.6

Data are the mean±SD. There is no significant difference.

Taurine concentration in the skeletal muscle was significantly higher in the EDL muscle of taurine-treated groups (ET) with respect to the untreated rats (EN). The results of GC and SOL were the same as with EDL (Fig. 25.2).

The duration of running time to exhaustion was 98.8 ± 17.5min. in taurine-administrated groups (ET) and 73.8 ± 8.2min. in non-taurine treated groups (EN). So, there was significant difference between taurine-administrated groups and non-taurine treated groups.

25.3.3 Study 3

Urinary excretions of creatinine were significantly decreased in T500 group at 24 hours after exercise. Urinary excretions of creatinine were significantly decreased in both T100 and T500 groups at 48 hours after exercise (Table 25.3). Urinary excretions of creatine were significantly decreased in both T100 and T500 groups at 24 and 48 hours after exercise (Table 25.4).

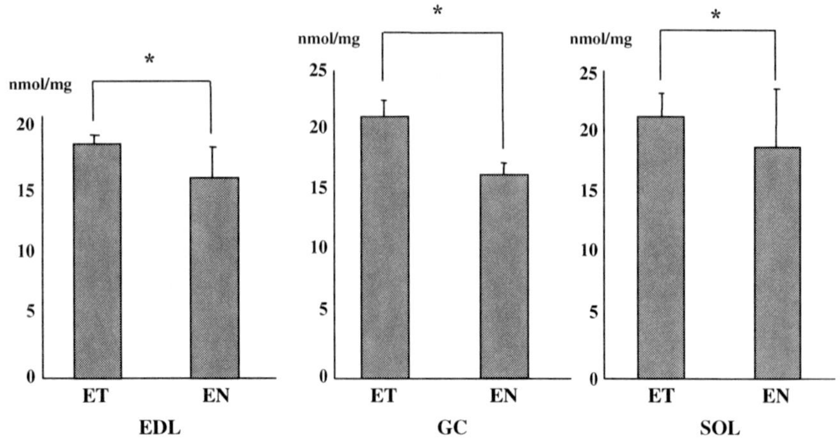

Fig. 25.2 Sketetal muscle taurine concentration in Study 2. Data are the mean±SD.*: There is significant difference ($P < 0.05$)

Table 25.3 Urinary excretions of creatinine in Study 3

Group	Pre.Ex.	Ex.24h	Ex.48h
CNT	57.2 ± 20.5	120.3 ± 41.0	119.1 ± 27.7
T100	62.4 ± 17.4	93.5 ± 29.2	78.9 ± 27.9*
T500	49.5 ± 15.0	81.4 ± 21.0*	77.0 ± 7.9*

Data are the mean±SD. *: There is significant difference ($P< 0.05$).

Table 25.4 Urinary excretions of creatine in Study 3

Group	Pre.Ex.	Ex.24h	Ex.48h
CNT	1.5 ± 0.7	4.7 ± 3.9	3.5 ± 1.6
T100	2.0 ± 0.7	2.2 ± 0.9*	1.5 ± 0.9*
T500	1.5 ± 0.6	2.1 ± 1.0*	1.6 ± 0.4*

Data are the mean±SD. *: $P< 0.05$.

Urinary excretions of 3-MH were significantly decreased in T500 group at 48 hours after exercise (Table 25.5). It seems that 500 mg/kg/day taurine administration group were more effective than 100 mg/kg/day group in Study 3.

Table 25.5 Urinary excretions of 3-MH in Study 3

Group	Pre.Ex.	Ex.24h	Ex.48h
CNT	36.2 ± 21.8	59.0 ± 25.5	68.6 ± 19.6
T100	35.3 ± 16.6	58.6 ± 26.3	46.5 ± 33.3
T500	33.8 ± 18.3	40.4 ± 19.7	34.5 ± 8.4*

Data are the mean±SD *: $P< 0.05$.

25.4 Discussion

We studied effects of taurine administration on taurine concentration in rat skeletal muscles on exercise in Study 1 and physical endurance in Study 2. And we investigated that oral administration of taurine has effects for maintaining taurine concentration in skeletal muscle on exercise and up-regulates physical endurance, suggesting that maintaining of taurine concentrations in skeletal muscles might be one of the explanations of this improved maximal performance on exercise.

Further, chronically lowered muscle taurine levels affect skeletal muscle function (Warskulat et al. 2004), and the cytoprotective role of taurine in exercise-induced muscle injury has been reported also. (Dawson et al. 2002; Zhang et al. 2004)

Whether taurine supplementation would minimize such increases in urinary taurine excrtion reflecting muscle damage with exercise is an interesting scientific question and merits investigation (Cuisinier et al. 2002). The next purpose is to investigate the effects of taurine administration for the exercise-induced muscle fatigue.

As rat urinary excretions of creatinine, creatine, 3-MH are thought to reflect muscle damage on exercise, we studied rat urinary excretions of creatinine, creatine, 3-MH after treadmill running, with and without taurine administration in Study 3. As shown in Tables 25.3, 25.4, 25.5, rat urinary excretions of creatinine, creatine, 3-MH after treadmill running were significantly decreased with peroral taurine administration.

The results of Study 3 clarify that oral taurine administration is considered to reduce the exercise-induced myotrauma also.

Furthermore, the effect of taurine on muscle atrophy induced by synthetic glucocorticoid dexamethason (DEX) has been published (Uozumi et al. 2006). They reported that treatment with taurine at 2mM protected cells from DEX-induced atrophic response. And they speculated that an increase in intracellular taurine content through taurine transporter (TauT) plays an important role in maintenance of normal muscle morphology.

Thus, effects of taurine administration on exercise in some aspect have been recently studied, while effects of taurine for the exercise remain uncertain. We are going to study effects of taurine on exercise, fatigue (including CNS fatigue), some kinds of stress or DNA damages, or aging.

Acknowledgments We express our appreciation to the Animal Resource Center, University of Tsukuba for animal humane care.

References

Baum M, Weiss M (2001) The influence of a taurine containing drink on cardiac parameters before and after exercise measured by echocardiography. Amino Acids 20:75–82

Camerino DC, Tricarico D, Pierno S, Desaphy JF, Liantonio A, Pusch M, Burdi R, Camerino C, Fraysse B, De Luca A (2004) Taurine and Skeletal Muscle Disorders. Neurochem Res 29: 135–142

Cuisinier C, Michotte De Welle J, Verbeeck RK, Poortmans JR, Ward R, Sturbois X, Francaux M (2002) Role of taurine in osmoregulation during endurance exercise. Eur J Appl Physiol 87:489–495

Cuisinier C, Ward RJ, Francaux M, Strbois X, De Witte P (2001) Changes in plasma and urinary taurine and amino acids in runners immediately and 24 h after a marathon. Amino Acids 20: 13–23

Dawson R, Biasetti M, Messina S, Dominy J (2002) The cytoprotective role of taurine in exercise-induced muscle injury. Amino Acids 22:309–324

Matsuzaki Y, Miyazaki T, Miyakawa S, Bouscarel B, Ikegami T, Tanaka N (2002) Decreased taurine concentration in skeletal muscles after exercise for various durations. Med Sci Sports Exerc 34:793–797

Uozumi Y, Ito T, Takahashi K, Matsuda T, Mohri T, Kimura Y, Fujio Y, Azuma J (2006) Myogenic induction of taurine transporter prevents dexamethasone-induced muscle atrophy. Adv Exp Med Biol 583:265–270

Warskulat U, Flögel U, Jacoby C, Hartwig HG, Thewissen M, Merx MW, Molojavyi A, Heller-Stilb B, Schrader J, Häussinger D (2004) Taurine transporter knockout depletes muscle taurine levels and results in severe skeletal muscle impairment but leaves cardiac function uncompromised. FASEB J 18:577–579

Zhang M, Izumi I, Kagamimori S, Sokejima S, Yamagami T, Liu Z, Qi B (2004) Role of taurine supplementation to prevent exercise-induced oxidative stress in healthyyoung men. Amino Acids 26:203–207

Chapter 26
Characterization of Myogenic Differentiation under Endoplasmic Reticulum Stress and Taurine Treatment

Hasuk Song, Hyemin Kim, Taesun Park, and Dong-Hee Lee

Abstract Cells undergo apoptosis when they are subjected to prolonged ER stress (ERS). Excessive lipid deposition causes ERS in adipocytes; however, it rarely serves as a stress factor that triggers apoptosis. This strongly implies that an anti-ERS mechanism may exist in differentiating adipocytes. We used 3T3L1 (adipocytes) and C2C12 (myocytes) to probe for a potential anti-ERS mechanism. After cells were induced to adipogenesis or myogenesis, they were treated with the ERS inducer, tunicamycin. After tunicamycin-mediated ERS, the expression of the key molecular chaperone, Bip, increased in both cell lines. The chaperone, GRP94, responded differently to extended tunicamycin treatment, with protein levels remaining largely unchanged in 3T3L1 cells but falling in C2C12 cells. In terms of CHOP expression, C2C12 cells contained higher levels than 3T3-L1 cells. When GRP94 expression was reduced by siRNAs, CHOP expression increased. Considering the high levels of GRP94 in 3T3L1 cells under ERS, the small rate of apoptosis in 3T3L1 cells might result from the downregulation of CHOP mediated by GRP94. When C2C12 cells were pretreated with taurine, GRP94 levels appeared to increase and CHOP expression decreased. These results strongly imply that taurine may play an important role in promoting GRP94 expression and blocking the progression to apoptosis under ERS through the inhibition of CHOP upregulation.

Abbreviations *ERS*, endoplasmic reticulum stress; *Bip*, binding protein or GRP78; *CHOP*, C/EBP homologous protein

H. Song (✉)
[1]Department of Life Sciences, University of Seoul, Seoul, Korea

J. Azuma et al. (eds.), *Taurine 7*, Advances in Experimental Medicine and Biology 643,
DOI 10.1007/978-0-387-75681-3_26,

26.1 Introduction

Nascent secretory and transmembrane proteins complete their functional structure by post-translational modification, folding, and oligomerization in the endoplasmic reticulum (ER). Due to ER's stringent quality control system, only correctly folded proteins exit the ER and unfolded or misfolded proteins are retained and ultimately degraded. During endoplasmic reticulum stress (ERS), nascent polypeptides lose opportunities for proper structural folding. Cells increase the expression of molecular chaperones when the ER suffer a critical setback in physiological homeostasis during ERS (Bando et al. 2004; Kadowaki et al. 2004; Rao et al. 2002; Lee, 2001). Under those circumstances, cells immediately respond to ERS by initiating protein unfolding and reducing protein expression to cope with cellular stress and disorder (Puthalakath et al. 2007; Ni and Lee 2007).

Excessive lipid deposition in the obese animal triggers ERS in cells (Lane and Tang 2005). When cells are exposed to prolonged ERS, however, the cells synthesize more lipids. Increased lipid synthesis can be explained by the overexpression of the SREBP factor in the majority of cell types. Apoptosis is usually evident during prolonged ERS (Misra and Pizzo 2005; Özcan et al. 2004). However, no definite explanation is available for the linkage between lipid deposition, ERS, and apoptosis (Sheu et al. 2007). In established cell lines, ERS is rapidly induced by tunicamycin, a powerful ERS inducer that blocks post-translational modification. Progression of ERS and ensuing apoptosis is inhibited when tunicamycin is removed.

Numerous biochemical and physiological stimuli can interfere with myogenic differentiation. c-fos inhibits myogenesis by blocking MyoD expression (Nakanishi et al. 2005; Sordella et al. 2003). Okadaic acid inhibits myogenesis by suppressing MyoD expression and promoting the expression of the Id factor. ERS initially stimulates microtubule formation in skeletal muscle (Kee et al. 2007) but eventually serves as a stress factor leading to apoptosis.

Taurine is known to reduce ERS by regulating osmolality. Taurine is known to promote the recovery of muscle from damage, and prevent physiological deterioration in cells under stressful conditions. To cope with cellular stresses, taurine ensures homeostasis and osmotic balance within cells. At the cellular level, taurine controls functional gating of ion channels involved in intercellular ion trafficking. However, no concrete molecular mechanism is available explaining the ability of taurine to counteract the detrimental effects of ERS.

To characterize the potential anti-apoptotic mechanism and the effect of ER stress on cellular differentiation in mouse myoblasts, we treated C2C12 cells with the ER stress inducer, tunicamycin. In the present study, treatment with taurine coincided with the upregulation of GRP94, an anti-apoptotic molecular chaperone and the downregulation of CHOP, a pro-apoptotic factor (Zinszner et al. 2007). Based on GRP94 and CHOP expression, taurine appears to exert an inhibitory effect on the progression of apoptosis. When the cells were treated with taurine prior to ER stress, the level of GRP94 significantly increased relative to other ER stress markers, PDI and GRP78. The significant upregulation of GRP94 suggests that taurine acts to specifically retard apoptosis induced under prolonged ER stress.

26.2 Methods

26.2.1 Cell Culture and Treatment

Cells (3T3-L1) were cultured in DMEM containing 10% bovine calf serum (BCS) in 5% CO_2 at 37°C. Two days after complete confluence (Day 0), adipogenic differentiation was induced by adding MDI mixture [0.5 mM isobutyl-1-methyxanthine (MIX, Sigma), 1 μg/ml insulin (Sigma) and 1 μM dexamethasone (Sigma)] in DMEM containing 10% fetal bovine serum. After 48 hours, the medium was replaced with DMEM FBS was replaced with DMEM containing only 2% FBS, and changed with fresh medium every other day. ER stress was induced by treating well-differentiated cultured cells with tunicamycin.

26.2.2 Western Blotting

Western blotting for ERS-inducible markers of GRP78, GRP94, and CHOP was carried out using a polyclonal antibody (Santa Cruz). Dilution factors were 1:500 and 1:1000 for GRP78 and GRP94, respectively. After electrophoresis on a 10% polyacrylamide gel, proteins were transferred onto nitrocellulose membrane over night. Membranes were blocked with 5% non-fat milk and probed with antibody in the presence of 3% non-fat milk. Appropriate secondary antibody and enhanced chemiluminescence (Roche) were used to develop the blots.

26.2.3 Flow Cytometry Analysis

When a single cell suspension was achieved, cells were pelleted at 500 g for 5 min and washed in PBS containing 0.1% sodium azide and 2% FCS. The cell precipitant was diluted in staining buffer that included blocking antibody. The sample contained 10^7 cells/ml and was equilibrated to room temperature for 3 minutes. After primary antibody was added, cells were briefly vortexed, incubated in the dark for 30 min on ice, and washed copiously in staining buffer twice. Each secondary antibody was added and incubated on ice in the dark for 1 hour. Using FACS buffer, cells were washed twice and incubated on ice for 10 minutes before flow cytometry analysis.

26.2.4 siRNA Inhibition Experiment

Commercially available GRP94 siRNA and transfection kit was purchased from Santa Cruz Biotech. RNA and transfection solution was mixed by gently tipping and then incubated for 30 minutes at room temperature. Cells were washed with 2 ml of siRNA media before transfection. Approximately one million cells were placed in one ml of siRNA transfection media containing the siRNA and incubated for 6 hours at 37°C in a CO_2 incubator. The transfected cells were transferred to appropriate culture dishes for the experiment.

26.3 Results

26.3.1 Expression of ER Stress Marker During Myogenesis

The expression of GRP78 and GRP94 was analyzed by Western blotting. Differentiating C2C12 cells were subjected to ER stress by treatment with 5 mM tunicamycin. Figure 26.1 shows that GRP78 expression increased when the ERS treatment period was increased, a finding typical for ER stress. GRP94 showed the opposite expression pattern. Under myogenic differentiation, GRP94 expression showed the highest level in the absence of tunicamycin, but the level of expression decreased when ERS treatment was extended up to 3 hours. Little or no expression of GRP94 was observed after three hours of tunicamycin treatment.

26.3.2 CHOP Expression During Adipogenesis and Myogenesis

Differences in molecular chaperone expression were evident in differentiating myocytes and adipocytes during ER stress conditions. Figures 26.2 and 26.3 show dissimilar expression patterns for CHOP protein. In developing myocytes, GRP94 expression rapidly decreased but CHOP expression increased during ERS.

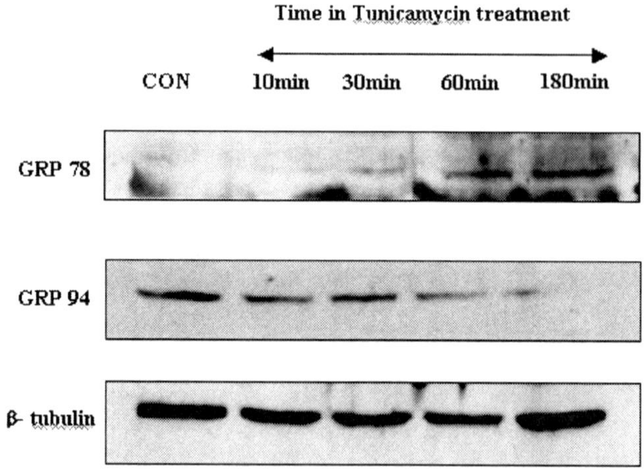

Fig. 26.1 Expression of GRP78 and GRP94 in differentiating C2C12 cells. Cells were treated with 5 mM tunicamycin for the indicated time periods. GRP78 expression increased when the tunicamycin treatment period was increased while GRP94 expression declined such that minimal levels of GRP94 were present after three hours

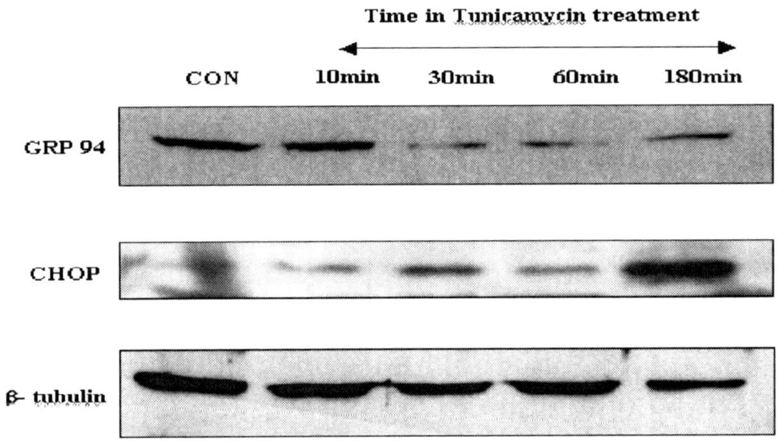

Fig. 26.2 Developing C2C12 cells showed a very sensitive response to tunicamycin treatment. While GRP94 expression decreased, CHOP expression rapidly increased when ERS treatment was extended to 3 hours

26.3.3 Comparison of Apoptotic Progression in 3T3L1 and C2C1 Cells

When the degree of apoptosis was measured using flow cytometry, differences in the degree of apoptosis was observed between differentiating myocytes and adipocytes.

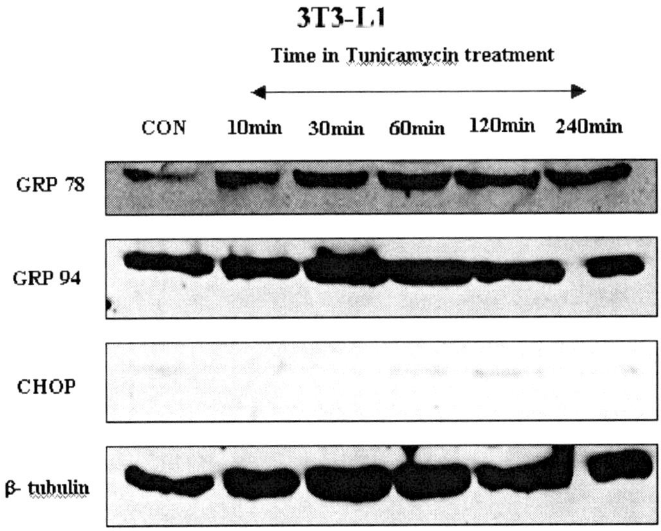

Fig. 26.3 Differentiating 3T3-L1 cells exhibited a steady state level of GRP94 expression despite prolonged ERS treatment. Little or no CHOP expression was observed as a result of ERS treatment

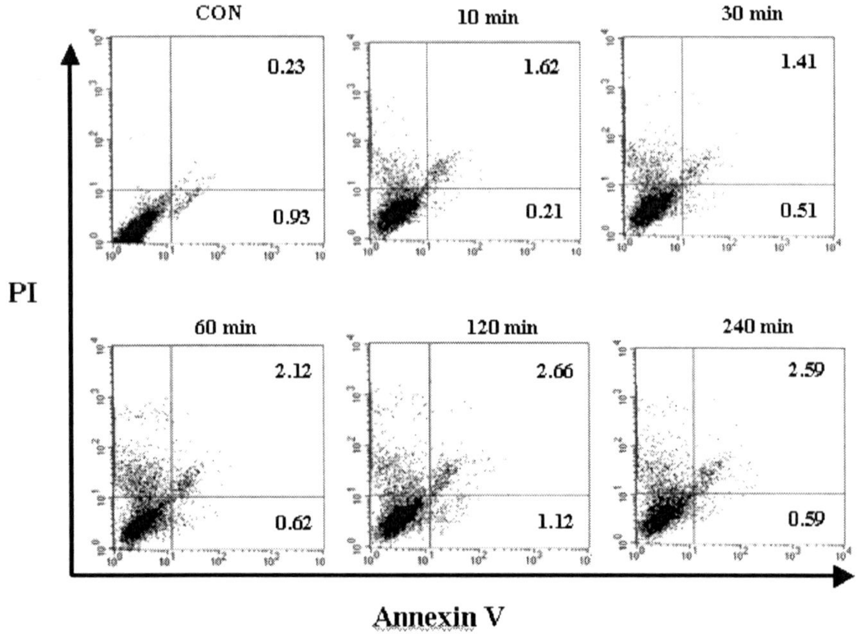

Fig. 26.4 The extent of apoptotic progression was measured after 3T3-L1 cells were treated with 5 mM tunicamycin for the indicated times. Despite prolonged treatment with tunicamycin, FACS analysis shows that the cells did not undergo any significant degree of apoptosis in 3T3-L1 cells

Tunicamycin caused little apoptosis among the 3T3-L1 cells (Fig. 26.4). However, C2C12 cells exhibited a significant level of apoptosis during ER stress (Fig. 26.5).

26.3.4 GRP94 Expression Causes Reduction in Pro-Apoptotic Marker, CHOP, with siRNA Treatment and Taurine Treatment

GRP94 and CHOP showed an opposite expression pattern. Figure 26.6 confirmed that GRP94 inhibited CHOP expression, such that CHOP expression was augmented when GRP94 was downregulated following siRNA transfection.

To examine the effect of taurine, cells were treated with taurine (0-50 mM) prior to ER stress. GRP94 and CHOP expression was compared with untreated controls. Table 26.1 shows the relative expression of the two proteins after taurine treatment. Taurine increased GRP94 expression in a concentration dependent manner. CHOP expression decreased when cells were pretreated with taurine. Based on the relative expression ratios of GRP94 and CHOP, CHOP was more sensitive to increasing concentrations of taurine.

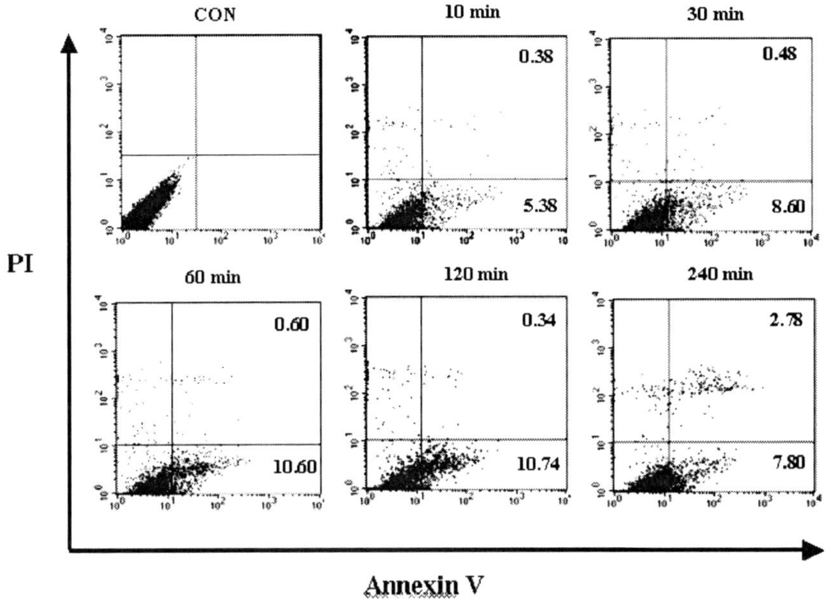

Fig. 26.5 ER stress was applied to C2C12 cells by treatment with 5 mM tunicamycin. Apoptotic progression was measured using flow cytometry. In contrast to 3T3-L1 cells, C2C12 cells exhibited a significant level of apoptosis following tunicamycin treatment

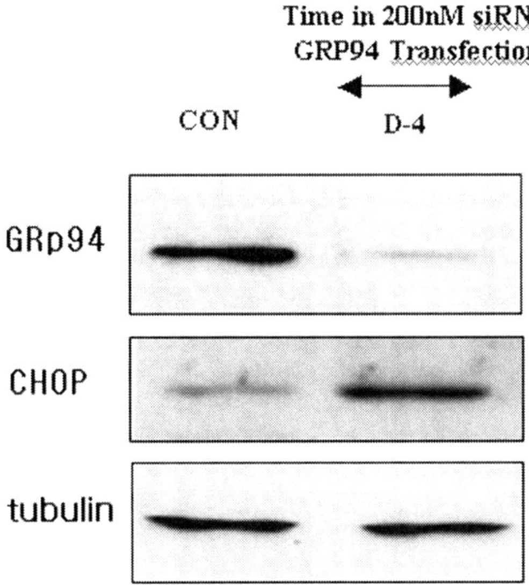

Fig. 26.6 GRP94 expression was downregulated four days (D-4) following siRNA. CHOP expression was upregulated when GRP94 was downregulated by siRNA

Table 26.1 Effect of taurine on the expression of GRP94 and CHOP

Taurine (mM)	GRP94	CHOP	[CHOP]/[GRP94]
0[a]	100.0	100.0	100.0
1[a]	112.5	85.1	75.64
5	125.3	78.6	62.72
10	152.9	63.5	41.53
50	138.7	50.3	36.26

Protein levels following taurine treatment was quantified by densitometry. For each protein, the measurements were standardized by fixing the taurine-free treatment as 100.0(%). Each value represents an average of two measurements.

26.4 Discussion

Taurine is known to play an important role in protecting muscle cells from impairment under various types of damaging conditions including ERS. In the present study, ERS caused a significant level of strain, especially in C2C12 cells, where it led to apoptosis. Different sensitivity has been observed in the two types of cell lines used in this study. Unlike 3T3-L1 cells, C2C12 cells exhibited sensitivity to tunicamycin by promoting apoptosis.

During ERS, the major difference between the two cell lines is the expression level of GRP94. Upon initiation of ERS, differentiating C2C12 cells responded to the stress by decreasing the expression of GRP94 while the expression of GRP94 remained unchanged in 3T3-L1 cells following similar treatment. This result implies that differentiating adipocytes are less sensitive to ERS agents and do not undergo apoptosis despite ongoing exposure to ERS including excessive deposition of lipid. In other words, GRP94 levels are maintained despite increasing concentrations of tunicamycin.

There are three major apoptotic ERS-linked pathways that are mediated by JNK, caspase-12 and CHOP, respectively. CHOP is the most responsive molecule when GRP94 is downregulated by siRNA. A controversy could emerge as to whether CHOP is the sole targeting molecule of GRP94. Since CHOP is induced by DNA damaging agents in addition to ERS, the effect of GRP94 overexpression might affect other cellular stress pathways. Based on the siRNA experiment, however, CHOP is the sole responsive molecule regulating the knockdown of GRP94. Induction of CHOP expression in the absence of GRP94 strongly suggests that the lack of apoptosis in 3T3-L1 cells results from the steady presence of GRP94 despite increasing degrees of ERS.

Taurine may exert a therapeutic effect directly or indirectly on the expression on GRP94, which modulates CHOP expression. Further study is necessary to elucidate whether knockdown of CHOP exerts a similar or even more inhibitory effect on apoptosis than GRP94. No study has been published to date that taurine consumption causes any harm. To enhance the beneficial effect of taurine against ERS, taurine should be consumed continuously in advance of certain types of strain and stress, including ER stress.

26.5 Conclusion

This study reports that taurine reduces the severity of ERS and apoptotic progression in myocytes. This observation can be explained on the basis of the promotion of GRP94 expression by taurine treatment. The resulting increase in GRP94 expression mediates a downregulation in CHOP expression, which serves as a triggering molecule for apoptosis. In addition, this study negates the potential use of taurine as an anti-obesity agent through the stimulation of adipocyte apoptosis, as taurine would increase the level of GRP94 and downregulate the expression of CHOP, which would tend to be anti-apoptotic.

Acknowledgments This research was supported by 2007 BioGreen21 from KRDA. The authors wish to express their great appreciation for the financial support.

References

Bando Y, Katayama T, Aleshin AN, Manabe T, Tohyama M (2004) GRP94 reduces cell death in SH-SY5Y cells perturbated calcium homeostasis. Apoptosis 9:501–508
Kadowaki H, Nishitoh H, Ichijo H (2004) Survival and apoptosis signals in ER stress: the role of protein kinases. J Chemi Neuro 28:93–100
Kee HJ, Kim JR, Nam KI, Park HY, Sin S, Kim JC, Yohei S, Masahide T, Jeong MH, Kim N, Kim KK, Kook H (2007) Enhancer of polycomb1, a novel homeodomain only protein-binding partner, induces skeletal muscle differentiation. J Biol Chem 282:7700–7709
Lane MD, Tang Q-Q (2005) From multipotent stem cell to adipocyte. Clinical & Mole Tera 73:476–477
Lee AS (2001) The glucose-regulated proteins: stress induction and clinical applications. Trends Biochem Sci 26:504–510
Misra UK, Pizzo SV (2005) Upregulation of GRP78 and antiapoptotic signaling in murine peritoneal macrophages exposed to insulin. J Leukoc Biol 78:187–194
Nakanishi K, Sudo T, Morishima N (2005) Endoplasmic reticulum stress signaling transmitted by ATF6 mediates apoptosis during muscle development. J Cell Biol 106:555–560
Ni M, Lee AS (2007) ER chaperones in mammalian development and human diseases. FEBS Lett 581:3641–3651
Özcan U. Cao Q, Erkan Yilmaz Lee A, Glimcher H, Hotamisligil GS. (2004) Endoplasmic reticulum stress links obesity insulin action type 2 diabetes. Science 306:457–461
Puthalakath H, O'Reilly LA, Gunn P, Lee L, Kelly PN, Huntington ND, Hughes PD, Michalak EM, McKimm-Breschkin J, Motoyama N, Gotoh T, Akira S, Bouillet P, Strasser A (2007) ER Stress triggers apoptosis by activating BH3-Only protein Bim. Cell 129:1337–1349
Rao RV, Peel A, Logvinova A, Rio GD, Hermel E, Yokota T, Goldsmith PC, Ellerby LM, Ellerby HM, Bredesen DE (2002) Coupling endoplasmic reticulum stress to the cell death program: role of the ER chaperone GRP78. FEBS Lett 514:122–128
Sheu ML, Liu SH, Lan KH (2007) Honokiol induces Calpain-mediated glucose-regulated protein-94 cleavage and apoptosis in human gastric cancer cells and reduces tumor growth. PLoSone 10:e1096
Sordella R, Jiang W, Chen G. C, Curto M, Settleman J (2003) Modulation of Rho GTPase signaling regulates a switch between adipogenesis and myogenesis. Cell 113:147–158
Zinszner H, Kuroda M, Wang XZ, Batchvarova N, Lightfoot RT, Remotti H, Stevens JL, Ron D (2007) CHOP is implicated in programmed cell death in response to impaired function of the endoplasmic reticulum. Genes Dev 12:982–995

Part IV
Gastroenteric and Hepatic Effects of Taurine

Chapter 27
Dietary Taurine Attenuates Dextran Sulfate Sodium (DSS)-induced Experimental Colitis in Mice

Makoto Shimizu, Zhaohui Zhao, Yoko Ishimoto, and Hideo Satsu

Abstract Effects of dietary taurine on the experimental colitis induced by dextran sulfate sodium (DSS) were studied. C57BL/6 mice administrated taurine or placebo for 5 days were given 3% DSS to induce acute. The colitis was as-sessed using indices such as diarrhea/bleeding scores, colon length change, histological score and tissue myeloperoxidase (MPO) activity. Further, tis-sue mRNA levels of interleukin (IL)-1β, tumor necrosis factor (TNF)-α and macrophage inflammatory protein (MIP)-2, were determined by real-time PCR. Taurine supplementation significantly attenuated the severity of diarrhea, colon shortening, histological score, MPO activity elevation and abnormal MIP-2 gene expression, indicating that taurine prevents DSS-induced colitis. Taurine also inhibited the TNF-α-induced secretion of IL-8 (a hu-man homologue of MIP-2) from human intestinal epithelial Caco-2 cells. Inhibition of chemokine secretion from intestinal cells may be involved in the mechanisms underlying the cytoprotective function of taurine in the intestinal epithelium.

Abbreviations *DSS*, dextran sulfate sodium; *MPO*, myeloperoxidase; *IL*, interleukin, *TNF-α*, tumor necrosis factorα; *MIP-2*, macrophage inflammatory protein 2

27.1 Introduction

In previous studies using human intestinal epithelial cell lines, we ob-served that intestinal taurine transporter (TAUT) activity was regulated by a variety of factors, including extracellular taurine concentration (Satsu et al. 1997), osmotic pressure (Satsu et al. 1999), lysophosphatidylcholine (Ishizuka et al. 2000; Ishizuka et al. 2002), and inflammatory cytokines, such as TNF-α and IL-1β (Mochizuki et al. 2002, Mochizuki et al. 2005). Up-regulation of TAUT by the inflammatory cytokines, followed by an increase in the intracellular taurine concentration, was particularly interesting because it indicated that the intestinal epithelial cells could

M. Shimizu (✉)
Department of Applied Biological Chemistry, The University of Tokyo, Tokyo, Japan

J. Azuma et al. (eds.), *Taurine 7*, Advances in Experimental Medicine and Biology 643,
DOI 10.1007/978-0-387-75681-3_27, © Springer Science+Business Media, LLC 2009

adaptively respond to the inflammatory cytokines by increasing the amount of TAUT in the cell membrane to raise the intracellular concentration of taurine. In other words, taurine can play a cytoprotective role under inflammatory conditions.

The cytoprotective and anti-inflammatory properties of taurine have been reported in immune cells. Taurine reacts with hypochlorous acid to produce taurine chloramine, which has shown to be a powerful anti-inflammatory agent by suppressing nuclear factor-κB (NF-κB) (Kanayama et al. 2002). However, little is known about the anti-inflammatory and cy-toprotective function of taurine in intestinal epithelial cells. Son et al. (1998) reported that taurine prevents trinitrobenzene sulfonic acid (TNBS)-induced colitis in rats (Son et al. 1998), and Satsu et al. reported that taurine protects epithelial cell damage caused by co-culturing human intestinal epithelial Caco-2 cells with human macrophage-like THP-1 cells (Satsu et al. 2006). The present study was undertaken to investigate whet-her taurine suppresses gut inflammation in an in vivo inflammatory bowel disease (IBD) model induced by dextran sulfate sodium (DSS).

27.2 Anti-Inflammatory Effects of Taurine on the IBD Model

27.2.1 Effect of Taurine on the Disease Activity Index of DSS-Induced Colitis, Histopathological Changes and Tissue Mye-loperoxidase (MPO) Activity

Female C57BL/6 mice (68 weeks old) were divided into 4 groups (6 mice in each group), and were given 2% taurine in water with or without 3% DSS (Fig. 27.1). The weight of each mouse and water intake were measured daily. Diarrhea scoring and bleeding scoring were performed on day 10 according to the method previously described (Cooper et al. 1993). Thereafter, the mice were sacrificed by cervical dislocation. The colon was removed from each mouse, and the length of the colon was measured. Two sections (1cm in length for each) were dissected from the proximal and distal colon. These were fixed in 10% formalin neutral buffer for histological assessment. Three segments from each specimen were sectioned, stained and scored. Inflammation and crypt damage were assessed for the H&E-stained sections by using a modification of the validated scoring scheme (Cooper et al. 1993; Matsuura et al. 2005). Tissue MPO activity was determined by standard enzymatic procedures (Krawisz et al. 1984) with minor changes (Zhao et al. 2006).

Compared with the control mice maintained on normal water, mice maintained on water supplemented with 2% taurine exhibited normal food and water intake (~3.7 mL/day, mouse), as well as weight gain. Five days following the addition of 3% DSS to the drinking water body weight fell 5%. Taurine supplementation significantly retarded DSS-induced weight loss (data not shown). It also significantly attenuated the diarrhea symptom, and appeared to diminish the severity of fecal bleeding induced by DSS (Fig. 27.2A and 27.2.B, respectively). Furthermore,

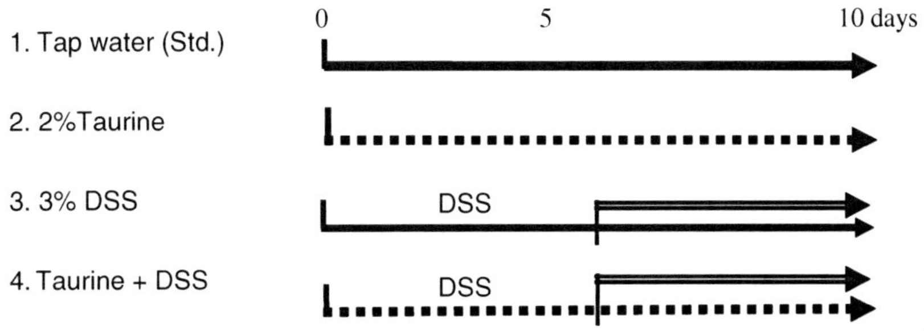

Fig. 27.1 Experimental design for DSS-induced IBD in mice. Six mice/group

taurine supplementation significantly inhibited DSS-mediated shortening of colon length (Fig. 27.2.C).

The severity of colonic mucosal damage and inflammation evaluated by histopathology is shown in Fig. 27.3. DSS induced a mild to moderate inflammatory infiltrate and crypt damage in the proximal colon. Taurine appeared to attenuate the inflammatory response, in particular DSS-mediated crypt damage. The total colitis scores for the DSS and DSS-Tau groups were 11.8 ± 2.0 and 8.8 ± 1.5, respectively (Fig. 27.3A). Although these scores were not statistically different, they revealed a trend implying a beneficial effect of taurine against DSS-induced colitis. To quantitatively evaluate the inflammatory changes in the colon, MPO activity of whole colonic tissue was determined (Fig. 27.3B). MPO activity of DSS-treated colonic

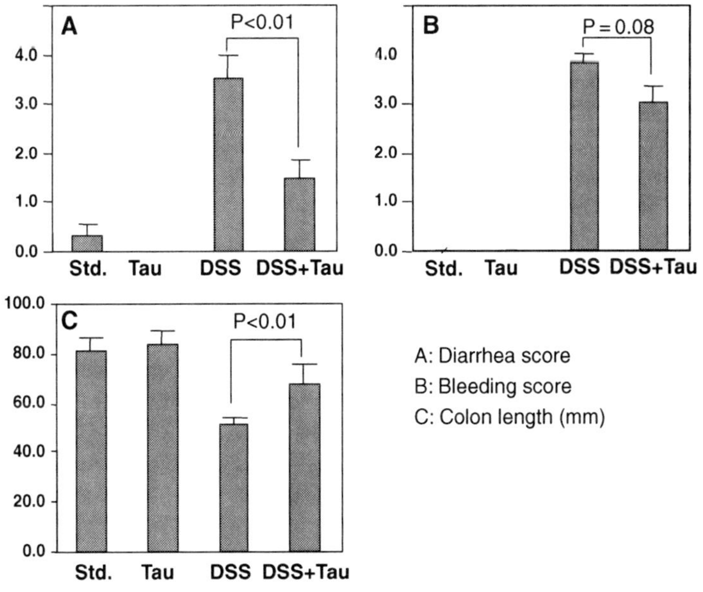

Fig. 27.2 Effect of taurine on the disease activity index of DSS-induced colitis

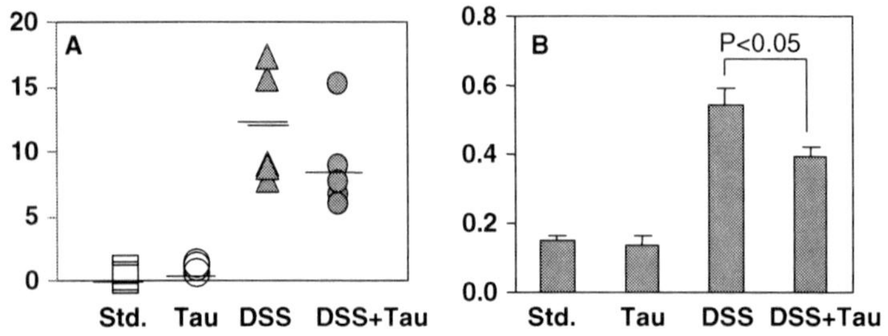

Fig. 27.3 Histological score (**A**) and MPO activity (**B**) of colonic tissue

tissue was 3.6-fold greater than that of normal tissue. Taurine treatment led to a 27% decrease in MPO activity compared to that of the DSS plus taurine-treated colonic tissue. TNBS- and DSS-induced colitis animal models are the most widely used for studying Chron's disease (CD) and ulcerative colitis (UC), respectively (Elson et al. 1995; Okayasu et al. 1990). Previous studies (Kim et al. 2006, Son et al. 1998) have demonstrated a therapeutic effect of taurine on TNBS-induced colitis in rats. Our present study further showed the beneficial effect of taurine against DSS-induced colitis in mice, suggesting that taurine would be useful for the treatment of both CD and UC.

27.2.2 Effect of Taurine on Expression of Pro-inflammatory Cytokines in the DSS-Treated Colon

RNA was extracted from tissue homogenate using the guanidium thiocyanate-phenol-chloroform method. Total poly A+ mRNA was subsequently purified from total RNA using Oligotex-dT30-super (Takara, Japan). The mRNA levels for TNF-α, IL-1β, MIP-2 were determined by real-time PCR. Total mRNA was then subjected to reverse transcription with the QuantiTect Reverse Transcription kit (Qiagen, Japan). The resulting complementary DNA was amplified using a QuantiTect SYBR Green real-time PCR kit (Qiagen, Japan). The reaction mixture was incubated for 15 min at 95°C, and then subjected to 50 amplification cycles, which included denaturation at 95°C for 15 s, an annealing step at 59°C (TNF-α and IL-1β) or 57°C (the others) for 15 s, and extension at 72°C for 15 s. Quantification of mRNA was performed using a comparative method, ΔΔCT method (Pfaffl et al. 2002 2002). Gene expression levels of the target molecules were finally normalized by using two housekeeping genes, β–actin and GAPDH.

The expression of MIP-2, TNF-α, and IL-1β was significantly increased in DSS-treated mice. The increase in MIP-2 mRNA levels was significantly reduced by taurine supplementation (Fig. 27.4A), whereas IL-1β and TNF-α mRNA levels were not significantly affected (Fig. 27.4.B, C). MIP-2, a mouse homologue of human IL-8, is a chemokine that plays a crucial role in mediating the infiltration of neutrophils into the intestinal mucosa. Taurine reduced MPO activity in DSS-treated

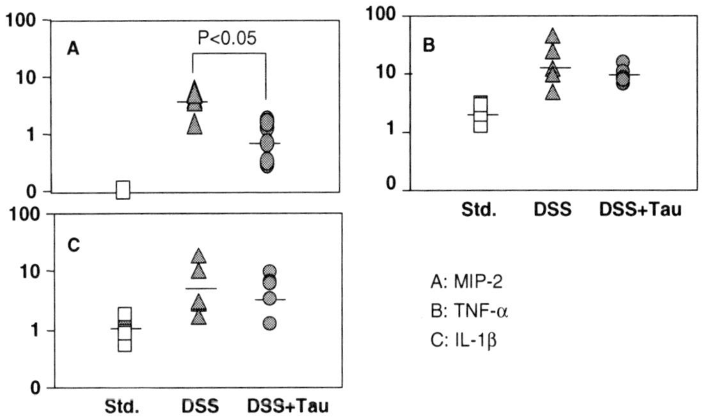

Fig. 27.4 Effect of taurine on the mRNA levels of pro-inflammatory cytokines in the DSS-treated colon

colonic mucosa (Fig. 27.4A), an observation that would explain the suppression of neutrophil infiltration by taurine treatment (Fig. 27.3B).

27.3 In Vitro Effect of Taurine on IL-8 Secretion by Intestinal Epithelial Cells

To verify the anti-inflammatory effects of taurine in vitro, we determined whether taurine could suppress TNF-α-induced IL-8 secretion by human intestinal Caco-2 cells. Caco-2 cells were pre-cultured on 24-well plates for 14 days (Mochizuki et al. 2002), before being treated with taurine for 3 h followed by TNF-α (50 ng/mL) for either 1h (mRNA determination) or 24 h (protein determination by ELISA). IL-8 mRNA was extracted and quantified by real-time PCR. Adding taurine to the medium significantly inhibited IL-8 secretion and decreased IL-8 mRNA levels of the Caco-2 cells in a dose-dependent manner (Fig. 27.5A, B). The taurine-mediated decrease in mRNA levels of IL-8 in vitro (Fig. 27.5B) and MIP-2 in vivo (Fig. 27.4A) suggests that regulation of chemokine production is involved in the beneficial effect of taurine on IBD.

Taurine reacts with hypochlorous acid to produce taurine chloramine which exhibits anti-inflammatory activity by depressing NF-κB and down-regulating pro-inflammatory mediators, such as TNF-α, PGE2, and COX2 in both rodent and human leukocytes (SchullerLevis and Park 2004; Kanayama et al. 2002). The anti-inflammatory properties of taurine chloramine are considered to be a major mechanism underlying the beneficial effect of taurine against LPS-induced acute lung injury in sheep (Elson et al. 1995) and TNBS-induced colitis in rat (Son et al. 1998). Our in vitro study showed that taurine inhibited TNF-α-induced secretion of IL-8 from the Caco-2 cell line. Because the Ca-co-2 cells, which produce no MPO,

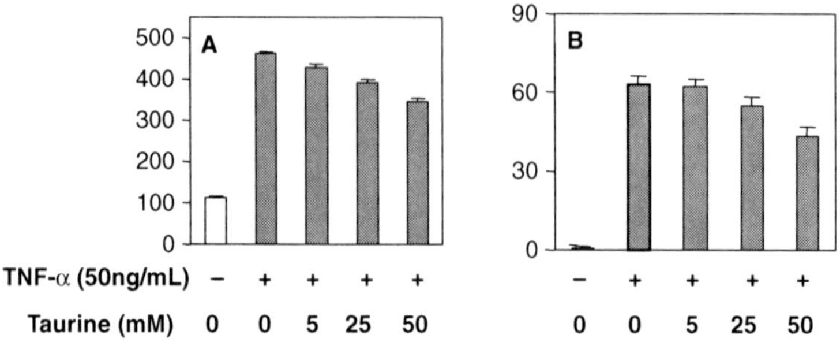

Fig. 27.5 Effect of taurine on IL-8 secretion (A; pg/ml) and mRNA levels (B; rela-tive value) in Caco-2 cells induced by TNF-α treatment

cannot convert taurine into taurine chloramine, our results indicate that intact taurine may also have anti-inflammatory potential and thus contribute to its beneficial effect against DSS-induced colitis (Zhao et al. 2007).

27.4 Conclusion

Taurine prevented DSS-induced colitis in mice. The inhibitory effect of taurine on the secretion of MIP-2 from intestinal epithelial cells would represent one of the cytoprotective mechanisms of taurine. Although taurine chloramine is considered to play a major role in preventing inflammatory diseases by down-regulating pro-inflammatory mediators, our in vitro study suggests that taurine itself also exerts some anti-inflammatory action.

References

Cooper HS, Murthy SN, Shah RS, Sedergran DJ (1993) Clinicopathologic study of dextran sulfate sodium experimental murine colitis. Lab Invest 69:238–249

Elson CO, Sartor RB, Tennyson GS, Riddell RH (1995) Experimental models of inflamma-tory bowel disease. Gastroenterology 109:1344–1367

Ishizuka K, Kanayama A, Satsu H, Miyamoto Y, Furihata K, Shimizu M (2000) Identifica-tion of a taurine transport inhibitory substance in sesame seeds. Biosci Biotechnol Bio-Chem 64: 1166–1172

Ishizuka K, Miyamoto Y, Satsu H, Sato R, Shimizu M (2002) Characteristics of lysophosp-hatidylcholine in its inhibition of taurine uptake by human intestinal Caco-2 cells. Biosci Biotechnol Biochem 66:730–736

Kanayama A, Inoue J, SugitaKonishi Y, Shimizu M, Miyamoto Y (2002) Oxidation of Ikappa Balpha at methionine 45 is one cause of taurine chloramine-induced inhibition of NFkappa B activation. J Biol Chem 277:24049–24056

Kim H, Jeon H, Kong H, Yang Y, Choi B, Kim YM, Neckers L, Jung Y (2006) A molecular mech-anism for the anti-inflammatory effect of taurineconjugated 5-aminosalicylic acid in inflamed colon. Mol Pharmacol 69:1405–1412

Krawisz JE, Sharon P, Stenson WF (1984) Quantitative assay for acute intestinal inflamma-tion based on myeloperoxidase activity. Assessment of inflammation in rat and hams-ter models. Gastroenterology 87:1344–1350

Matsuura M, Okazaki K, Nishio A, Nakase H, Tamaki H, Uchida K, Nishi T, Asada M, Kawasaki K, Fukui T, Yoshizawa H, Ohashi S, Inoue S, Kawanami C, Hiai H, Tabata Y, Chiba T (2005) Therapeutic effects of rectal administration of basic fibroblast growth factor on experimental murine colitis. Gastroenterology 128:975–986

Mochizuki T, Satsu H, Shimizu M (2002) Tumor necrosis factor alpha stimulates 0 taurine uptake and transporter gene expression in human intestinal Caco-2 cells. FEBS Lett 517:92–96

Mochizuki T, Satsu H, Shimizu M (2005) Signaling pathways involved in tumor necrosis factor alpha-induced up-regulation of the taurine transporter in Caco-2 cells. FEBS Lett 579: 3069–3074

Okayasu I, Hatakeyama S, Yamada M, Ohkusa T, Inagaki Y, Nakaya R I, Hatakeyama S, Yamada M, Ohkusa T, Inagaki Y, Nakaya R (1990) A novel method in the induction of reliable experimental acute and chronic ulcerative colitis in mice. Gastroenterology 98:694–702

Pfaffl MW, Horgan GW, Dempfle L (2002) Relative expression software tool (REST) for groupwise comparison and statistical analysis of relative expression results in real-time PCR. Nucleic Acids Res 30:e3–e6

Satsu H, Ishimoto Y, Nakano T, Mochizuki T, Iwanaga T, Shimizu M (2006) Induction by activated macrophagelike THP1 cells of apoptotic and necrotic cell death in intestinal epithelial Caco-2 monolayers via tumor necrosis factoralpha. Exp Cell Res 312:3909–3019

Satsu H, Miyamoto Y, Shimizu M (1999) Hypertonicity stimulates taurine uptake and transporter gene expression in Caco-2 cells. Biochim Biophys Acta 1419:89–96

Satsu H, Watanabe H, Arai S, Shimizu M (1997) Characterization and regulation of taurine transport in Caco-2, human intestinal cells. J Biochem (Tokyo) 121:108–127

SchullerLevis GB, Park E (2004) Taurine and its chloramine: modulators of immunity. Neu-rochem Res 29:117–126

Son MW, Ko JI, Doh HM, Kim WB, Park TS, Shim MJ, Kim BK (1998) Protective effect of taurine on TNBS-induced inflammatory bowel disease in rats. Arch Pharm Res 21:531–536

Zhao ZH, Hyun JS, Satsu H, Kakuta S, Shimizu M (2006) Oral exposure to cadmium chlo-ride triggers an acute inflammatory response in the intestines of mice, initiated by the overexpression of tissue macrophage inflammatory protein2 mRNA. Toxicology Let-ters 164:144–154

Zhao Z, Satsu H, Fujisawa M, Hori M, Ishimoto Y, Totsuka M, Nambu A, Kakuta S, Ozaki H, Shimizu M (2008) Attenuation by dietary taurine of dextran sulfate sodium-induced colitis in mice and of THP-1-induced damage to intestinal Caco-2 cell monolayers. Amino Acids 35(1):217-24.

Chapter 28
Protective Effect of Taurine against Nitrosative Stress in the Stomach of Rat with Water Immersion Restraint Stress

Ning Ma, Takeshi Sasaki, Hiromi Sakata-Haga, Ken-ichi Ohta, Ming Gao, Shosuke Kawanishi, and Yoshihiro Fukui

Abstract In the present study, we examined by immunohistochemistry the formation and distribution of 8-nitroguanine, a sensitive marker of nitrosative DNA damage, in rat stomach of rats subjected to water immersion restraint stress (WIR). WIR induced an increase in 8-nitroguanine content of gastric gland epithelium. 8-Nitroguanine immunoreactivity, which was observed mainly in the nuclei of stomach epithelium, increased with the severity of inflammation. Expression of iNOS was also observed in the inflammatory cells of lamina propria. Therefore, it is logical that iNOS-mediated nitrosative stress must participate in the development of ulcers through apoptotic cell death linked to the formation of 8-nitroguanine during chronic inflammation. Taurine administration attenuated stress-induced gastric mucosal injury. These results demonstrate that nitrosative stress participates in stress-mediated ulcer formation. Taurine exerts a prophylactic effect against mucosal lesions of the stomach caused by stress. This effect of taurine may have a potential clinical benefit in preventing gastritis associated with stress.

Abbreviations *WIR*, water immersion restraint stress

28.1 Introduction

Taurine (2-aminoethanosulfonic acid) is present in high concentration in the brain, liver, kidney, and heart of mammals, where it represents one of the most abundant free amino acids in the body and is considered a conditionally semi-essential amino

N. Ma (✉)
Department of Anatomy and Developmental Neurobiology, Institute of Health Biosciences, The University of Tokushima Graduate School, Tokushima Japan

J. Azuma et al. (eds.), *Taurine 7*, Advances in Experimental Medicine and Biology 643,
DOI 10.1007/978-0-387-75681-3_28,

acid (Huxtable 1992). It plays an important role in several essential biological processes, including neuromodulation, osmoregulation and protection against oxidative stress. Despite its wide distribution, the biochemical and physiologic function of taurine is not fully understood. Yet it is widely accepted that taurine functions as a scavenger of endogenously generated hypochlorous acid (HOCl) (Marquez and Dunford 1994). Taurine chloramine, which is formed from the reaction of taurine with HOCl, acts as a cellular signaling molecule to downregulate the expression of inflammatory agents, such as inducible nitric oxide synthase, tumor necrosis factor- α and cyclooxygenase-2 (Liu et al. 1998). It is well established that tissues undergoing inflammation are at increased risk of cell damage. Reactive nitrogen species (RNS) and reactive oxygen species (ROS) are considered to play key roles in inflammation mediated pathology (Helmut and Jagadeesan 2006). Excess nitric oxide (NO) is produced from inflammatory cells via the expression of inducible nitric oxide synthase (iNOS). Overproduction of NO contributes to the generation of various RNSs, such as Nox and peroxynitrite ($ONOO^-$). RNSs can mediate the formation of 8-nitroguanine, a marker of nitrosative DNA damage. Evidence continues to accumulate showing that ROS and RNS are involved in stomach disorders (Ma et al. 2004). Several studies demonstrate that taurine treatment prevents tissue damage in various model of inflammation (Hung 2006; Zeybek et al. 2006). It has been suggested that the beneficial effect of taurine against indomethacin-induced gastric mucosal damage involves its ability to serve as an antioxidant (Son et al. 1996). These reports led us to propose the hypothesis that stress-induced RNSs cause nitrosative DNA damage and subsequent cell death in the mucosa of the stomach, and are responsible for the pathogenesis of digestive abnormalities. Accordingly, we examined nitrosative DNA damage and the resulting pathological changes in gastric mucosa with the aim of clarifying the mechanism underlying taurine-mediated cytoprotection.

The formation and distribution of 8-nitroguanine, a sensitive marker of nitrosative DNA damage (Akaike et al. 2003; Ma et al. 2004, 2006), was assessed using immunohistochemistry and histological changes of the gastric mucosa in the stomach of rats subjected to water immersion restraint stress.

28.2 Methods

28.2.1 Chemicals and Materials

8-Nitroguanine was purchased from Biolog Life Science Institute (Bremen, Germany). Mouse monoclonal anti-iNOS, anti-H+, K+-ATPase and anti-TNF-α antibodies were purchased from Calbiochem-Novabiochem (Darmstadt, Germany) and Stressgen Biotechnologies (Victoria, BC, Canada), respectively. Alexa 594-labeled goat antibody against rabbit IgG and Alexa 488-labeled goat antibody against mouse IgG were obtained from Molecular Probes (Eugene, Oregon, USA). We purchased a cellulofine GCL-2000m column from Seikagaku Kogyo, Tokyo, Japan.

28.2.2 Animals and Experimental Design

Male Wistar rats, weighing 220–250 g, were purchased from SLC (Hamamatsu, Japan). All animal procedures were conducted according to the Guide for the Care and Use of Laboratory Animals, and were reviewed by the Institutional Animal Care and Use Committee of The University of Tokushima. Great care was taken to minimize the number of animals used and their suffering. Rats are housed individually in a room with a 12 hour dark-light cycle and with central air conditioning (24°C temperature). Rats were allowed free access to water and pellet food. Before the experiment, rats were deprived of food for 24 hours, but were allowed free access to water. Rats were restrained in a wire cage and immersed up to the depth of the xiphoid process in a 23 ± 0.5°C water bath for 2 hours to produce WIR stress-induced gastric mucosal lesion as described by Takagi (Takagi and Okabe 1968). Four experimental groups were designed. In the stress group (WIR group, $n = 12$), rats were applied to WIRS for hours in one time (series A) or 2 hours of WIRS applied every day for up to 5 days (series B) at 23°C ± 0.5°C. After the experiment, the gastric samples were fixed 6 (12 hours (series A) and 7 days (Series B). Control rats (Control group, $n = 12$) were placed on the same container for 2 hours of one time or 2 hours of every day for up to 5 days. Rats were killed under ether anesthesia after application of WIR at which point their stomachs were removed. The observation of lesions recognized as linear breaks (erosions) at the mucosal surface of the glandular part and the length (mm) and the width (mm) of erosion on the gastric mucosa were measured under a stereoscopic microscope. The extent of the lesion (lesion index) is expressed as the sum of the area of the erosions per stomach. Rats received an intraperitoneal injection of taurine (WIR-taurine-IP group, 300 mg kg^{--1}) was dissolved in isotonic saline 60 minutes before the onset of WIR stress. In a separate experiment, rats received an oral administration of taurine (WIR-taurine-PO group 300 mg/kg dissolved in isotonic saline) 60 minutes prior to the onset of WIR stress. Stomachs were placed in a 4.0% paraformaldehyde/0.1% glutaraldehyde fixative for one night, and embedded in paraffin. Five-μm thick sections were mounted on albumin-coated slides.

28.2.3 Production of Anti-8-Nitroguanine Antibody

Anti-8-nitroguanine polyclonal antibody was produced by the modified method of Akaike et al. (2003). 8-Nitroguanosine was incubated with sodium metaperiodate for 20 minutes at room temperature and then conjugated with RSA for 1 hour followed by incubation with sodium borohydride for 1 hour. The conjugate was dialyzed overnight against 150 mM NaCl. 8-Nitroguanine-aldehyde-RSA conjugate mixed with Freund's complete adjuvant was injected in rabbit by intracutaneous administration. After 4 weeks of the immunization, the same antigen was given and the blood was taken 10 days later. We immobilized 8-nitroguanine in a cellulofine GCL-2000 m column (Seikagaku Kogyo), and then purified the antibody by affinity chromatography. Specificity of the purified antibody was examined by a dot immunobinding assay and absorption test (Pinlaor et al. 2004).

28.2.4 Double Immunofluorescence Staining of 8-Nitroguanine and iNOS or H^+, K^+ – ATPase

Double immunofluorescence labeling studies of 8-nitroguanine, iNOS and H^+, K^+-ATPase in the stomach were performed as described previously (Ma et al. 2004). Briefly, paraffin sections (6 μm thickness) were deparaffinized and microwaved twice in 5% urea for 5 minutes. The sections were incubated with 1% skim milk for 30 minutes. Then, the sections were incubated with either rabbit polyclonal anti-8-nitroguanine antibody (2 μg/ml), mouse monoclonal anti-H^+, K^+-ATPase (3 μg/ml) antibody or mouse monoclonal anti-iNOS antibody (5 μg/ml) overnight at room temperature. Finally, they were incubated with Alexa 594-labeled goat antibody against rabbit IgG and Alexa 488-labeled goat antibody against mouse IgG (1:400) for 3 hours. The immunostained sections were examined under an inverted Laser Scan Microscope (Olympus, Japan).

28.2.5 Immunohistochemical Study of Taurine and TNF-α Expression

The sections were incubated with rabbit polyclonal anti-taurine antibody (1:500) followed by incubation with goat anti-rabbit IgG-HRP (1:200) and analyzed using a microscope. Sections are visualized with 3,3-diaminobenzidine tetrahydrochloride as chromogen. In certain experiments, immunohistochemical analysis of taurine, H^+, K^+-ATPase or TNF-α expression were carried out in the stomach sections by indirect immunofluorescence.

28.3 Results

28.3.1 Morphological and Histological Alterations in WIR Rat Gastric Mucosa with or Without Administration of Taurine

Gastric mucosal lesions in rats undergoing WIR stress were observed 12 hours after onset of stress, with no gastric mucosal lesions found in rats without WIR stress (control group). The development of gastric mucosal lesions in rats subjected to WIR stress for 2 hours was significantly attenuated by pre-administration of taurine at a dose of 300 mg kg^{-1} by intraperitoneal injection or by oral administration. Semi-quantitative histological damage scores in the presence or absence of taurine are summarized in Fig. 28.1. The damage scores for one time WIR were 5.00 ± 1.34 mm^2 in the WIR group, 0.86 ± 1.86 mm^2 in the WIR-taurine-PO group, and

0.13 ± 0.29 mm^2 in the WIR-taurine-IP group; the damage scores for five-day WIR were 18.32 ± 2.42 mm^2 in the WIR group, 2.42 ± 1.03 mm^2 in the WIR-taurine-PO group, and 2.23 ± 0.94 mm^2 in the WIR-taurine-IP group. The histological damage score in the WIR group was elevated in the gastric mucosa ($P < 0.01$) compared to those of the WIR-taurine-IP and WIR-taurine-PO groups. Taurine treatment significantly reduced the microscopic damage score of the stomach relative to the one time WIR and 5 day WIR-induced damage groups (Fig. 28.1). Gastric mucosa was morphologically intact in WIR rats administrated taurine. Nevertheless, numerous hemorrhagic erosions were observed in WIR rats without administration of taurine.

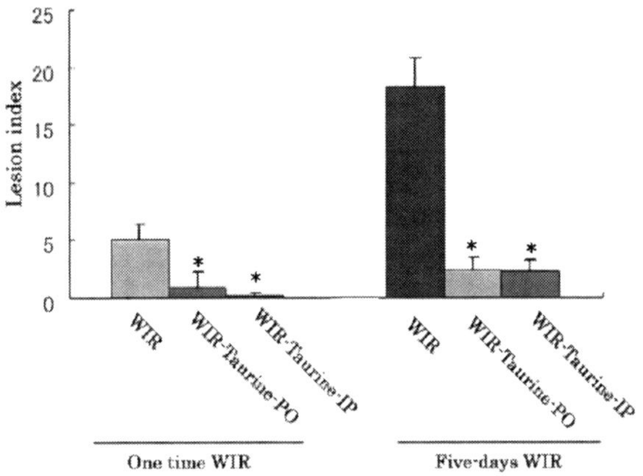

Fig. 28.1 Effect of pre-administration of taurine on gastric mucosal lesion development in rats subjected to WIR stress and five day WIR stress. Each microscopic scoring damage value represents the mean ± SD of 12 animals. $^*P < 0.01$ compared with the WIR group based on the Student's t-test

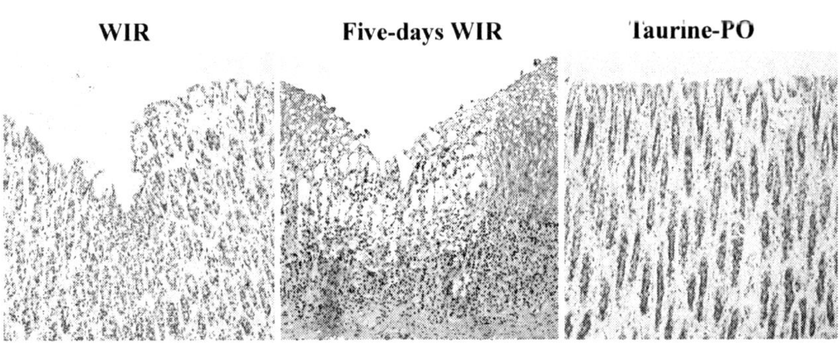

Fig. 28.2 WIR group: A pronounced disruption of the upper cell layer and lamina propria is observed. Dilated gastric glands, severe congestion of blood vessels and inflammatory cell infiltration are present. WIR-five-day group: Severe degeneration of surface epithelium and dilated gastric glands, significant congestion of blood vessels and inflammatory cell infiltration are observed. Taurine administration group: Seemingly normal stomach mucosa with surface epithelium, gastric pits and gastric glands are observed

Histological studies showed pronounced cell injury in both the epithelial layer and lamina propria of the WIR rat stomach. Severe degeneration of surface epithelium and gastric glands with dilatations, prominent congestion of blood vessels, and inflammatory cell infiltration were present in the WIR group (Fig. 28.2). However, almost normal fundic mucosa with gastric pits, mucus and glandular cells were observed in the pre-taurine administration groups (Fig. 28.2).

28.3.2 8-Nitroguanine Formation and iNOS Expression in the Stomach of WIR Rats

We produced a specific anti-8-nitroguanine antibody using an 8-nitroguanine-RSA conjugate. The purified antibody reacted with 8-nitroguanine conjugates but the immunoreactivity disappeared when the antibody was preincubated with 8-nitroguanine. However, immunoreactivity with the 8-nitroguanine conjugate did not disappear when the antibody was preincubated with 3-nitrotyrosine, guanosine, 8-oxodG, deoxyguanosine, 8-bromoguanosine, and xanthosine (see data in Pinlaor, et al. 2004).

8-Nitroguanine formation in the gastric mucosa of WIR rats pretreated with or without taurine is shown in Fig. 28.3. Intense immunoreactivity of 8-nitroguanine was detected in gastric gland epithelial cells of rats treated with WIR for five days. On the other hand, the WIR stressed rat pretreated with taurine exhibited little or no immunoreactivity in gastric gland epithelial cells. 8-Nitroguanine formation was observed mainly in the nuclei of the labeled epithelial cells, suggesting that DNA is modified. H, K^+-ATPase immunoreactivity, a marker for parietal cells, colocalized with that of 8-nitroguanine. Even after one time WIR stress treatment, weak immunoreactivity of 8-nitroguanine formation was observed in the cytoplasm of the parietal cells.

28.3.3 Effect of Taurine on Gastric Mucosal Lesion Development and Changes in Gastric Mucosal Taurine Distribution in Rats with WIR Stress

To assess the extent and distribution of taurine in the gastric gland after stress, immunoreactivity of taurine was examined in stomach sections (Fig. 28.4). Intense staining of parietal cells was noted in gastric sections of the control group (non-WIR group). However, WIR treatment markedly reduced the amount or the number of taurine containing cells of the gastric gland. Moreover, pre-administration of taurine had no effect on intracellular taurine content of the parietal cells. Interestingly, stress decreased the number of taurine immunoreactive cells of the epithelium, indicating that taurine is released from parietal cells during WIR (Fig. 28.4).

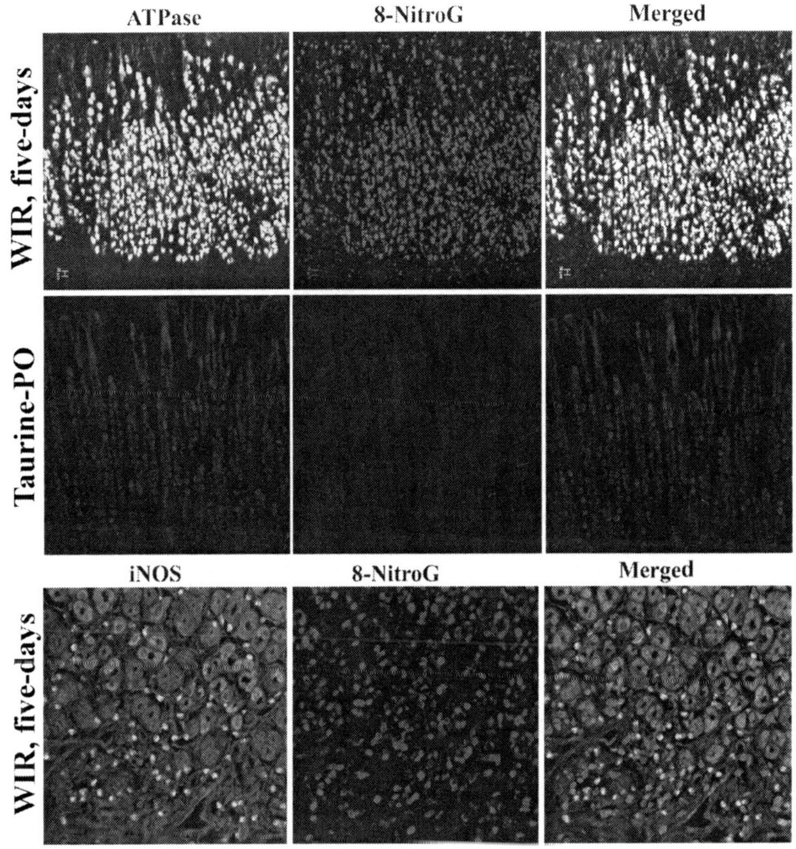

Fig. 28.3 WIR stress induced 8-nitroguanine formation in gastric mucosa. Elevated 8-nitroguanine immunoreactivity was observed in cell nuclei of gastric mucosa of rats with five-day WIR. The distribution pattern of 8-nitroguanine immunoreactive cells is similar to that of H^+, K^+-ATPase positive cells. iNOS immunoreactivity was detected in the cytoplasm of many infiltrated cells of the lamina propria

28.3.4 *Effect of Taurine on TNF-α Expression of Rat Gastric Mucosa Following WIR Stress*

TNF-α was expressed in some inflammatory cells of the gastric mucosa. The number of cells staining positive for TNF-α in gastric mucosa was significantly increased by stress, with the effect being prominent in the superficial portion of the mucosa. The increase in gastric mucosal TNF-α immunoreactive cells was significantly attenuated in rats pretreated with taurine (Fig. 28.5). Interestingly, endothelial cells and epithelial cells did not stain for TNF-α.

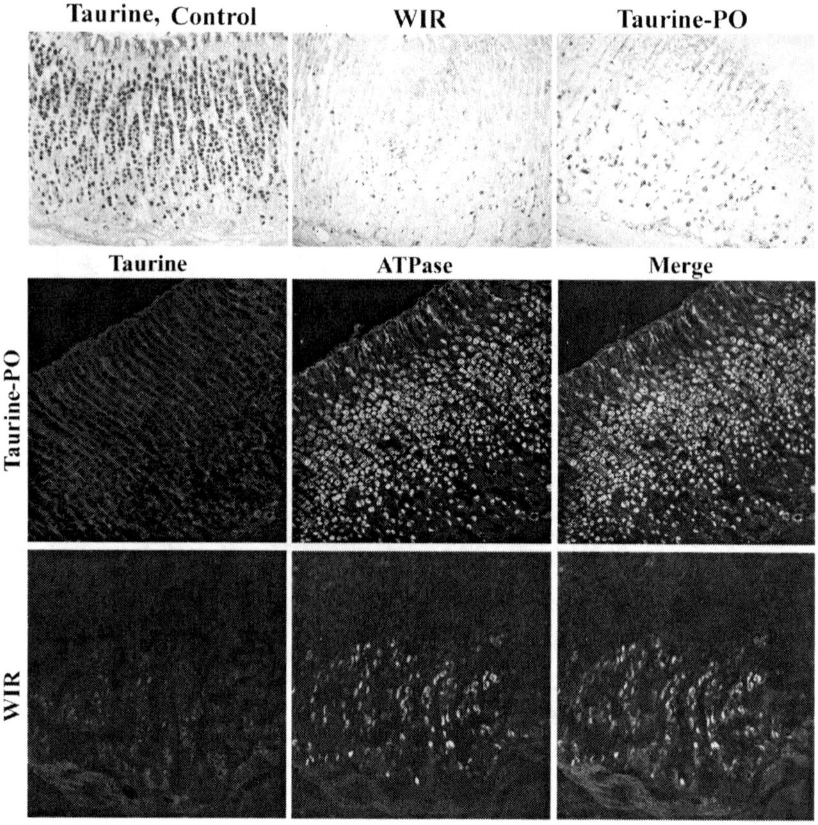

Fig. 28.4 Taurine immunoreactivity in gastric mucosa and double immunofluorescence labeling with taurine antibody and H^+, K^+-ATPase antibody. After WIR stress, taurine content in parietal cells disappeared

28.4 Discussion

It is well known that oxygen-free radicals, primarily superoxide anion (O_2) and hydroxyl radical (OH·), play important roles in the pathogenesis of acute gastric lesions induced by experimental stress (Yasukawa et al. (2004). WIR used in this study has been widely used to induce gastric lesions in animals, providing a model system for studying stress-induced ulcers in humans (Takagi and Okabe 1968; Uramoto et al. 1990). In the present study, we demonstrated an increase in 8-nitroguanine content of gastric gland epithelium following water immersion restraint stress. 8-Nitroguanine immunoreactivity was observed mainly in the nuclei of stomach epithelium. It is noteworthy that 8-nitroguanine increased, with the amount being dependent on the severity of inflammation. Expression of iNOS was also observed in the inflammatory cells of the lamina propia, indicating that the formation of 8-nitroguanine was mediated by NO production linked

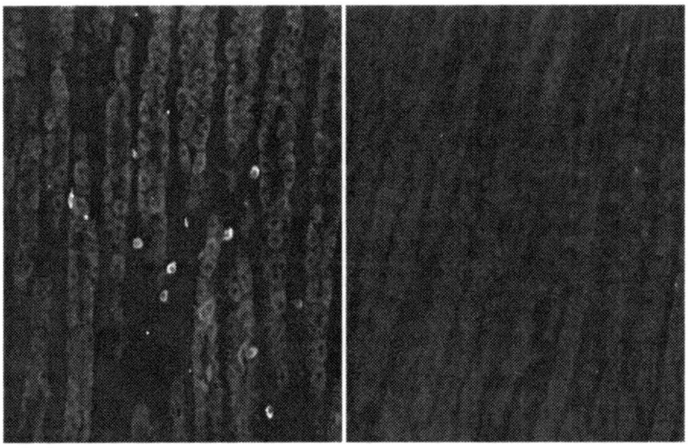

Fig. 28.5 Expression of TNF-α in gastric mucosa 12 hours after application of WIR. Application of stress increased the number of TNF-α-positive cells, especially in the superficial portion of the mucosa. Administration of taurine before stress reduced TNF-α expression. Endothelial cells and epithelial cells did not stain for TNF-α

to iNOS expression. In agreement with Hamaguchi et al. (2001), exposure of rats to WIR exhibited histological changes, such as severe degeneration of surface epithelium and of gastric glands with dilatations, prominent congestion of blood vessels, and inflammatory cell infiltration. These results suggest that nitrosative stress is caused by an inflammatory response. In addition, significant 8-nitroguanine formation was observed in five-day WIR stressed rats. Therefore, it is logical to conclude that iNOS-mediated nitrosative stress participates in the development of ulcers, largely by causing apoptosis through the formation of 8-nitroguanine during chronic inflammation. It is noteworthy that the present study was the first to demonstrate that nitrosative stress participates in ulcer formation during stress. The mechanism underlying the formation of 8-nitroguanine during inflammation is as follows. Superoxide anion radical (O_2^-) is generated from macrophages (Crowther et al. 2004), and neutrophils, and is then dismutated to hydrogen peroxide (H_2O_2). However, 8-nitroguanine formation requires the generation of NO, which is mainly derived from iNOS in mononuclear cells, including macrophage. NO reacts with O_2^- to produce peroxynitrite ($ONOO^-$) (Wink and Michel 1998), which reacts with guanine to form 8-nitroguanine. Because 8-nitroguanine formed in DNA is extremely unstable, it readily undergoes depurination, resulting in the formation of an apurinic site. The resulting apurinic site can lead to a G:C > T:A transversion (Loeb and Preston 1986; Kawanishi et al. 2006). It is been reported that expression of iNOS is increased in gastric mucosa of WIR rats (Nishida et al. 1998). Our finding that iNOS was expressed in macrophages found in the lamina propria of the WIR group is supported by these reports.

Our previous report demonstrates that high concentrations of taurine are present in parietal cells of gastric mucosa (Ma et al. 2003). In this study, WIR treatment markedly reduced either the levels of taurine in each cell or the number of taurine containing parietal cells. Interestingly, stress decreases the number of taurine immunoreactive epithelial cells, indicating that taurine is released from parietal cells during WIR. The mechanism of stress-induced taurine release is not fully understood. Taurine plays an important role in cytoprotection against ischemia-reperfusion and hypoxia (Kingston et al. 2004). Taurine efflux may also contribute to the regulatory volume decrease mediated by ion channels and triggered as a response to ischemia-induced cell swelling. Released taurine can in turn react with $HOCl^-$ produced by activated leukocytes forming taurine chloramine, a reactant that inhibits the production of NO, TNF-α, and other proinflammatory mediators (Barua et al. 2001).

Tumor necrosis factor-α is a proinflammatory cytokine that mediates ischemia-reperfusion induced gastric mucosal injury and that strongly stimulates neutrophil adherence by inducing the synthesis and expression of adhesion molecules on endothelial cells and neutrophils. In fact, WIR-induced gastric inflammation is accompanied by an increase in the expression of TNF-α (Kwiecien et al. 2002). In the present study, taurine administration resulted in a decrease in TNF-α immunoreactivity in the gastric mucosa of WIR-stressed rats. This finding indicates that taurine likely inhibits neutrophil infiltration into the gastric mucosa by suppressing TNF-α production.

28.5 Conclusion

In conclusion, taurine administration attenuated stress-induced gastric mucosal injury, likely via its ability to prevent not only oxidative stress but also nitrosative stress. Taking into account the role of local elevations in NO from the inducible isoform of NOS expressed in inflamed tissue, the data presented here provide evidence to support a possible protective effect of taurine treatment before and during the development of inflammation in the stomach. These results suggest that taurine exerts a prophylactic effect against stress-mediated mucosa lesions of the stomach. This effect of taurine could serve as potential therapy for the prevention of gastritis associated with stress.

References

Akaike T, Okamoto S, Sawa T, Yoshitake J, Tamura F, Ichimori K, Miyazaki K, Sasamoto K, Maeda H (2003) 8-nitroguanosine formation in viral pneumonia and its implication for pathogenesis. Proc Natl Acad Sci USA 100:685–690

Barua M, Liu Y, Quinn MR (2001) Taurine chloramines inhibit inducible nitric oxide synthase and TNF-α gene expression in activated alveolar macrophages: decreased NF-κB kinase activity. J Immunol 167:2275–2281

Crowther JE, Kutala VK, Kuppusamy P, Ferguson JS, Beharka AA, Zweier JL, McCormac FX, Schlesinger LS (2004) Pulmonary surfactant protein a inhibits macrophage reactive oxygen intermediate production in response to stimuli by reducing NADPH oxidase activity. J Immunol 172:6866–6874

Hamaguchi M, Watanabe T, Higuchi K, Tominaga K, Fujiwara Y, Arakawa T (2001) Mechanisms and roles of neutrophil infiltration in stress-induced gastric injury in rats. Dig Dis Sci 46: 2708–2715

Helmut B, Jagadeesan N (2006) Chronic inflammation and oxidative stress in the genesis and perpetuation of cancer: role of lipid peroxidation DNA damage and repair Langenbecks. Arch Surg 391:499–510

Hung CR (2006) Effect of taurine on gastric oxidative stress and hemorrhagic erosion in brain ischemic rats. Chinese J Physiol 49:152–159

Huxtable RJ (1992) Physiological actions of taurine. Physiol Rev 72:101–163

Liu Y, Tonna-Demasi M, Park E, Schuller-Levis G, Quinn MR (1998) Taurine chloramines inhibits production of nitric oxide and prostaglandin E2 in activated C6 glioma cells by suppressing inducible nitric oxide synthase and cyclooxygenase-2 expression. Molecular Brain Res 59: 189–195

Loeb LA, Preston BD (1986) Mutagenesis by apurinic/apyrimidinc sites. Annu Rev Genet 20: 201–230

Kawanishi S, Hiraku Y, Pinlaor S, Ma N (2006) Oxidative and nitrative DNA damage in animals and patients with inflammatory diseases to inflammation-related carcinogeneis. Biol Chem 387:365–372

Kingston R, Kelly CJ, Murray P (2004) The therapeutic role of taurine in ischaemia-reperfusion injury. Cur Pharm Des 10:2401–2410

Kwiecien S, Brzozowski T, Konturek SJ (2002) Effects of reactive oxygen species action on gastric mucosa in various models of mucosal injury. J Physiol Pharmacol 53:39–50

Ma N, Tagawa T, Hiraku Y, Murata M, Ding X, Kawanishi S (2006) 8-Nitroguanine formation in oral leukoplakia a premalignant lesion. Nitric Oxide 14:137–143

Ma N, Adachi Y, Hiraku Y, Horiki N, Horiike S, Imoto I, Pinlaor S, Semba R, Kawanishi S (2004) Accumulation of 8-nitroguanine in human gastric epithelium induced by Helicobacter pylori infection. Biochem Biophys Res Commun 319:506–510

Ma N, Ding X, Miwa T, Semba R (2003) Immunohistochemical localization of taurine in the rat stomach. Adv Exp Med Biol 526:229–236

Marquez LA, Dunford HB (1994) Chlorination of Taurine by Myeloperoxidase. J Biol Chem 269:7950–7956

Nishida K, Ohta Y, Ishiguro I (1998) Contribution of NO synthases to neutrophil infiltration in the gastric mucosal lesions in rats with water immersion restraint stress. FEBS Lett 425:243–248

Pinlaor S, Hiraku Y, Ma N, Yongvanit P, Semba R, Oikawa S, Murata M, Sripa B, Sithithaworn P, Kawanishi S (2004) Mechanism of NO-mediated oxidative and nitrative DNA damage in hamsters infected with Opisthorchis viverrini: a model of inflammation-mediated carcinogenesis. Nitric Oxide 11:175–183

Son M, Kim HK, Kim WB, Yang J, Kim BK (1996) Protective effect of taurine on indomethacin-induced gastric mucosal injury. Adv Exp Med Biol 403:147–155

Takagi K, Okabe S (1968) The effects of drugs in the stress ulcer in rats. Jpn J Pharmacol 18:9–18

Uramoto H, Ohno T, Ishihara T (1990) Gastric mucosal protection induced by restraint and water-immersion stress in rats. Jpn J Pharmacol. 54:287–298

Wink DA, Michel JB (1998) Chemical biology of nitric oxide: insights into regulatory cytotoxic and cytoprotective mechanisms of nitric oxide. Free Radic Biol Med 25:434–456

Yasukawa K, Kasazaki K, Hyodo F (2004) Non-invasive analysis of reactive oxygen species generated in rats with water immersion restraint-induced gastric lesions using in vivo electron spin resonance spectroscopy. Free Radic Res 38:147–155

Zeybek A, Ercan F, Cetinel S, Cikler E, Saglam B, Sener G (2006) Taurine ameliorates water avoidance stress-induced degenerations of gastrointestinal tract and liver. Dig Dis Sci 51:1853–1861

Chapter 29
Taurine Feeding Inhibits Bile Acid Absorption from the Ileum in Rats Fed a High Cholesterol and High Fat Diet

Naomichi Nishimura, Tatsuro Yamamoto, and Toru Ota

Abstract We have previously reported that taurine-mediated reductions in plasma cholesterol in cholesterol-fed rats were caused by increased excretion of bile acids into feces. In the present study, we examined the effect of taurine on bile acid secretion into bile and the absorption of bile acids from the ileum. Male Wistar rats were divided into 2 groups, one group that was fed a diet containing 1% cholesterol (HC diet) and the other group fed a HC diet supplemented with 1% taurine for 2 weeks. Bile acid concentrations in the mesenteric blood, the distal ileum and the colorectal digesta were determined. Mesenteric bile acid concentration in the distal ileum was significantly lower in rats fed the taurine containing diet than in those fed only the HC diet. Colorectal, but not distal ileal, bile acid concentration was significantly higher in rats fed the taurine diet than in those fed the HC diet. However, the secretion of bile acids into the bile was similar in the two groups. These results suggest that the absorption of bile acids from the distal ileum to the rectum is inhibited by taurine.

Abbreviations *HC*, the diet containing 1% cholesterol

29.1 Introduction

Taurine has a plasma cholesterol-lowering effect in rats fed a high cholesterol diet (Yokogoshi et al. 1999; Yamanaka et al. 1985; Murakami et al. 1999). Taurine feeding induces increased fecal bile acid excretion (Yamanaka et al. 1985; Nishimura et al. 2003). Since bile acid excretion constitutes a major route of elimination of cholesterol from the body (Turley and Dietschy 1988), we believe that the taurine-mediated increase in fecal bile acid excretion may be the primary factor underlying the plasma cholesterol-lowering effect of taurine.

Bile acid excretion is thought to depend on cholesterol catabolism, bile acid secretion into bile and the absorption of bile acids from the distal ileum.

N. Nishimura (✉)
Department of Nutritional Sciences, Faculty of Health and Welfare Science, Nayoro City University, Hokkaido, Japan

J. Azuma et al. (eds.), *Taurine 7*, Advances in Experimental Medicine and Biology 643,
DOI 10.1007/978-0-387-75681-3_29, © Springer Science+Business Media, LLC 2009

Therefore taurine seems to act on one or more of these processes. We and Yokogoshi et al. (1999) showed in a previous study that taurine induced hepatic gene expression of cholesterol 7α-hydroxylase, which is a rate-limiting enzyme of cholesterol catabolism (Yokogoshi et al. 1999; Nishimura et al. 2003). These data suggest that taurine enhances cholesterol catabolism, thereby increasing bile acid excretion. We recently demonstrated that plasma VLDL+LDL-cholesterol concentration negatively correlated with fecal bile acid excretion (unpublished data). But fecal bile acid excretion didn't correlate with CYP7A1 activity (unpublished data). These results suggest that the increase in bile acid excretion by the taurine fed rat may involve factors in addition to enhanced cholesterol catabolism. Therefore, we examined whether taurine affects bile acid secretion into bile and the absorption of bile acids from the distal ileum. Few studies have examined the effect of taurine on the secretion and absorption of bile acids.

29.2 Methods

29.2.1 Animals and Diets

The study was approved by the Nayoro City University Animal Use Committee. The animals were maintained in accordance with the guidelines for the care and use of laboratory animals, Nayoro City University.

Male Wistar rats weighing 100 g were obtained from Japan SLC (Hamamatsu, Japan). They were housed in individual cages containing stainless steel screen bottoms in a room maintained at 23 ± 1°C with lighting from 0700 to 1900. Rats were acclimated by feeding a 20% casein diet containing 10 g/kg for 14 d before feeding test diets in order to raise the serum cholesterol concentration, the composition of which was, in g/kg, as follows: casein, 200; α-cornstarch, 362; sucrose, 181; corn oil, 50; lard, 100; AIN-93G mineral mix(Reeves et al. 1993), 35; AIN-93 vitamin mix (Reeves et al. 1993), 10; choline chloride, 2.0; cholesterol, 10; and cellulose, 50.

After the induction period, the rats were divided into 2 groups of six each based on body weight and serum cholesterol concentration. The rats received HC diets with or without 10 g/kg taurine supplementation for 14 days. Taurine was supplied by Taisho Pharmaceutical Co., Ltd. (Tokyo, Japan).

29.2.2 Sampling Procedures

We collected blood samples from the tail veins on days −14, −7, 0 (start of taurine administration), 3, 7, 10 and 14 in experiment 1, and on days 0 and 14 in experiment 2 for determination of the total plasma cholesterol concentration. Feces were collected to determine bile acid concentration. They were lyophilized, weighed and stored at −40°C until analysis of bile acids.

At the end of the experimental period, rats were anesthetized by intraperitoneal injection of Nembutal (sodium pentobarbital 50 mg/kg body wt). In experiment 1, we collected blood samples (0.5 mL) from the mesenteric vein in the distal ileum (distal 5 cm) and placed the samples in a microtube containing heparin on ice. After bleeding was halted, we collected bile via the catheter for 60 minutes and stored them in a microtube on ice. In experiment 2, we collected the digesta from the distal ileum (distal 10 cm) and colorectum (distal 5 cm) for bile acid analysis. Plasma was separated by centrifugation at $1,200 \times g$ for 20 min at 4°C.

29.2.3 Biochemical Analyses

Plasma cholesterol concentration was determined by an enzymatic method using a commercial kit (Cholesterol E-test, Wako, Osaka, Japan). Fecal bile acids were extracted from lyophilized feces using 40 vol of 0.5 mol/L ethanolic KOH at 80°C for 2 h (Moundras et al. 1997) and enzymatically analyzed by a slight modification of the 3 α-hydroxysteroid dehydrogenase assay of Sheltawy and Losowsky (1975). Lithocholic acid was used as the standard substrate.

Extracted fecal bile acid was chemically deconjugated (1.2 N NaOH, 125°C, 1 h), and the resulting unconjugated bile acids were isolated by solvent extraction into n-hexane. They were then analyzed by capillary gas chromatography as methyl ester trimethylsilyl derivatives using a Shimadzu GC-17A gas chromatograph equipped with a flame ionization detector and a CP-Sil 5 CB (25 m $\times$ 0.25 mm; Chrompack, Inc.) column. Helium was used in the gas flow at 130 kPa, with a split ratio of 100:1. The temperature was programmed as follows: initial temperature of 100°C(2 min), increased to 278°C at a rate of 32°C/min, and kept at 278°C for 23 min. Identification of the bile acids was done using authentic standards.

29.2.4 Statistical Analysis

Values in the text are means $\pm$ SEM. Mean values obtained for the HC and the taurine groups were compared using the Student's t test when variances of each group were equal. All statistical analyses were performed using SPSS software (version 14.0J; Tokyo, Japan). Differences were considered statistically significant when the calculated P value was less than 0.05.

29.3 Results

29.3.1 Effect of Taurine on the Secretion of Bile Acids and Bile Acid Absorption from the Distal Ileum (Experiment 1)

No significant differences in either body weight gain or food intake were observed between the HC and taurine groups (data not shown). Figure 29.1 shows the time-dependent changes in plasma cholesterol concentration and fecal bile acid excretion.

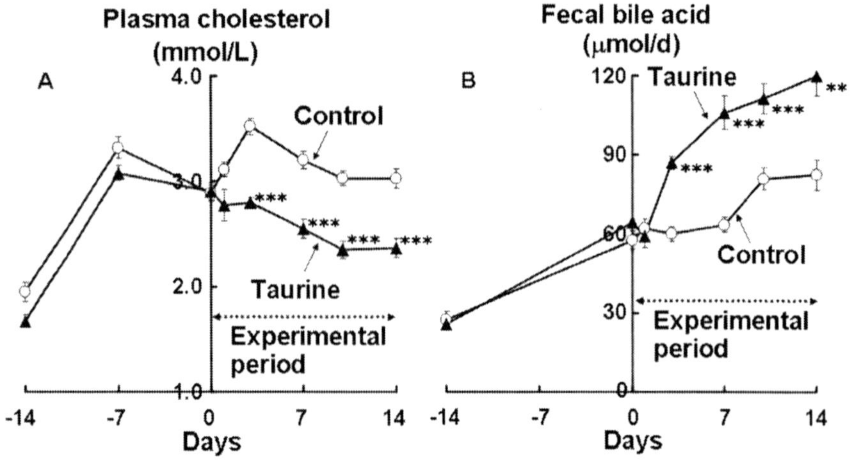

Fig. 29.1 Change in plasma cholesterol concentration and fecal bile acid excretion in rats fed the control diet and taurine diet. Values represent means ±SEM($n = 6$). Day 0 shows the start of taurine administration. Control: rats fed the diet containing 1% cholesterol and 10% lard (HC diet); Taurine: rats fed the HC diet supplemented with 1% taurine. Asterisks indicate significant difference from the corresponding values of the control group at $P < 0.01(**)$ and $P < 0.001(***)$

Plasma cholesterol concentrations after the 3rd-day of the experiment was significantly lower in rats fed the taurine diet than in those fed the HC diet. After 3 d of feeding, the taurine-fed rats also showed a significant elevation in fecal bile acid excretion compared with the rats fed the HC diet. However fecal weight did not differ in the two groups (data not shown).

Table 29.1 shows the effect of taurine on the secretion of bile and biliary bile acids and the bile acid concentration in the mesenteric vein. The secretion of bile and biliary bile acids did not differ in the 2 groups. Bile acid concentration in the mesenteric vein was significantly lower in the taurine-fed rats than in those fed the HC diet.

Table 29.1 Effect of taurine on bile acid concentration in mesenteric vein and bile

	Bile			Mesenteric vein
	Bile secretion mL/h	Bile acid μmol/mL	Bile acid secretion μmol/hr	Plasma bile acid μmol/L
Control	0.600 ± 0.012	21.0 ± 2.2	12.5 ± 1.2	224 ± 31
Taurine	0.544 ± 0.029	23.0 ± 2.0	12.5 ± 1.2	111 ± 27*

Values represent means ± SEM ($n = 6$). Control: rats fed the diet containing 1% cholesterol and 10% lard (HC diet); Taurine: rats fed the HC diet supplemented with 1% taurine. Asterisks denote significant difference from the corresponding values of the control group at $P = 0.016$.

Table 29.2 Bile acid concentration in ileum, colorectum and feces of rats fed taurine

	Distal ileal digesta μmol/g	Colorectal digesta μmol/g	Feces μmol/g
Control	131±22	64.2±2.9	69.1±1.6
Taurine	137±30	82.5±9.0	89.3 ± 2.8***

Values represent means ± SEM ($n = 6$). Control: rats fed the diet containing 1% cholesterol and 10% lard (HC diet); Taurine: rats fed the HC diet supplemented with 1% taurine. Asterisks denote significant differences from the corresponding values of the control group at $P < 0.001$(***).

29.3.2 *Effect of Taurine on Bile Acid Concentration and Composition in the Distal Ileal and Colorectal Digesta (Experiment 2)*

No significant differences in body weight gain and food intake were observed between the HC and taurine groups. Plasma cholesterol concentration was significantly lower in rats fed the taurine diet (2.53 ± 0.07mmol/L) than in those fed the HC diet (3.48 ± 0.11mmol/L). Fecal bile acid excretion was significantly higher in the taurine fed rats (114 ± 15μmol/24h) than in those fed the HC diet (87.1 ± 5.2μmol/24h). However fecal weight did not differ in the two groups (data not shown).

Table 29.2 shows the effect of taurine on bile acid concentration in the distal ileal digesta, colorectal digesta and feces. Bile acid concentration in the distal ileal digesta did not differ in the 2 groups, however, the bile acid concentration in the colorectal digesta tended to be higher in the taurine-fed rats than in those fed the HC diet. Additionally, the bile acid concentration in feces was significantly higher in the taurine-fed rats.

Table 29.3 reveals the effect of taurine on bile acid composition in the distal ileal and colorectal digesta. No bile acids in the ileal digesta were affected by taurine feeding. Colorectal chenodeoxycholic acid and lithocholic acid concentrations were significantly higher in the taurine-fed rats than in those fed the HC diet. Colorectal

Table 29.3 Bile acid composition in distal ileal and colorectal digesta of rats fed taurine

	CA μmol/g	CDCA μmol/g	DCA μmol/g	MCA μmol/g	LCA μmol/g	UDCA μmol/g
Distal ileum						
Control	48.8±7.6	6.76±1.75	4.37±0.56	51.6±8.4	0.415±0.193	3.00±0.55
Taurine	45.5±8.8	7.67±2.69	4.59±0.96	50.2±10.1	0.850±0.392	2.80±0.62
Colorectum						
Control	33.6±2.5	3.74±0.19	6.65±0.44	0.64±0.08	6.09±0.29	8.97±1.13
Taurine	49.4±8.8	5.74 ± 0.71**	7.69±0.93	1.64±0.63	9.93 ± 0.89*	11.9±1.7

Values represent means ± SEM ($n = 6$). Control: rats fed the diet containing 1% cholesterol and 10% lard (HC diet); Taurine: rats fed the HC diet supplemented with 1% taurine. Asterisks denote significant difference from the corresponding values of the control group at $P < 0.05$(*) and $P < 0.01$(**).

cholic acid concentration tended to be higher in the taurine-fed rats. Sixty percent of the increase in colorectal bile acid concentration by taurine feeding consisted of cholic acid.

29.4 Discussion

The excretion of bile acids into feces constitutes a major route of elimination of cholesterol from the body(Turley and Dietschy 1988), therefore, increased fecal bile acid excretion contributes to the reduction in plasma cholesterol concentration. In the present study, taurine administration reduced plasma cholesterol concentration and increased bile acid excretion into the feces. We previously demonstrated similar results (Nishimura et al. 2003; Chen et al. 2003). Masuda and Horisaka (1986) also reported that fecal bile acid excretion increased in taurine-fed rats. Therefore, the taurine-mediated increase in bile acid excretion into feces results in a reduction in plasma cholesterol concentration.

The extent of bile acid excretion into feces depends on the balance between bile acid secretion into the bile and the absorption of bile acids. In the present study, the secretion of biliary bile acids was not affected by taurine feeding, whereas the bile acid concentration in the mesenteric vein of the distal ileum was reduced by taurine. Moreover, we showed that the bile acid concentration in the colorectal digesta, but not in the distal ileal digesta, was greater in the taurine-fed rats. We also reported that the bile acid concentration in the cecal and colonic digesta was elevated by taurine feeding in rats (Chen et al. 2003). These results suggest that the inhibition of bile acid absorption from the distal ileum, but not increased bile acid secretion, is involved the taurine-mediated increase in excretion of bile acids into feces.

Conjugated bile acids are absorbed from the distal ileum via an apical sodium bile acid transporter (Shneider et al. 1995). The rate of absorption of bile acids from the distal ileum varies among the various bile acids, cholic acid, chenodeoxycholic acid, and deoxycholic acid etc (Aldini et al. 1992; Aldini et al. 1996). Conjugated taurocholic acid is absorbed faster than other conjugated bile acids (Aldini et al. 1992). On the other hand, unconjugated bile acids are absorbed from the distal ileum and the colon by passive transport, although unconjugated cholic acid is not readily absorbed compared with the other unconjugated bile acids (Aldini et al. 1992). In the present study, cholic acid was the bile acid with the greatest increase in colorectal levels following taurine feeding. This may indicate that taurine feeding affects the conjugation of bile acids, resulting in reduced absorption from the distal ileum and the colon. Masuda and Horisaka demonstrated that taurine suppressed the elevation in glycine conjugated bile acid content by cholesterol feeding in rats. It also reduced the glycine/taurine-conjugation ratio among the biliary bile acids (Masuda and Horisaka 1986). However, this change in the conjugation ratio of bile acids cannot account for the inhibition of bile acid absorption shown in the present study. These discrepancies require further investigation to explain the relationship between the inhibition of bile acid absorption and changes in the conjugation type, unconjugated bile acid content and the composition of the bile acids.

29.5 Conclusion

In summary, we found that taurine inhibits bile acid absorption from the distal ileum, resulting in enhanced bile acid excretion in feces. This effect of taurine on bile acid absorption probably depends on the composition of the biliary bile acids.

Acknowledgments I'd like to thank Dr. H. Yokogoshi and Dr. H. Oda for their valuable advice. We also wish to thank the Taisho Pharmaceutical Co., Ltd. for their generous financial assistance.

References

Aldini R, Roda A, Lenzi PL, Ussia G, Vaccari MC, Mazzella G, Festi D, Bazzoli F, Galletti G, Casanova S, et al. (1992) Bile acid active and passive ileal transport in the rabbit: effect of luminal stirring. Eur J Clin Invest 22:744–750

Aldini R, Roda A, Montagnani M, Cerre C, Pellicciari R, Roda E (1996) Relationship between structure and intestinal absorption of bile acids with a steroid or side-chain modification. Steroids 61:590–597

Chen W, Nishimura N, Oda H, Yokogoshi H (2003) Effect of taurine on cholesterol degradation and bile acid pool in rats fed a high-cholesterol diet. Adv Exp Med Biol 526:261–267

Masuda M, Horisaka K (1986) Effect of taurine and homotaurine on bile acid metabolism in dietary hyperlipidemic rats. J Pharmacobiodyn 9:934–940

Moundras C, Behr SR, Remesy C, Demigne C (1997) Fecal losses of sterols and bile acids induced by feeding rats guar gum are due to greater pool size and liver bile acid secretion. J Nutr 127:1068–1076

Murakami S, Kondo-Ohta Y, Tomisawa K (1999) Improvement in cholesterol metabolism in mice given chronic treatment of taurine and fed a high-fat diet. Life Sci 64:83–91

Nishimura N, Umeda C, Oda H, Yokogoshi H (2003) The effect of taurine on the cholesterol metabolism in rats fed diets supplemented with cholestyramine or high amounts of bile acid. J Nutr Sci Vitaminol 49:21–26

Reeves PG, Nielsen FH, Fahey GC Jr (1993) AIN-93 purified diets for laboratory rodents: final report of the American Institute of Nutrition ad hoc writing committee on the reformulation of the AIN-76A rodent diet. J Nutr 123:1939–1951

Sheltawy MJ, Losowsky MS (1975) Determination of faecal bile acids by an enzymic method. Clin Chim Acta 64:127–132

Shneider BL, Dawson PA, Christie DM, Hardikar W, Wong MH, Suchy FJ (1995) Cloning and molecular characterization of the ontogeny of a rat ileal sodium-dependent bile acid transporter. J Clin Invest 95:745–754

Turley SD, Dietschy JM (1988) In Arias IM, Jakoby WB, Popper H, Schachter D, Shafritz DA (eds) The Liver: biology and pathobiology, Raven Press, New York, pp 617–641

Yamanaka Y, Tsuji K, Ichikawa T, Nakagawa Y, Kawamura M (1985) Effect of dietary taurine on cholesterol gallstone formation and tissue cholesterol contents in mice. J Nutr Sci Vitaminol 31:225–232

Yokogoshi H, Mochizuki H, Nanami K, Hida Y, Miyachi F, Oda H (1999) Dietary taurine enhances cholesterol degradation and reduces serum and liver cholesterol concentrations in rats fed a high-cholesterol diet. J Nutr 129:1705–1712

Chapter 30
The Protective Effect of Taurine Against Hepatic Damage in a Model of Liver Disease and Hepatic Stellate Cells

Teruo Miyazaki, Bernard Bouscarel, Tadashi Ikegami, Akira Honda, and Yasushi Matsuzaki

Abstract Taurine plays a protective role against free radicals and toxins in various cells and tissues. However, the effect of taurine on hepatic injury and fibrosis developed by activated hepatic stellate cells (HSC) and myofibroblast-like cells is not fully understood. We investigated the effects of taurine on the hepatic fibrogenesis and damage in rats and isolated HSC. Rats were divided into a normal and two CCl_4-induced liver damage (LD) groups, one untreated and the other maintained for 5 weeks on a 2% taurine diet. The HSC isolated from a normal rat were cultured either for a day only or for an additional 3–6 days with ~50 mM taurine. LD rats maintained on the taurine diet were resistant to CCl_4-induced loss of taurine from the liver. The liver of the LD rats were also protected against histological damage, fibrosis, significant reductions in oxidative stress markers (LPO and 8-OHdG) and hepatic fibrogenic factors ($TGF-\beta_1$ mRNA, hydroxyproline, α-SMA). Proliferation, oxidative stress, and fibrogenesis were significantly inhibited in HSC by treatment with taurine. Thus, supplementation with taurine should be considered as a therapeutic approach to lessen the severity of oxidative stress-induced liver injury and hepatic fibrosis.

Abbreviations *HSC*, hepatic stellate cells; CCl_4, carbon tetrachloride, *LPO*, lipid hydroperoxide; $TGF-\beta_1$, transforming growth factor$-\beta_1$; α-*SMA*, α-smooth muscle actin protein

---

T. Miyazaki (✉)
The George Washington University, District of Columbia, USA

J. Azuma et al. (eds.), *Taurine 7*, Advances in Experimental Medicine and Biology 643,
DOI 10.1007/978-0-387-75681-3_30, © Springer Science+Business Media, LLC 2009

30.1 Introduction

Taurine, 2-amino ethylsulfonic acid, is the most abundant amino acid in mammalian tissues, including heart (Awapara 1956), skeletal muscle (Jacobsen and Swimth 1968), nerve, brain, and liver (Jacobsen and Swimth 1968; Huxtable 1980). The plasma and intracellular concentration of taurine are considerably higher than those of other amino acids; plasma: 200~250μM), heart: 25~30mM (Timbrell et al.1995), lung: 11~27mM (Timbrell et al. 1995), neutrophils: 50~100mM (Fukuda et al. 1982; Green 1991), retina: 50~ 70mM (Sturman 1993). The high intracellular levels of taurine are maintained by active uptake and endogenous synthesis ,mainly in the liver where taurine is an endproduct of sulfur amino acid catabolism (Hosokawa et al. 1990; Kaisaki et al. 1995; Reymond et al. 1996; Tappaz et al. 1999). The biosynthesis of taurine in the liver is limited, particularly in the case of liver disease, where the biosynthetic ability might be reduced. Therefore, exogenous uptake via the diet is largely responsible for maintaining the high intracellular concentration of taurine that account for its many physiological and pharmacological roles, including cellular plasma membrane stabilization (Pasantes et al. 1985), osmoregulation (Nieminen et al. 1988), neuromodulation and neurotransmission (Davison and Kaczmarec 1971; Kuriyama 1980; Kanner and Nurit 1994), anti-oxidation (Nakashima et al. 1990; Waterfield et al. 1993a, b; Timbrell et al. 1995), detoxification (Huxtable 1992) and conjugation to bile acids in the liver (Sjovall 1959; Danielsson 1963).

We have demonstrated the effects of taurine on oxidative stress induced by exercise and/or by hepatotoxin in rats. In exercised rats, oral taurine administration reduced oxidative stress (lipid peroxidation, oxidized glutathione) and diminished the decrease in taurine content of skeletal muscle (Miyazaki et al. 2004a). Hepatic taurine concentration was significantly decreased in liver disease mediated by a free radical inducer CCl_4 (Miyazaki et al. 2004b). Previous studies have shown that taurine treatment inhibits hepatic damage induced by CCl_4-mediated oxidative stress (Vohra and Hui 2001; Dinçcer et al. 2002) and thioacetamide-mediated toxicity (Balkan et al. 2001). Indeed, the anti-oxidant activity of taurine has been reported in chick heart (Azuma et al. 1987), hamster lung (Wang et al. 1991; Gordon et al. 1992), and the rat central nervous system (Vohra and Hui 2001), as well as in human lymphoblastoid cells (Pasantes et al. 1984), neutrophils (Wright et al. 1985), and blood cells in culture (Koyama et al. 1992). Therefore, these results show that taurine serves as an anti-oxidant to prevent liver disease.

Oxidative stress and cytokine tranforming growth factor-beta$_1$ (TGF$-\beta_1$) participate closely (Li and Friedman 1999). TGF$-\beta_1$ is mainly released from inflammatory cells, most likely Kupffer cells and neutrophils. It is widely accepted that TGF$-\beta_1$ activates and transforms hepatic stellate cells (HSC; Ito cells, lipocytes, vitamin A-storing cells) into myofibroblast-like cells. The transformed HSC secrete the amino acid, hydroxyproline, which is abundant in collagen-structured and extracellular matrix (ECM) fibrotic tissue. In turn, the HSC transformation induces the augmentation of ECM synthesis (Schafer et al. 1987; Nakatsukasa et al. 1990; Gressner 1996). Transformed HSC characteristically express alpha-smooth muscle actin (α-SMA) (Enzan et al. 1994). Increased ECM synthesis leads to the progression of fibrosis. Therefore, the inhibition of hepatic damage potentially diminishes both inflammatory cell infiltration (*i.e.* Kupffer cells and neutrophils) and platelet aggregation. Consequently, the secretion of TGF−β_1 by those inflammatory cells should be reduced, the activation and transformation of myofibroblasts should be decreased and hepatic fibrosis/cirrhosis should diminish (Friedman 1993).

In a previous study, Dinçcer et al. (2002) observed that acute taurine treatment reduced the degree of CCl_4-mediated lipid peroxidation and liver necrosis. However, there is little information on the effect of taurine on chronic hepatic damage and oxidative stress mediated by repetitive administration of CCl_4. Furthermore, the effect of taurine on the activation of HSC remains unclear. Therefore, in the present study, we proposed to examine the effects of taurine on hepatic fibrosis and oxidative damage in a hepatic damage rat model induced by chronic CCl_4 administration and on fibrogenesis and oxidative stress in isolated rat HSC.

30.2 Materials and Methods

30.2.1 Animal Model Experiment

30.2.1.1 Hepatic Disease Rat Model

Male 6-week old rats (Sprague-Dawley; Japan SLC, Shizuoka, Japan) were divided randomly into 3 groups: normal ($N = 8$), liver disease (LD: $N = 18$), and LD treated with taurine (LDT: $N = 18$). The rats in the LDT group were given 2% taurine mixed with standard diet. After 1 week, the rats in the LD and LDT groups were subcutaneously injected 50% CCl_4 olive oil solution as 1 mL/kg body weight twice a week for 5 weeks. As in other studies, they also received 0.05% phenobarbital dissolved in the drinking water (Matsuzaki et al. 2002b; Miyazaki et al. 2003a, b; Miyazaki et al.2004b; Miyazaki et al. 2005; Zhang et al. 2006). Blood and liver were collected from the anesthetized animals. Taurine content was measured by amino acid analysis (Matsuzaki et al. 2002a, 2002b). All rats were kept at 21–25°C under 12-hour dark/light cycles, and received humane care in accordance with *The Guidelines of the University of Tsukuba for the Care of Laboratory Animals.*

30.2.1.2 Fibrosis Analysis

For histological analysis, the liver tissue was embedded on paraffin for Hematoxin & Eosin (HE) and silver staining, and immunohistrochemical (IHC) staining for $\alpha - SMA$. Hepatic fibrosis was quantified from positive silver staining as described in a previous study (Ludwig 1993).

Hepatic hydroxyproline content was used as a measure of fibrosis. Homogenized liver was incubated with 6N HCl at 105°C for 18 hours and centrifuged to get bottom's layer. After drying the pellet, the sample was incubated with chloramine-T solution at 50°C for 90 min, and the hepatic hydroxyproline concentration was determined by measuring the absorbance at 540 nm by a 96-well plate reader.

TGF$-\beta_1$ mRNA content in the liver samples was quantified by real-time PCR. After reverse-transcription of RNA, hepatic TGF$-\beta_1$ mRNA content was detected using TaqMan® Universal PCR Master Mix (Perkin Elmer) and an Automated Sequence Detection System (Perkin Elmer). The temperature profile and the sequence of primers and probes for TGF$-\beta_1$ and β-actin (internal standard) are as follows: at 50°C for 2 min, at 95°C for 10 min, 40 cycles of at 95°C for 15 sec and at 60°C for 1 min, TGF$-\beta_1$ probe; 5'-AGT GGC TGA ACC AAG GAG ACG GAA TAC-3', TGF$-\beta_1$ sense primer; 5'-CGC CTG CAG AGA TTC AAG TCA A-3', TGF$-\beta_1$ anti-sense primer; 5'-GTC GGT TCA TGT CAT GGA TGG T-3', β-actin probe; 5'-TTT GAC ACC TTC AAC ACC CCA GCC A-3', β-actin sense primer; 5'-CGT GAA AAG ATG ACC CAG ATC A-3', β-actin anti-sense primer; 5'-ACA CAG CCT GGA TGG CTA CGT A-3'. The 5' and 3' terminals in both probes were FAM and TAMRA labeled, respectively. The β-actin probe was designed and provided by Nippon Flour Mills Co., Ltd. (Tokyo, Japan).

The expression of $\alpha - SMA$ protein was determined by Western blotting of homogenized liver.

30.2.1.3 Oxidative Stress Markers

The concentration of lipid hydroperoxide (LPO) and 8-hydroxy-2'-deoxyguanosine (8-OHdG) in serum and liver were used to assess lipid peroxidation and oxidative DNA damage, respectively. The LPO concentration was measured as hydroperoxides utilizing a redox reaction with ferrous ions (Mihaljevic et al. 1996) as described in the LPO assay kit (Cayman chemical company, Ann Arbor, MI). The 8-OHdG concentration was determined using the ELISA detection kit (8-OHdG Check; Japan Institute for the Control of Aging, Shizuoka, Japan).

30.2.2 Hepatic Satellite Cells Experiment

30.2.2.1 Isolation and Culture of HSC

HSC were isolated from normal Sprague-Dawley rat (Japan SLC) by the two-step collagenase perfusion method as described in our previous study (Zhang et al. 2006). Isolated HSC were plated on a 96-well plate at 1×10^4 cells/well and in 35mm plastic culture dishes at 5×10^5 cells/dish and incubated with DMEM containing 10% fetal bovine serum and 100 U/mL penicillin/streptomycin. HSC were cultured at 37°C in a humidified atmosphere of $5\% CO_2$. Activation of HSC was preliminary confirmed after $3 \sim 7$ days of plating, as determined by the loss of vitamin A autofluorescence and the increased expression of $\alpha - SMA$. After 24 hours in culture, the HSC were cultured for an additional 3 days (condition A) and 6 days (condition C) with 0, 25, and 50mM (Tau-0, -25, and -50) taurine, or for an additional 3 days with taurine followed by 3 days of culture without taurine (condition B). The culture medium was changed every 2 days. The HSC were then harvested. In addition, culture medium before and after the completion of protocols B and C was kept for assays of LPO and hydroxyproline.

30.2.2.2 HSC Proliferation

Cell growth and DNA synthesis were measured by the MTT assay and the $5-bromo-2'$-deoxyuridine (BrdU) cell proliferation assay (Merck Biosciences, Darmstadt, Germany), respectively. Signaling protein expression associated with cell proliferation, such as the total and phosphorylated forms of MAPKs and Akt were determined by Western blotting. The antibodies used were raised against either total or the phosphorylated forms: Thr^{202}/Tyr^{204} p44/42 MAPK, Thr^{180}/Tyr^{182} p38 MARK, Thr^{183}/Tyr^{185} stress activated protein kinase/c-Jun N-terminal kinase (SAPK/JNK) and Ser^{473} Akt, respectively (Cell Signaling Technology, VA).

30.2.2.3 Transforming of HSC, Fibrogenesis, and Oxidative Stress Assays

In the harvested HSC subjected to condition A, the expression levels of $\alpha - SMA$ protein and TGF$-\beta_1$ mRNA was detected by Western blotting and real-time PCR, respectively, as described above. The LPO and hydroxyproline concentrations were determined in the culture medium collected before and after incubation with taurine.

30.2.3 Statistical Analysis

All data are presented as the means $\pm SD$. Significant differences were determined by the unpaired student's *t*-test, the one-way ANOVA post hoc Fisher's PLSD test or the Mann-Whitney *U*-test. Statistical significance was set at $p< 0.05$.

30.3 Results

30.3.1 CCl_4-Exposed Rats Treated with Taurine

30.3.1.1 Blood Biochemical Analysis and Taurine Concentration

Serum aspartate aminotransferase (AST) and alanine aminotransferase (ALT) were significantly increased in the LD group compared to that of the normal and LDT group, although ALT levels of the LDT were significantly higher than those of the normal group (Fig. 30.1). Total serum bilirubin content was increased eight-ten-fold in the LD and LDT groups compared to that of the normal, however there was no significant difference between the LD and LDT groups. Furthermore, serum albumin levels in the LD and LDT groups tended to be lower than those of the normal group; there was no significant difference between the LD and LDT groups. Serum taurine concentration of the LD group was unchanged relative to the normal, but levels

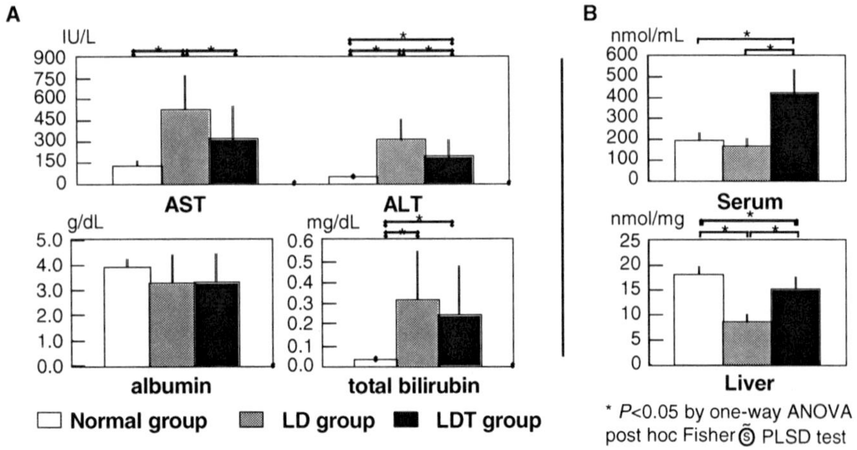

Fig. 30.1 Blood biochemical analyses and taurine concentration in serum and liver of CCl_4-treated rats with or without dietary taurine supplementation. **A**: Serum analysis of liver function, **B**: Taurine concentration in serum and liver. Data represent means ±SD

of the LDT groups were increased more than two-fold compared to those of the normal and LD groups (Fig. 30.1B). On the other hand, hepatic taurine content was significantly decreased in the LD group compared to that of the normal group; oral taurine administration preserved hepatic taurine levels against depletion observed in the LD group.

30.3.1.2 Histological Hepatic Pathology

Based on HE staining, CCl_4-administered rat livers underwent fibrosis, marked fatty degeneration, necrosis, cellular inflammation and infiltration of inflammatory cells including macrophages and lymphocytes. In the LD group, this damage was found throughout the liver and fibrotic infiltrations appeared in both the pericentral and periportal regions, with bridging fibrosis seen (Fig. 30.2). In the LDT group, hepatic necrosis and infiltration of inflammatory cells were inhibited compared to the LD group, particularly in the pericentral region. Similarly, the development of hepatic fibrosis was diminished in the pericentral region, the degree of bridging fibrosis was

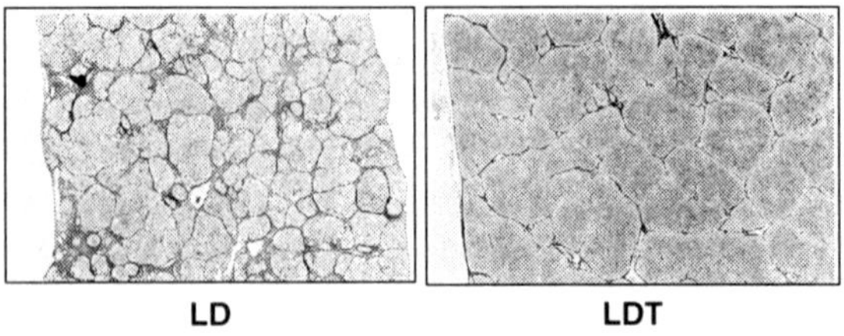

Fig. 30.2 Light micrographs of silver stained liver. Objective was ×5

Table 30.1 Parameters of fibrogenesis and oxidative stress in serum and liver tissue

Parameter	tissue	(unit)	Normal group	LD group	LDT group
Hydroxyproline	liver	(μg/L)	16.10 ± 6.04	304.83 ± 128.77[a]	212.27 ± 81.49[ab]
TGF−β_1 mRNA	liver		0.03 ± 0.01	0.98 ± 0.25[a]	0.45 ± 0.25[ab]
LPO	serum	(μmol/L)	1.11 ± 0.68	12.40 ± 5.09[a]	3.04 ± 1.91[ab]
	liver	(μmol/L)	0.88 ± 0.59	31.70 ± 11.35[a]	14.86 ± 14.2[ab]
8-OHdG	serum	(μmol/L)	1.58 ± 0.55	15.27 ± 4.7[a]	11.71 ± 4.41[ab]
	liver	(ng/g)	1.76 ± 0.39	9.27 ± 2.48[a]	6.88 ± 0.95[ab]

TGF−β_1 mRNA is expressed relative to β-actin mRNA. a and b show significant difference ($P < 0.05$) compared to the normal and LD groups, respectively, by one-way ANOVA post hoc Fisher's PLSD test. Data represent means±SD.

significantly reduced, and the architecture of the hepatic parenchyma and lobules was preserved. The fibrotic area quantified from the positive silver staining was 21.6 ± 9.1% in the LD group and 8.0 ± 5.9% in the LDT group, although there was no significant difference between the groups. Based on Western blot analysis and IHC staining, the number and intensity of α-SMA of the LD group was markedly increased compared to those of the LDT group.

30.3.1.3 Fibrogenesis and Oxidative Stress Markers

Hepatic hydroxyproline content and TGF−β_1 mRNA levels of the LD and LDT groups were significantly increased compared to those of the normal group, with the levels being significantly lower in the LDT group than in the LD group (Table 30.1). Likewise, the LPO and 8-OHdG content of serum and liver was significantly increased in both the LD and LDT groups compared to that of the normal group, but content was significantly lower in the LDT group than in the LD group.

30.3.2 Primary HSC Cultured with Taurine

30.3.2.1 HSC Proliferation and Signaling Pathways of Cell Growth

Cell growth of HSC assessed by the MTT assay was significantly decreased in cells cultured with 50 mM taurine compared to those cultured without taurine (Fig. 30.3). There was also a significant difference between cells incubated with medium containing 25 mM taurine for 6 days (condition C) compared to those cultured without taurine. Similarly, based on the BrdU assay, DNA synthesis of the 50 mM taurine group was significantly reduced compared to that of the 0 mM taurine group. For conditions B and C, DNA synthesis was significantly lower in the 50 mM taurine group than in the 25 mM taurine group. HSC incubated with taurine according to condition A showed no significant difference in total and phosphorylated forms of the MAPKs and Akt.

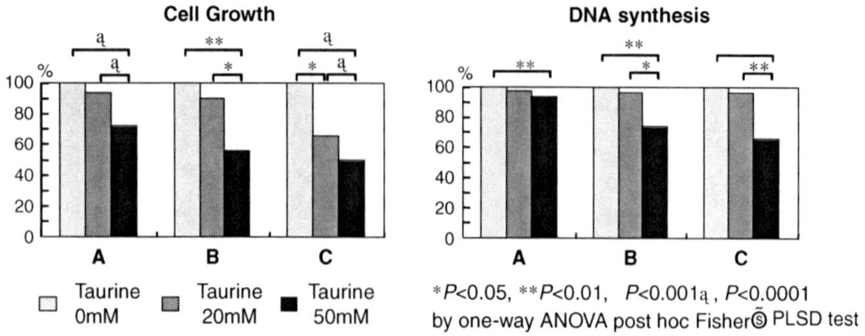

Fig. 30.3 Cell proliferation of primary cultured HSC treated with taurine. Cell growth and DNA synthesis were analyzed by MMT and BrdU assays. **Condition A**: cultured with taurine for 3 days, **Condition B**: cultured with taurine for 3 days followed by 3 days without taurine, **Condition C**: cultured with taurine for 6 days. Data represent means±SD

30.3.2.2 Fibrogenesis and Oxidative Stress Factors of Cultured HSC

When incubated according to condition A, TGF$-\beta_1$ mRNA content of TAU-25 (0.02 ± 0.01 fold compared to β-actin) and -50 (0.03 ± 0.01) was significantly decreased compared to that of Tau-0 (0.04 ± 0.03, $P < 0.05$). However, there was no significant difference between the Tau-25 and -50 groups. The LPO and hydroxyproline content of culture medium collected before (Pre) and after treatment with taurine for 3 days (condition A) was higher in the Tau-0 (0.48±0.07 g/L and 7.16±3.85 μ mol/L, respectively) than in the Pre group (0.04±0.01, $P < 0.01$ and 1.92 ± 0.39, $P < 0.05$, respectively). The hydroxyproline concentration (g/L) was significantly lower in the Tau-50 (0.38 ± 0.08, $P < 0.05$) than in the Tau-0 group. The LPO concentration (μ mol/L) in the Tau-25 (4.84 ± 2.53, $P < 0.05$) and -50 (3.13±2.50, $P < 0.01$) groups was significantly decreased compared to those of the Tau-0 group. There was no significant difference between the hydroxyproline and LPO concentrations of the Tau-25 and -50 groups.

30.4 Discussion

In the present study, taurine treatment prevented the decline in hepatic taurine content of the CCl_4 treated rat. Taurine treatment also protected the liver against oxidative stress-induced damage trigger by repetitive CCl_4 administration in rats. Particularly noteworthy was the reduction in hepatic fibrosis of the CCl_4-administered rats treated with taurine. Furthermore, in the cultured HSC study, taurine inhibited the transformation of HSC while reducing oxidative stress. These results showed that fibrosis was directly inhibited by taurine through attenuation of oxidative stress-mediated transformation of HSC to myofibroblast-like cells. However, the mechanism of taurine-mediated modulation of cell proliferation of HSC remains unclear, but does not appear to involve the signaling pathways of MAPKs and

Akt. Therefore, there must be some other mechanism that can mediate the effect of taurine on HSC proliferation. Furthermore, the mechanism underlying taurine-mediated protection against liver disease must involve multiple factors because taurine diminished diverse types of hepatic damage, including hepatocyte necrosis, fatty degeneration, infiltration of inflammatory cells and fibrosis. Therefore, taurine acts through multiple and complex mechanisms, including cell membrane stabilization, osmoregulation, detoxification, anti-apoptosis, modulation of bile acid conjugation/excretion, cholesterol excretion, anti-inflammation, anti-oxidation, inhibition of fibrogenesis and cytokine secretion, inhibition of HSC transformation and inhibition of cytokine/autocrine action. In conclusion, taurine supplementation should be considered as a therapy to lessen the severity of oxidative stress-induced hepatic injury and fibrosis.

Acknowledgments This study was supported by a research grant from Taisho Pharmaceutical Co., Ltd. Japan. This study has been previously published in part in the *Journal of Hepatology* in 2005.

References

Awapara J (1956) The taurine concentration of organs from fed and fasted rats. J Biol Chem 218(2):571-6.

Azuma J, Hamaguchi T, Ohta H, Takihara K, Awata N, Sawamura A, Harada H, Tanaka Y, Kishimoto S (1987) Calcium overload-induced myocardial damage caused by isoproterenol and by adriamycin: possible role of taurine in its prevention. Adv Exp Med Biol 217:167–179

Balkan J, Dogru-Abbasoglu S, Kanbagli O, Cevikbas U, Aykac-Toker G, Uysal M (2001) Taurine has a protective effect against thioacetamide-induced liver cirrhosis by decreasing oxidative stress. Hum Exp Toxicol 20:251–254

Danielsson H (1963) Present status of research on catabolism, excretion of cholesterol. Adv Lipid Res 1:335–385

Davison AN, Kaczmarec LK (1971) Taurine-a possible neurotransmitter? Nature 234:107–108

Dinçer S, Özenirler S, Öz E, Akyol G, Özogul C (2002) The protective effect of taurine pretreatment on carbon tetrachloride-induced hepatic damage – A light and electron microscopic study. Amino Acids 22:417–426

Enzan H, Himeno H, Iwamura S, Saibara T, Onishi S, Yamamoto Y, Hara H (1994) Immunohistochemical identification of Ito cells and their myofibroblastic transformation in adult human liver. Virchows Arch 424:249–256

Friedman SL (1993)Seminars in medicine of the Beth Israel Hospital, Boston. The cellular basis of hepatic fibrosis. Mechanisms and treatment strategies. N Engl J Med 328:1828–1835

Fukuda K, Hirai Y, Yoshida H, Nakajima T, Usui T (1982) Free amino acid content of lymphocytes and granulocytes compared. Clin Chem 28:1758–1761

Gordon RE, Heller RF, Heller RF (1992) Taurine protection of lungs in hamster models of oxidant injury: a morphologic time study of paraquat and bleomycin treatment. Adv Exp Med Biol 315:319–328

Green TR, Fellman JH, Eicher AL, Pratt KL (1991) Antioxidant role and subcellular location of hypotaurine and taurine in human neutrophils. Biochim Biophys Acta 1073:91–97

Gressner AM (1996) Transdifferentiation of hepatic stellate cells (Ito cells) to myofibroblasts: a key event in hepatic fibrogenesis. Kidney Int Suppl 54:S39–S45

Hosokawa Y, Matsumoto A, Oka J, Itakura H, Yamaguchi K (1990) Isolation and characterization of a cDNA for rat liver cysteine dioxygenase. Biochem Biophys Res Commun 168:473–478

Huxtable RJ (1980) Does taurine have a function? Fed Proc 39:2678–2679
Huxtable RJ (1992) Physiological actions of taurine. Physiol Rev 72:101–163
Jacobsen JG, Swimth LH (1968) Biochemistry and physiology of taurine and taurine derivatives. Physiol Rev 48:424–511
Kaisaki PJ, Jerkins AA, Goodspeed DC, Steele RD (1995) Cloning and characterization of rat cysteine sulfinic acid decarboxylase. Biochim Biophys Acta 1262:79–82
Kanner BI, Nurit KD (1994) Structure and function of sodium-coupled neurotransmitter transporters. Cell Physiol Biochem 4:174
Koyama I, Nakamura T, Ogasawara M, Nemoto M, Yoshida T (1992) The protective effect of taurine on the biomembrane against damage produced by the oxygen radical. Adv Exp Med Biol 315:355–359
Kuriyama K (1980) Taurine as a neuromodulator. Fed Proc 39:2680–2684
Li D, Friedman SL (1999) Liver fibrogenesis and the role of hepatic stellate cells: new insights and prospects for therapy. J Gastroenterol Hepatol 14:618–633
Ludwig J (1993) The nomenclature of chronic active hepatitis:an obituary. Gastroenterol 105: 274–278
Matsuzaki Y, Miyazaki T, Miyakawa S, Bouscarel B, Ikegami T, Tanaka N (2002a) Decreased taurine concentration in skeletal muscles after exercise for various durations. Med Sci Sports Exerc 34:793–797
Matsuzaki Y, Miyazaki T, Ohkoshi N, Miyakawa S, Bouscarel B, Tanaka N (2002b) Degeneration of skeletal muscle fibers in the rat administrated carbon tetrachloride:similar histological findings of the muscle in a 64-year-old patient of LC with muscle cramp. Hepatol Res 24:368–378
Mihaljevic B, Katusin-Razem B, Razem D (1996) The reevaluation of the ferric thiocyanate assay for lipid hydroperoxides with special considerations of the mechanistic aspects of the response. Free Radic Biol Med 21:53–63
Miyazaki T, Ikegami T, Zhang Y, Honda A, Doy M, Matsuzaki Y (2005) The prevention mechanisms of taurine on hepatic damage and fibrosis via antioxidative stress. Jpn Pharmacol Ther 33:S105
Miyazaki T, Matsuzaki Y, Ikegami T, Miyakawa S, Doy M, Tanaka N, Bouscarel B (2004a) Optimal and effective oral dose of taurine to prolong exercise performance in rat. Amino Acids 27:291–298
Miyazaki T, Matsuzaki Y, Ikegami T, Miyakawa S, Doy M, Tanaka N, Bouscarel B (2004b) The harmful effect of exercise on reducing taurine concentration in the tissues of rats treated with CCl_4 administration. J Gastroenterol 39:557–562
Miyazaki T, Matsuzaki Y, Karube M, Bouscarel B, Miyakawa S, Tanaka N (2003a) Amino acid ratios in plasma and tissues in a rat model of liver cirrhosis before and after exercise. Hepatol Res 27:230–237
Miyazaki T, Matsuzaki Y, Karube M, Miyakawa S, Tanaka N (2003b) Clinical importance of taurine maintenance on liver disease. Gastroenterology 37:558–562
Nakashima T, Seto Y, Nakajima T, Shima T, Sakamoto Y, Cho N, Sano A, Iwai M, Kagawa K, Okanoue T, et al. (1990) Calcium-associated cytoprotective effect of taurine on the calcium and oxygen paradoxes in isolated rat hepatocytes. Liver 10:167–172
Nakatsukasa H, Nagy P, Evarts RP, Hsia CC, Marsden E, Thorgeirsson SS (1990) Cellular distribution of transforming growth factor-beta 1 and procollagen types I, III, and IV transcripts in carbon tetrachloride- induced rat liver fibrosis. J Clin Invest 85:1833–1843
Nieminen ML, Tuomisto L, Solatunturi E, Eriksson L, Paasonen MK (1988) Taurine in the osmoregulation of the Brattleboro rat. Life Sci 42:2137–2143
Pasantes MH, Wright CE, Gaull GE (1984) Protective effect of taurine, zinc and tocopherol on retinol-induced damage in human lymphoblastoid cells. J Nutr 114:2256–2261
Pasantes MH, Wright CE, Gaull GE (1985) Taurine protection of lymphoblastoid cells from iron-ascorbate induced damage. Biochem Pharmacol 34:2205–2207
Reymond I, Sergeant A, Tappaz M (1996) Molecular cloning and sequence analysis of the cDNA encoding rat liver cysteine sulfinate decarboxylase (CSD). Biochim Biophys Acta 1307: 152–156

Schafer S, Zerbe O, Gressner AM (1987) The synthesis of proteoglycans in fat-storing cells of rat liver. Hepatology 7:680–687

Sjovall J (1959) Dietary glycine and taurine on bile acid conjugation in man; bile acids and steroids 75. Proc Soc Exp Biol Med 100:676–678

Sturman JA (1993) Taurine in development. Physiol Rev 73:119–147

Tappaz M, Bitoun M, Reymond I, Sergeant A (1999) Characterization of the cDNA coding for rat brain cysteine sulfinate decarboxylase: brain and liver enzymes are identical proteins encoded by two distinct mRNAs. J Neurochem 73:903–912

Timbrell JA, Seabra V, Waterfield CJ (1995) The in vivo and in vitro protective properties of taurine. Gen Pharmacol 26:453–462

Vohra BP, Hui X (2001) Taurine protects against carbon tetrachloride toxicity in the cultured neurons and in vivo. Arch Physiol Biochem 109:90–94

Wang QJ, Giri SN, Hyde DM, Li C (1991) Amelioration of bleomycin-induced pulmonary fibrosis in hamsters by combined treatment with taurine and niacin. Biochem Pharmacol 42:1115–1122

Waterfield C.J, Mesquita M, Parnham P, Timbrell J.A (1993a) Taurine protects against the cytotoxicity of hydrazine, 1,4-naphthoquinone and carbon tetrachloride in isolated rat hepatocytes. Biochem Pharmacol 46:589–595

Waterfield CJ, Turton JA, Scales MD, Timbrell JA (1993b) Reduction of liver taurine in rats by beta-alanine treatment increases carbon tetrachloride toxicity. Toxicology 29:7–20

Wright CE, Lin TT, Lin YY, Sturman JA, Gaull GE (1985) Taurine scavenges oxidized chlorine in biological systems. Prog Clin Biol Res 179:137–147

Zhang Y, Ikegami T, Honda A, Miyazaki T, Bouscarel B, Rojkind M, Hyodo I, Matsuzaki Y (2006) Involvement of integrin-linked kinase in carbon tetrachloride-induced hepatic fibrosis in rats. Hepatology 44:612–622

Chapter 31
Taurine Depletion by β-Alanine Inhibits Induction of Hepatotoxicity in Mice Treated Acutely with Carbon Tetrachloride

Dalwoong Choi, Sun Ju Kim, Do Young Kwon, Sun Young Lee, and Young Chul Kim

Abstract We examined the effect of taurine depletion on hepatic sulfur-containing amino acid metabolism and carbon tetrachloride-induced acute liver injury. Mice were supplemented with β-alanine (3%) in drinking water for one week. β-Alanine intake significantly reduced hepatic taurine levels, but did not influence *S*-adenosylmethionine, *S*-adenosylhomocysteine, glutathione levels or methionine adenosyltransferase activity in liver. However, hepatic cysteine levels were significantly elevated by β-alanine administration. Hepatotoxicity caused by carbon tetrachloride (50 μ*l*/kg, ip) in mice fed β-alanine was decreased, as determined by changes in serum aspartate aminotransferase, alanine aminotransferase and sorbitol dehydrogenase activities. Hepatic glutathione and taurine levels after a carbon tetrachloride challenge were markedly increased by β-alanine exposure. The results suggest that enhanced availability of cysteine for synthesis of glutathione and/or taurine may account for the hepatoprotective effects of β-alanine against carbon tetrachloride-induced acute liver injury.

Abbreviations *MAT* methionine adenosyltransferase; *GSH*, glutathione; *GSSG*, glutathione disulfide; *SAM*, *S*-adenosylmethionine; *SAH*, *S*-adenosylhomocysteine; *C*β*S*, cystathionine β-synthase; $C\gamma L$, cystathionine γ-lyase; *CYP*, cytochrome P450; *AST*, aspartate aminotransferase; *ALT*, alanine aminotransferase; *SDH*, sorbitol dehydrogenase; *GES*, guanidinoethane sulfonate

31.1 Introduction

Taurine or 2-aminoethanesulfonic acid is a β-amino acid that is one of the most abundant free amino acids in mammalian tissues. Its biochemical and physiological functions are thought to include antioxidation, detoxification, osmoregulation,

---

D. Choi (✉)
College of Health Science, Korea University, Korea

J. Azuma et al. (eds.), *Taurine 7*, Advances in Experimental Medicine and Biology 643, DOI 10.1007/978-0-387-75681-3_31,

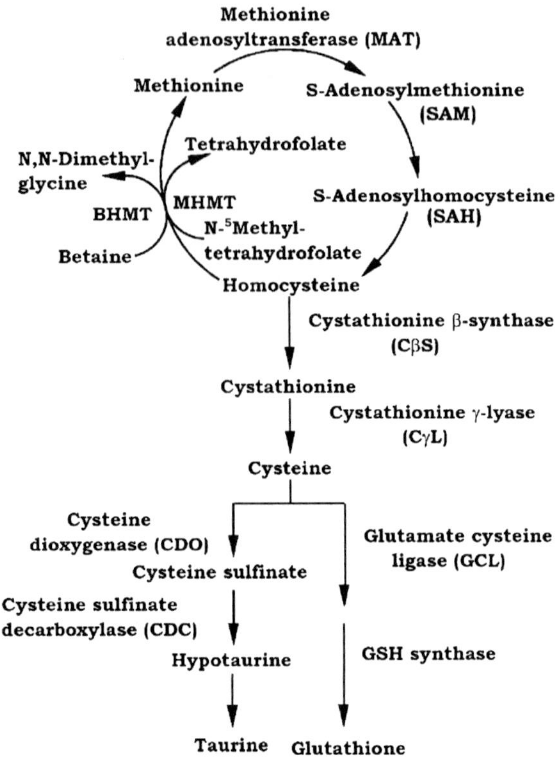

Fig. 31.1 Metabolic reactions in the transsulfuration pathway (Adapted from Kim and Kim 2005)

membrane stabilization and neuromodulation (Huxtable 1992). Taurine is both ingested directly from foodstuffs and synthesized in liver as an end product of sulfur amino acid metabolism. Sulfur amino acid metabolism primarily proceeds via the transsulfuration pathway, which results in the transfer of sulfur from methionine to serine to form cysteine (Fig. 31.1). The first step in methionine metabolism is the formation of *S*-adenosylmethionine (SAM), which is catalyzed by methionine adenosyltransferase (MAT). SAM serves as a methyl donor for various biological methylation reactions, and the co-product of transmethylation, *S*-adenosyl homocysteine (SAH), is hydrolyzed to yield homocysteine which is either remethylated to methionine or condensed with serine to form cystathionine. Transsulfuration of homocysteine to cysteine via cystathionine is mediated by consecutive actions of cystathionine β-synthase (CβS) and cystathionine γ-lyase ($C\gamma$L). Cysteine is irreversibly metabolized to yield either taurine, inorganic sulfate, or glutathione (GSH). Therefore, utilization of cysteine for generation of GSH and for generation of taurine are competitive and may be regulated by the need for GSH in biological systems (Stipanuk et al. 1992).

Carbon tetrachloride (CCl_4) is a model hepatotoxicant, whose toxicity involves the generation of a reactive free radical metabolite by CYP2E1. Several studies have shown that taurine may modulate the hepatotoxicity of CCl_4 (Nakashima et al. 1982; Vohra and Hui, 2001; Dincer et al. 2002; Erman et al. 2004; Miyazaki et al. 2005). Since taurine has a minimal effect on hepatic CYP activity (Matsuda et al. 2002; unpublished observations in this laboratory), most investigators attributed the protective effect of taurine against CCl_4 hepatotoxicity to its antioxidant activity. However, the oxygen radical scavenging activity of taurine *per se* is negligible (Aruoma et al. 1988; unpublished observation in this laboratory). Therefore, the mechanism of protection against CCl_4 hepatotoxicity provided by taurine remains unclear.

In this study we determined the hepatotoxicity of CCl_4 in mice supplemented with β-alanine. This β-amino acid competitively inhibits taurine reuptake from the renal tubular fluid into proximal tubular cells (Shaffer and Kocsis 1981). It was felt that the effect of taurine on CCl_4-induced acute liver injury could be examined by assessing the effect of taurine depletion on CCl_4-induced hepatotoxicity.

31.2 Methods

31.2.1 Animals and Treatments

Male ICR mice, weighing 20-25 g, were obtained from Dae-Han Laboratory Animal (Seoul, Korea). The use of animals was in compliance with the guidelines established by the Animal Care Committee of this institute. Animals were acclimated to temperature (22 ± 2°C) and humidity (55 ± 5%) in controlled rooms with a 12-hr light/dark cycle (light: 0700-1900, dark 1900-0700). Laboratory chow and tap water were allowed *ad libitum*. β-Alanine (3%) was added daily to regular tap water for one week before CCl_4 (50μ*l*/kg, ip) treatment. CCl_4 was dissolved in corn oil. Mice were sacrificed 24 hr following CCl_4 treatment.

31.2.2 Measurement of Hepatotoxicity

Activities of aspartate aminotransferase (AST) and alanine aminotransferase (ALT) in serum were determined using the method of Reitman and Frankel (1957). The method of Gerlach (1983) was used for the measurement of sorbitol dehydrogenase (SDH) activity.

31.2.3 Measurement of Sulfur-Containing Metabolites and Enzyme Activities

Livers were homogenized in a four-fold volume of cold 1 M perchloric acid with 2 mM EDTA. Denatured protein was removed by centrifugation at 10,000 x g for 10 min. Total GSH content was determined using the method of Neuschwander-Tetri and Roll (1989). An HPLC system equipped with a fluorescence detector and a 3.5 μm Symmetry C18 column (4.6 × 75 mm; Waters, Milford, MA, U.S.A.) was employed. Cysteine and cystine were quantified by the acid-ninhydrin method (Gaitonde 1967). An HPLC method was used for the determination of *S*-adenosylmethionine (SAM) and *S*-adenosylhomocysteine (SAH) (She et al. 1994). The supernatant was directly applied to an HPLC equipped with a UV detector and a TSK-GEL ODS-80TM column (4.6 × 250 mm; Tosoh, Tokyo, Japan). The method of Ide (1997) was used to quantify hypotaurine and taurine. Hypotaurine and taurine were derivatized with *O*-phthalaldehyde/ 2-mercaptoethanol before injection into an HPLC equipped with a fluorescence detector and a 3.5μm Kromasil C18 column (4.6 × 100 mm; Eka Chemicals, Bohus, Sweden).

Livers were homogenized in a three-fold volume of a buffer consisting of 0.154 M KCl/50 mM Tris-HCl and 1 mM EDTA (pH 7.4). The homogenate was centrifuged at 10,000 g for 20 min. The supernatant fraction was further centrifuged at 104,000 g for 60 min. The 104,000 g supernatant fraction was used to determine enzyme activities. Activity of methionine adenosyltransferase (MAT) was estimated by quantifying SAM and SAH production (Kim et al. 2003). Two different methionine concentrations, 0.1 mM and 5.0 mM, were utilized in the incubation mixture. Activity of NADPH-dependent GSH reductase was determined employing an enzymatic recycling method (Smith et al. 1988). GSH peroxidase activity was assayed by using hydrogen peroxide as a substrate (Lawrence and Burk 1976).

31.3 Results and Discussion

Hepatic taurine levels were reduced to approximately 60% of control values in mice supplemented with β-alanine for a week, but cysteine concentration in liver was significantly increased (Table 31.1). It is known that depletion of taurine by β-alanine

Table 31.1 Effect of β-alanine administration on major sulfur-containing metabolites and enzyme activities in liver [a]

		Control	β-Alanine
SAM (nmol/g liver)		112.3 ± 7.8	125.2 ± 9.3
SAH (nmol/g liver)		32.3 ± 2.5	27.9 ± 2.1
Cysteine (nmol/g liver)		78.7 ± 3.3	121.6 ± 9.2**
GSH μ mol/g liver		7.4 ± 0.3	7.9 ± 0.8
Taurine (μ mol/g liver)		12.2 ± 1.7	7.6 ± 0.8*
Hypotaurine (nmol/g liver)		97.6 ± 8.6	71.5 ± 9.5
GSSG reductase (units/mg protein)		0.09 ± 0.01	0.10 ± 0.004
GSH peroxidase (units/mg protein)		0.49 ± 0.03	0.54 ± 0.04
MAT	(5 m M)	0.18 ± 0.02	0.24 ± 0.04
(nmol/min/mg protein)	(0.1 m M)	0.04 ± 0.001	0.04 ± 0.01

[a]Mice were provided with drinking water containing 3% β-alanine for one week. Each value represents the mean ± S.E.M. for five mice.

*,**Significantly different from control at $P < 0.05, 0.01$, respectively (Student's *t*-test).

administration is associated with elevated elimination of taurine in the urine (Shaffer and Kocsis 1981). The present results suggest that the need to replenish the taurine pool accounts for the elevation in hepatic cysteine content of the β-alanine-fed mouse. Cysteine is both synthesized in the transsulfuration pathway and taken up from the systemic circulation. As seen in Table 31.1, neither the activity of MAT nor the levels of SAM in liver were changed by β-alanine administration, although the immediate precursors of cysteine, homocysteine and cystathionine, were not determined. The mechanism responsible for the elevation in hepatic cysteine content by β-alanine remains to be examined.

An acute dose of CCl_4 significantly elevated serum AST, ALT and SDH activities (Table 31.2). Hepatic GSH concentrations were not affected, but taurine levels were reduced to approximately 30 % of the control. In contrast, hepatic cysteine was markedly elevated. β-Alanine administration inhibited the induction of CCl_4-mediated hepatotoxicity, as determined by the attenuated elevation in serum enzyme activities. Hepatic levels of both GSH and taurine in the taurine deficient level were increased after CCl_4 treatment. It is suggested that the β-alanine-induced increase in the availability of cysteine (Table 31.1) is responsible for the elevation of hepatic GSH and taurine levels in mice challenged with CCl_4.

The induction of CCl_4 toxicity is attributed to the generation of a free radical, CCl_3, by reactions mediated mostly by CYP2E1. The free radical binds covalently to lipids and proteins, resulting in structural damage of membranes and inhibition of a variety of enzymes. The reactive species also initiates lipid peroxidation by attacking unsaturated fatty acids leading to the generation of organic free radicals, which may in turn react with O_2 to form peroxides and other cytotoxic metabolites (Recknagel et al. 1989).

It has been shown that taurine intake may inhibit the induction of hepatotoxicity by CCl_4in animals (Nakashima et al. 1982; Vohra and Hui, 2001; Dincer et al. 2002; Erman et al. 2004; Miyazaki et al. 2005). Because taurine does not alter hepatic CYP activities (Matsuda et al. 2002; unpublished observation in this laboratory) and has no significant direct antioxidant activity (Aruoma et al. 1988; unpublished

Table 31.2 Effect of β-alanine administration on changes in serum enzyme activities and sulfur-containing amino acid metabolites in mice treated with CCl_4^a

	Control	CCl_4	β-Alanine + CCl_4
AST (units/ml)	72 ± 11	5226 ± 773[###]	3281 ± 119[*]
ALT (units/ml)	39 ± 8	12888 ± 1182[###]	5964 ± 409[***]
SDH (units/ml)	12 ± 4	11869 ± 588[###]	7747 ± 731[**]
Cysteine (nmol/g liver)	78.7 ± 3.3	115.0 ± 2.4[##]	96.5 ± 3.8[**]
GSH (μmol/g liver)	7.4 ± 0.3	7.9 ± 1.2	11.4 ± 0.4[*]
Taurine (μmol/g liver)	12.2 ± 1.7	3.8 ± 0.5[##]	6.7 ± 0.5[**]

[a]Mice were provided with drinking water containing 3% β-alanine for one week before CCl_4 treatment. Each value represents the mean ± S.E.M. of five mice.
[##,###]Significantly different from control at $P < 0.01$, 0.001, respectively (Student's *t*-test).
[*,**,***]Significantly different from mice treated with CCl_4 only at $P < 0.05$, 0.01, 0.001, respectively.

observation in this laboratory), the mechanism of hepatoprotection provided by taurine still remains unclear. Interestingly, several studies have demonstrated that administration of a taurine-deleting agent may alleviate the toxicity of xenobiotics in various tissues. Guanidinoethane sulfonate (GES), an amidino analog of taurine, decreased the hepatotoxicity of methylene dianiline (Seabra and Timbrell 1997), and the toxicological changes in liver and heart induced by monocrotaline (Yan and Huxtable 1996). The mechanism of protection provided by GES was not studied, but in both studies an elevation of hepatic GSH was observed. Also β-alanine was shown to provide cardioprotection from reperfusion injury (Allo et al. 1997), and to decrease the toxic effects of β-amyloid in rat brain endothelial cells (Preston et al. 1998), suggesting that this substance might reduce oxidative damage in biological systems. We also observed a reduction of bacterial lipopolysaccharide-induced hepatotoxicity in rats fed with β-alanine (Kim and Kim 2002).

Taurine is a major product in the transsulfuration pathway, and its depletion in the body is suspected to result in significant changes in the metabolism of sulfur-containing substances in liver. In fact increased methionine levels in various tissues were demonstrated in rats treated with GES (Marnela and Kontro 1984; Marnela et al. 1984). β-Alanine was also shown to raise the urinary excretion of cysteine and homocysteine (Kerai et al. 2001), and the hepatic GSH and taurine concentrations after CCl_4 treatment (Waterfield et al. 1993), which are compatible with the present results. In this study the hepatoprotective effect of β-alanine against CCl_4 could be accounted for by the increased supply of cysteine for production of taurine and/or GSH, both known to have important roles in the maintenance of normal physiology biochemistry of the liver. These findings also suggest that the consequence of taurine depletion is not specific for taurine, but may induce significant changes in the metabolism of various biologically important sulfur-containing substances. The extent of the alteration in the transsulfuration reactions induced by taurine depletion and its physiological significance remain to be investigated.

References

Allo SN, Bagby L, Schaffer SW (1997) Taurine depletion a novel mechanism for cardioprotection from regional ischemia. Am J Physiol Heart Circ Physiol 273:H1956–1961

Aruoma OI, Halliwell B, Hoey BM, Butler J (1988) The antioxidant action of taurine hypotaurine and their metabolic precursors. Biochem J 256:251–255

Dincer S, Ozenirler S, Oz E, Akyol G, Ozogul C (2002) The protective effect of taurine pretreatment on carbon tetrachloride-induced hepatic damage – a light and electron microscopic study. Amino Acids 22:417–426

Erman F, Balkan J, Cevikbas U, Kocak-Toker N, Uysal M (2004) upbetaine or taurine administration prevents fibrosis and lipid peroxidation induced by rat liver by ethanol plus carbon tetrachloride intoxication. Amino Acids 27:199–205

Gaitonde MK (1967) A spectrophotometric method for the direct determination of cysteine in the presence of other naturally occurring amino acids. Biochem J 104:627–633

Gerlach U (1983) Sorbitol dehydrogenase. In: HU Bergmeyer (ed) Methods of enzymatic analysis, vol 3. Verlag Chemie Weinheim, pp 112–117

Huxtable RJ (1992) Physiological actions of taurine. Physiol Rev 72:101–163

Ide T (1997) Simple high-performance liquid chromatographic method for assaying cysteine-sulfinic acid decarboxylase activity in rat tissue. J Chromatogr B Biomed Sci Appl 694:325–332
Kerai MD, Waterfield CJ, Kenyon SH, Asker DS, Timbrell JA (2001) The effect of taurine depletion by upbeta-alanine treatment on the susceptibility to ethanol-induced hepatic dysfunction in rats. Alcohol Alcohol 36:29–38
Kim SK, Kim YC (2002) Attenuation of bacterial lipopolysaccharide-induced hepatotoxicity by upbetaine or taurine in rats. Food Chem Toxicol 40:545–549
Kim SK, Kim YC (2005) Effects of upbetaine supplementation on hepatic metabolism of sulfur-containing amino acids in mice. J Hepatol 42:907–913
Kim SK, Seo JM, Jung YS, Kwak HE, Kim YC (2003) Alterations in hepatic metabolism of sulfur-containing amino acids induced by ethanol in rats. Amino Acids 24:103–110
Lawrence RA, Burk RF (1976) Glutathione peroxidase activity in selenium-deficient rat liver. Biochem Biophys Res Commun 71:952–958
Marnela KM, Kontro P (1984) Free amino acids and the uptake and binding of taurine in the central nervous system of rats treated with guanidinoethanesulphonate. Neuroscience 12:323–328
Marnela KM, Kontro P, Oja SS (1984) Effects of prolonged guanidinoethanesulphonate administration on taurine and other amino acids in rat tissues. Med Biol 62:239–244
Matsuda H, Kinoshita K, Sumida A, Takahashi K, Fukuen S, Fukuda T, Takahashi K, Yamamoto I, Azuma J (2002) Taurine modulates induction of cytochrome P450 3A4 mRNA by rifampicin in the HepG2 cell line. Biochim Biophys Acta 1593:93–98
Miyazaki T, Karube M, Matsuzaki Y, Ikegami T, Doy M, Tanaka N, Bouscarel B (2005) Taurine inhibits oxidative damage and prevents fibrosis in carbon tetrachloride-induced hepatic fibrosis. J Hepatol 43:117–125
Nakashima T, Taniko T, Kuriyama K (1982) Therapeutic effect of taurine administration on carbon tetrachloride-induced hepatic injury. Jpn J Pharmacol 32:583–589
Neuschwander-Tetri BA, Roll FJ (1989) Glutathione measurement by high-performance liquid chromatography separation and fluorometric detection of the glutathione-orthophthalaldehyde adduct. Anal Biochem 179:236–241
Preston JE, Hipkiss AR, Himsworth DT, Romero IA, Abbott JN (1998) Toxic effects of β-amyloid(25-35) on immortalised rat brain endothelial cell:protection by carnosine homocarnosine and β-alanine. Neurosci Lett 242:105–108
Recknagel RO, Glende EA, Dolak JA, Waller RL (1989) Mechanisms of carbon tetrachloride toxicity. Pharmacol Ther 43:139–154
Reitman S, Frankel SA (1957) Colorimetric method for determination of serum glutamic oxaloacetic and glutamic pyruvic transaminases. Am J Clin Pathol 28:56–63
Seabra V, Timbrell JA (1997) Modulation of taurine levels in the rat liver alters methylene dianiline hepatotoxicity. Toxicology 122:193–204
Shaffer JE, Kocsis JJ (1981) Taurine mobilizing effects of upbeta alanine and other inhibitors of taurine transport. Life Sci 28:2727–2736
She QB, Nagao I, Hayakawa T, Tsuge H (1994) A simple HPLC method for the determination of S-adenosylmethionine and S-adenosylhomocysteine in rat tissues: the effect of vitamin B6 deficiency on these concentrations in rat liver. Biochem Biophys Res Commun 205:1748–1754
Smith IK, Vierheller TL, Thorne CA (1988) Assay of glutathione reductase in crude tissue homogenates using 55'-dithiobis(2-nitrobenzoic acid). Anal Biochem 175:408–413
Stipanuk MH, Coloso RM, Garcia RA, Banks MF (1992) Cysteine concentration regulates cysteine metabolism to glutathione sulfate and taurine in rat hepatocytes. J Nutr 122:420–427
Vohra BP, Hui X (2001) Taurine protects against carbon tetrachloride toxicity in the cultured neurons and in vivo. Arch Physiol Biochem 109:90–94
Waterfield CJ, Turton JA, Scales MD, Timbrell JA (1993) Reduction of liver taurine in rats by -alanine treatment increases carbon tetrachloride toxicity. Toxicology 77:7–20
Yan CC, Huxtable RJ (1996) Effects of taurine and guanidinoethane sulfonate on toxicity of the pyrrolizidine alkaloid monocrotaline. Biochem Pharmacol 51:321–329

Chapter 32
Effect of Taurine on Alcoholic Liver Disease in Rats

Gaofeng Wu, Jiancheng Yang, Changmian Sun, Xinhong Luan, Jiao Shi, and Jianmin Hu

Abstract To investigate the effect of taurine on alcoholic liver disease in rats, male Wistar rats were administered alcohol intragastrically for 3 months. The effect of β-alanine-mediated taurine depletion and taurine administration on the development of alcoholic liver disease was examined. It was found that taurine administration produced lower levels of aspartate aminotransferase and alkaline aminotransferase than that of the untreated group. In addition, the levels of hepatic total protein, glutathione and superoxide dismutase were higher in the taurine treated groups than in the untreated control or the taurine depleted group, while hepatic malondialdehyde content exhibited the opposite effect. Moreover, the content of hepatic hydroxyproline, serum hyaluronic acid, interleukin-2, interleukin-6, tumor necrosis factor-α and laminin were all decreased in the taurine treated group. The pathological changes showed that the percentage of fatty degeneration and inflammation in the taurine group were less than that of the control, taurine depleted and automatic recovery groups. These in-vivo findings demonstrate that hepatic disease caused by chronic alcohol consumption can be prevented and reversed by administration of taurine.

Abbreviations *ALD*, alcoholic liver disease; *ALT*, alkaline aminotransferase; *AST*, aspartate aminotransferase; *GSH*, glutathione; *SOD*, superoxide dismutase; IL-2 or 6, interleukin-2 or 6; *Hyp*, hydroxyproline; *TNF*-α, tumor necrosis factor-α; *HA*, hyaluronic acid; *LN*, laminin

32.1 Introduction

Liver is considered to be the main organ responsible for the oxidation of alcohol. Drinking too much alcohol can lead to serious damage to the liver. It has been reported that the mortality rate among youths caused by alcoholic liver disease

G. Wu (✉)
College of Animal Science and Veterinary Medicine, Shenyang Agricultural University, Shenyang, China

J. Azuma et al. (eds.), *Taurine 7*, Advances in Experimental Medicine and Biology 643,
DOI 10.1007/978-0-387-75681-3_32, © Springer Science+Business Media, LLC 2009

(ALD) constitute 80% of total hepatic-linked mortality (Liu and Zhang 2004). In China, with the rise in living standards, the rate of ALD has progressively increased.

Taurine, a sulfur-containing β-amino acid, is the major free intracellular amino acid present in many human and animal tissues (Huxtable 1992). There is significant evidence that taurine exerts physiologic and pharmacologic functions, such as maintaining normal structure and function of the hematological system, immune system, reproductive system, visual system, cardiovascular system and nervous system. Since taurine is synthesized mainly in the liver, the effects of taurine on the alimentary system have assumed particular attention. Kerial et al. have demonstrated a protective effect of taurine against hepatic steatosis and lipid peroxidation when co-administered with alcohol for 28 days (Kerial et al. 1998). In China, Chen et al. (1999) reported that taurine protects against liver damage caused by multiple insults, although they did not study ALD. The present study was to investigate whether taurine has both preventive and curative effects on ALD, and elucidate the mechanism underlying the cytoprotection.

32.2 Methods and Materials

32.2.1 Experimental Animals and Treatments

Six weeks old male Wistar rats weighing 125 to 150 g and acclimatized for 7 days before experiment were used in the study. Animals were maintained under a controlled condition of light (12 hr of light, 12 hr of dark) and temperature (23 ± 2°C), and were given free access to food (commercial standard rat chow) and water.

32.2.2 Chemicals

Sixty degree alcohol was purchased from Shenyang LaoLong kou distillery. Taurine and pyrazole were purchased from Sigma Chemical Company (St. Louis, MO, USA). Reagent kits of reduced glutathione (GSH), malondialdehyde (MDA), superoxide dismutase (SOD), alanine aminotransferase (ALT), aspartate aminotransfer (AST), nitrogen monoxidum (NO), inducible nitricoxide synthase (iNOS) and hydroxyproline (Hyp) were purchased from Nanjing Jiancheng Bioengineering Institute. Interleukin-2 (IL-2), interleukin-6 (IL-6) and tumor necrosis factor-α (TNF-α) radioimmunoassay kits were purchased from Beijing chemclin biotech co. ltd. Hyaluronic acid (HA) and laminin (LN) radioimmunoassay kits were purchased from Beijing North Institute of Biological Technology (BNIBT in short) (Pan Jia Mao A20, Feng Tai District).

32.2.3 Experimental Design

The experiment was divided into two arms. In one experiment (the preventive experiment), 50 Wistar rats were randomly divided into 5 groups ($n = 10$). The control group (C) was fed commercial standard rat chow, and received an intragastric administration of physiologic saline every morning. Alcohol and pyrazole at dosages of 8g/(kg • d) and 24 mg/(kg • d) respectively were co-administered intragastrically to rats of the model group (M) every morning. These rats were fed a high-fat diet (commercial standard rat chow were dipped into corn oil). Other groups were treated with alcohol and pyrazole in addition to being maintained on tap water containing 1% β-alanine to cause endogenous taurine depletion (D). Still other rats (belonging to the preventive group) were maintained on tap water containing either 1% or 2% taurine (PI, PII). After 12 weeks on alcohol and taurine, the rats were euthanatized and blood and hepatic tissue were collected for biochemical analysis and histological tests.

In the curative experiment, the ALD model rats were randomly divided into 3 groups (n=10). The rats were allowed free access to commercial standard rat chow. The rats in the automatic recovery group (A) were given tap water, while the taurine treatment groups (TI, TII) were given water containing 1% and 2% taurine respectively. After 4 weeks of treatment, rats were euthanatized and blood and hepatic tissue were collected for biochemical analysis and histological tests.

32.2.4 Biochemical Analysis

Serum was separated by centrifugation (+4°C, 1500 r/min, 15 min). Serum concentrations of ALT and AST were determined by a calorimetry method based on appropriate reagent kits.

Liver tissue was taken and rapidly put into ice-cold saline to prepare homogenate. Hepatic content of TP, GSH, MDA and SOD were determined by a colorimetry method based on appropriate reagent kits.

Serum concentrations of HA, LN, IL-2, IL-6 and TNF-α were determined by a radioimmunoassay based on reagent kits. Hepatic content of Hyp was determined by a colorimetry method based on the appropriate reagent kit.

Hepatic content of NO and iNOS were determined by a colorimetry method based on the appropriate reagent kits.

32.2.5 Histological Analysis

Hepatic tissue from the left hepatic lobe was fixed in 10% (v/v) phosphate buffered formalin solution (PH 7.0) and embedded in paraffin wax. Sections were cut into 5 μm and stained with haematoxylin and eosin. The adipohepatic percentage was the ratio of adipohepatic area to the area of total visual field. Classify the degree of fatty degeneration according to the proportion of hepatocytes containing fat droplets.

32.2.6 *Statistic Analysis*

Data were presented as the mean ± SE and significant differences were determined by Duncan's multiple range test using SPSS 12.0 statistical analysis software. *P* values less than 0.05 were considered significant.

32.3 Results

32.3.1 *Hepatic Index*

Table 32.1 shows that the taurine-treated and normal animals exhibited smaller hepatic indices than the model and taurine depleted groups following the 3 months of alcohol and pyrazole consumption. Four weeks after addition of taurine to the water, the hepatic index significantly decreased in the 2% taurine group compared with the automatic recovery group ($p < 0.05$).

32.3.2 *Serum Analysis*

As shown in Table 32.1, serum concentrations of ALT and AST were significantly raised by alcohol treatment ($P < 0.05$), especially when endogenous taurine was depleted by β-alanine administration ($P < 0.01$). However, the levels of serum ALT and AST were significantly decreased when 2% taurine was co-administered with alcohol and was maintained for 4 weeks after alcohol withdrawal ($P < 0.05$).

32.3.3 *Hepatic Analysis*

As shown in Table 32.2, hepatic content of TP and GSH were significantly decreased by alcohol treatment, but they were significantly increased when 2% taurine was administered both preventively and curatively compared with the model group.

Table 32.1 Effects of taurine on hepatic index and serum concentrations of ALT and AST

Group		Hepatic index (%)	AST (U/L)	ALT (U/L)]
	C	2.67 ± 0.03^a	91.86 ± 3.87^a	41.36 ± 1.13^a
Preventive	M	3.37 ± 0.02^b	175.87 ± 9.53^b	57.75 ± 2.46^b
experiment	D	4.37 ± 0.13^c	194.13 ± 4.02^b	75.29 ± 2.98^c
	PI	3.10 ± 0.06^b	135.67 ± 5.53^c	53.24 ± 2.91^b
	PII	2.83 ± 0.15^a	103.22 ± 3.04^a	44.22 ± 1.96^a
	C	2.43 ± 0.07^a	87.83 ± 2.49^a	42.41 ± 0.08^a
Curative	A	3.07 ± 0.18^b	170.37 ± 9.92^b	72.83 ± 2.38^b
experiment	TI	2.70 ± 0.01^{ab}	145.30 ± 5.02^b	67.79 ± 0.52^b
	TII	2.60 ± 0.03^a	110.31 ± 10.00^a	47.80 ± 2.54^a

The hepatic index is the ratio of liver weight to body weight. *AST*: aspartate aminotransfer, *ALT*: alanine aminotransferase.Results represent means ± SE (n=5). Superscripts reveal significant differences between two groups within the same experiment ($P < 0.05$).

Table 32.2 Effects of taurine on hepatic lipid peroxidation in rats

Group		GSH (mmol/mg · pro)	MDA (nmol/mg · pro)	SOD (u/mg · pro)	TP (g/L)
Preventive experiment	C	29.90 ± 0.27^{a}	1.63 ± 0.01^{ac}	336.03 ± 13.75^{ac}	1.35 ± 0.07^{a}
	M	22.47 ± 1.64^{b}	2.33 ± 0.11^{b}	259.88 ± 20.55^{a}	1.06 ± 0.03^{b}
	D	16.93 ± 0.08^{c}	2.48 ± 0.18^{b}	257.47 ± 18.13^{a}	0.80 ± 0.06^{c}
	PI	27.42 ± 1.15^{a}	1.90 ± 0.07^{a}	335.62 ± 28.86^{ac}	1.14 ± 0.02^{b}
	PII	28.71 ± 1.51^{a}	1.47 ± 0.10^{c}	367.20 ± 39.11^{c}	1.31 ± 0.06^{a}
Curative experiment	C	29.03 ± 0.47^{a}	1.82 ± 0.12^{a}	229.03 ± 4.04^{a}	2.02 ± 0.04^{a}
	A	20.08 ± 0.06^{b}	2.74 ± 0.12^{b}	205.61 ± 2.83^{b}	1.60 ± 0.10^{b}
	TI	22.99 ± 0.14^{c}	2.45 ± 0.05^{bc}	217.60 ± 0.29^{a}	1.88 ± 0.06^{a}
	TII	27.91 ± 0.27^{d}	2.16 ± 0.06^{ac}	226.27 ± 5.20^{a}	2.02 ± 0.10^{a}

GSH: glutathione, *MDA*: malondialdehyde, *SOD*: superoxide dismutase, *TP*: total protein. Results represent means ± SE ($n = 5$). Superscripts represent significant differences between two groups within the same experiment ($P < 0.05$).

Hepatic levels of MDA were significantly increased by alcohol treatment, especially in the taurine depleted group ($P < 0.01$). However, 2% taurine treatment during alcohol consumption significantly lowered hepatic content of MDA compared with the model group, as well as in the curative experiment. Alcohol treatment also caused a decrease in hepatic content of SOD. However, administration of 2% taurine significantly increased hepatic content of SOD in the preventive group compared with the model group ($P < 0.05$).

32.3.4 Hepatic Levels of NO and iNOS Analysis

As shown in Table 32.3, the hepatic levels of NO and iNOS were significantly raised by alcohol treatment ($P < 0.05$), especially in the β-alanine treated group ($P < 0.01$). When 2% taurine was administered preventively and curatively hepatic levels of NO and iNOS were significantly decreased compared with the model group ($P < 0.05$).

32.3.5 Hepatic Fibrosis Analysis

Hepatic content of Hyp and serum concentrations of HA and LN were raised by alcohol treatment, with the level of serum LN being partially elevated relative to the control group ($P < 0.05$, Table 32.3). But all the three indices were significantly raised when endogenous taurine was depleted by β-alanine administration ($P < 0.05$). The results showed that the levels of hepatic Hyp, serum HA and serum LN were significantly decreased when 2% taurine was administered both preventively and curatively (Table 32.3).

Table 32.3 Effects of taurine on hepatic content of NO, iNOS and the fibrosis index in rats

Group		NO (μmol/mgprot)	iNOS (U/mgprot)	Hyp (μg/mgprot)	HA (μg/L)	LN (μg/L)
Preventive experiment	C	4.29 ± 0.40^{ac}	1.53 ± 0.10^{a}	0.77 ± 0.05^{a}	546.81 ± 2.68^{a}	44.28 ± 3.87^{a}
	M	5.56 ± 0.28^{b}	2.28 ± 0.07^{b}	0.92 ± 0.07^{a}	690.25 ± 48.99^{ab}	65.68 ± 1.99^{bc}
	D	5.26 ± 0.22^{b}	3.01 ± 0.22^{c}	1.18 ± 0.09^{b}	765.89 ± 34.11^{b}	73.93 ± 1.82^{c}
	PI	5.10 ± 0.21^{ab}	2.18 ± 0.03^{b}	0.86 ± 0.02^{a}	609.33 ± 15.87^{ac}	51.84 ± 11.84^{ab}
	PII	3.60 ± 0.13^{c}	1.27 ± 0.12^{a}	0.83 ± 0.04^{a}	501.69 ± 82.32^{c}	44.59 ± 2.90^{a}
Curative experiment	C	5.51 ± 0.01^{a}	1.02 ± 0.01^{a}	0.50 ± 0.01^{a}	570.76 ± 68.69^{a}	40.72 ± 0.72^{a}
	A	6.48 ± 0.39^{b}	1.59 ± 0.04^{b}	0.59 ± 0.02^{b}	787.13 ± 12.87^{b}	84.84 ± 4.30^{b}
	TI	6.02 ± 0.00^{ab}	1.25 ± 0.01^{c}	0.52 ± 0.02^{a}	640.96 ± 96.26^{ab}	65.38 ± 2.50^{bc}
	TII	5.30 ± 0.23^{a}	1.23 ± 0.00^{c}	0.49 ± 0.02^{a}	509.49 ± 0.00^{a}	55.07 ± 7.68^{ac}

NO: nitric oxide, *iNOS*: inducible nitric oxide synthase, *Hyp*: hydroxyproline, *HA*: Hyaluronic acid, *LN*: laminin. Results represent means ± SE ($n = 5$). Superscripts represent significant differences between two groups within the same experiment ($P < 0.05$).

32.3.6 Serum Cytokine Concentrations

As shown in Table 32.4, the serum concentrations of IL-2, IL-6 and TNF-α in the model and the taurine depleted groups were all remarkably higher than values of the control group, ($P < 0.05$). The above indices were remarkably decreased when 2% taurine was administered preventively and curatively ($P < 0.05$).

32.3.7 Histological Analysis

The results of hepatic histological examination showed that the livers of rats treated with alcohol and β-alanine were enlarged, with hyperaemic surfaces, tight teguments and granulations on the surface. Optical micro-scopic examination showed that the rats treated with alcohol had developed significant steatosis and contained inflammatory cell infiltrations, with significant engorged hepatic cells, loose cyto-

Table 32.4 Effects of taurine on correlation cytokine of rats

Group		IL-2 (ng/ml)	IL-6 (pg/ml)	TNF-α (ng/ml)
Preventive experiment	C	2.48 ± 0.12^{a}	80.22 ± 4.27^{a}	0.70 ± 0.15^{a}
	M	4.75 ± 0.97^{bc}	130.02 ± 6.16^{bc}	1.18 ± 0.07^{bc}
	D	5.18 ± 0.49^{b}	147.31 ± 19.35^{c}	1.28 ± 0.04^{b}
	PI	3.12 ± 0.35^{a}	107.32 ± 22.05^{ac}	0.94 ± 0.22^{ac}
	PII	2.53 ± 0.18^{a}	86.35 ± 13.94^{ab}	0.71 ± 0.08^{a}
Curative experiment	C	3.04 ± 0.04^{a}	90.00 ± 30.00^{a}	0.99 ± 0.04^{a}
	A	4.68 ± 0.15^{b}	154.21 ± 13.25^{b}	1.31 ± 0.05^{b}
	TI	4.02 ± 0.02^{ab}	130.62 ± 21.60^{ab}	1.19 ± 0.10^{ab}
	TII	3.33 ± 0.17^{a}	95.61 ± 24.73^{a}	1.02 ± 0.01^{a}

IL-2: interleukin-2, *IL-6*: interleukin-6, *TNF*-α: tumor necrosis factor-α. Results represent means ± SE ($n = 5$). Superscripts represent significant differences between two groups within the same experiment ($P < 0.05$).

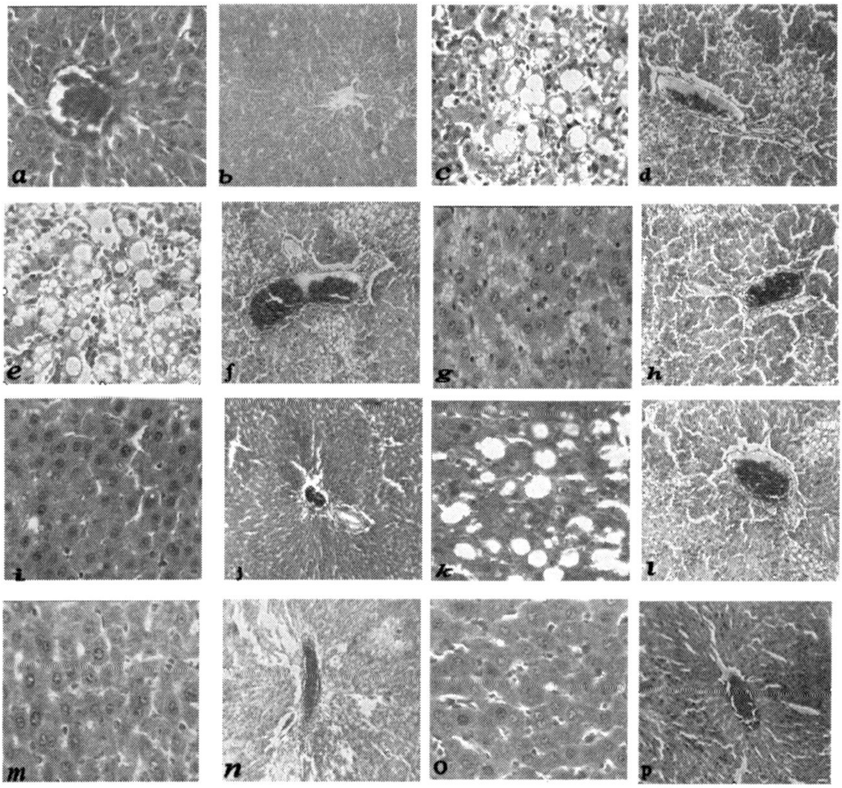

Fig. 32.1 Histological changes of the liver. These photomicrographs of hepatic cells and central veins show different degrees of steatosis, thrombosis, inflammatory cell infiltration and collagenous fibre hyperplasia in each group. **a and b**: control group. **c and d**: model group. **e and f**: taurine depleted group. **g and h**: 1% taurine prevention group. **i and j**: 2% taurine prevention group. **k and l**: automatic recovery group. **m and n**: 1% taurine treatment group. **o and p**: 2% taurine treatment group. HE stain. a,c,e,g,i,k,m,o ×400. b,d,f,h,j,l,n,p ×100

plasm and many fat vacuoles in the cell plasma. The nuclei were located on the periphery of the cells (Fig. 32.1c). The liver cell cords of the model and the taurine depleted groups lacked normal orderly structure, with the collagenous fibers enlarged and the hepatic lobules separated (Fig. 32.1d, f). The degree of steatosis, inflammation and fibroplasia were remarkably reduced when treated with taurine preventively (Fig. 32.1g, h, i, j) and curatively (Fig. 32.1m, n, o, p). But in automatic recovery group, there were also fat vacuoles localized in the cytoplasm, the cell nuclei were dispersed to the periphery of the cells and the central veins were in disorder (Fig. 32.1k, l).

Table 32.5 shows a hepatic steatosis classification chart and adipohepatic percentage data. In the preventive experiment, the adipohepatic percentage of the model group was significantly larger than that of the control group ($P < 0.05$). However, when the rat was treated with taurine at the time of alcohol consumption,

Table 32.5 Effects of taurine on pathologic changes of the liver

Group		Steatosis classification	Adipohepatic percentage
Preventive experiment	C	1	0.04 ± 0.04^{a}
	M	6	27.36 ± 0.16^{b}
	D	6	37.38 ± 2.42^{c}
	PI	5	19.06 ± 0.27^{d}
	PII	4	9.28 ± 0.30^{e}
Curative experiment	C	1	0.04 ± 0.04^{a}
	A	5	18.74 ± 0.24^{b}
	TI	4	5.97 ± 0.47^{c}
	TII	1	0.56 ± 0.07^{a}

Results represent means ± SE ($n = 5$). Superscripts represent significant difference between two groups within the same experiment ($P < 0.05$).

the adipohepatic percentage was significantly reduced compared with the model group ($P < 0.05$). In the curative experiment, the adipohepatic percentage was significantly reduced by taurine administration when compared with the automatic recovery group ($P < 0.05$).

32.4 Discussion

The co-administration of alcohol and pyrazole combined with a high fat diet leads to ALD, associated with prolonged alcohol metabolism, hepatic cell membrane dysfunction and liver injury. The present study describes a model of ALD exhibiting the damage mentioned above (Zhou et al. 1986).

Taurine is a β-type sulfur amino acid that is widely distributed in tissues, cells and body fluids of human beings and animals. As a conditional essential amino acid of humans and most animals, taurine has been used for many years in the treatment of hepatopathologies, such as chronic active hepatitis, however, its preventive and curative applications on ALD have received little attention. The data showing that the hepatic index and serum concentrations of ALT and AST were significantly increased, while the hepatic content of total protein was significantly decreased in the model and taurine depleted groups may result from alcohol-mediated lipid accumulation in the hepatic cells, hepatic cell hypertrophy and membrane damage. Under those conditions, hepatic ALT and AST are released into the blood. It is known that protein is mainly synthesized in liver and when the liver is injured, protein synthesis is adversely affected. Through this mechanism alcohol could cause a decrease in hepatic protein content, as shown in the present experiments. However, taurine administration would appear to reduce the degree of fat deposition, inhibit hepatic cellular swelling and attenuate the increase in the hepatic index mediated by alcohol. Based on the sensitive index of liver dysfunction, namely, serum ALT, AST and hepatic TP, taurine treatment partially prevents ALD.

Chronic alcohol feeding leads to the proliferation of smooth surfaced endoplasmic reticulum. Because many of the enzymes involved in the synthesis of triglycerides and phospholipids are associated with the endoplasmic reticulum, the lipid-synthesizing capacity of the liver could be enhanced. As alcohol is also metabolized by the microsomal enzyme system, it interferes with lipid metabolism. Microsomal alcohol oxidation may interfere with lipid metabolism by generating oxygen radicals such as O^{2-} and HO^{-}, which initiate a cascade of lipid peroxidation and damage cell membranes (Polavarapu et al. 1998). There is evidence that taurine exhibits antioxidant activity, regulates calcium, modulates osmotic pressure and stabilizes membranes (Trachtman et al. 1993; Wang et al. 2002). In this study, both alcohol and β-alanine significantly increased hepatic lipid peroxidation, while taurine significantly increased hepatic levels of GSH and SOD while decreasing hepatic levels of MDA. The results indicate that taurine may protect against free radical damage, elevate cytoprotective factors (reducing agent) and reduce damaging factors (oxidative products and radicals) (Banks et al. 1991). Thus, taurine reduces alcohol-induced hepatic steatosis and lipid peroxidation (Kerial et al. 1999).

Nitric oxide plays an important role in the process of alcohol hepatic damage. Under physiological conditions, NO catalyzed by endothelial NOS (eNOS) regulates the microcirculation, thereby inhibiting oxidative stress and reducing the capability of generating free radicals formed by peroxidases. But when the liver is damaged by alcohol, large amounts of NO are formed by inducible NOS (iNOS), aggravating oxidative damage and disturbing the microcirculation. There is evidence that taurine has no significant effect on the expression of iNOS in damaged lung tissue caused by pneumosilicosis. However, the present study found that hepatic content of NO and iNOS was significantly elevated by alcohol, but taurine treatment could reduce the synthesis of NO by inhibiting iNOS expression.

Alcohol treatment resulted in alcoholic hepatic fibrosis, increasing interstitial extracellular matrix (ECM) through a decrease in the degradation of ECM. Because Hyp is the main constituent of collagen protein, the degree of hepatic fibrosis can be estimated by hepatic Hyp content. Meanwhile, liver fibrosis resulted in abnormal proliferation and insufficient degradation of protein constituents, such as collagen protein, proteoglycan and glycoproteinogen. In this study, the abnormal increase of serum HA, LN and hepatic Hyp is consistent with the previous conclusion that taurine can inhibit hepatic fibrosis caused by CCl4 (Chen et al. 1999; Wei et al. 2004). The results indicate that taurine can inhibit the proliferation of hepatic stellate cells and collagenoblasts, the synthesis of collagen and the deposition of hepatic ECM. In this way, taurine minimize the degree of hepatic fibrosis. In addition, the values of serum IL-2, IL-6 and TNF-α obtained from this experiment showed that taurine administration inhibits the abnormal increase in cytokine content, indicating that taurine minimizes ALD by regulating the immunological network.

Histological observations showed that taurine not only prevents but actually reverses the pathomorphological changes of ALD, such as decreases in fat deposition and inflammatory cell infiltration.

32.5 Conclusion

In conclusion, this study demonstrated that taurine has both preventive and curative effects on ALD and may have therapeutic relevance.

Acknowledgments This study was supported by a grant from the Education Department of Liaoning Province, China. The authors thank Dr. Masugi Nishihara, Tokyo University (Japan) for reviewing the manuscript and giving valuable advice.

References

Banks MA, Porter DW, Martin WG, Castranova V (1991) Ozone induced lipid peroxidation and membrane leakage in isolated rat alveolar macrophages: protective effects of taurine. J Nutr Biochem 2: 308–313

Chen YX, Li S, Zhang XR (1999) Inhibition effects of taurine on hepatic fibrosis in rats caused by CCL4. Chinese J Digest 19:185–187

Huxtable RJ (1992) Physiological action of taurine. Physiol Rev 72:101–163.

Kerial MDJ, Waterfield CJ, Kenyon SH, Asker DS, Timbrell JA (1998) Taurine: Protective properties against ethanol-induced hepatic steatosis and lipid peroxidation during chronic ethanol consumption in rats. Amino Acids 15:53–76

Kerial MDJ, Waterfield CJ, Kenyon SH, Asker DS, Timbrell JA (1999) Reversal of ethanol-induced hepatic steatosis and lipid peroxidation by taurine: a study in rats. Alcohol and Alcoholism 34:529–541

Liu S, Zhang MX (2004) Research summarize of alcoholic liver disease. Chinese Arch. Chinese Med 22:85–86

Polavarapu R, Spitz DR, Sim JE, Follansbee MH, Oberley LW, Rahemtulla A, Nanji AA (1998) Increase lipid peroxidation and impaired antioxidant enzyme function is associated with pathological liver injury in experimental alcoholic liver disease in rats fed diets high in corn oil and fish oil. Hepatology 27: 1317–1323

Trachtman H, Futterwert C, Robert B (1993) Taurine prevents glucose-induced lipid peroxidation and increased collagen production in cultured rat mesangial cells. Biochem Biophys 191:759

Wang CH, Zhang XK, Miu YM (2002) Protective effects of taurine on hepatic mitochondria in the process of ischemia reperfusion in rats. Chinese J Integ Trad Western Med Liver Dis 2:351–354 and 357

Wei X, Liang J, Mao DW (2004) Influence of taurine on the formation of hepatic fibrosis in rats. Chinese J Integ Trad Western Med Digest 12:6–7

Zhou SQ, Jia CY, Gao DS (1986) Pathomorphologic observations on hepatic damages in rats caused by experimental chronic alcoholic intoxicate. Chinese J Pathol 15:136–139

Part V
Effect of Taurine on Bone

Chapter 33
Does Taurine Deficiency Cause Metabolic Bone Disease and Rickets in Polar Bear Cubs Raised in Captivity?

Russell W. Chesney, Gail E. Hedberg, Quinton R. Rogers, Ellen S. Dierenfeld, Bruce E. Hollis, Andrew Derocher, and Magnus Andersen

Abstract Rickets and fractures have been reported in captive polar bears. Taurine (TAU) is key for the conjugation of ursodeoxycholic acid (UDCA), a bile acid unique to bears. Since TAU-conjugated UDCA optimizes fat and fat-soluble vitamin absorption, we asked if TAU deficiency could cause vitamin D malabsorption and lead to metabolic bone disease in captive polar bears. We measured TAU levels in plasma (P) and whole blood (WB) from captive and free-ranging cubs and adults, and vitamin D_3 and TAU concentrations in milk samples from lactating sows. Plasma and WB TAU levels were significantly higher in cubs vs captive and free-ranging adult bears. Vitamin D in polar bear milk was 649.2±569.2 IU/L, similar to that found in formula. The amount of TAU in polar bear milk is 3166.4±771 nmol/ml, 26-fold higher than in formula. Levels of vitamin D in bear milk and formula as well as in plasma do not indicate classical nutritional vitamin D deficiency. Higher dietary intake of TAU by free-ranging cubs may influence bile acid conjugation and improve vitamin D absorption.

Abbreviations *TAU*, taurine; *UCDA*, ursodeoxycholic acid; *P*, plasma; *WB*, whole blood; *MBD*, metabolic bone disease

33.1 Introduction

The zoo world has been puzzled by a spate of rickets, fractures and metabolic bone disease (MBD) in polar bear cubs raised in captivity (Kenny et al. 1999; Hedberg and Chesney 2004; Hedberg et al. 2007). Many of these cubs have been hand-reared

R.W. Chesney (✉)
The University of Tennessee Health Science Center, USA

J. Azuma et al. (eds.), *Taurine 7*, Advances in Experimental Medicine and Biology 643,
DOI 10.1007/978-0-387-75681-3_33, © Springer Science+Business Media, LLC 2009

because of maternal neglect. The main features of MBD in polar bears are single or multiple fractures (19 fractures in 12 bears) after a trivial fall and radiographic evidence of rickets (Lin et al. 2005). Antebrachial fractures of a forelimb are the most common breaks encountered. The etiology and pathogenesis of MBD is as yet unclear.

Hand-reared bear cubs are often fed a cow milk-based formula, called Esbilac, developed as a milk replacer for puppies, alone or in combination with a standard cow milk-based full-term human infant formula. More than 50% of cubs reared in captivity over the past 25 years have failed to survive beyond three months (Hedberg et al. 2004). At birth, polar bear cubs weigh approximately 600 g, and remain in the den with their mother until about four months of age. Cubs may continue to nurse for up to two years.

Vitamin D deficiency has been considered a major factor in causing rickets and MBD in cubs (Johnston and Cutchins 1985; Kenny et al. 1998; Kenny et al. 1999; Lin et al. 2005). Nonetheless, cubs are fed formula(s) with 675–950 IU/L (17–24 μg) of vitamin D, an amount known to prevent rickets in most mammalian species (Chesney 2003). The quantity of calcium and phosphorus in formula is comparable to that found in polar bear milk (Derocher et al. 1993; Kenny et al. 1998). Serum values of 25-hydroxyvitamin D in 36 captive and 56 free-ranging polar bears were not different (approximately 350 nmol/mL) (Kenny et al. 1998). Thus, while calcium, phosphorous or vitamin D deficiency cannot be ruled out as etiologic factors in MBD, the possibility of vitamin D malabsorption should be considered. Supporting this hypothesis is the finding that polar bears require ingestion of vitamin D to maintain stores, as their black skin (with limited 7-dehydrocholesterol), thick coats, and limited arctic sunshine prevent synthesis of adequate amounts of vitamin D by the photocutaneous route (Kenny et al. 1998).

Vitamin D is fat-soluble and requires the presence of conjugated bile acids to initiate the formation of micelles that enhance the diffusion of fats through the unstirred layer of the small intestine (Hagey et al. 1993; Zamboni et al. 1993; Hofmann 1999). Being the most intense carnivore among bears, the polar bear conjugates its bile acids only with TAU (Hagey et al. 1993; Hofmann 1999; Sacco and Valkenburgh 2004). These tauro-conjugates are especially effective in the absorption of fat-soluble vitamins.

The polar bear, like all bears, synthesizes a unique primary bile acid, ursodeoxycholic acid (UDCA), which, when TAU-conjugated, is the most efficient bile salt in terms of fat absorption (Hagey et al. 1993; Zamboni et al. 1993; Hofmann 1999). A pertinent question is whether the polar bear cub, which requires dietary sources of vitamin D, also needs adequate TAU to optimally conjugate UDCA and other bile acids. Accordingly, not only the vitamin D content of polar bear milk or cow milk-based formula is relevant, but also the TAU content of the diets of polar bear cubs should be evaluated.

We therefore measured the plasma concentration of TAU in polar bear cubs and in captive and free-ranging adult bears, as well as the TAU and vitamin D levels in polar bear milk. The milk values were then compared with the measured or referenced values for TAU and vitamin D content in cow milk-based formula.

33.2 Methods

Blood samples were obtained from two wild-caught cubs, 14 captive adults and 50 free-ranging adult polar bears on Spitzbergen and other Svalbard islands, a chain of Norwegian islands above the Arctic Circle (77 – 79° latitude). Milk samples were obtained from 16 lactating sows with cubs. Quantitation of vitamin D_3 and 25-hydroxyvitamin D_3 in milk samples was performed using previously described methods that include methanol:methylene chloride extraction, precipitation with cold methanol and ether, and an alkaline buffer wash. The samples were then passed through a silica chromatography cartridge and the eluted vitamins subjected to normal-phase and reverse-phase high performance liquid chromatography. Final quantitation was by competitive protein binding assay (Hollis 1983; Hollis 2003).

Analysis of TAU in plasma and milk samples was done by precipitation of proteins followed by ion exchange chromatography. Quantification was determined colorimetrically using ninhydrin for color development (Hedberg et al. 2007).

33.3 Results

33.3.1 TAU in Plasma and Whole Blood

The concentrations of TAU obtained from polar bears are shown in Fig. 33.1. Plasma TAU in captive bears was lower (99 ± 16 nmol/ml) than in the two wild-caught cubs (237 ± 10 nmol/ml), $p < 0.02$. The whole blood values, which include TAU leakage from white blood cells and platelets, were 453 ± 8 nmol/ml in wild-caught cubs and 253 ± 36 nmol/ml in captive bears ($p < 0.05$). TAU content was measured over the next four years in samples from the now captive cubs, and the values fell to levels comparable to other captive bears (plasma 101.6 ± 17.9 nmol/ml, $n = 5$; whole blood 258.2 + 31.6 nmol/ml, $n = 9$). No appreciable differences in plasma

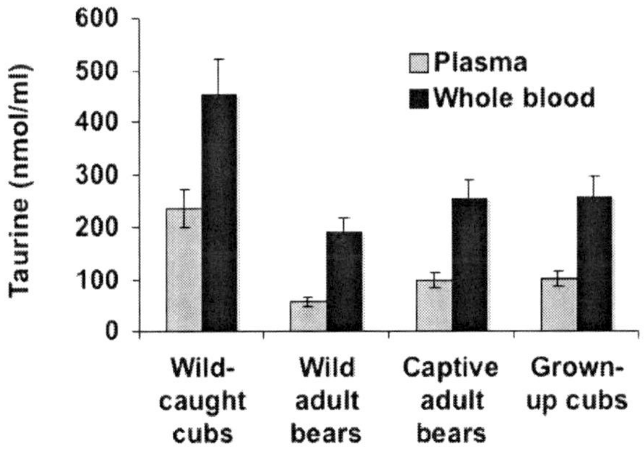

Fig. 33.1 Plasma and whole blood TAU values in free-ranging and captive polar bears

or whole blood TAU content were evident with regard to sex or age of the captive animals (Fig. 33.1).

33.3.2 Free-Ranging Bears

Plasma and whole blood TAU levels were measured in 50 adult free-ranging bears, which consume a diet composed primarily of seals. Plasma TAU was 55.9 ± 39.4 nmol/ml ($n = 50$), and whole blood TAU was 189.5 ± 53.2 nmol/ml ($n = 40$). These values are not different from values in captive bears in United States zoos consuming a prepared diet.

33.3.3 Vitamin D in Milk

It has been suggested that metabolic bone disease in polar bear cubs might be related to reductions in vitamin D in the diet. Preliminary results from selective data of vitamin D_3 in polar bear milk ($n = 10$) from samples at early lactation (3–11 months) months was 18.4 ± 15.8 ng/ml or 738.4 ± 632.5 IU/L (Table. 33.1). This is not different from the 675 to 950 IU/L (17–24 nmol/ml) of vitamin D in both Esbilac and Enfamil formulas ingested by cubs in captivity (Hedberg et al. 2007). Hence, the concentrations of vitamin D, and, presumably, the intake of this fat-soluble vitamin, do not differ in free-ranging and hand-reared captive cubs.

33.3.4 TAU in Milk

The formula liquid Esbilac fed captive bears contains 120 nmol/ml TAU (Hedberg et al. 2007). In contrast, the concentration of TAU in the milk of 16

Table 33.1 Vitamin D and TAU levels in milk and milk-based formulas

Milk source	Vitamin D (IU/L)	Vitamin D (ng/ml)	TAU (nmol/ml)	TAU mg/L
Human	75	1.875	360–400	45–50
Enfamil	400	10	319.5	40
Cow	40	1	< 10	1
	suppl. with 400	suppl. with 10		
Esbilac (liquid)	950	23.75	120	9.6
(solid)	2633	65.83		
Polar bear (with 3–11-month-old cubs, $n = 10$)	738.4	18.46	3166	390
Polar bear (with 3–24-month-old cubs, $n = 16$)	649	16.23	3069	380

free-ranging lactating bears was 3166.4±771 nmol/ml, a 26-fold difference ($p < 0.002$, Table. 33.1). Accordingly, the values of TAU in the gut, a site for the enterohepatic circulation of bile acids and salts, would be remarkably different.

33.4 Discussion

Polar bears in captivity have an increased frequency of antebrachial and other fractures as compared to free-ranging bears (Lin et al. 2005). The affected bears may have subnormal serum 25-hydroxyvitamin D (25(OH)D): indeed, two of three bears with fractures had circulating 25(OH)D levels under 135 nmol/ml. The fractures occurred when bears were pushed or fell into moats, while playing, or slipping on ice. They were usually antebrachial fractures, and the affected limbs were not able to tolerate weight (Lin et al. 2005). In addition, a number of cubs that were hand-reared have developed rickets (Anonymous 1985; Kenny et al. 1999; Engeli et al. 2005; Lin et al. 2005). In Lin et al. (2005), three bears with antebrachial fractures were less than three years old and had metabolic bone disease. In general, MBD has been related to very low 25(OH)D values in the sera of affected, captive bears. Interestingly, the average 25(OH)D level in seven captive polar bears was 108 ± 37 nmol/L versus 360 ± 135 nmol/L in 56 adult free-ranging bears. Some rachitic cubs have extremely low 25(OH)D levels, less than 20 nmol/L (Kenny et al. 1999).

Researchers have tended to focus their attention on the role of vitamin D deficiency in the pathogenesis of MBD and rickets: such an approach seems appropriate for a number of reasons. Polar bear cubs den during the winter, when sunshine exposure is the most limited (Kenny et al. 2004). Moreover, newborn cubs weigh from 0.1 to 0.3% of their adult weight, so there is a huge requirement for vitamin D in order to mineralize rapidly growing bone (Kenny et al. 2004; Kenny 2007). The serum values for 25(OH)D in two rachitic cubs in the Denver Zoo were 15.5 and 18.3 nmol/L (Kenny et al. 2004): clearly deficient and well below the normal value of serum (approximately 350 nmol/L) in either captive or free-ranging adult bears.

It is well established that free-ranging polar bears ingest arctic marine mammals, such as the ringed seal or beluga whale, which have extremely high levels of vitamin D in their blubber (157–747 ng vitamin D/g fat). Seals and whales acquire vitamin D from their own diets of fish, which contain 605 to 2472 ng vitamin D per gram of liver (Kenny et al. 2004). Interestingly, marine mammals that feed on phytoplankton and zooplankton, such as the walrus and bowhead whale, have very little vitamin D in their blubber. The polar bear appears dependent upon its diet as a source of vitamin D. In the food chain, the polar bear eats the ringed seal or beluga whale, which has ingested whitefish, salmon, sculpin, arctic char, and other arctic fish (Kenny et al. 1998).

The photocutaneous synthesis of vitamin D_3 from 7-dehydrocholesterol in the skin of polar bears is limited (Kenny et al. 1998). Their skin is black and covered with thick fur; only the tip of the nose is exposed. Skin biopsies from captive polar bears had only 0.11 ± 0.03 nmol/cm^2 of 7-dehydrocholesterol, as compared to

1.28 nmol/cm^{2}in humans (Kenny et al. 1998). Hence, the polar bear is especially dependent on dietary vitamin D.

The finding of extremely reduced or significantly low values of 25(OH)D in bears with MBD raises support for the role of vitamin D deficiency in this situation. Since previous studies (Kenny et al. 1998) and the present study indicate that polar bear milk contains roughly 10-fold higher values of vitamin D than human milk, the supply of vitamin D in the diet of free-ranging denned cubs seems adequate. The vitamin D content of the formulas used in hand-rearing cubs (17–24 nmol/ml) is similar to that found in polar bear milk (18.4 $\pm$ 15.8 ng/ml), and would seem to be adequate.

The observation in this study that plasma and whole blood TAU levels were high in cubs born in the wild suggested a role for TAU in the absorption of vitamin D. It is well established that TAU is necessary for the conjugation of bile acids, especially ursodeoxycholic acid, a bile acid almost exclusively found in ursids (Hagey et al. 1993; Zamboni et al. 1993; Hofmann 1999; Sacco and Valkenburgh 2004). Moreover, these tauro-conjugates have been shown to be particularly effective in the absorption of fat-soluble vitamins (Hagey et al. 1993; Zamboni et al. 1993; Sacco and Valkenburgh 2004). The high TAU content in the diet of free-ranging bears results in high TAU levels in the milk of lactating sows. This study shows that the TAU content of polar bear milk is approximately 26-fold higher than in the cow milk-based formulas (Esbilac) fed hand-reared cubs. It is also 10-fold higher than in Enfamil formula. We speculate that the greater availability of TAU in the diet of cubs in the wild serves to enhance bile acid conjugation and the absorption of vitamin D and vitamin D_3, which can then be 25-hydroxylated in the liver, resulting in higher levels of circulating 25(OH)D.

Just as domestic cats and zoo felids require particular attention to the TAU content of their diets, (Hedberg et al. 2007), attention to the TAU content of the diets of developing polar bears in captivity is appropriate. The high level of TAU in polar bear milk as compared to cow milk-based formulas makes it logical to supplement the diet of captive bears, especially if they are being hand-reared. A prospective trial seems indicated.

33.5 Conclusion

In conclusion, TAU appears likely to play a role in the pathogenesis of MBD in polar bears reared in captivity. Serum or plasma 25(OH)D has been shown to be extremely low in bears with rickets or fractures. The vitamin D_3 content of polar bear milk and cow milk-based formulas are not different. A major difference, however, is the 10- to 26-fold higher level of TAU in polar bear milk versus cow milk-based formula. This increased dietary TAU content likely results in higher levels of tauro-conjugated bile salts, which enhances absorption of vitamin D_3 and other fat-soluble vitamins. TAU supplementation of diets currently used to hand-rear polar bear cubs seems worthwhile, but will require a prospective study to determine if the fracture rate in captive polar bears is reduced.

References

Anonymous (1985) Nursery notes 2.0 polar bear (*Ursus maritimus*). Zool Soc San Diego:23

Chesney RW (2003). Renal tubular acidosis: rickets associated with renal tubular acidosis. In: Behrman R, Kliegman RM Jenson HB (eds) Nelson Textbook of Pediatrics, W. B. Saunders, Orlando, p 1762

Derocher AE, Andriashek D, Arnould JPY (1993) Aspects of milk composition and lactation in polar bears. Can J Zool 71:561-567

Engeli E, Lin RC, Goodrich LR, Prowten AW (2005) Fractures of the radius and ulna secondary to possible vitamin D deficiency in captive polar bears (*Ursus maritimus*). Polar Bears International. http://www.polarbearsinternational.org/pbhc/fractures.htm

Hagey LR, Crombie DL, Espinosa E, Carey MC, Igimi H, Hofmann AF (1993) Ursodeoxycholic acid in the *Ursidae*: biliary bile acids of bears, pandas, and related carnivores. J Lipid Res 34:1911–1917

Hedberg GE, Chesney RW (2004) A preliminary investigation of taurine in zoo carnivores. Feline Conserv Fed 48, http://www.felineconservation.org

Hedberg GE, Dierenfeld ES, Chesney RW, Rogers QR (in press) Speculations on pathogenesis of metabolic bone disease in captive Polar Bears (*Ursus maritimus*) with links to taurine status In: Clauss M, Fidgett A, Huisman T, Jansen G, Jorgensen M, Nijboer J, (eds). Zoo Animal Nutrition V. Fürth, Germany: Filander Verlag

Hedberg GE, Dierenfeld ES, Rogers QR (2007) Taurine and zoo felids: Considerations of dietary and biological tissue concentrations. Zoo Biol 26(6):517–531

Hedberg GE, Dunker F, Chesney RW (2004) Pathogenesis of metabolic bone disease in captive polar bears (*Ursus maritimus*). International Polar Bear Husbandry Conference. San Diego, CA

Hofmann AF (1999) The continuing importance of bile acids in liver and intestinal disease. Arch Intern Med 159:2647–2658

Hollis BW (1983) Individual quantitation of vitamin D_2, vitamin D3, 25-hydroxyvitamin D_2, and 25-hydroxyvitamin D_3 in human milk. Anal Biochem 131:211–219

Hollis BW (2003) Detection of Vitamin D and Its Major Metabolites. San Diego, Academic Press.

Johnston G, Cutchins J (1985) Andy Bear - A Polar Cub Grows Up at the Zoo. New York, William Morrow & Co., Inc

Kenny DE (2007) 25-(OH)-D_3, calcium, phosphorus, and alkaline phosphatase content in sera from captive polar bear (*Ursus maritimus*) cubs at the Denver Zoological Foundation. Zoo Biol 26:167–173

Kenny DE, Irlbeck NA, Chen TC, Lu Z, Holick MF (1998) Determination of vitamins D, A, and E in sera and vitamin D in milk from captive and free-ranging polar bears (*Ursus maritimus*) and 7-dehydrocholesterol levels in skin from captive polar bears. Zoo Biol 17:285–293

Kenny DE, Irlbeck NA, Eller JL (1999) Rickets in two hand-reared polar bear (*Ursus maritimus*) cubs. J Zoo Wildl Med 30:132–140

Kenny DE, O 'Hara TM, Chen TC, Lu Z, Tian X, Holick MF (2004) Vitamin D content in Alaskan arctic zooplankton, fishes, and marine mammals. Zoo Biol 23:33–43

Lin RC, Engeli E, Prowten AW, Erb HN, Ducharme NG, Goodrich LR (2005) Antebrachial fractures in four captive polar bears (*Ursus maritimus*). Vet Surg 34:358–65

Sacco T, Valkenburgh BV (2004) Ecomorphological indicators of feeding behaviour in the bears (*Carnivora: Ursidae*). J Zool:41–54

Zamboni G, Piemonte G, Bolner A, Antoniazzi F, Dall'Agnola A, Messner H, Gambaro G, Tato L (1993) Influence of dietary taurine on vitamin D absorption. Acta Paediatr 82:811–815

Chapter 34
The Preventive Effect of Fermented Milk Supplement Containing Tomato (Lycopersion Esculentum) and Taurine on Bone Loss in Ovariectomized Rats

Sun Hee Cheong and Kyung Ja Chang

Abstract This study was conducted to examine the preventive effects of fermented milk supplement with tomato and taurine on bone loss in ovariectomized rats. Twenty-seven female Sprague-Dawley rats weighing 200g were ovariectomized and then were randomly assigned to three groups (OVX, ovariectomized and fed a low calcium diet; OVX-C, ovariectomized control; OVX+TS, ovariectomized and fed fermented milk supplement with tomato). Each group was fed a low calcium diet for three weeks. After three weeks, the OVX group was sacrificed and the other groups were fed an experimental diet for 4 weeks. Length, weight and calcium levels of the femur and lumbar in the OVX+TS group were significantly elevated compared to those of the OVX group. Bone mass, breaking force and bone mineral density (BMD) of femur of the OVX+TS group were significantly greater than that of the OVX-C group. Serum alkaline phosphatase activity (ALP) was highest in the OVX-C group. Therefore, our findings suggest that fermented milk supplemented with tomato and taurine might improve bone health in postmenopausal osteoporotic rats.

Abbreviations *OVX*, ovariectomized and fed low calcium diet; *OVX-C*, ovariectomized control; *OVX+TS*, ovariectomized and fed fermented milk supplement with tomato; *BMD*, bone mineral density; *ALP*, alkaline phosphatase

34.1 Introduction

Osteoporosis is a major public health problem characterized by a decrease in bone mineral density and by bone loss. The development of osteoporosis can also be related to inadequate calcium intake, breast feeding, age at menarche, first

S.H. Cheong (✉)
Department of Food and Nutrition, Inha University, Incheon, Korea

J. Azuma et al. (eds.), *Taurine 7*, Advances in Experimental Medicine and Biology 643, DOI 10.1007/978-0-387-75681-3_34,

pregnancy, the number of pregnancies and obesity (Lee et al. 2004; McClung et al. 2004). Ovarian hormone deficiency is a major risk factor for osteoporosis. Numerous antiresorptive agents, such as estrogen, calcitonin, and anabolic steroids, may prevent further bone loss in postmenopausal osteoporosis although they cannot restore bone mass that has already been lost. Moreover, some agents have been reported to have undesirable side effects (Raisz 1993). Thus it is desirable to develop functional food products with fewer undesirable side effects. In the last decade, numerous studies have reported that several amino acids contribute to bone mineralization. In particular, taurine, which is localized in matrices of the bone, can regulate osteoblast metabolism through an antiosteopenic effect and through inhibition of bacteria-stimulated osteoclast formation *in vitro* (D'Eufemia et al. 2007).

Tomatoes and tomato-based products are rich sources of lycopene, an antioxidant carotenoid reported to be more stable and a more potent singlet oxygen quenching agent than other carotenoids (Canene-Adams et al. 2005). These phytochemicals reduce the risk of developing several chronic diseases, such as cardiovascular disease and prostate cancer (Hwang and Bowen 2004; Shi et al. 2005). Recent studies have shown that the dietary antioxidant lycopene reduces oxidative stress and the levels of bone turnover markers in postmenopausal women, and may be beneficial in reducing the risk of osteoporosis (Kim et al. 2003; Rao et al. 2003). However, the role of lycopene and tomatoes in osteoporosis has not been investigated.

Therefore, this study was conducted to investigate the effect of fermented milk supplement with tomato and taurine on the bone of ovariectomy-induced osteoporotic female rats.

34.2 Materials and Methods

34.2.1 Animals and Diet

Twenty-three female Sprague-Dawley rats weighing 200g were supplied from Biolink (Seoul, Korea), and were kept in shoebox cages in a room with controlled temperature (22±2°C), humidity (65±5%), and 12 h light-dark cycle (06:00∼18:00). All rats were fed a commercial chow diet for one week. The animals were ovariectomized and then were randomly divided into three groups (OVX group, ovariectomized and fed low calcium diet and taurine; OVX-C group, ovariectomized and fed standard diet of basic AIN-76 and taurine; OVX+TS group, ovariectomized and then fed fermented milk supplement with tomato and taurine). All of the rats were fed a low calcium diet for three weeks. After three weeks the rats of the OVX group were sacrificed while those of the other groups were fed either the standard diet alone or the standard diet containing fermented milk supplemented with tomato for 4 weeks (Table 34.1). Diets and water were provided *ad libitum.* During the experimental period, body weight was measured once per week and food intake was measured every day.

Table 34.1 Composition of experimental diet (g/kg diet)

	Experimental groups[1] (g/kg diet)		
Ingredients	OVX	OVX-C	OVX + TS
Casein	200	185	185
dl-Methionine	3	3	3
Corn starch	615.87	628.12	628.12
Corn oil	50	50	50
-Cellulose	50	50	50
Choline-chloride	2	2	2
Vitamin mix.	10	10	10
Ca-, P-free mineral mix.	35	35	35
Fe-citrate	0.19	0.19	0.19
Taurine	15	15	15
$CaHPO_4 \cdot 2H_2O$	6.74	6.74	6.74
$CaCO_3$	2.8	2.8	2.8
KH_2PO_4	12.15	12.15	12.15
Fermented milk supplement with tomato[2]			4ml

[1]OVX; ovariectomized and fed low calcium diet and taurine, OVX-C; ovariectomized and then fed basal diet and taurine, OVX+TS; ovariectomized and then fed basal diet + fermented milk supplement with tomato and taurine.[2]In our previous study, we carried out a nutritional survey of elderly, urban Koreans and then developed fermented milk containing tomato to prevent senility. In order to supply the recommended dietary allowance for elderly Koreans, we prepared fermented milk containing tomato. The amount of the fermented milk supplement containing tomato was estimated in the Korean National Health and Nutrition Examination Survey (Ministry of Health and Welfare 2001) to represent an average daily dietary intake for a person weighing 65 ~ 70 kg. The fermented milk supplement containing tomato substituted for casein and corn starch and contained 14.15 g carbohydrate, 4.11 g protein, 0.14 g fat, 1.00 g crude fiber, 2.25 mg cholesterol, 526.50 mg calcium, 309.35 mg phosphorous, 0.70 mg iron, 60.80 mg sodium 203.24 mg potassium, 0.49 mg zinc, 35.45 μg vitamin A, 0.56 μg, retinol, 137.75 μg β-carotene, 0.07 mg vitamin B_1, 0.20 mg vitamin B_2, 0.11 mg vitamin B_6, 2.86 mg vitamin C 200IU vitamin D_3, 5.74 mg vitamin E, 0.55 mg niacin, and 3.79 μg folic acid. The calorie content of the fermented milk supplement containing tomato was 74.83 kcal per 150 ml.

34.2.2 Blood Sampling and Chemical Analysis

After 8 weeks of feeding the experimental diets, the rats were sacrificed. The liver, femur and lumbar of each rat were removed and weighed. Blood samples were obtained through heart puncture and serum was separated at 3000 x g for 20 min. Serum were stored at –70°C until analysis. The calcium and phosphorus concentration of the serum, femur and lumbar were analyzed by atomic absorption spectrophotometer (PERKIN ELMER, Analyst 100, 422.7 nm, 340 nm, respectively). Serum ALP activity and rat osteocalcin concentration were measured using commercial kits (Youngdong Pharmaceutical Co., Korea, ELISA kits, USA, respectively). The weight and length of the femur and lumbar were measured using caliper. The strength of the right femur was measured using Instron (Tensilon/UTM-4-100, TOYO BALDWIN Co. LTD). The ash of the left femur was obtained after being placed in a 600°C furnace for 6 hours, and bone mass was calculated as follows:

Bone mass = ash weight / bone length

The BMD of the femur was measured by dual energy X-ray absorptionmetry (DEXA), (Lunar Radiation corp. Madison, Wisconsin, U.S.A).

34.2.3 Statistical Analysis

Statistical analysis of the data was carried out using the SPSS 12.0 program. All results were expressed as means ± SD and statistical significance between the groups was assessed by using analysis of variance (ANOVA) and Duncan's multiple range test. The level of significance used was $p < 0.05$.

34.3 Results

34.3.1 Body Weight and Food Intake

Body weight and food intake are shown in Table 34.2. The body weight and body weight gain in the OVX+TS group was significantly greater than that observed in the OVX group ($p<0.05$, $p<0.05$, respectively). Food intake in the OVX+TS group was significantly elevated compared to the other groups ($p<0.05$). However, the food efficiency ratio in the OVX-C group was significantly reduced compared to the OVX group ($p<0.05$).

34.3.2 Liver, Kidney and Relative Liver Weight

Liver, kidney weight and relative liver weight are shown in Table 34.3. Liver weight in the OVX-C group was significantly reduced compared to the other groups

Table 34.2 Effect of fermented milk supplement containing tomato on body weight, food intake and food efficiency ratio

	Experimental groups					
Measure	OVX (n =5)		OVX-C (n =9)		OVX + TS (n =9)	
Initial body weight (g)	221.6	±2.8$^{N.S.}$	222.1	±7.4	220.3	±8.9
Final body weight (g)	257.8	±3.4^{a}	276.5	±2.9^{b}	280.9	±12.3^{b}
Body weight gain (g)	36.2	±1.9^{a}	54.4	±3.7^{b}	60.6	±8.2^{b}
Food intake (g/day)	19.9	±1.2^{a}	19.5	±1.7^{a}	21.2	±5.8^{b}
FER (%)	6.21	±1.1^{a}	5.75	±1.0^{b}	6.01	±2.1^{ab}

Each value is the mean ± SD. OVX, ovariectomized and fed low calcium diet; OVX-C, ovariectomized and then fed basal diet; OVX+TS, ovariectomized and then fed basal diet + fermented milk supplement containing tomato. Food efficiency ratio (FER, %) was calculated by the formula: FER (%) = [Total body weight gain (g) / total food intake (g)] × 100. Values containing the same superscripts or N.S. are not significantly different at $p < 0.05$ by Duncan's multiple range test.

Table 34.3 Effect of fermented milk supplement containing tomato on liver and kidney weight

	Experimental groups		
Measure	OVX (n =5)	OVX-C (n =9)	OVX + TS (n =9)
Liver weight/100g BW (g)	7.73 ± 0.68^{a}	6.61 ± 0.58^{b}	7.18 ± 0.33^{a}
Relative liver weight (%)	$2.69 \pm 0.08^{N.S.}$	2.34 ± 0.05	2.56 ± 0.10
Kidney weight/100g BW (g)	1.85 ± 0.09^{ab}	1.70 ± 0.13^{b}	1.92 ± 0.16^{a}

Each value is the mean ± SD. OVX, ovariectomized and fed low calcium diet; OVX-C, ovariectomized and then fed basal diet; OVX+TS, ovariectomized and then fed basal diet + fermented milk supplement containing tomato. Relative liver weight was calculated from the percent of liver weight to body weight. Values with the same superscripts or N.S. are not significantly different at $p < 0.05$ by Duncan's multiple range test.

(p<0.05). For the experimental period, kidney weight in the OVX+TS group was significantly elevated compared to that of the OVX-C group (p<0.05).

34.3.3 The Length and Weight of Femur and Lumbar

The length and weight of femur and lumbar of the group fed the supplement containing tomato and taurine are shown in Table 34.4. Femur and lumbar length in the OVX+TS group were significantly elevated compared to the OVX and OVX-C groups (p<0.05). Femur and lumbar weight and calcium level in OVX+TS group

Table 34.4 Bone length, weight and calcium and phosphorus content of the right femur and fourth-lumbar and serum in rats

	Experimental groups		
Measure	OVX (n =5)	OVX-C (n = 9)	OVX + TS (n = 9)
Length(cm)			
Femur	3.52 ± 0.13^{a}	3.62 ± 0.12^{a}	3.66 ± 0.08^{b}
Lumbar	3.40 ± 0.51^{a}	3.76 ± 0.11^{b}	4.02 ± 0.16^{b}
Weight (g)			
Femur	0.79 ± 0.03^{a}	0.85 ± 0.07^{ab}	0.86 ± 0.03^{b}
Lumbar	1.22 ± 0.15^{a}	1.80 ± 0.30^{b}	1.88 ± 0.01^{b}
Ca content (mg/cm^3)			
Serum	7.74 ± 1.04^{NS}	8.88 ± 0.17	8.89 ± 0.10
Femur	180.39 ± 22.63^{a}	248.84 ± 14.87^{b}	248.58 ± 25.23^{b}
Lumbar	75.69 ± 23.80^{a}	107.42 ± 16.90^{b}	126.26 ± 13.75^{b}
P content (mg/cm^3)			
Serum	15.58 ± 2.08^{NS}	12.30 ± 3.09	8.57 ± 3.39
Femur	66.33 ± 9.68^{NS}	71.07 ± 8.52	63.58 ± 5.98
Lumbar	72.43 ± 3.58^{NS}	71.62 ± 5.39	76.90 ± 2.67

Each value is the mean ± SD. Values with the same superscripts or N.S. are not significantly different at $p < 0.05$ by Duncan's multiple range test. OVX, ovariectomized and fed low calcium diet; OVX-C, ovariectomized and then fed basal diet; OVX+TS, ovariectomized and then fed basal diet + fermented milk supplement containing tomato.

Table 34.5 Breaking force of femur, bone mineral density, bone mass, serum osteocalcin and serum alkaline phosphatase activity (ALP)

	Experimental groups		
Measure	OVX (n =5)	OVX-C (n =9)	OVX + TS (n =9)
Breaking force of femur (N/m^2)	11.19 ± 1.18^a	12.35 ± 1.92^{ab}	13.7 ± 1.35^b
Bone mineral density (g/cm^2)	0.173 ± 0.000^a	0.182 ± 0.002^{ab}	0.187 ± 0.003^b
Bone mass (g/cm)	0.08 ± 0.002^a	0.09 ± 0.001^a	0.13 ± 0.001^b
Serum osteocalcin (μg/ml)	10.4 ± 1.09^{NS}	9.91 ± 1.25	10.1 ± 1.90
Serum ALP (U/l)	36.61 ± 3.65^a	76.61 ± 1.10^c	52.64 ± 6.42^b

Each value is the mean ± SD. Values with the same superscripts or N.S. are not significantly different at $p < 0.05$ by Duncan's multiple range test. OVX, ovariectomized and fed low calcium diet; OVX-C, ovariectomized and then fed basal diet; OVX+TS, ovariectomized and then fed basal diet + fermented milk supplement containing tomato.

were significantly elevated compared to the OVX group (p<0.05, p<0.05, respectively). These results indicate that the treatment with fermented milk containing tomato appeared to prevent bone loss in the femur and lumbar vertebra. This beneficial effect may be due to the presence of natural components and phytochemicals in tomato, such as lycopene, β-carotene, lutein, zeaxanthin, phytoene, phytofluene, ascorbic acid and tomatine (Canene-Adams et al. 2005).

34.3.4 Breaking Force, Bone Mineral Density, Bone Mass, Serum Osteocalcin and ALP Activity

The effect of the fermented milk supplement containing tomato on the breaking force, bone mineral density, bone mass, serum osteocalcin and ALP activity is shown in Table 34.5. The breaking force and BMD of the OVX-TS group was significantly elevated compared to the OVX-C group (p<0.05). Bone mass in the OVX+TS group was significantly elevated compared to the other two groups (p<0.05). ALP activity, which is a marker enzyme of osteoblasts involved in bone mineralization, was highest among in the OVX-C group (p<0.05).

34.4 Discussion

This study was conducted to investigate the effect of a fermented milk supplement containing tomato and taurine on the bone of ovariectomy-induced osteoporotic female rats. We showed that daily supplementation of fermented milk containing tomato (4 ml/kg diet) and taurine may prevent osteoporosis of postmenopausal elderly women. In this study, the length, weight and calcium level of femur and lumbar in the OVX+TS group were significantly elevated compared to those of the OVX and OVX-C groups. These results indicate that fermented milk supplemented with tomato (4 ml/kg diet) appeared to prevent bone loss in the femur and lumbar vertebra. Moreover, the breaking force and BMD of the OVX-TS group was significantly elevated compared to those of the OVX-C group. In general, antioxidants have been shown to inhibit free radical production and bone resorption (Ries et al. 1992). However, Leveille et al. (1997) reported no association between total intake of vitamin E or β-carotene and hip bone density among postmenopausal women. In addition, the bone mass of the OVX+TS group was significantly elevated compared to the other groups. However, the ALP activity was highest in the OVX-C group. Rao et al. (2003) reported that lycopene, the antioxidant carotenoid found in tomatoes, inhibits mineral resorption by interfering with osteoclast formation and the production of ROS *in vivo*. Also, a positive association of lumbar spine bone mass with dietary β-carotene intake was observed in postmenopausal women (Wattanapenpaiboon et al. 2003). A recent study has shown that β-cryptoxanthin affects calcium content and ALP activity in rat femoral-diaphyseal and femoral-metaphyseal tissues *in vitro*. However, another study reported that lycopene had no anabolic effect on bone calcification (Yamaguchi and Uchiyama 2003). Because of this diversity of opinion, further investigation is warranted to clarify the regulation of osteoporosis in animal and human studies.

34.5 Conclusion

In conclusion, our results suggest that daily supplementation of fermented milk containing both tomato (4 ml/kg diet) and taurine may prevent osteoporosis of elderly postmenopausal women. Although the efficacy of the dietary supplement containing tomato and taurine in improving bone health was shown in the postmenopausal model rat, there are limitations to extrapolating these results to postmenopausal osteoporotic women. Therefore, clinical or community trials should be conducted in postmenopausal osteoporotic women maintained long-term on the supplement containing tomato and taurine.

Acknowledgments This study was partially supported by a grant of the Korea Health 21 R and D Project, Ministry of Health and Welfare, Republic of Korea (HMP-00-CH-17-0016). We thank the Dong-A Pharmaceutical Co., which donated taurine and contributed to technical advice.

References

Canene-Adams K, Campbell JK, Zaripheh S, Jeffery EH, Erdman JW (2005) The tomato as a functional food. J Nutr 135:1226–1230

D'Eufemia P, Finocchiaro R, Zambrano A, Tetti M, Ferrucci V, Celli M (2007) Reduction of plasma taurine level in chidren affected by osteogenesis imperfecta during bisphosphonate therapy. Biomed Pharmacother 61:235–240

Hwang ES., Bowen PE (2004) Effects of tomatoes and lycopene on prostate cancer prevention and treatment. J Korean Soc Food Sci Nutr 33:455–462
Kim L, Rao AV, Rao LG (2003) Lycopene II–effect on osteoblasts: the carotenoid lycopene stimulates cell proliferation and alkaline phosphatase activity of SaOS-2 cells. J Med Food 6:79–86
Lee YB, Lee HJ, Kim KS, Lee JY, Nam SY, Cheon SH, Sohn HS (2004) Evaluation of the preventive effect of isoflavone extract on bone loss in ovariectomized rats. Biosci Biotechnol Biochem 68:1040–1045
Leveille SG, LaCroix AZ, Koepsell TD, Beresford S, Belle GV, Buchner DM (1997) Do dietary antioxidants prevent postmenopausal bone loss? Nutrition Research 17:1261–1269
McClung MR, Wasnich RD, Hosking DJ, Christiansen C, Ravn P, Wu M, Mantz AM, Yates J, Ross PD, Santora AC (2004) Prevention of postmenopausal bone loss:six-year results from the early postmenopausal intervention cohort study. J Clin Endocrinol Metab 89:4879–4885
Raisz LG (1993) Bone cell biology:new approaches and unanswered questions. J Bone Miner Res 8:S457–S465
Rao LG, Krishnadev N, Banasikowska K, Rao AV (2003) Lycopene I–effect on osteoclasts: lycopene inhibits basal and parathyroid hormone-stimulated osteoclast formation and mineral resorption mediated by reactive oxygen species in rat bone marrow cultures. J Med Food 6:69–78
Ries WL, Key LL, Rodriguiz RM (1992) Nitroblue tetrazolium reduction and bone resorption by osteoclasts in vitro inhibited by a manganese-based superoxide dismutase mimic. J Bone Min Res 7:931–939
Shi J, MacNaughton L, Kakuda Y, Bettger W, Teung D, Jiang Y (2004) Bioavailability of lycopene from tomato products. J Food Sci Nutr 9:98–106
Wattanapenpaiboon N, Lukito W, Wahlqvist ML, Strauss BJ (2003) Dietary carotenoid intake as a predictor of bone mineral density. Asia Pac J Clin Nutr 12:467–473
Yamaguchi M, Uchiyama S (2003) Effect of carotenoid on calcium content and alkaline phosphatase activity in rat femoral tissues in vitro: the unique anabolic effect of β-cryptoxanthin. Biol Pharm Bull 26:1188–1191

Chapter 35
The Effects of Dietary Taurine Supplementation on Bone Mineral Density in Ovariectomized Rats

Mi-Ja Choi and Nancy M. DiMarco

Abstract This study was performed to evaluate the effect of a diet rich in taurine (2.0 g/100 g) on bone metabolism in ovariectomized (OVX) rats. All rats were fed deionized water during the experimental period. Bone mineral density (BMD) and bone mineral content (BMC) of spine and femur were measured. Serum and urinary calcium and phosphorus content were determined. The levels serum osteocalcin and alkaline phosphatase (ALP) were used to assess bone formation. The rate of bone resorption was measured by the deoxypyridinoline (DPD) crosslink immunoassay and corrected for creatinine. Urinary Ca and P excretion, serum osteocalcin content, and the crosslink value were not significantly different between the Sham groups. The taurine supplemented, Sham group had higher spinal and femur BMC than those of the untreated control group, but the difference was not statistically significant. However, the taurine supplemented, Sham group had significantly higher spine and femur BMC per weight than those of the untreated control group. Within the OVX group, the taurine supplemented group had a lower crosslink value than the casein group. The taurine supplemented, OVX group had higher femur bone mineral content per weight than those of the control, OVX group, but the difference was not statistically significant. A study examining the long-term effect of taurine supplementation in humans is warranted.

Abbreviations *OVX*, ovariectomized; *Sham*, sham operated

35.1 Introduction

Osteoporosis is defined as a disease characterized by loss of bone mass, accompanied by microarchitectural deterioration of bone tissue, which leads to an unacceptable increase in the risk of fracture. Osteoporosis is well recognized as a major public health problem. Osteoporosis and low bone mass are currently estimated to be a major public health threat for almost 44 million US men and women aged 50 and older, representing 55% of the population in that age range (Jeri 2005). More than

---

M.-J. Choi (✉)
Department of Food and Nutrition, Keimyung University, Daegu, Korea

J. Azuma et al. (eds.), *Taurine 7*, Advances in Experimental Medicine and Biology 643,
DOI 10.1007/978-0-387-75681-3_35,

40% of postmenopausal women (Chrischilles et al. 1990) and up to 25% of men (Nguyen 1996)will sustain osteoporotic fractures, which will result in substantial expense (Dolan and Torgerson 1998; Stafford et al. 2004) morbidity (Melton 2003) and mortality (Melton 2003; Jalava et al. 2003). The causes of osteoporosis are multiple (Raisz 1999; Heaney 1993). The facts that the organic matrix in bone is mainly composed of protein and that most of the bone mineral content is calcium suggests that the important nutrients for bone health are protein and calcium (Ilich and Kerstetter 2000). Although a great deal of attention has been given to the importance of calcium intake, much less is known about the effects of other nutrients on bone, although recent reports have supported the importance of potassium, magnesium, vitamin K, and fruit and vegetables (Tucker et al. 1999; Booth et al. 2000). Women are at higher risk than men of developing osteoporosis as a result of naturally lower peak bone mass and rapid bone loss after menopause. Estrogen replacement therapy in postmenopausal women reduces the risk of osteoporosis or coronary heart disease (CHD) in part by modulating serum cholesterol. However, estrogen replacement therapy and cholesterol-lowering pharmacologic agents may be accompanied by side effects. On the other hand, taurine is thought to be quite safe and there is little concern about the side effects of excessive intake of taurine (Furukawa et al. 1991).

Taurine could act either directly or indirectly by enhancing growth factor production (Boujendar et al. 2002). According to Boujendar et al. (2002) taurine supplementation of the maternal diet restored normal serum insulin-like growth factor II (IGF-II) expression in islet cells of fetuses of low protein-fed rats (Boujendar et al. 2002). Interest has been expressed in the relation between skeletal maintenance and age-related decreases in serum insulin-like growth factor 1 (IGF-I) concentrations (Jensen et al. 2002). Recombinant IGF-I increases bone formation activity in postmenopausal women (Ghiron et al. 1995; Ebeling et al. 1993; Grinspoon et al. 1995, suggesting that increasing IGF-I levels may help restore bone mass. IGF-I production is markedly affected by the intake of nutrients (Jensen et al. 2002).

The major hormones that regulate tissue growth and metabolism all have a major influence on skeletal growth and remodeling, including the growth hormone-insulin-like growth factor (GH-IGH) system. The GH-IGH system determines body size and regulates the distribution of body fat, lean body mass, and bone modeling and remodeling after epiphyseal closures (Sjogren et al. 1999). GH can stimulate IGF production not only in the liver but also in other target organs, including bone. The GH-IGF system stimulates both resorption and formation (Yakar et al. 2002). Taurine could act either directly or indirectly by enhancing growth factor production (Boujendar et al. 2002). Human growth hormone (hGH) has been shown in some studies to have anabolic effects on bone (Tangpricha et al. 2006) but the effects of taurine on bone are unknown. Taurine is a sulfur-containing amino acid that is best known for its conjugation with bile acids, but it is also involved in the coordination of nerve function, the stabilization of the cell membrane, detoxification, antioxidant reactions and the modulation of osmotic pressure (Huxtable 1992; Sturman 1993). The present study investigated the relationship between taurine content and bone. Influence of diet on postmenopausal bone loss is not well understood. Most work has focused on calcium and vitamin D or a few isolated nutrients, but little work has focused on taurine. The aim of this study was to determine whether there is an association between taurine supplementation and the indexes of bone health (bone resorption markers and BMD) in the estrogen deficient rat (OVX).

35.2 Methods

35.2.1 Animals

Forty female Sprague-Dawley rats (body weight 200±5 g, 9 weeks old age) were randomly divided into two groups. One group was ovariectomized (OVX) while the other group received a sham operation (SHAM). For a 6 week period, each rat group was further divided into control and taurine supplemented (2.0%) dietary groups.

35.2.2 Diet

The dietary supply of vitamins, minerals, and protein was in accordance with the recommended dietary allowances for rats from the American Institute of Nutrition (AIN-93; Reeves et al. 1993) as shown in Table 35.1. All rats were fed an experimental diet and provided deionized water ad libitum for 6 weeks.

Table 35.1 Composition of experimental diets (g/100 g diet)

Ingredients	Control	Taurine
Casein[1]	20	20
Corn starch[2]	52.9486	52.9486
Sucrose	10	10
Soybean oil[3]	7	7
Cellulose[4]	5	5
Min-mix[5]	3.5	3.5
Vit-mix[6]	1.0	1.0
L-cystine	0.30	0.30
Choline[7]	0.25	0.25
Tert-butyl hydroquinone	0.0014	0.0014
Taurine[8]	–	2.0

[1]Lactic Casein, 30mesh, New Zealand Dairy Board, Willington, N.Z.
[2]Corn Starch, Doosan Co. 234-17 Maam-Ri, Bubal-Eup, Inchon-City, Kyunggi-Do.
[3]Soybean oil, CJ CheilJedang Co. Seoul, Korea.
[4]Cellulose, supplied by SIGMA Chemical Company.
[5]Mineral Mixture (AIN-93G), supplied by U.S. CORNING Laboratory Services Company. TEKLAD TEST DIETS, Madison.
[6]Vitamin mixture (AIN-93), supplied by U.S. CORNING Laboratory Services Company. TEKLAD TEST DIETS, Madison, Wisconsin.
[7]Choline, supplied by SIGMA Chemical Company.
[8]Taurine, Dong-A Pharm. Co. Ltd. 434-4 Moknae-dong, Ansan-City, Kyunggi-Do.

35.2.3 Bone and Bone Markers Determination

Bone mineral density (BMD) and bone mineral content (BMC) of spine and femur were measured using PIXImus (GE Lunar Co, Wisconsin, USA). Serum alkaline phosphatase activity (ALP), osteocalcin and urinary DPD crosslink values were measured as markers of bone formation and resorption. Bone resorption was calculated by measuring urinary excretion of deoxypyridinoline and bone formation by measuring serum osteocalcin.

35.2.4 Statistic Analysis

The statistical significance of differences among the groups was evaluated by two-way ANOVA, using a computer software package (version 9.13, SAS Institute Inc, Cary, NC). Individual comparisons were made by Duncan's multiple range test using the ANOVA. Differences were considered to be significant at $p < 0.05$. Data are expressed as means ± SD.

35.3 Results

35.3.1 Weight Gain and FER

The results of this study indicate that body weight gain was higher in the OVX groups than in the SHAM groups regardless of diet (Table 35.2). Food intake and the food efficiency ratio were not significantly different between the groups (Table 35.3).

Table 35.2 The effect of diet on body weight and weight gain in rats

	Sham		Ovx	
	Control	Taurine	Control	Taurine
Initial weight (g)	204.7 ± 9.4[1,a]	201.5 ± 6.3[a]	211.5 ± 6.7[a]	204.1 ± 7.6[a]
Final weight (g)	276.6 ± 21.9[a]	278.8 ± 16.7[a]	317.9 ± 26.3[b]	329.2 ± 22.6[b]
Weight gain (g)	71.9 ± 13.2[a]	77.30 ± 13.9[a]	106.38 ± 13.2[b]	119.10 ± 11.2[b]

[1]Mean ± SD
Values with different superscripts within a given row are significantly different at $p < 0.05$ by Duncan's multiple range test.

Table 35.3 The effect of diet on food intake and food efficiency ratio (FER) in rats

	Sham		Ovx	
	Control	Taurine	Control	Taurine
Food intake (g)	17.60 ± 1.09[1,a]	18.23 ± 1.09[a]	18.89 ± 2.26[a]	21.09 ± 1.77[a]
FER	0.09 ± 0.01[a]	0.10 ± 0.01[a]	0.13 ± 0.05[a]	0.14 ± 0.01[a]

[1]Mean ±SD
Values with different superscripts within a given row are significantly different at $p < 0.05$ by Duncan's multiple range test.

35.3.2 Serum Ca, P, Alkaline Phosphatase and Osteocalcin

Serum calcium and phosphorus content of animals are presented in Table 35.4. Serum calcium and phosphorus content were unaffected by ovariectomy or taurine supplementation.

Table 35.4 The effect of diet on serum Ca and P in rats

	Sham		Ovx	
	Control	Taurine	Control	Taurine
Ca (mg/dl)	9.14 ± 0.45[1,a]	9.22 ± 0.18[a]	9.88 ± 0.24[a]	9.78 ± 0.90[a]
P (mg/dl)	7.74 ± 0.85[a]	6.68 ± 0.19[a]	7.90 ± 0.68[a]	6.86 ± 1.25[a]

[1]Mean ± SD
Values with different superscripts within a given row are significantly different at $p < 0.05$ by Duncan's multiple range test.

Serum ALP was significantly higher in the OVX groups than in the SHAM groups. The values for the taurine, OVX group was significantly higher than those of the control, OVX group (Table 35.5).

Bone formation was determined by measuring serum osteocalcin. And the content of serum osteocalcin was not significantly different between the groups (Table 35.5).

Table 35.5 The effect of diet on serum alkaline phosphatase (ALP) and osteocalcin content of rats

	Sham		Ovx	
	Control	Taurine	Control	Taurine
ALP (u/l)	169.4 ± 371.3[1,a]	260.0 ± 31.4[bc]	202.3 ± 18.6[ab]	292.8 ± 16.1[c]
Ostcocalcin (ng/ml)	0.31 ± 0.16[a]	0.29 ± 0.11[a]	0.32 ± 0.01[a]	0.38 ± 0.03[a]

[1]Mean ± SD
Values with different superscripts within a given row are significantly different at $p < 0.05$ by Duncan's multiple range test.

Table 35.6 The effect of diet on urine Ca and P in rats

	Sham		Ovx	
	Control	Taurine	Control	Taurine
Ca (mg/day)	0.27 ± 0.20[1,a]	0.31 ± 0.18[a]	0.26 ± 0.13[a]	0.32 ± 0.19[a]
P(mg/day)	10.59 ± 2.53[a]	5.80 ± 2.18[b]	12.13 ± 4.74[a]	8.37 ± 2.49[ab]

[1]Mean ± SD
Values with different superscripts within a given row are significantly different at $p < 0.05$ by Duncan's multiple range test.

Table 35.7 The effect of diet on urine deoxypyridinoline, creatinine and crosslink value of rats

	Sham		Ovx	
	Control	Taurine	Control	Taurine
DPD (nM)	646.5 ± 311.4[1,a]	683.2 ± 381.9[a]	880.3 ± 487.5[a]	993.8 ± 456.7[a]
creatinine (mM)	4.20 ± 1.71[a]	7.60 ± 7.11[a]	4.44 ± 2.09[a]	8.82 ± 4.97[a]
Crosslink value (nM/mM)	150.4 ± 26.8[a]	109.3 ± 36.9[a]	207.8 ± 73.5[b]	121.5 ± 22.7[a]

Mean ± SD
Values with different superscripts within a given row are significantly different at $p < 0.05$ by Duncan's multiple range test.

Table 35.8 The effect of diet on spinal BMD, BMC, BMD per weight and BMC per weight of rats

	Sham		Ovx	
	Control	Taurine	Control	Taurine
SBMD(g/cm^2)	0.151 ± 0.013[1,ab]	0.158 ± 0.018[a]	0.136 ± 0.011[b]	0.136 ± 0.008[b]
SBMD(g/cm^2) /Wt(kg)	0.55 ± 0.07[a]	0.57 ± 0.10[a]	0.44 ± 0.10[b]	0.49 ± 0.02[b]
SBMC(g/cm^2)	0.494 ± 0.066[a]	0.563 ± 0.057[a]	0.440 ± 0.035[b]	0.437 ± 0.046[b]
SBMC(g/cm^2) /Wt(kg)	1.80 ± 0.27[ba]	2.04 ± 0.32[a]	1.52 ± 0.21[b]	1.69 ± 0.18[ab]

[1]Mean ± SD
Values with different superscripts within a given row are significantly different at $p<0.05$ by Duncan's multiple range test.

Table 35.9 The effect of diet on femur BMD, BMC, femur BMD per weight and BMC per weight of rats

	Sham		Ovx	
	Control	Taurine	Control	Taurine
FBMD(g/cm^2)	0.196 ± 0.007[1,a]	0.202 ± 0.011[a]	0.192 ± 0.011[a]	0.195 ± 0.012[a]
FBMD(g/cm^2) /Wt(kg)	0.70 ± 0.06[a]	0.73 ± 0.08[a]	0.60 ± 0.12[b]	0.60 ± 0.03[b]
FBMC(g/cm^2)	0.388 ± 0.018[a]	0.409 ± 0.027[a]	0.380 ± 0.014[a]	0.399 ± 0.035[a]
FBMC(g/cm^2) /Wt(kg)	1.30 ± 0.10[ab]	1.52 ± 0.18[a]	1.22 ± 0.15[b]	1.29 ± 0.05[ab]

[1]Mean ± SD
Values with different superscripts within a given row are significantly different at $p < 0.05$ by Duncan's multiple range test.

35.3.3 Urine Ca, P, Deoxypyridinoline, Creatinine and Crosslink Value

Urinary Ca excretion was not significantly different between the experimental groups. But urinary P excretion was significantly decreased in rats fed taurine. Serum calcium and phosphorus content of the SHAM groups were not different from those of the ovarectomized rats (Table 35.6).

Bone resorption was calculated by measuring urinary excretion of deoxypyridinoline, and bone formation by measuring serum osteocalcin. The crosslink value was increased in the ovariectomy group fed the control diet. However, the crosslink value was significantly decreased in the ovarietomized, taurine group (Table 35.7).

35.3.4 Spine and Femur BMD, BMC, BMD per Weight and BMC per Weight

Spine BMD of the ovariectomized groups was significantly lower than that of the SHAM groups. Spine BMD and BMC divided by body weight appears to have a higher BMD (7.5%) and significantly higher BMC (4.5%) in the taurine group, which indicates that taurine has a positive influence on spine bone mineral density and bone mineral content (Table 35.8).

Femur BMC divided by body weight appears to have a larger BMC in the taurine group in ovx rats (Table 35.9).

35.4 Discussion

The influence of nutrition on bone health remains largely undefined because most studies have focused on calcium intake. The treatment of osteoporosis remains a major challenge, despite an increasing array of therapeutic agents, including bisphosphonates, hormone replacement therapy, and selective estrogen receptor modulators. Despite widespread use, however, these agents all rely on decreasing osteoclastic absorption of bone. The most potent bone-inducing factors are growth factors, such as bone morphogenetic proteins (Edwards et al. 2000).

Taurine is found in bone tissue, but its function is not fully understood. Therefore, in this study using ovariectomized (OVX) rats we examined the effect of taurine on ovarian hormone deficiency-induced bone loss. Within the OVX group, the taurine supplemented subgroup had a lower crosslink value than the casein subgroup. Thus, taurine supplementation seems to decrease bone resorption. The taurine supplemented subgroup had higher spinal BMC and femur BMC than those of the control, Sham group, although the difference was not statistically significant. However, the taurine supplemented group had significantly higher spinal and femur BMC per weight than those of the control, OVX group. Moreover, spinal BMD (7.5%) and BMC (4.5%) divided by body weight appears to be greater in the taurine group, indicating that taurine has a positive influence on spinal bone mineral density and bone mineral content. Femur BMC divided by body weight appears to have a larger BMC (5.7%) in the taurine group, indicating that taurine has a positive influence on femur bone mineral content.

The present study suggests that BMD may improve upon taurine supplementation of postmenopausal women, as long as they met the currently recommended intake of calcium and vitamin D. Further research is needed to determine whether a similar association exists in OVX rats consuming a less nutrious diet, such as less calcium intake. Although there is little information on the influence of dietary intake on bone metabolism markers, several theories may help to explain our findings.

35.5 Conclusion

Within the OVX group, the taurine supplemented rats tended to have higher femur bone mineral content per weight than those of control rats, although the difference was not statistically significant. Clearly, a study on the long-term effect of taurine supplement in humans is warranted, focusing in part on the effect of taurine on the characteristics of bone.

Acknowledgments We thank the Dong-A Pharmaceutical Co., which donated taurine.

References

Booth SL, Tucker KL, Chen H, Hannan MT, Gagnon DG, Cupples LA, Wilson PWF, Ordovas J, Schaefer EJ, Dawson-Hughes B, Kiel DP (2000) Dietary vitamin K intakes are associated with hip fracture but not with bone mineral density in elderly men and women. Am J Clin Nutr 71:1201–1208

Boujendar S, Reusens B, Merezak S, Ahn M T, Arany E, Hill D, Remacle C (2002) Taurine supplementation to a low protein diet during foetal and early postnatal life restores a normal proliferation and apoptosis of rat pancreatic islets. Diabetologia 45:856–866

Chrischilles EA, Butler CD, Davis CS, Wallace RB (1990) A model of lifetime steoporosis impact. Arch Intern Med 151:2026–2032

Dolan P, Torgerson DJ (1998) The cost of treating osteoporotic fractures in the United Kingdom female population. Osteoporos Int 8:611–617

Ebeling PR, Jones JD, O'Fallon WM, Janes CH, Riggs BL. (1993) Short-term effects of recombinant human insulin-like growth factor I on bone turnover in normal women.J Clin Endocrinol Metab 77:1384–1387

Edwards CJ, Hart DJ, Spector TD (2000) Oral ststins and increased bone-mineral density in postmenopausal women. Lancet 355; 2218–2219

Furukawa S, Katto M, Kouyama H, Nishida I, Kikumori M, Taniguchi Y, Toda T., Araki H (1991) Repeated dose toxicity study of intravenous treatment with taurine for 13 weeks and recovery test for 5 weeks in rats. Jap Phar Thera 19:275–306

Ghiron L, Thompson JL, Holloway L (1995) Effects of recombinant insulin-like growth factor-I and growth hormone on bone turnover in elderly women. J BoneMiner Res 10:1844–1852

Grinspoon SK, Baum HBA, Peterson S, Klibanski A. (1995) Effects of rhIGF-I administration on bone turnover during short-term fasting. J Clin Invest 96:900–906

Heaney RP (1993) Nutritional factors in osteoporosis. Annu Rev Nutr 13:287–316
Huxtable RJ (1992) Physiological actions of taurine. Physiol. Rev 72:101–163
Ilich JZ, Kerstetter JE (2000) Nutrition in bone health revisited: a story beyond calcium.J Am Coll Nutr 19:715–737
Jalava T, Sarna S, Pylkkanen L, Mawer B, Kamichael-Davies JA, Adams J, Francis RM, Robinson J, Euge (2003) Association between vertebral fracture and increased mortality in osteoporotic patients. J Bone Miner Res 18:1254–1260
Jensen C, Holloway L, Block G, Spiller G, Gildengorin G, Gunderson E, Butterfield G, Marcus R (2002) Long-term effects of nutrient intervention on markers of bone remodeling and calciotropic hormones in late-postmenopausal women. Am J Clin Nutr 75:1114–1120
Jeri WN (2005) Supplement: women and micronutrients:Addressing the gap throughout the life cycle: Osteoporosis: the role of micronutrients. Am. J. Clinical Nutrition 81:1232S–1239S
Melton LJ III (2003) Adverse outcomes of osteoporotic fractures in the general population. J Bone Miner Res 18:1139–1141
Nguyen TV, Eisman JA, Kelly PJ, Sambrook PN (1996) Risk factors for osteoporoticfractures in elderly men. Am J Epidemiol 144:255 263
Raisz LJ (1999) Osteoporosis: current approaches and future prospects in diagnosis, pathogenesis and management. J Bone Miner Res 17:79–89
Reeves PG, Nielsen FH, Fahey GC (1993) AIN-93 purified diets for laboratory rodents: final report of the American Institute of Nutrition Ad Hoc Writing Committee on the Reformulation of the AIN 76 A Rodent Diet. J Nutr 123:1939–1951
Sjogren K, Liu JL, Blad K (1999) Liver-derived insulin-like growth factor I (IGH-I) is the principal source of IGF-Iin blood but is not required for postnatal body growth in mice. Proc Natl Acad Sci USA 96:7088–7092
Stafford RS, Brieling RL, Hersh AL (2004) National trends in osteoporosis visits and osteoporosis treatment . Arch Intern Med 164:1525–1530
Sturman, G. A (1993) Taurine in development. Physiol. Rev 73:119–147
Tangpricha M, Luo C, Fernandez-Estivariz LH, Gu N, Bazargan JM, Klapproth SV, Sitaraman JR, Galloway LM, Leader T R, Ziegler (2006) Growth hormone favorably affects bone turnover and bone mineral density in patients with short bowel syndrome undergoing intestinal rehabilitation. JPEN 30:480–486
Tucker KL, Hannan MT, Chen H, Cupples LA, Wilson PWF, Kiel DP (1999) Potassium, magnesium, and fruit and vegetable intakes are associated with greater bone mineral density in elderly men and women. Am J Clin Nutr 69:727–736
Yakar S, Rosen CJ, Beamer WG (2002) Circulating levels of IGF-1 directly regulatebone growth and density. J ClinInvest 110:771–781

Part VI
Effect of Taurine on Diabetes and Obesity

Chapter 36
Taurine Supplementation and Pancreatic Remodeling

Abdeslem El Idrissi, Latifa Boukarrou, and William L'Amoreaux

Abstract Taurine is a semi-essential sulphur containing amino acid derived from methionine and cysteine metabolism. Taurine has several biological processes such as hypoglycemic action, antioxidation, and detoxification. In this study we evaluated the role of taurine in pancreatic islets development, since the endocrine pancreas undergoes significant modifications during neonatal life. Histological examination of the pancreas from taurine-fed mice revealed no histological abnormalities in the endocrine or exocrine parts of the pancreas. However, supplementation of taurine in the drinking water resulted in a drastic and significant increase in the number of islets per section. Furthermore, islets size was significantly larger. We hypothesize that supplementation of taurine, which is important for the development of the endocrine pancreas may reduce cytokine-induced apoptosis in pancreatic beta cells.

The endocrine pancreas undergoes significant modifications during neonatal life and apoptosis is an important mechanism in this remodeling. We suggest that alteration of this remodeling process during this period of time, when a fine balance between cell replication and cell death is critical, would affect the development of the pancreatic islets of Langerhans, and could have important effects on the pancreatic cell mass and the endocrine function.

Abbreviations *PCNA*, proliferating cell nuclear antigen; *Tau*, taurine; *GAD*, glutamic acid decarboxylase; *iNOS*, inducible nitric oxide synthase; *IGF*, insulin-like growth factor; *WT*, wild type controls

36.1 Introduction

Taurine (2-aminoethanesulfonic acid) is a sulfur-containing amino acid. It is one of the most abundant free amino acids in many excitable tissues, including the brain, skeletal and cardiac muscles. Physiological actions of taurine are widespread and include bile acid conjugation, detoxification, membrane stabilization, osmoregulation, neurotransmission, and modulation of cellular calcium levels (Lambardini

A. El Idrissi (✉)
Department of Biology, City University of New York Graduate School, NewYork

J. Azuma et al. (eds.), *Taurine 7*, Advances in Experimental Medicine and Biology 643,
DOI 10.1007/978-0-387-75681-3_36,

1985; Solis et al. 1988; Foos and Wu 2002; Saransaari and Oja 2000; Schaffer et al. 2000). Furthermore, taurine plays an important role in modulating glutamate and GABA neurotransmission (Militante and Lombardini 1998; El Idrissi and Trenkner 1999, 2004). We have previously shown that taurine prevents excitotoxicity in vitro primarily through modulation of intracellular calcium homeostasis (El Idrissi and Trenkner 1999). In neurons, calcium plays a key role in mediating glutamate excitotoxicity. Taurine is added to milk formula and in solution for parenteral nutrition of premature babies to prevent retinal degeneration and cholestasis (Huxtable 1992; Lourenco, Camilo 2002). More recently, it has been shown that gestational taurine is able to prevent pancreatic alterations induced by gestational malnutrition especially low-protein diet (Dahri et al. 1991; Cherif et al. 1996; Merezak et al. 2001; Boujendar et al. 2002). In addition, taurine administration during gestation delays the mean onset time of diabetes in NOD mice (Arany et al. 2004); whereas taurine supplementation on dams fed with normal diet produces weak glucose intolerance, and increases islet sensitivity to cytokines in offspring (Merezak et al. 2001). Moreover, taurine plays a role in glucose metabolism in adults (Hansen 2001; Franconi et al. 2006).

As a potent anti-oxidant, taurine has been shown to have a protective effect on the pancreas by preventing or scavenging free radicals38. Previous reports propose the islets from taurine treated mice had almost double the number of cells positive for proliferating cell nuclear antigen (PCNA). This increase proliferation was accompanied by a reduction in the incidence of apoptosis in islet cells, and also a significant increase in the number of islet cells immunopositive for IGF-II (Arany et al. 2004). Peak of islet cell apoptosis is maximal in the rat pancreas 14 days after birth and is temporally associated with a fall in the islet cell expression of IGF-II (Petrik et al. 1998). IGF-II was shown to function as an islet survival factor in vitro. The induction of islet cell apoptosis in vivo may involve an increased expression of inducible nitric oxide synthase (iNOS) within β cells. Interestingly, taurine as been shown to be a potent inhibitor of iNOS (Liu et al. 1998). Similarly, Scaglia et al. (1997) have shown decreased replication and increased incidence of apoptosis in the β cells in the presence of IGF-II. These data show that the endocrine pancreas undergoes significant modification during neonatal life and that apoptosis is an important mechanism in this remodeling. Disregulation of this remodeling process during this period of time when a fine balance between cell replication and cell death determines the development of the islets of Langerhans in the pancreas could have significant impact on pancreatic cell mass and endocrine function. Taurine may improve pancreatic health by decreasing synthesis of nitric oxide (Nakaya et al. 2000) through inhibition of iNOS (Boujendar et al. 2002).

36.2 Methods

36.2.1 Quantification of Size and Number of Pancreatic Islets

Two months old mice were perfused with 4% paraformaldehyde and pancreas were isolated attached the pyloric region of the stomach and the duodenum. The initial part of the duodenum served to orient the pancreas for the sectional plane. Tissue was cryoprotected with 30% sucrose and cryosectioned at a thickness of 15 μm. Sections were stained with hematoxylin and eosin. Contiguous sections were stained with propidium iodide to visualize condensed chromatin as an indication of apoptotic cell death. microscopy was performed by histologist unaware of the treatment conditions.

36.2.2 Statistic Analysis

Statistical significance was determined by Student's t-test. Each value was expressed as the mean ± SEM. Differences were considered statistically significant when the calculated P value was less than 0.05.

36.3 Results

36.3.1 Taurine Supplementation Increases the Size and Number of the Islets of Langerhans

In this study we examined the effects of taurine supplementation on pancreatic remodeling. Mice were supplemented with taurine (0.05%) in drinking water at the age of four weeks and remained on this diet for an additional for weeks. When mice were two months old, pancreas were removed and processed for histology. Pancreata were dissected attached to the pyloric region of the stomach and the duodenum. The initial part of the duodenum served to orient the pancreas for the sectional plane. The pancreas of mice is not encapsulated in connective tissue, and it was difficult to isolate the pancreas without the surrounding adipose tissue of the peritoneum.

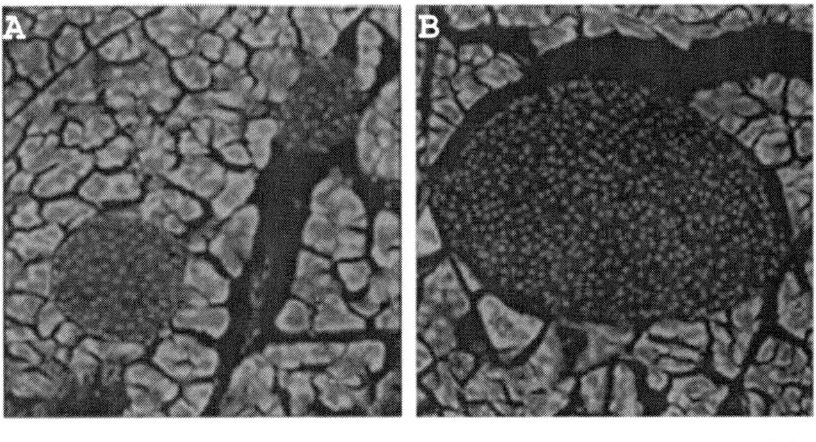

Fig. 36.1 Sections of the pancreas stained with propidium iodide showing the serous acini and islets of Langerhans. **A**. Representative islet from control pancreas. **B**. this is the relative size of an islet from a taurine-fed mouse

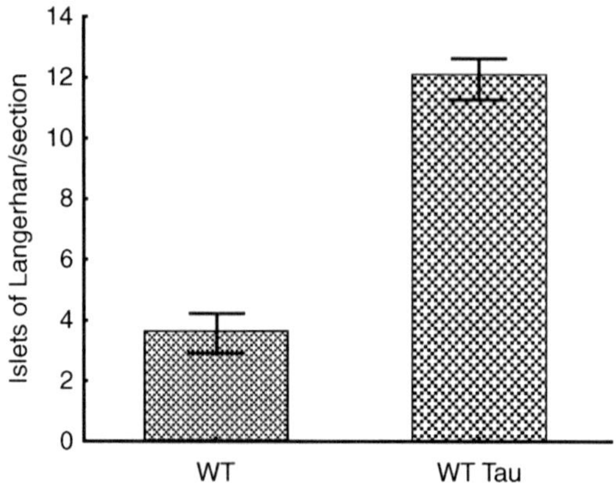

Fig. 36.2 Taurine induces an increase in the number of islets in the pancreas. All mice were 2 months old. Taurine (0.05%) was supplemented in the drinking water for 4 weeks. control, *n*=4; Tau, *n*=5. Pancreata were crysectioned (15 um) and stained with hematoxylin and Eosin. Each pancreas yielded approximately 150 sections. All pancreata were cut in the longitudinal plane. Supplementation of taurine to mice caused a significant increase in the size and number of islets ($p< 0.001$)

Each pancreas yielded approximately 150 sections. Quantification of the number of islets was based on determining the number of islets per section. Histological examination of pancreas from taurine-fed mice revealed a significant increase in their size (Fig. 36.1). The pancreas from these mice did not seem to be enlarged and the exocrine serous acini were of normal histology when compared to the control pancreas. Interestingly however, the number of islets was significantly increased (Fig. 36.2). On average, a pancreas from control mice yielded approximately 4 islets per section, whereas a pancreas of taurine-fed mice contained more than 10 islets. The number of islets per section was not uniform throughout the pancreas, but differences in islets size and number between controls and taurine-fed mice was maintained proportionally throughout the different regions of the pancreas.

36.4 Discussion

Histological examination of the pancreas revealed that taurine-fed mice had a significant increase in the size of islets of Langerhans when compared to controls. The overall size of the pancreas was not affected. There were no histological abnormalities in the endocrine or exocrine parts of the pancreas. Surprisingly, supplementation of taurine in the drinking water resulted in a drastic and significant increase in the number of islets per section. Previously, it has been reported that the islets from taurine treated mice had almost double the number of cells immunopositive for proliferating cell nuclear antigen (PCNA). This increase proliferation was accompanied by a reduction in the incidence of apoptosis in islet cells, and also a significant increase in the number of islet cells immunopositive for IGF-II (Arany et al. 2004). We have supplemented taurine in the drinking water shortly after weaning of mice. It has been shown that a peak of islet cell apoptosis is maximal in the pancreas 14 days after birth and is temporally associated with a fall in the islet cell expression of IGF-II (Petrik et al. 1998). IGF-II was shown to function as an islet survival factor *in vitro*. The induction of islet cell apoptosis *in vivo* may involve an increased expression of inducible nitric oxide synthase (iNOS) within ß cells. Interestingly, taurine as been shown to be a potent inhibitor of iNOS (Liu et al. 1998). Similarly, Scaglia et al. (1997) have shown decreased replication and increased incidence of apoptosis in the ß cells in the presence of IGF-II. These data show that the endocrine pancreas undergoes significant modification during neonatal life and that apoptosis is an important mechanism in this remodeling. Dysregulation of this remodeling process during this period of time when a fine balance between cell replication and cell death determines the development of the islets of Langerhans in the pancreas and could have important effects on the pancreatic cell mass and the endocrine function.

36.5 Conclusion

In summary, this study shows that supplementation of taurine in the drinking water resulted in a significant increase in the size and number of the islets of Langerhans. These histological effects of taurine on the pancreas are consistent with the hypoglycemic effects of taurine and may have implication in diabetes.

Acknowledgments We thank Ekaterina Zavyalova, George Malliaros, Candice Cruz and Labentina Shala for helping with the histological evaluation and staining of tissue. This work was supported by PSC-CUNY and CSI.

References

Arany E, Strutt B, Romanus P, Remacle C, Reusens B, Hill DJ (2004) Taurine supplement in early life altered islet morphology, decreased insulitis and delayed the onset of diabetes in non-obese diabetic mice. Diabetologia 47:1831–1837

Boujendar S, Reusens B, Merezak S, Ahn MT, Arany E, Hill D, Remacle C (2002) Taurine supplementation to a low protein diet during foetal and early postnatal life restores a normal proliferation and apoptosis of rat pancreatic islets. Diabetologia 45:856–866

Cherif H, Reusens B, Dahri S, Remacle C, Hoet JJ (1996) Stimulatory effects of taurine on insulin secretion by fetal rat islets cultured in vitro. J Endocrinol 151:501–506

Dahri S, Snoeck A, Reusens-Billen B, Remacle C, Hoet JJ (1991) Islet function in offspring of mothers on low protein diet during gestation. Diabetes 40:115–120

El Idrissi A, Trenkner E (1999) Growth factors and taurine protect against excitotoxicity by stabilizing calcium homeostasis and energy metabolism. J Neurosci 19:9459–9468

El Idrissi A, Trenkner E (2004) Taurine as a modulator of excitatory and inhibitory neurotransmission. Neurochem Res 1:189–197

Foos T, Wu JY (2002) The role of Taurine in the central nervous system and the modulation of intracellular calcium homeostasis. Neurochem Res 27:21–26

Franconi F, Loizzo A, Ghirlanda G, Seghieri G (2006) Taurine supplementation and diabetes mellitus. Curr Opin Clin Nutr Metab Care 9:32–36

Hansen SH (2001) The role of taurine in diabetes and the development of diabetic complications, Diabet Metab Res Rev 17:330–346

Huxtable RJ (1992) Physiological actions of taurine. Physiol Rev 72:101–163

Lambardini JB (1985) Effects of taurine on calcium ion uptake and protein phosphorylation in rat retinal membrane preparations. J Neurochem 45:268–275

Liu Y, Tonna-DeMasi M, Park E, Schuller-Levis G, Quinn MR (1998) Taurine chloramine inhibits production of nitric oxide and prostaglandin E2 in activated C6 glioma cells by suppressing inducible nitric oxide synthase and cyclooxygenase-2 expression. Brain Res Mol Brain Res 59:189–195

Lourenco R, Camilo ME (2002) Taurine: a conditionally essential amino acid in humans? An overview in health and disease. Nutr Hosp 17:262–270

Merezak S, Hardikar AA, Yajnik CS, Remacle C, Reusens B (2001) Intrauterine low protein diet increases fetal beta-cell sensitivity to NO and IL-1 beta: the protective role of taurine. J Endocrinol 171:299–308

Militante JD, Lombardini JB (1998) Pharmacological characterization of the effects of taurine on calcium uptake in the rat retina. Amino Acids 15:99–108

Nakaya Y, Minami A, Harada N, Sakamoto S, Niwa Y, Ohnaka M (2000) Taurine improves insulin sensitivity in the Otsuka Long-Evans Tokushima Fatty rat, a model of spontaneous type 2 diabetes. Am J Clin Nutr 71:54–58

Petrik J, Arany E, McDonald TJ, Hill DJ (1998) Apoptosis in the pancreatic islet cells of the neonatal rat is associated with a reduced expression of insulin-like growth factor II that may act as a survival factor. Endocrinology 139:2994–3004

Scaglia L, Cahill CJ, Finegood DT, Bonner-Weir S (1997) Apoptosis participates in the remodeling of the endocrine pancreas in the neonatal rat. Endocrinology 138:1736–1741

Saransaari P, Oja SS (2000) Taurine and neuronal cell damage. Amino Acids 19:509–526

Schaffer S, Takahashi K, Azuma J (2000) Role of osmoregulation in the actions of taurine. Amino Acids 19:527–546

Solis JM, Herranz AS, Erreras O, Lerma J, Martin del Rio R (1988) Does taurine act as an osmoregulatory substance in the rat brain. Neurosci Lett 91:53–58

Chapter 37
The Effects of Taurine, Taurine Homologs and Hypotaurine on Cell and Membrane Antioxidative System Alterations Caused by Type 2 Diabetes in Rat Erythrocytes

Davekanand Gossai and Cesar A. Lau-Cam

Abstract This study compared taurine, aminomethanesulfonic acid, homotaurine and hypotaurine for the ability to modify indices of oxidative stress and membrane damage associated with type 2 diabetes. In the study, male Goto-Kakizaki and Wistar-Kyoto rats were allowed free access to a high fat and normal diet, respectively, for 9 weeks. At the end of week 8, half of the animals in each group received a daily intraperitoneal dose of a sulfur compound (0.612 M/kg) for 5 days and, 24 hr after the last treatment, blood samples were withdrawn by cardiac puncture to obtain plasma and erythrocyte fractions for biochemical analyses. Relative to control values, taurine and its congeners reduced membrane damage, the formation of intracellular malondialdehyde and oxidized glutathione, and the decreases in reduced glutathione and antioxidative enzyme activities in diabetic erythrocytes. Except for a few isolated instances, all test compounds were equiprotective.

Abbreviations *GK*, Goto Kakizaki rats; *WKY*, Wistar-Kyoto rats; *RBC*, erythrocyte; *TAU*, taurine; *AMSA*, aminomethanesulfonic acid; *HTAU*, homotaurine; *HYTAU*, hypotaurine

37.1 Introduction

Chronic hyperglycemia and oxidative stress are key players in the onset and progression of diabetes mellitus (Kaneto et al. 2004; Maiese et al. 2007). In diabetes, oxidative stress is manifested by the overproduction of reactive oxygen and nitrogen species (ROS and RNS), an increase in reactive carbonyl compounds, the lowering of antioxidant defenses and altered metabolic rates (Baynes and

D. Gossai (✉)
Department of Pharmaceutical Sciences, College of Pharmacy and Allied Health Professions, St. John's University, Jamaica, NY, USA

J. Azuma et al. (eds.), *Taurine 7*, Advances in Experimental Medicine and Biology 643,
DOI 10.1007/978-0-387-75681-3_37,

Thorpe 1999). Potential sources of ROS in diabetes are the autoxidation of sugars, glycated proteins and glycated lipids, xanthine oxidase, NAD(P)H oxidase, and the mitochondrial respiratory chain (Baynes and Thorpe 1999). The accumulation of advanced glycation end products (AGEs) and lipid peroxidation (LPO) products may jointly contribute to the pathogenesis and development of complications of diabetes (Ceriello 2003; Gallou et al. 1994).

On the premise that a balance between oxidant and antioxidant species will minimize ROS production and, hence, prevent complications of diabetes, numerous compounds with antioxidant properties have been tested in diabetic humans and experimental animals for their ability to control hyperglycemia, preserve β-cell morphology and function, improve insulin sensitivity, and maintain the patency of intracellular antioxidant defenses (Ceriello 2003; Da Ros et al. 2004; Evans 2007).

Among natural compounds, taurine (TAU) is among those extensively investigated for effects on diabetes-related biochemical, morphological and functional alterations, including insulin secretion (Kulakowski and Maturo 1984), blood glucose (Kulakowski and Maturo 1984; Tas et al. 2007), blood lipids (Goodman and Shihabi 1990; Tas et al. 2007), formation of lipid peroxidation (LPO) products (Goodman and Shihabi 1990; Tas et al. 2007; Trachtman et al. 1995) and advanced protein glycation products (Trachtman et al. 1995) in spontaneous and pharmacologically-induced diabetes.

The main purpose of the present study was to compare compounds representing an analog (hypotaurine) or homolog (aminomethanesulfonic acid, homotaurine) with the parent molecule in terms of protective effects against membrane damage and changes in enzymatic and nonenzymatic defenses in erythrocytes (RBCs) from type 2 diabetic rats.

37.2 Materials and Methods

37.2.1 Animals

Male Goto-Kakizaki (GK) rats 200–250 g, were obtained from Taconic Farms Inc., Germantown, NY. For 9 weeks, the animals had free access to a high fat pelleted diet (Purina Rat Chow 13004, Ralston Purina Co., St. Louis, MO) and tap water. Male Wistar-Kyoto (WKY) rats, similar in weight to the GK rats and serving as the control group, were maintained for 9 weeks on a standard pelleted diet (Purina Rat Chow 5008, Ralston Purina Co., St. Louis, MO) and tap water.

37.2.2 Treatment Solutions and Treatments

The solutions of TAU, aminomethanesulfonic acid (AMSA), hypotaurine (HYTAU) (from Sigma Chemical Company, St. Louis, MO, USA) and homotaurine (HMTAU) (from Acros Organics, Pittsburgh, PA, USA) were prepared in distilled water. All treatments were performed by the intraperitoneal (IP) route, once daily, at a dose of 0.612 M (200–300 mg)/kg/2 mL, starting with week 8, for 5 consecutive days.

37.2.3 Blood Samples and RBC Suspension

One day after the last treatment with a sulfur compound, blood samples were withdrawn by cardiac puncture into heparinized test tubes, and centrifuged at 700 $\times$ g and 4°C for 10 min to separate the plasma from the erythrocytes (RBCs). The plasma sample was used for the assay of extracellular hemoglobin (Hb) and lactate dehydrogenase (LDH). The RBC pellet was put through two cycles of washings with ice-cold isotonic phosphate buffered saline, pH 7.4, (PBS) solution followed by brief centrifugation at 700 $\times$ g and 4°C, and suspended in PBS supplemented with 5 mM of glucose (PBSG) to a hematocrit of 20%. This suspension was used for the assay of malondialdehyde (MDA), Hb, LDH, reduced glutathione (GSH), oxidized glutathione (GSSG), catalase (CAT), glutathione peroxidase (GPX), and superoxide dismutase (SOD).

37.2.4 Assays in Plasma and RBCs

The Hb released to the plasma *in vivo* or to the extracellular medium after exposing a RBC suspension to 30% H_2O_2 (in a 1:4 ratio) was measured with a commercial assay kit (Procedure No. 525-A, Sigma Chemical Co.), and its concentration was reported as mg/dL. The activity of LDH released to the plasma *in vivo*, or to the extracellular medium after exposing a RBC suspension to water (in 1:4 ratio), was measured using a commercial assay kit (Procedure No. 228 UV, Sigma Chemical Co.), and its activity was expressed in units per liter (U/L).

37.2.5 Assays in RBCs

The concentration of MDA was measured as TBA reactive substances (TBARS) by the end-point assay method of Buege and Aust (1978) and expressed as nM/g Hb. The intracellular concentrations of GSH and GSSG were measured fluorometrically by the method of Hissin and Hilf (1976), and expressed as μM/g Hb. The CAT activity was measured as described by Aebi (1984) and calculated as U/g Hb. The GPX activity was measured by the spectrophotometric method of Flohé and Günzler (1984) and reported as μM NADPH/min/g Hb. The CuZn SOD activity was measured spectrophotometrically by the kinetic method of Misra and Fridovich (1972), and was reported as U/g Hb.

37.2.6 Statistical Analysis of the Data

The results are reported as the mean ± SEM for n = 3. Intergroup comparisons were performed by Student's t-test, followed by one-way analysis of variance (ANOVA), and Tukey's *post hoc* test. Values were considered to be significantly different from each other at $p < 0.05$.

37.3 Results

The results of the assays for nonenzymatic and enzymatic parameters of oxidative stress along with relevant ratios are presented in Tables 37.1, 37.2 and 37.3.

The level of plasma Hb in diabetic rats was two-fold higher than that of control rats ($p < 0.001$). Without exceptions, a treatment with a sulfur compound resulted in a dramatic decrease in plasma Hb (by 51–53%, $p < 0.001$) relative to untreated diabetic rats. None of the test compounds exhibited intrinsic hemolytic activity. Whereas diabetes increased the ratio of plasma Hb to RBC Hb to a significant extent (2.02, $p < 0.001$) relative to a control value (0.49), a treatment with a sulfur compound abolished the effect (ratio from 0.45 to 0.54) (Table 37.1). RBCs from diabetic rats contained less Hb than RBCs from normal rats, but after a treatment with a sulfur compound they released quantities of Hb that were just above (by 1–4%) the control values, and from 108 to 114% higher than those of RBCs from untreated diabetic rats ($p < 0.001$) (Table 37.1).

RBCs from diabetic rats showed about 32% less LDH activity than RBCs from nondiabetic rats ($p < 0.05$) (Table 37.1). However, following a treatment with a

Table 37.1 Effects of TAU and structurally-related compounds on plasma and RBC values of LDH and Hb from type 2 diabetic rats

Treatment	P-LDH U/L	R-LDH U/L	P-Hb mg/dL	R-Hb mg/dL	P/R-LDH	P/R-Hb
Normal	54.02± 2.56	135.05± 4.60	610.0± 28.0	1240.0± 25.0	0.40± 0.05	0.59± 0.08
Diabetic	95.42± 2.69***	91.36± 2.65***	1218.0± 20.0$^{*,+++}$	602.5± 32.5***	1.04± 0.04***	2.02± 0.09***
+ AMSA	57.69± 3.58^{+++}	130.26± 4.02^{+++}	576.0± 34.5^{+++}	1292.0± 34.5$^{*,+++}$	0.44± 0.01^{+++}	0.45± 0.07$^{*,+++}$
+ HMTAU	50.85± 2.98^{+++}	137.26± 4.33^{+++}	570.0± 24.0^{+++}	1284.5± 20.5$^{*,+++}$	0.37± 0.05^{+++}	0.44± 0.08$^{*,+++}$
+ HYTAU	50.68± 4.01^{+++}	136.21± 2.91^{+++}	597.0± 16.0^{+++}	1281.0± 28.5$^{*,+++}$	0.37± 0.06^{+++}	0.47± 0.09$^{*,+++}$
+ TAU	54.95± 2.30^{+++}	134.98± 3.67^{++}	630.0± 28.5$^{*,+++}$	1172.5± 36.5$^{*,+++}$	0.41± 0.08^{+++}	0.54± 0.04^{+++}

[a]Values represent the mean ± SEM for $n = 3$.
[b]Statistical comparisons were vs. normal rats at $^{*}p < 0.05$, $^{**}p < 0.01$ and $^{***}p < 0.001$; and vs. diabetic rats at $^{+}p < 0.05$, $^{++}p < 0.01$ and $^{+++} < 0.001$.
[c]P-LDH, Plasma LDH; R-LDH, RBC LDH; P-Hb, Plasma Hb; R-Hb, RBC Hb; P/R-LDH, Plasma LDH/RBC LDH ratio; P/R-Hb, Plasma Hb/RBC Hb ratio.

Table 37.2 Effects of TAU and structurally-related compounds on the values of nonenzymatic indices of oxidative stress for RBCs from type 2 diabetic rats

Treatment	MDA μM/g Hb	GSH μM/g Hb	GSSG μM/g Hb	GSH/GSSG
Normal	11.33 ± 0.83	2.66 ± 0.24	$4.14 \pm 0.46^{*}$	0.64 ± 0.08
Diabetic	$25.02 \pm 0.57^{***}$	$2.06 \pm 0.18^{*}$	$4.79 \pm 0.15^{+}$	$0.43 \pm 0.07^{***}$
+ AMSA	$10.65 \pm 0.65^{+++}$	$2.59 \pm 0.17^{+}$	$4.17 \pm 0.20^{+}$	$0.62 \pm 0.04^{+++}$
+ HMTAU	$9.11 \pm 0.44^{*,+++}$	$2.49 \pm 0.18^{+}$	$4.05 \pm 0.13^{+}$	$0.61 \pm 0.07^{+++}$
+ HYTAU	$8.65 \pm 0.75^{*,+++}$	$2.76 \pm 0.20^{+}$	$3.98 \pm 0.19^{+}$	$0.69 \pm 0.08^{+++}$
+ TAU	$10.84 \pm 0.95^{+++}$	$2.38 \pm 0.16^{+}$	$4.16 \pm 0.15^{+}$	$0.57 \pm 0.08^{*,++}$

[a]Values represent the mean ± SEM for $n = 3$.
[b]Statistical comparisons were vs. normal rats at $^{*}p < 0.05$, $^{**}p < 0.01$ and $^{***}p < 0.001$; and vs. diabetic rats at $^{+}p < 0.05$, $^{++}p < 0.01$ and $^{+++}p < 0.001$.

Table 37.3 Effects of TAU and structurally related compounds on enzymatic indices of oxidative stress for RBCs from type 2 diabetic rats

Treatment	CAT U/g Hb	SOD U/L	GPX μM NADPH/min/ g Hb	CAT/SOD	SOD/GPX
Normal	1091 ± 122	1507 ± 102	144 ± 5.98	0.72 ± 0.05	10.47 ± 1.65
Diabetic	$796 \pm 105^{**}$	$832 \pm 105^{***}$	$103 \pm 3.65^{**}$	$0.96 \pm 0.07^{***}$	$8.08 \pm 1.54^{**}$
+ AMSA	$1094 \pm 105^{++}$	$1062 \pm 105^{**,++}$	$136 \pm 3.80^{++}$	$1.03 \pm 0.04^{***}$	$7.81 \pm 1.62^{***}$
+ HMTAU	$1075 \pm 119^{++}$	$1225 \pm 106^{**,++}$	$131 \pm 2.46^{*,++}$	$0.88 \pm 0.06^{***}$	$9.36 \pm 1.65^{*,++}$
+ HYTAU	$1101 \pm 103^{++}$	$1238 \pm 105^{**,++}$	$128 \pm 2.98^{*,++}$	$0.89 \pm 0.06^{***}$	$9.65 \pm 1.45^{*,++}$
+ TAU	$1064 \pm 105^{++}$	$1169 \pm 108^{**,++}$	$119 \pm 3.07^{**,++}$	$0.91 \pm 0.08^{***}$	$9.83 \pm 1.76^{*,++}$

[a]Values represent the mean ± SEM for $n = 3$.
[b]Statistical comparisons were vs. normal rats at $^{*}p < 0.05$, $^{**}p < 0.01$ and $^{***}p < 0.001$; and vs. diabetic rats at $^{+}p < 0.05$. $^{++}p < 0.01$ and $^{+++}p < 0.001$.

sulfur compound the intracellular activity of LDH in diabetic RBCs became either about equal to (HMTAU, HYTAU, TAU, $p < 0.01$) or just below (AMSA, $p < 0.05$) the control value (Table 37.1). Diabetes increased the plasma LDH activity by 77% over the control value ($p < 0.001$). This effect was attenuated significantly ($p < 0.01$) by a treatment with a sulfur compound in the decreasing order HMTAU (by 53%) > HYTAU (by 47%)>TAU (by 42.5%)>AMSA (by 40%). Diabetes also caused a drastic increase in the ratio of plasma LDH to RBC LDH (1.04, $p < 0.001$ vs. control), an effect that was effectively counteracted by a treatment with a sulfur compound (ratio from 0.37 to 0.44) (Table 37.1).

Diabetic RBCs contained less GSH (–23%, $p < 0.05$) and more GSSG (+16%, $p < 0.05$) than normal RBCs (Table 37.2). The GSH loss was effectively counteracted by a sulfur-containing compound ($p < 0.05$ relative to diabetes alone), with the loss amounting to only 3% and 6% in the presence of AMSA and HMTAU, in that order, and exceeding control values by 1% and 4% in the presence of TAU and HYTAU, respectively (Table 37.2). Likewise, a sulfur-containing compound had a normalizing effect on the increase in GSSG seen with diabetes, which became either equal to (AMSA, TAU) or just below (HMTAU, HYTAU) the control value ($p < 0.05$ vs. diabetic group). Diabetes lowered the GSH/GSSG ratio from a normal value of 0.64 to one equal to 0.43 ($p < 0.01$) Following a treatment with a sulfur compound the ratio ranged for 0.57–0.69 ($p < 0.01$) (Table 37.2).

The MDA value of diabetic RBCs was almost 2-fold higher than that of normal rats. The formation of MDA induced by diabetes was significantly decreased ($p < 0.001$) by a sulfur compound in the order HYTAU (65.5%) • HMTAU (64%)>AMSA, TAU (–57%) (Table 37.2).

In diabetic rats, the RBC the activities of CAT, CuZn SOD and GPX were reduced by about 27% ($p < 0.05$), 45% ($p < 0.01$) and 29% ($p < 0.01$) relative to normal values (Table 37.3). A treatment with a sulfur-containing compound raised these activities by 34–37.5% ($p < 0.01$), 28–49% ($p < 0.01$), and 6–17%, respectively, above diabetic values. HYTAU and HMTAU were somewhat more potent than TAU and AMSA in their effect on CAT and SOD; and AMSA was slightly better than HYTAU and HMTAU on GPX. Diabetes raised the CAT/SOD ratio of normal RBCs from 0.72 to 0.96 (Table 37.3). Except for AMS (ratio ~1.0), the remaining sulfur compounds lowered the diabetic CAT/SOD ratio in the potency order HMTAU (0.88) > HYTAU (0.89) > AMSA, TAU (0.91) (Table 37.3). The SOD/GPX activity ratio of normal RBCs was reduced by diabetes by about 23% (10.5 vs. 8.08, $p < 0.01$); but increased following a treatment with a sulfur compound to between 8.55 (AMSA) and 9.83 (TAU) (Table 37.3).

37.4 Discussion

In the present study, RBCs from GK rats, a rat strain genetically-predisposed to develop type 2 diabetes spontaneously, and from age, weight – and gender-matched WKY rats were biochemically examined. Relative to RBCs from WKY rats, those from GK rats exhibited a higher TBARS and a lower GSH content. The depletion of cellular GSH may be a consequence of its interaction with α-oxoaldehydes formed from the autoxidation or the metabolism of plasma carbohydrates by glycolysis and the pentose phosphate pathway (Beard et al. 2003). The increase in GSSG in diabetic RBCs may reflect an increased utilization of GSH to cope with the oxidative stress of diabetes. Similar results have been reported by other laboratories for RBCs from diabetic subjects (e.g., Donma et al. 2002; Tas et al. 2007). Likewise, the higher ratios of plasma/RBC ratios for Hb and LDH in GK rats relative to WKY rats point to a greater disturbance of the RBC membrane integrity in the former than in the latter rats, possibly because of increased LPO.

Treating diabetic rats with TAU for 5 days led to significant protection against membrane and biochemical alterations induced by diabetes. Thus, TAU kept the intracellular GSH content of diabetic RBCs equal to that of normal RBCs. In spite of the seemingly greater availability of GSH created by TAU, the GSSG content was surprisingly not different from that found in control RBCs. Although the reason for this contrasting effect is not obvious at this time, it may reflect the ability of TAU to stimulate the recycling of GSSG to GSH. Future experiments aimed at measuring the activities of glutathione reductase and glutathione transferase may help to clarify this possibility. Furthermore, TAU lowered MDA formation and prevented the leakage of both Hb and LDH from RBCs. The former effect may be partly the result of a direct interaction of the amino group of TAU with MDA in a manner analogous to the nonenzymatic glycation of exposed amino groups in proteins observed during hyperglycemic states (Ogasawara et al. 1994), since *in vitro* experiments have found TAU to readily bind to acetaldehyde at physiological pH (Ogasawara et al. 1994). In terms of the second effect, TAU is probably acting on the cell surface as a membrane stabilizer. This effect could be the result of a reversible ion-pair type of interaction between the zwitterion form of this β-amino acid and charged proteins (Schaffer and Azuma 1992) or phospholipids heads on the external leaf of the RBC membrane (Huxtable and Sebring 1986). As a result, the membrane packing density may increase to render the membrane more resistant to attack by ROS, MDA and other forms of extracellular insult (Huxtable 1992). The likelihood of this mechanism is supported by the results of an earlier study from this laboratory that found protection of RBCs by TAU against membrane damage by exogenous oxidants to disappear when the RBCs were washed with PBSG pH 7.4 just before the addition of the oxidant (Pokhrel and Lau-Cam 2000).

Two additional compounds, one representing the immediate lower (AMSA) and the other the immediate higher (HMTAU) homolog of TAU, were tested alongside the parent molecule. In general, both compounds offered the same pattern of protection against diabetes-related alterations of nonenzymatic parameters of oxidative stress as that derived from TAU. With the possible exception of effects on MDA formation, plasma LDH leakage and SOD activity, for which HMTAU was slightly more effective than TAU, the three compounds exhibited equipotent actions against all other biochemical alterations caused by diabetes.

In diabetic rats, HYTAU, the sulfinate analog of TAU, was more effective than TAU in preventing MDA formation and leakage of Hb and LDH into the vascular compartment; and about equal to TAU in most other tests. These findings are, in part, unexpected since HYTAU is regarded to be a better antioxidant than TAU by virtue of its readily oxidizable sulfinate moiety (Fellman and Roth 1985) and a greater radical scavenging ability *in vitro* (Aruoma et al. 1988; Pitari et al. 2000; Tadolini et al. 1995). However, as found here, under *in vivo* conditions the two compounds appear to display equivalent antioxidant potencies not anticipated on structural grounds. In this context, when the present results are interpreted in conjunction with those found earlier in this laboratory (Pokhrel and Lau-Cam 2000), it would appear that even though both the sulfonic and amino groups contribute to the antioxidant activity of TAU and its immediate homologs under *in vivo* conditions, a different picture emerges when these compounds are tested in nonbiological samples. Indeed, while ethanesulfonic acid (ESA), the deaminated analog of TAU, and TAU protected rat RBCs against the damaging actions of an exogenous oxidant to about the same extent *in vitro*, ESA was less effective than TAU in protecting RBCs against the same oxidant in rats (Pokhrel and Lau-Cam 2000).

The RBC is endowed with CAT, SOD and GPX activities for pro-tection against oxidative damage. Hence, it would be expected that in the face of oxidative stress these enzymes will be activated to scavenge harmful H_2O_2 or ROS, and that changes in the ratios of CAT/SOD and GPX/SOD will serve as indicators of protective efficiency by these enzymes against oxidative stress (Pitari et al. 2000). In this regard, decreases in enzyme activities and in enzyme ratios will signify overwhelming oxidative stress and a lack of enzymatic responsiveness, respectively, situations that will increase the risk of oxidative damage. Although the present findings agree with those of most published reports (e.g., Donma et al. 2002; Tas et al. 2007) in that diabetes lowers the RBC activities of SOD, CAT and GPX, a great deal of variation seems to exist regarding enzyme values reported by different laboratories. For example, one study found the RBC activities of SOD, GPX and CAT to be equal among healthy subjects, type 2 diabetics and diabetics with retinopathy (Gürler et al. 2000); and in another the RBC activity of CAT was increased while those of GPX (Memisogullari et al. 2003) and SOD (Djordjevic et al. 2004) were decreased. Yet, another laboratory reported a higher than normal CAT and GPX activity and a normal SOD activity in RBCs from type 2 diabetic patients with microvascular complications (Kesavulu et al. 2000). The higher CAT/SOD and lower SOD/GPX ratio of RBCs from diabetic rats relative to ratios of RBCs from normal or diabetic rats treated with a sulfur compound probably reflect the higher degree of oxidative state in the former than in the latter rats.

37.5 Conclusions

TAU and its related compounds protected RBCs against diabetes-induced alterations in enzymatic and nonenzymatic indices of oxidative stress and against membrane susceptibility to oxidative damage Although TAU and its related compounds provided protective effects consonant with true antioxidant activity, their differences in potencies appeared to vary within a narrow range of one another and to be more dependent on differences in oxidation state of the sulfur-containing functionality than on differences in carbon chain length.

References

Aebi H (1984) Catalase *in vitro*. Methods Enzymol 105:121–126

Aruoma OI, Halliwell B, Hoey BM, Butler J (1988) The antioxidant action of taurine, hypotaurine and their metabolic precursors Biochem J 256:251–255

Baynes JW, Thorpe SR (1999) Role of oxidative stress in diabetic complications. A new perspective on an old paradigm. Diabetes 48:1–9

Beard KM, Shangari N, Wu B, O'Brien PJ (2003) Metabolism, not autoxidation plays a role in α-oxoaldehyde- and reducing sugar-induced erythrocyte GSH depletion: relevance for diabetes mellitus. Mol Cell Biochem 252:331–338

Buege JA, Aust SD (1978) Microsomal lipid peroxidation. Methods Enzymol 52:302–310

Ceriello A (2003) New insights on oxidative stress and diabetic complications may lead to a "causal" antioxidant therapy. Diabetes Care 26:1589–1596

Da Ros R, Assaloni R, Ceriello A (2004) Antioxidant therapy in diabetic complications:what is new? Curr Vasc Pharmacol 2:335–341

Donma O, Yorulmaz E, Pekel H, Suyugul N (2002) Blood and lens lipid peroxidation and antioxidant status in normal individuals, senile and diabetic cataractous patients. Curr Eye Res 25: 9–16

Djordjevic A, Spasic S, Jovanovic-Galovic A, Djordjevic R, Grubor-Lajsic G (2004) Oxidative stress in diabetic pregnancy: SOD, CAT and GSH-Px activity and lipid peroxidation. J Matern Fetal Neonatal Med 16:367–372

Evans JL (2007) Antioxidants: do they have a role in the treatment of insulin resistance? Indian J Med Res 125:355–372

Fellman JH, Roth ES (1985) The biological oxidation of hypotaurine to taurine: hypotaurine as an antioxidant. In: Oja SS, Ahtee L, Kontro P, Paasonen MK (eds) Taurine: biological actions and clinical perspectives, Alan R. Liss, New York, pp 71–82

Flohé L, Günzler WA (1984) Assays of glutathione peroxidase. Methods Enzymol 105: 114–121

Gallou G, Ruelland A, Campion L, Maugendre D, Le Moullec N, Legras B, Allannic H, Cloarec L (1994) Increase in thiobarbituric acid-reactive substances and vascular complications in type 2 diabetes mellitus Diabetes Metab 20:258–264

Goodman HO, Shihabi ZK (1990) Supplemental taurine in diabetic rats: effects on plasma glucose and triglycerides. Biochem Med Metab Biol 43:1–9

Gürler B, Vural H, Yilmaz N, Oguz H, Satici A, Aksoy N (2000) The role of oxidative stress in diabetic retinopathy. Eye 14 (Pt 5):730–735

Hissin PJ, Hilf R (1976) A fluorometric method for determination of oxidized and reduced glutathione in tissues. Anal Biochem 74:214–226

Huxtable RJ, Sebring LA (1986) Towards a unifying theory for the actions of taurine. Trends Pharmacol Sci 7:481–485

Huxtable RJ (1992) Physiological actions of taurine. Pharmacol Rev 72:101–163

Kaneto H, Nakatani Y, Kawamori D, Miyatsuka T, Matsuoka T (2004) Involvement of oxidative stress and the JNK pathway in glucose toxicity. Rev Diabet Stud 1:165–174

Kesavulu MM, Giri R, Kameswara R.B, Apparao C (2000) Lipid peroxidation and antioxidant enzyme levels in type 2 diabetics with microvascular complications. Diabetes Metab 26: 387–392

Kulakowski EC, Maturo J (1984) Hypoglycemic properties of taurine: not mediated by enhanced insulin release. Biochem Pharmacol 33:2835–2838

Maiese K, Morhan SD, Chong ZZ (2007) Oxidative stress biology and cell injury during type 1 and type 2 diabetes mellitus. Curr Neurovasc Res 4:63–71

Memisogullari R, Taysi S, Bakan E, Capoglu I (2003) Antioxidant status and lipid peroxidation in type II diabetes mellitus. Cell Biochem Funct 21:291–296

Misra HP, Fridovich I (1972) The role of superoxide anion in the autoxidation of epinephrine and a simple assay for superoxide dismutase. J Biol Chem 247:3170–3175

Ogasawara M, Nakamura T, Koyama I, Nemoto M, Yoshida T (1994) Reactivity of taurine with aldehydes and its physiological role. In: Huxtable R, Michalk DV (eds) Taurine in health and disease, Plenum Press, New York, pp 71–78

Pitari G,Dupre S, Sprinto A, Antoni G (2000) Hypotaurine protection on cell damage by singlet oxygen. In: Della Corte L, Huxtable RJ, Sgaragli G, Tipton KF (eds) Taurine 4: taurine and excitable tissues, Kluwer Academic/Plenum Publishers, New York, pp 157–162

Pokhrel KP, Lau-Cam C.A (2000) *In vitro* and *in vivo* effects of taurine and structurally related sulfur-containing compounds against phenylhydrazine-induced oxidative damage to erythrocytes. In: Della Corte L, Huxtable RJ, Sgaragli G, Tipton KF (eds) Taurine 4: taurine and excitable tissues, Kluwer Academic/Plenum Publishers, New York, pp 503–522

Schaffer SW, Azuma J (1992) Myocardial physiological effects of taurine and their significance. In: Lombardini JB, Schaffer SW, Azuma J (eds) Taurine: nutritional value and mechanisms of action, Plenum Press, New York, pp 105–120

Tadolini B, Pintus G, Pinna GG, Bennardini F, Franconi F (1995) Effects of taurine and hypotaurine on lipid peroxidation. Biochem Biophys Res Commun 213:820–826

Tas S, Sarandol, E, Ayvalik SZ, Serdar Z, Dirican M (2007) Vanadyl sulfate, taurine, and combined vanadyl sulfate and taurine treatments in diabetic rats: effects on the oxidative and antioxidative systems. Arch Med Res 38:276–283

Trachtman H, Futterweit D, Maesaka J, Ma C, Valderrama E, Fuchs A, Tarectecan AA, Rao PS, Sturman JA, Boles TH et al. (1995) Taurine ameliorates chronic streptozocin-induced diabetic nephropathy in rats Am J Physiol 269:F429–F438

Chapter 38
The Effects of Taurine, Hypotaurine, and Taurine Homologs on Erythrocyte Morphology, Membrane Fluidity and Cytoskeletal Spectrin Alterations Due to Diabetes, Alcoholism and Diabetes-Alcoholism in the Rat

Davekanand Gossai and Cesar A. Lau-Cam

Abstract Taurine (TAU) and compounds representing a TAU analog (hypotaurine = HYTAU) or homolog (aminomethanesulfonic acid = AMSA, homotaurine = HMTAU) were tested for their counteracting effects against alterations in erythrocyte (RBC) morphology, membrane fluidity and cytoskeletal spectrin distribution due to diabetes, alcoholism and diabetes-alcoholism in male Goto-Kakizaki rats (made diabetic with a high fat diet and alcoholic upon feeding on a flavored alcohol solution) and Wistar-Kyoto rats (serving as controls). Both diabetes and alcoholism changed the RBC discoidal biconcave shape to a spiculated one, lowered membrane fluidity, and caused spectrin to become marginalized. While AMSA and HYTAU returned the RBC shape to normal, HMTAU made it only discoidal, and TAU was without effect. All test compounds, but TAU, maintained the membrane fluidity normal; and HYTAU and AMSA, but not TAU or HMTAU, kept spectrin uniformly distributed. The noted effects were correlated with compound structure and RBC values for malondialdehyde and cholesterol/phospholipid ratio.

Abbreviations *TAU*, taurine; *HYTAU*, hypotaurine; *HMTAU*, homotaurine; *AMSA*, aminomethanesulfonic acid; *RBCs*, erythrocytes

38.1 Introduction

Alcoholism and diabetes mellitus are two independent and unrelated metabolic disorders sharing the common ability of being able to modify biological membranes both physically and chemically. As a result, these two metabolic disorders can

D. Gossai (✉)
Department of Pharmaceutical Sciences, College of Pharmacy and Allied Health Professions, St. John's University, Jamaica, New York, USA

J. Azuma et al. (eds.), *Taurine 7*, Advances in Experimental Medicine and Biology 643,
DOI 10.1007/978-0-387-75681-3_38, © Springer Science+Business Media, LLC 2009

separately induce changes in cell membrane composition, structure and function and, hence, affect membrane fluidity (Harris and Schroeder 1981; Watanabe et al. 1990), and cell shape, deformability and surface properties (Ballard 1997; McLawhon et al. 1987; Straface et al. 2002).

Membranes targeted by alcohol and diabetes are of ubiquitous occurrence and encompassing a wide range of cellular locations and types. In the case of alcohol, the effect on a membrane varies according to the duration of the exposure. On an acute basis, alcohol exerts a universal fluidizing action (Harris and Schroeder 1981) but, upon chronic consumption, it makes membranes less fluid (Watanabe et al. 1990) and tolerant to its fluidizing action *in vitro* (Taraschi et al. 1986). Oxidative stress, formation of reactive oxygen species (ROS) and lipid peroxidation (LPO) are among the factors that may play a role in membrane fluidification (Sergent et al. 2005) and rigidification (Banerjee et al. 1998; Benedetti et al. 1987; Eichenberger et al. 1982) by alcohol. In common with alcoholism, diabetes is also accompanied by the development of a state of oxidative stress leading to the generation of ROS but which, unlike alcoholism, is accentuated by hyperglycemia (Dave and Kalia 2007; Jain et al. 1989). Additional factors that may contribute to lowering membrane fluidity in chronic alcoholism and diabetes are changes in the type, proportion, and distribution of lipids component within the membrane bilayer (Benedetti et al. 1987; Lindi et al. 1998).

TAU and compound representing an analog (e.g., HYTAU) or homolog (e.g., AMSA, HMTAU) have displayed varying degrees of antioxidative properties under *in vitro* (Aruoma et al. 1988; Tadolini et al. 1995) and *in vivo* (Balkan et al. 2002) conditions. In addition, TAU has been found to lower diabetes-related hyperglycemia (Tokunaga et al. 1979) and to influence bilayer phospholipids mobility by directly interacting with biological membranes (Nakashima et al. 1996; Huxtable and Sebring 1986). Based on this evidence, the present study was undertaken in RBCs from normal rats, from rats chronically fed alcohol, and from rats genetically modified to spontaneously develop type 2 diabetes to determine whether these compounds can counteract alterations in morphology, fluidity and cytoskeletal spectrin distribution caused by chronic alcoholism and type 2 diabetes acting independently and, more significantly, concurrently.

38.2 Methods

38.2.1 Animals

Male Goto-Kakizaki (GK) rats 200–250 g, were from Taconic Farms Inc., Germantown, NY. During a 48 days period, rats in the diabetic group had free access to a pelleted high fat diet (Purina Rat Chow 13004, Ralston Purina Co., St. Louis, MO) and tap water. Rats in the alcohol group were maintained on a normal diet and tap water for 42 days. On day 43, these rats started to drink 40% alcohol (w/v) in 60% Kool-Aid® sweetened with SplendaÒ in place of tap water, and continued to do so for another 12 days. Rats in the diabetes plus alcohol group consumed the high fat diet for 55 days, tap water until day 42, and 40% alcohol from day 43 until day 55. Male Wistar-Kyoto (WKY) rats of similar age and weight to the GK rats, serving as the control group, were maintained for 48 days on a standard pelleted diet (Purina Rat Chow 5008, Ralston Purina Co., St. Louis, MO) and tap water. Feedings were on an *ad libitum* basis. Frank hyperglycemia was evident by day 48.

38.2.2 Treatment Solutions and Treatments

TAU and other sulfur-containing compounds were dissolved in phosphate buffer saline (PBS) pH 7.4, and administered by the intraperitoneal route at a dose of 0.612 M ($\sim 200 - 300$ mg)/kg/2 mL. Treatments were started on day 43 for rats in the diabetic group, and on day 50 for diabetic-alcoholic and alcoholic rats, and continued for another 4 days. Control rats received 2 mL of physiological saline in place of a sulfur compound solution.

38.2.3 Collection of Blood and Preparation of RBC Suspension

At 24 hr after the last treatment with a sulfur compound, blood was withdrawn by cardiac puncture into heparinized test tubes. After separating the plasma and buffy coat from the RBCs by centrifugation at $700 \times$ g and 4°C for 10 min, the RBC pellet was washed twice with ice-cold isotonic PBS pH 7.4 supplemented with 5 mM of glucose (PBSG) followed by centrifugation each time. The pellet of RBCs was used directly for the assay of malondialdehyde (MDA) or, alternatively, resuspended in PBSG to a hematocrit of 20% and stored on ice.

38.2.4 Assessment of Membrane Fluidity

The sample was prepared by incubating an aliquot of RBC suspension with an equal volume of 0.25 μM DPH in HEPES buffer pH 7.0 at 37°C for 1 hr. The fluorescence polarization (P) was measured at 25°C in the steady mode with a spectrofluorometer set at an excitation wavelength of 360 nm and emission wavelength of 450 nm. The value of P was calculated from: $P = [I_x - I_y]/[I_x + I_y]$, where I_x and I_y are the intensities of the polarized light emitted in parallel (vertically) and perpendicular (horizontally), respectively, to the direction of the incident polarized beam. The steady-state anisotropy (r_s was calculated from the value of P and the following equation: $r_s = 2P/3 - P$

38.2.5 Assessment of Cell Morphology

The RBCs were placed on a piece of filter paper, fixed with 1.5% glutaraldehyde in PBS pH 7.4 at 4°C for 3 hr, washed twice with PBS pH 7.4, dehydrated by passages through graded acetone solutions (30–60%) up to absolute acetone, critically point dried in CO_2, and sputter-coated with platinum. The RBCs were examined with a scanning electron microscope (SEM).

38.2.6 Assessment of Cytoskeletal Spectrin Distribution

RBCs for spectrin distribution studies were fixed with 3.7% formaldehyde in PBS pH 7.4 for 10 min, washed with PBS, and permeabilized with 0.5% Triton-X 100 in PBS for 5 min. After a treatment with goat serum, the RBCs were incubated at 37°C for 30 min intervals, first with rabbit anti-chicken spectrin and, next, with anti-rabbit IgG (whole molecule) FITC conjugate. The sample was rinsed with water, mounted on glycerol-PBS (2:1), and examined with a confocal microscope.

38.2.7 Assessment of LPO

The MDA present in the RBC pellet was measured as thiobarbituric acid reactive substances (TBARS) by the end-point assay method of Buege and Aust (1978). The amounts of MDA were expressed as nM/g Hb.

38.2.8 Statistical Analysis of the Data

The results are reported as the mean ± SEM for $n = 3$. Intergroup comparisons were made by Student's t-test, followed by one-way analysis of variance (ANOVA), and Tukey's *post hoc* test. Values were considered to be significantly different from each other at $p < 0.05$.

38.3 Results

38.3.1 Membrane Fluidity

Based on the anisotropy values presented in Fig. 38.1a–d, it is apparent that chronic alcoholism, type 2 diabetes and alcoholism-diabetes lowered the membrane fluidity of normal RBCs (Fig. 38.1a) to a significant extent ($p < 0.01$). This effect was greater in diabetic plus alcoholic (59% less) than in alcoholic (52% less) or diabetic (51% less) RBCs (Fig. 38.1b–d). Except for TAU (12% increase), AMSA, HMTAU and HYTAU did not alter the fluidity of normal RBCs. Also, while AMSA,

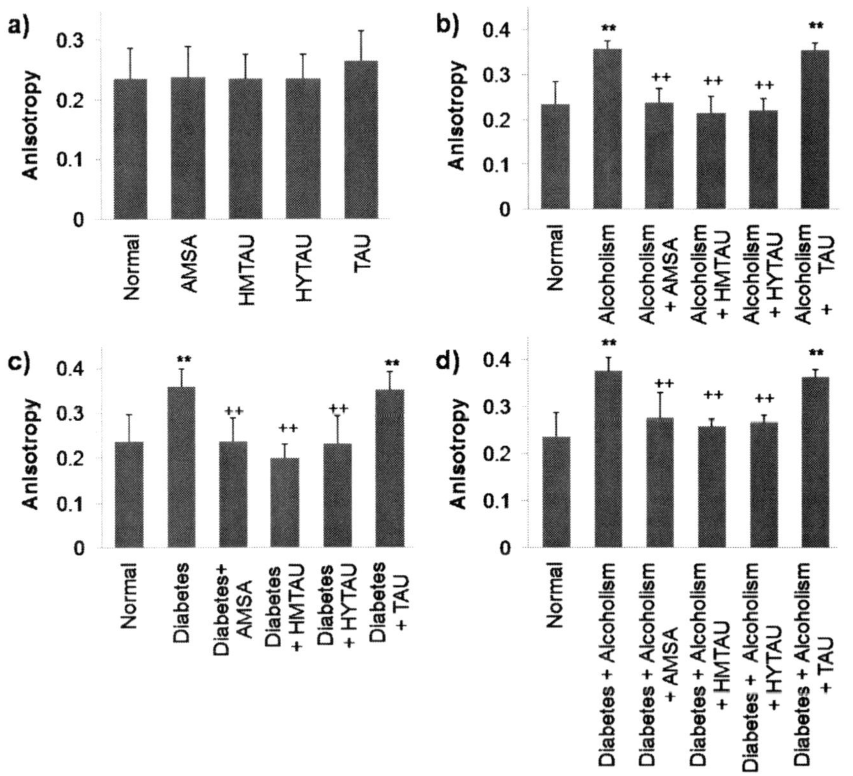

Fig. 38.1 Effect of TAU and structurally related compounds on the RBC membrane anisotropy values from (**a**) normal rats; (**b**) alcoholic rats; (**c**) diabetic rats, and (**d**) diabetic-alcoholic rats. Values represent the mean ± SEM for n =. Statistical comparisons were significantly different vs. normal RBCs at $**p < 0.01$; and vs. diabetic RBCs at $^{++}p < 0.01$

HMTAU and HYTAU attenuated the membrane-rigidifying action of alcohol (by 33–40%), diabetes (by 34–44%), and alcohol-diabetes (by 27–32%) in the potency order HMTAU>HYTAU>AMSA. On the other hand, TAU was found, in all cases, to lack a protective effect.

38.3.2 Morphology

RBCs from normal rats were discocytic and biconcave, and remained in this shape after a treatment with a sulfur compound (Fig. 38.2a). In contrast, alcoholic, diabetic and alcoholic-diabetic RBCs appeared echynocytic (Fig. 38.2b). Treating alcoholic rats with a sulfur compound led to cell shapes that varied according to the treatment. Thus, RBCs appeared ovalocytic or elongated after AMSA; rather thin, cup-like, and with slightly curled edges after HMTAU; and spherocytic after HYTAU or TAU (Fig. 38.2c). In contrast, all sulfur compounds but TAU returned the RBCs to their

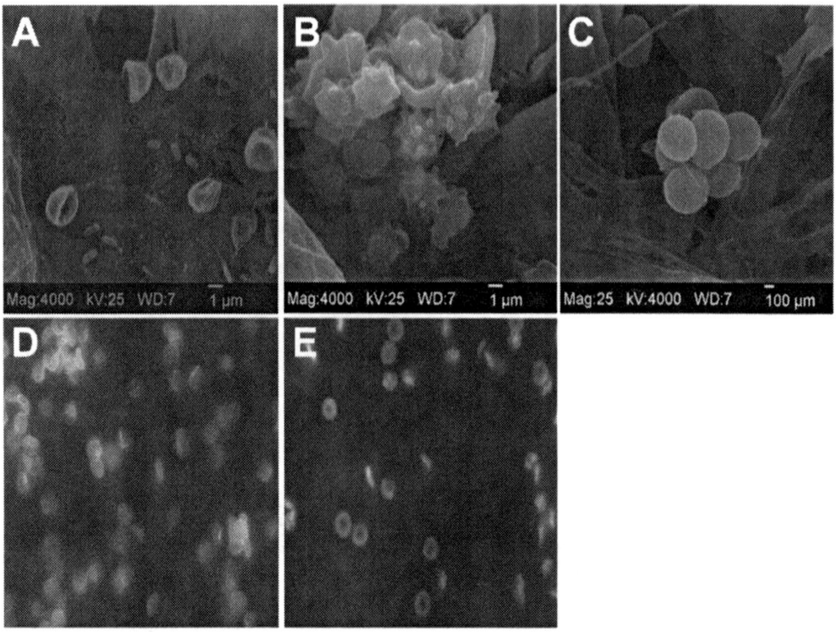

Fig. 38.2 Upper row shows SEM photomicrographs of (**A**) normal discocytic, (**B**) echinocytic; and (**C**) spherocytic RBCs at a magnification of ×4000. Lower row shows confocal photomicrographs of immunofluorescent spectrin (**D**) normally distributed and (**E**) marginalized within the RBCs

normal morphology. Relative to normal RBCs, those from alcoholic-diabetic rats appeared as discocytes after a treatment with HMTAU, or as spherocytes after a treatment with AMSA, HYTAU or TAU.

38.3.3 Spectrin Distribution

In normal RBCs spectrin was uniformly distributed throughout the cell surface, an appearance that was not modified by a treatment with a sulfur compound (Fig. 38.2d). In contrast, alcoholism, diabetes and alcoholism-diabetes caused spectrin to segregate towards the periphery as a fairly narrow fluorescent band (Fig. 38.2e). A treatment with either AMSA or HYTAU, but not one with HMTAU or TAU, normalized the distribution of spectrin in diabetic, alcoholic and diabetic-alcoholic

38.3.4 Production of MDA

None of the test compounds altered the MDA value of normal RBCs (Fig. 38.3a). In contrast, chronic alcoholism, diabetes and alcoholism-diabetes increased the baseline MDA value by ∼2-fold, 3-fold and 3.5 fold, respectively. A treatment with a

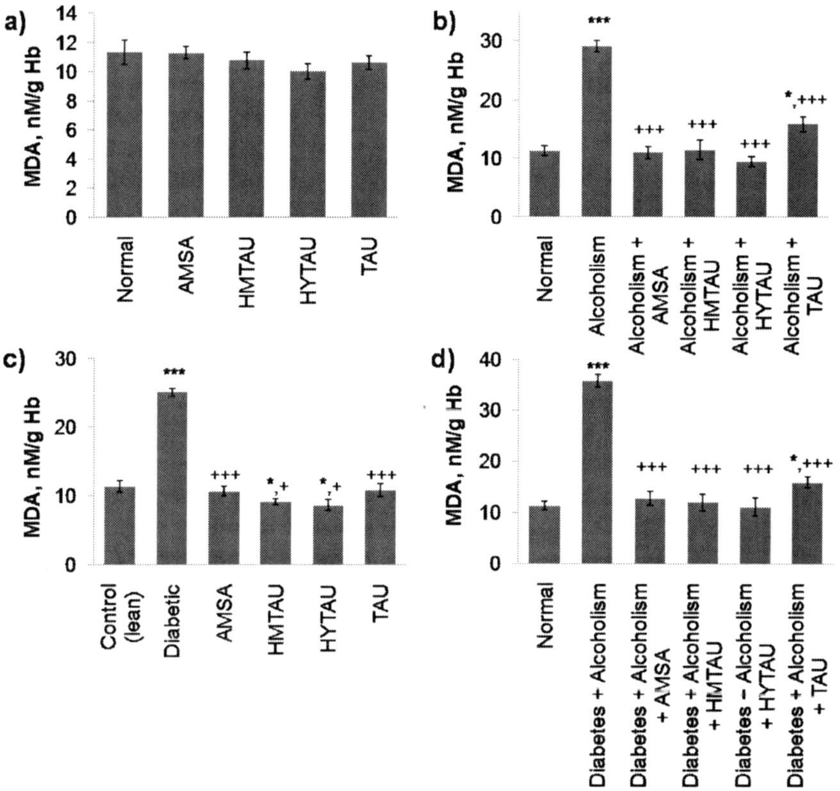

Fig. 38.3 Effect of TAU and structurally related compounds on the formation of MDA by RBCs from (**a**) normal rats; (**b**) alcoholic rats; (**c**) diabetic rats; and (**d**) diabetic-alcoholic rats. Values represent the mean ± SEM for $n = 3$. Statistical comparisons were significantly different vs. normal RBCs at $*p < 0.05$ and $***p < 0.001$; and vs. diabetic RBCs at $^{+}p < 0.05$, and $^{+++}p < 0.001$

sulfur compound attenuated these increases to a significant extent ($p < 0.001$), with the magnitude of the effect ranging from 52 to 69% (Fig. 38.3b–d). Overall, HYTAU was the most protective (65–69% decreases), TAU the least (52–57% decreases), and AMSA and HMTAU were equipotent in offering intermediate protection (61–65% decreases).

38.4 Discussion

38.4.1 Membrane Fluidity, MDA and Cholesterol/Phospholipids

The increase in steady anisotropy (r_s) for DPH in the membrane of diabetic, alcoholic or diabetic-alcoholic RBCs suggests a lowering in membrane fluidity. Based on currently available evidence, such an effect is probably the result of multiple factors and, thus, not amenable to a simple explanation. In spite of such a difficulty, two possible mechanisms for membrane rigidification were investigated here in an attempt to find an explanation for the counteracting effects noted with all the test compounds but TAU. The first one is based on the increase in oxidative stress, production of ROS, and eventual LPO of polyunsaturated membrane fatty acids that has been observed in both metabolic disorders (Parthiban et al. 1995; Watanabe et al. 1990). In accordance with the results reported by other laboratories, MDA, a secondary product of LPO, will be expected to cause membrane rigidification in RBCs from both chronic alcoholics and diabetics by reacting with amino groups of membrane proteins and phospholipids to promote their cross linking (Chiu et al. 1989). Alternatively, increased order among phospholipid acyl chains may be the result of their LPO to adjacent lipid radicals which, upon interacting with each other, will form covalent bonds (Eichenberger et al. 1982).

While the marked decrease in membrane fluidity verified here for diabetic and alcoholic RBCs, either as independent or co-occurring events, was effectively counteracted by AMSA, HMTAU or HYTAU, one with TAU was always without an effect. Hence, a direct correlation between protection against decrease in membrane fluidity by diabetes and chronic alcoholism and decreased MDA formation appears to only apply to AMSA, HMTAU and HYTAU, even when TAU was found to lower MDA formation by at least 3.8-fold relative to untreated alcoholic or diabetic RBCs. On the other hand, both chronic alcoholism (Benedetti et al. 1987) and diabetes (Banerjee et al. 1998) are reported to increase the cholesterol/phospholipid (C/PL) ratio in the RBC membrane. Measurement of this ratio in this laboratory (details not given here) have indicated that all sulfur compounds but TAU have the ability to normalize the C/PL ratio of RBCs from diabetic, alcoholic and diabetic-alcoholic rats. One possible reason for the failure of TAU to more effectively reverse diabetes- and alcoholism-induced membrane changes might be that after directly binding to neutral membrane phospholipids via loose ion-ion interactions (Huxtable and Sebring, 1986; Sebring and Huxtable 1986) to decrease the mobility of head groups of polar phospholipids, this effect is diminished in the presence of the high concentrations of calcium ions present in oxidatively-stressed RBCs (Nakashima et al. 1996; Palek and Lux 1983).

38.4.2 Cell Morphology

The morphology of RBCs from alcoholic, diabetic and alcoholic-diabetic rats was that described as echinocytic, i.e., exhibiting spur-like protrusions along their periphery. The spurs are ascribed to the incorporation of excessive amounts of cholesterol into the cell membrane, an effect that will cause areas of the membrane surface to expand by outward bulging but without a corresponding increase in cell volume (Ballard 1997; Cooper 1978). Hence, it quite likely that the same mechanism is operative in RBCs from diabetic rats since most studies find the ratio of cholesterol to phospholipids to be higher in these RBCs than in those from normal subjects (Benedetti et al. 1987; Lindi et al. 1998). Increases in membrane cholesterol in alcoholic and diabetic RBCs are believed to be the result of an increase in the rate of transfer of this sterol from the plasma to the RBC membrane (Daniels and Goldstein 1982), a process that in diabetics would be favored by high plasma levels of cholesterol-rich plasma lipoproteins (Watala and Winocour 1992).

HYTAU and TAU were found to be slightly more effective than either AMSA or HMTAU in preventing the formation of spurs and in inducing a spherocytic shape. In contrast, both AMSA and HMTAU failed to reestablish the discoidal shape. RBCs from rats treated with AMSA were either oval, somewhat elongated, or having a flat and uniconcave appearance; and those of rats receiving HMTAU were flat, uniconcave and curled at the edges. The flattened RBC shape has been associated with a modest elevation in membrane cholesterol (Ballard 1997); and spherocytosis has been related to losses of cholesterol, phospholipids, integral membrane proteins and some surface material (Palek and Lux 1983).

RBCs from diabetic rats differed from those from alcoholic rats in two respects. First, the normal discocytic morphology was preserved in RBCs from rats treated with both AMSA and HYTAU. Second, while RBCs treated with HMTAU exhibited the flattened appearance noted with alcoholic RBCs, those treated with TAU were echynocytic. At variance with RBCs from rats with only diabetes, those from alcoholic-diabetic rats were either of normal, discocytic (HMTAU) or spherocytic appearance. This finding suggests that diabetes could be playing an attenuating role on the morphological effects of alcohol when the two are present together.

38.4.3 Spectrin Distribution

By imparting shear stiffness to the RBC membrane, the cytoskeleton stabilizes cell shape and allows the RBC to acquire its typical discocytic biconcave shape. In an environment promoting oxidative stress, such as those fostered by diabetes and chronic alcoholism, the cytoskeletal network, comprised primarily of spectrin and actin, can undergo drastic oxidative modifications at spectrin-actin junction sites responsible for RBC deformability and shape maintenance, thus resulting in alteration of the microfilament system, uncoupling of the cytoskeleton from the lipid bilayer (Knowles et al. 1997) and the development of cell morphology changes and membrane rigidity (Malorni et al. 2000; Straface et al. 2002). In RBCs from alcoholic, diabetic and alcoholic-diabetic rats, the fluorescence of spectrin became marginalized, possibly as a result of oxidative protein damage, denaturation and/or cross linking (Malorni et al. 2000; Straface et al. 2002). Among the test compounds, AMSA and HYTAU, but not HMTAU and TAU, were able to counteract the displacing effect of diabetes and alcoholism on spectrin. This difference might stem from the manner how these compounds interact with components of the cytoskeletal network, as imposed by their carbon-chain length and/or oxidation state of their sulfur-containing functionality, to serve as antioxidants.

38.5 Conclusions

In spite of small differences in carbon chain length between TAU (C_2), AMSA (C_1) and HMTAU (C_3); and in the state of oxidation between these sulfonic acids and HYTAU, the sulfinic analog of TAU, these compounds differed in the type and extent of their effects on cell membrane fluidity, morphology and spectrin distribution alterations due to diabetes, alcoholism or diabetes-alcoholism. Generally, AMSA and HYTAU were more protective than HMTAU and TAU. Failure of TAU to normalize RBC membrane fluidity and spectrin distribution might be related to the way how this amino acid interacts with components of the membrane bilayer and underlining cytoskeleton. The membrane fluidity changes observed in metabolic disorders such as those examined were found to be more dependent on the C/PL ratio than on LPO within the RBC membrane.

References

Aruoma OI, Halliwell B, Hoey BM, Butlerv J (1988) The antioxidant action of taurine, hypotaurine and their metabolic precursors. Biochem J 256:251–255

Balkan J, Oztezcan S, Aykaç-Toker G, Uysal M (2002) Effects of added dietary taurine on erythrocyte lipids and oxidative stress in rabbits fed a high cholesterol diet. Biosci Biotechnol Biochem 66:2701–2705

Ballard HS (1997) The hematological complications of alcoholism. Alcohol Health Res World 21:42–52

Banerjee R, Nageshwari K, Puniyani RR (1998) The diagnostic relevance of red cell rigidity. Clin Hemorheol Microcirc 19:21–24

Benedetti A, Birarelli AM, Brunelli E, Curatola G, Ferretti G, Del Prete U, Jezequel AM, Orlandi F (1987) Modification of lipid composition of erythrocyte membranes in chronic alcoholism. Pharmacol Res Commun 19:651–662

Buege JA, Aust SD (1978) Microsomal lipid peroxidation. Methods Enzymol 52:302–310

Chiu D, Kuypers F, Lubin B (1989) Lipid peroxidation in human red cells. Semin Hematol 26: 257–276

Cooper RA (1978) Influence of increased membrane cholesterol on membrane fluidity and cell function in human red blood cells. J Supramol Struct 8:413–430

Daniels CK, Goldstein D (1982) Movement of free cholesterol from lipoproteins or lipid vesicles into erythrocytes Acceleration by ethanol in vitro. Mol Pharmacol 21:694–700

Dave GS, Kalia K (2007) Hyperglycemia induced oxidative stress in type-1 and type-2 diabetic patients with and without nephropathy. Cell Mol Biol53:68–78

Eichenberger K, Böhni P, Winterhalter KH, Kawato S, Richter C (1982) Microsomal lipid peroxidation causes an increase in the order of the membrane lipid domain. FEBS Lett 142:59–62

Harris RA, Schroeder F (1981) Ethanol and the physical properties of brain membranes. Fluorescence studies. Mol Pharmacol 20:128–137

Huxtable RJ, Sebring LA (1986) Towards a unifying theory for the actions of taurine. Trends Pharmacol Sci 7:481–485

Jain SK, McVie R, Duett J, Herbst JJ (1989) Erythrocyte membrane lipid peroxidation and glycosylated hemoglobin in diabetes. Diabetes 38:1539–1543

Knowles DW, Tilley L, Mohandas N, Chasis JA (1997) Erythrocyte membrane vacuolation: model for the molecular mechanism of protein sorting. Proc Natl Acad Sci USA 94:12969–12974

Lindi C, Montorfano G, Marciani P (1998) Rat erythrocyte susceptibility to lipid peroxidation after chronic ethanol intake. Alcohol 16:311–316

Malorni W, Straface E, Pagano G, Monti D, Zatterale A, Del Principe D, Deeva IB, Franceschi C, Masella R, Korkina LG (2000) Cytoskeleton alterations of erythrocytes from patients with Fanconi's anemia. FEBS Lett 468:125–128
McLawhon RW, Marikovsky Y, Thomas NJ, Weinstein RS (1987) Ethanol-induced alterations in human erythrocyte shape and surface properties: modulatory role of prostglandin E1. J Membrane Biol 99:73–78
Nakashima T, Shima T, Sakai M, Yama H, Mitsuyoshi H, Inaba K, Matsumoto N, Sakamoto Y, Kashima K, Nishikawa H (1996) Evidence of a direct action of taurine and calcium on biological membranes. Biochem Pharmacol 52:173–176
Palek J, Lux SE (1983) Red cell membrane skeletal defects in hereditary and acquired hemolytic anemias. Semin Hematol 20:189–224
Parthiban A, Vijayalingam S, Shanmugasundaram KR, Mohan R (1995) Oxidative stress and the development of diabetic complications – antioxidants and lipid peroxidation in erythrocytes and cell membrane. Cell Biol Int 19:987–993
Sebring LA, Huxtable RJ (1986) Low affinity binding of taurine to phospholiposomes and cardiac sarcolemmal. Biochim Biophys Acta 884:559–566
Sergent O, Pereira M, Belhomme C, Chevanne M, Huc L, Lagadic-Gossmann D (2005) Role for membrane fluidity in ethanol-induced oxidative stress of primary rat hepatocytes. J Pharmacol Exp Ther 313:104–111
Straface E, Rivabene R, Masella R, Santulli M, Paganelli R, Malorni W (2002) Structural changes of the erythrocyte as a marker of non-insulin dependent diabetes:protective effects of N-acetylcysteine. Chem Biophys Res Commun 290:1393–1398
Tadolini B, Pintus, G, Pinna GG, Bennardini F, Franconi F (1995) Effects of taurine and hypotaurine on lipid peroxidation. Biochem Biophys Res Commun 213:820–826
Taraschi, TF, Ellingson JS, Wu A, Zimmerman R, Rubin E (1986) Membrane tolerance to ethanol is rapidly lost after withdrawal: a model for studies of membrane adaptation. Proc Natl Acad Sci USA 83:3669–3673
Tokunaga H, Yoneida Y, Kuriyama K (1979) Protective actions of taurine against streptozotocin-induced hyperglycemia. Biochem Pharmacol 15:2807–2811
Watala C, Winocour PD (1992) The relationship of chemical modification of membrane proteins and plasma lipoproteins to reduced membrane fluidity of erythrocytes from diabetic patients. Eur J Clin Chem Clin Biochem 30:513–519
Watanabe H, Kobayashi A, Yamamoto T, Suzuki S, Hayashi H, Yamazaki N (1990) Alterations of human erythrocyte membrane fluidity by oxygen-derived free radicals and calcium. Free Radic Biol Med 8:507–514

Chapter 39
Effect of PTP1B Inhibitors and Taurine on Blood Lipid Profiles in Adolescent Obesity

Sun Hee Cheong, Hyeongjin Cho, and Kyung Ja Chang

Abstract The purpose of this study was to investigate the effect of protein tyrosine phosphatase 1B (PTP1B) inhibitors and taurine on blood lipid profiles of the obese adolescent. Three week-old thirty-six male Sprague-Dawley rats were randomly assigned to six groups (high fat diet (HFD) group, HFD group; high fat diet + taurine group, HFD + taurine (TR) group; high fat diet + PTP1B inhibitor A (A) group, HF + A group; high fat diet + PTP1B inhibitor B, HF + B group; high fat diet + PTP1B inhibitor A + taurine group, HF + A + TR group; and high fat diet + PTP1B inhibitor B + taurine group, HF+B+TR group). The body weight of the HF+A and HF + A + TR groups were significantly reduced compared to those of the other groups, the exception being the HF + B + TR group. Serum triglyceride (TG), total cholesterol (TC) and low density lipoprotein cholesterol (LDL-C) levels of the HF + A + TR group were significantly reduced compared to those of the other groups. However, serum high density lipoprotein cholesterol (HDL-C) levels in HF + B + TR group was significantly reduced compared those of the HFD and HF + TR groups. Therefore, these results suggest a possible effect of PTP1B inhibitors and taurine on blood triglyceride, total cholesterol and low density lipoprotein cholesterol in the obese adolescent.

Abbreviations *PTP1B*, protein tyrosine phosphatase; *TG*, triglyceride; *TC*, total cholesterol; *LDL-C*, low density lipoprotein cholesterol; *HDL-C*, high density lipoprotein cholesterol; *HF*; high fat diet, *TR*, taurine

39.1 Introduction

Obesity is the most common metabolic disease in developed nations. In particular, the increasing prevalence of obesity among children and adolescents is a major concern and is a predictive indicator of obesity in future adults (Bays 2004).

S.H. Cheong (✉)
Department of Food and Nutrition, Inha University, Incheon, Korea

J. Azuma et al. (eds.), *Taurine 7*, Advances in Experimental Medicine and Biology 643, DOI 10.1007/978-0-387-75681-3_39, © Springer Science+Business Media, LLC 2009

Taurine (2-aminoethane sulfonic acid) is a non-protein amino acid that is abundant in most mammalian tissues, cells and seafood, which has a variety of physiological and biochemical functions (Balkan et al. 2004; Tsuboyama-Kasaoka et al. 2007). In previous studies it has been reported that taurine has hypolipidemic, antiatherosclerotic and antiobesity effects (Obinata et al. 1996; Brøns et al. 2004; Zhang et al. 2004; Tsuboyama-Kasaoka et al. 2007). Moreover, the effect of taurine on atherogenesis induced by high fat diet in rats, a species which depends entirely on taurine for conjugation of bile acids, has been investigated (Sethupathy et al. 2002).

Leptin, the 16kDa protein product of the obese gene, is an adipocyte-derived hormone that is secreted in proportion to total body fat mass. In various human and rodent studies it has been reported that defective leptin signaling due to either leptin deficiency or dysfunctional leptin receptors leads to early onset of obesity (Korner and Aronne 2003; Lund et al. 2005). Protein tyrosine phosphatase 1B (PTP1B) is a key element in the negative regulation of the insulin signaling pathway and may play an important role in diabetes and obesity (Ahn et al. 2003; Bays 2004;Dubé and Tremblay 2005; Bharatham et al. 2007) Most obese individuals develop hyperleptinemia and leptin resistance. Mice lacking PTP1B are protected from diet-induced obesity and are hypersensitive to leptin, but the site and the mechanism underlying these effects remain controversial (Bence et al. 2006). Klaman et al. (2000) have reported that increases in insulin sensitivity, resistance to diet-induced obesity and augmentation in energy expenditure are characteristic features of the PTP1B-deficient mouse. In general, salicylic acid is known to be a weak competitive inhibitor of PTP1B, with an inhibition constant of 19.4 mM (Burke and Zhang 1998).

However, the effect of taurine and PTP1B inhibitors on the blood lipid profiles in the obese adolescent has received little attention. Therefore, this study was conducted in order to examine the effect of dietary taurine, the PTP1B inhibitor, methylenedisalicylic acid (MDSA) and a derivative of MDSA, on the blood lipid profile of high fat diet-induced obesity in rats.

39.2 Materials and Methods

39.2.1 Animals and Diet

Three week-old thirty-six male Sprague-Dawley rats were supplied from Biolink (Seoul, Korea), and were kept in shoe-box cages in a room with controlled temperature (23± 2°c), humidity (55 ± 5%), and light-dark-cycle (07:00–19:00). The rats were fed a commercial diet for one week. All rats were randomly assigned to six groups (high fat diet group, HFD group; high fat diet + taurine group, HF + TR group; high fat diet + PTP1B inhibitor A group, HF + A group; high fat diet + PTP1B inhibitor B, HF + B group; high fat diet + PTP1B inhibitor A + taurine group, HF + A + TR group; and high fat diet + PTP1B inhibitor B + taurine group, HF+B+TR group).

In this study, MDSA served as the PTP1B type A inhibitor while PTP1B type B inhibitor was a derivative of MDSA. The PTP1B inhibitor A was synthesized by treatment of methyl 3-phenylsalicylate with 1,3,5-trioxane, in the presence of glacial acetic acid at 95°C (Shrestha et al 2007). The PTP1B inhibitor A was dissolved in 1M NaOH and the PTP1B inhibitor B was dissolved in distilled water and given orally (administered 0.165 mg per 100 g diet). Rats were fed the experimental diets for 8 weeks (Table 39.1). Diets and water were provided *ad libitum*.

39.2.2 Blood Sampling and Chemical Analysis

After 8 weeks of feeding the experimental diets, rats were sacrificed. The weight of the liver, adipose tissue of the epididymis, the gastrocnemius and the psoas muscle were measured. The size of the abdominal adipose cell was examined by light

Table 39.1 Composition of experimental diet (g/100g diet)

	HFD[5]	HF+TR	HF+A	HF+B	HF+A+TR	HF+B+TR
Casein	20	20	20	20	20	20
DL-Methionine	0.3	0.3	0.3	0.3	0.3	0.3
Corn starch	55	55	55	55	55	55
Cellulose	5	5	5	5	5	5
Corn Oil	-	-	-	-	-	-
Beef tallow	15	15	15	15	15	15
Choline bitartrate	0.2	0.2	0.2	0.2	0.2	0.2
AIN Mineral Mix [1]	3.5	3.5	3.5	3.5	3.5	3.5
AIN Vitamin Mix [2]	1	1	1	1	1	1
PTP1B inhibitor (type A) [3](mg)	-	-	0.165	-	0.165	-
PTP1B inhibitor (type B) [4](mg)	- -	- -	- -	0.165	- -	0.165
Taurine(mg)	- -	1000	- -	- -	1000	1000
Calorie(kcal/100g diet)	435	435	435	435	435	435

[1]AIN Mineral Mixture(g/kg) ; Calcium phosphate dibasic 500 g, Sodium chloride 74 g, Potassium citrate monohydrate 220 g, Potassium sulfate 52 g, Magnesium oxide 24 g, Manganous carbonate(43-48%Mn) 3.5 g, Ferric citrate(16-17%Fe) 6 g, Zinc carbonate(70% ZnO) 1.6 g, Cupric carbonate(53-55% Cu) 0.3 g, Potassium iodate 0.01 g, Sodium selenite 0.01 g, Chromium potassium sulfate 0.55 g, Sucrose finely powdered 118 g. [2]AIN Vitamin Mixture(g/kg) ; Thiamine hydrochloride 600 mg, Riboflavin 600 mg, pyridoxine hydrochloride 700 mg, nicotinic acid 3 g, D-calcium pantothenate 1.6 g, Folic acid 200 mg, D-biotin 20 mg, Cyanocobalamin 1 mg Retinyl palmitate pre-mix(250,000IU/g) 1.6 g, DL-alpha-tocopherol acetate(250IU/g) 20 g, cholecalciferol(400,000IU/g) 250 mg, Menaquinone 5 mg Sucrose, finely powdered 972.9 g. [3]The PTP1B inhibitor A dissolved in 1M NaOH was given orally 0.165 mg per 100 g diet. [4]The PTP1B inhibitor B dissolved in distilled water was given orally (0.165 mg per 100 g diet). [5]Experimental rats were randomly assigned to six groups (high fat diet group, HFD group; high fat diet + taurine group, HF+TR group; high fat diet + PTP1B inhibitor A group, HF+A group; high fat diet + PTP1B inhibitor B, HF+B group; high fat diet + PTP1B inhibitor A + taurine group, HF+A+TR group; and high fat diet + PTP1B inhibitor B + taurine group, HF+B+TR group).

microscopy. Blood samples were obtained through heart puncture and serum was separated at 3000x g for 20 min. Serum samples were stored at −70°C until analysis. Serum triglyceride, total cholesterol and HDL-cholesterol levels were analyzed using methods described in commercial kits that involved enzymatic procedures employing a UV/VIS spectrophotometer at 546 nm, 500 nm and 500 nm, respectively (Youngdong Pharmaceutical Co., Seoul, Korea). Serum LDL-cholesterol was calculated by the formula (Friedewald et al 1972)

$$\text{LDL-cholesterol} = \text{Total cholesterol} - (\text{HDL-cholesterol} + \text{triglyceride}/5)$$

Also HDL-cholesterol/total cholesterol and LDL-cholesterol/HDL-cholesterol ratios were calculated.

39.2.3 Statistical Analysis

Statistical analysis of data was carried out using the SPSS 10.0 program. All results were expressed as means ± SE and statistical significance between the groups was assessed by using analysis of variance (ANOVA) and Duncan's multiple range test. The level of significance used was $p<0.05$.

39.3 Results

39.3.1 Final Body Weight, Adipose Tissue and Muscle Weight

Body weight, weight of the adipocytes of the epididymis, the liver, the gastrocnemius and the psoas muscle are shown in Table 39.2. Final body weight and the weight of adipose tissue of the epididymis in the HF+A and the HF+A+TR groups were significantly reduced compared to those of the other groups ($P<0.05$). Relative

Table 39.2 Final body weight, weight of adipose tissue of the epididymis, the liver, the gastrocnemius and the psoas muscle

Group	Final body weight (g)	Adipose tissue of epididymis (g)	Relative Liver (g)/100g BW	Gastrocnemius muscle (g)	Psoas muscle (g)
HFD	277.5 ± 4.5^{b}	3.8 ± 0.4^{cd}	11.6 ± 0.1^{a}	2.4 ± 0.1^{a}	1.9 ± 0.1^{a}
HF+TR	282.5 ± 9.2^{b}	4.0 ± 0.5^{d}	9.9 ± 1.3^{ab}	2.1 ± 0.1^{ab}	1.8 ± 0.1^{ab}
HF+A	258.8 ± 8.3^{a}	2.8 ± 0.3^{a}	8.8 ± 0.2^{b}	1.8 ± 0.4^{b}	1.4 ± 0.2^{b}
HF+B	288.8 ± 13.1^{b}	3.6 ± 0.2^{c}	11.0 ± 0.7^{c}	2.0 ± 0.2^{ab}	1.9 ± 0.1^{a}
HF+A+TR	252.5 ± 8.5^{a}	3.3 ± 0.5^{b}	9.4 ± 0.4^{bc}	2.0 ± 0.1^{ab}	1.7 ± 0.1^{ab}
HF+B+TR	269.0 ± 6.4^{ab}	3.6 ± 0.1^{c}	9.4 ± 0.2^{bc}	1.9 ± 0.4^{b}	1.7 ± 0.3^{ab}

Values are means ± SD. Means in each column not sharing a common superscript letter are significantly different at $p<0.05$ by Duncan's multiple range test.

liver weight in the HF+A group was significantly reduced compared to that of the other groups ($P<0.05$). The weight of the gastrocnemius and the psoas muscle in the HF+A group were significantly reduced compared to those of the HFD group ($P<0.05$).

39.3.2 Serum Lipid Profiles

The serum lipid profile of the various groups is shown in Table 39.3. Serum triglyceride and total cholesterol levels were significantly lower in the HF+A+TR and HF+B+TR groups than in the other groups ($p<0.05$). Serum LDL-cholesterol level was significantly higher in the HFD group than in the other groups ($p<0.05$). The serum HDL-cholesterol level and the ratio of the HDL-cholesterol to total cholesterol in the serum was significantly elevated in the HF +A, HF + B, HF + A + TR and HF + B + TR groups compared to those of the HFD and HFD+TR groups ($p<0.05$). The ratio of LDL-cholesterol to HDL-cholesterol in serum was significantly elevated in the HFD group compared to that of the other groups ($p<0.05$).

39.3.3 Adipose Cell Size

The size of adipocytes located around the epididymis is shown in Fig. 39.1. The size of adipocytes in the HF+B+TR group was different from those of the other groups. These results suggest that the size of adipocytes may reflect the status of obesity. In a previous study, it was reported that taurine supplementation plays an important role in cardiovascular disease prevention in overweight or obese subjects (Sethupathy et al. 2002).

Table 39.3 Effect of PTP1B inhibitors and taurine on serum lipid profiles

	TG[1] (mg/dl)	TC (mg/dl)	LDL-C (mg/dl)	HDL-C (mg/dl)	HDL-C /TC	LDL-C /HDL-C
HFD	222.7±25.7^{a}	134.1±23.1^{a}	83.4±2.9^{a}	30.7±8.4^{a}	0.2±0.0^{a}	2.9±0.1^{a}
HF+TR	174.7±5.7^{b}	102.3±16.2^{b}	58.1±8.7^{b}	46.3±3.2^{a}	0.5±0.1^{a}	1.4±0.2^{b}
HF+A	132.9±23.4^{c}	88.3±5.6^{bc}	50.9±3.5^{b}	54.0±9.8^{b}	0.6±0.1^{b}	1.0±0.1^{b}
HF+B	111.3±11.2^{bcd}	94.1±3.6^{bc}	57.4±5.1^{b}	72.1±8.5^{c}	0.8±0.1^{bc}	0.9±0.3^{b}
HF+A+TR	83.5±7.9^{de}	76.3±8.6^{c}	45.7±7.3^{b}	69.5±9.8^{c}	1.0±0.0^{bc}	0.7±0.0^{b}
HF+B+TR	76.7±7.1^{de}	84.9±2.6^{bc}	53.1±3.9^{b}	82.4±6.2^{d}	1.0±0.2^{bcd}	0.7±0.1^{b}

Values are means ± SD. Means in each column not sharing a common superscript letter are significantly different at $p<0.05$ by Duncan's multiple range test.[1]TG: Triglyceride; TC: Total cholesterol; HDL-C: High density lipoprotein-cholesterol; LDL-C: Low density lipoprotein-cholesterol; HDL-C/TC: The ratio of HDL-cholesterol to total cholesterol; LDL-C/HDL-C: The ratio of LDL-cholesterol to HDL-cholesterol.

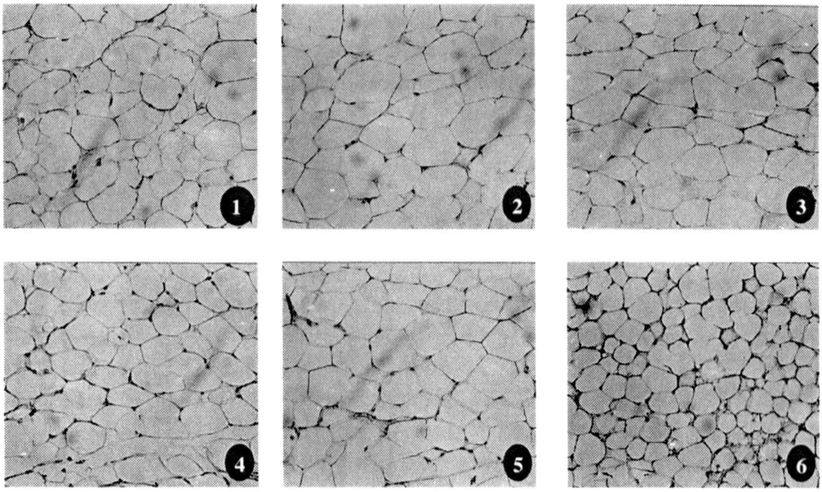

Fig. 39.1 Size of adipocytes of the epididymis examined by light microscopy (Bar =100 μm) Experimental rats were randomly assigned to six groups (1: high fat diet group, HFD group; 2: high fat diet + taurine group, HF+TR group; 3: high fat diet + PTP1B inhibitor A group, HF+A group; 4: high fat diet + PTP1B inhibitor B, HF+B group; 5: high fat diet + PTP1B inhibitor A + taurine group, HF+A+TR group; 6: high fat diet + PTP1B inhibitor B + taurine group, HF+B+TR group)

39.4 Discussion

This study was conducted in order to examine the effect of dietary taurine and PTP1B inhibitors on the blood lipid profile of high fat diet-induced obese rats. In this study, we showed that taurine and two kinds of PTP1B inhibitors, a PTP1B type A inhibitor and a type B inhibitor reduced blood triglyceride, total cholesterol, low density lipoprotein cholesterol and adipocyte size in adolescent obesity. In this study, taurine and the PTP1B type A inhibitor influenced final body weight and the weight of adipose tissue of the epididymis. Zhang et al. (2004) have previously reported that body weight was reduced following treatment with 3 g taurine. However, Kishida et al. (2003) reported that body weight and liver weight was not affected by taurine in rats fed a diet of corn oil and coconut oil. In this study, the weights of the gastrocnemius and the psoas muscle in the HF+A group were significantly reduced compared to those of the HFD group. Klaman et al. (2000) have reported that PTP1B deficient mice exhibited a significant increase in insulin sensitivity, as evidenced by the increase in insulin-stimulated glucose uptake in skeletal muscle. In this study, serum triglyceride, total cholesterol and LDL-cholesterol levels were significantly lower in the HF+A+TR and HF+B+TR groups than in the other groups. However, serum HDL-cholesterol levels and the ratio of HDL-cholesterol to total cholesterol in the serum was significantly higher in the HF+B+TR group than in the HFD and HFD+TR groups. In a previous study, it was reported that the plasma concentration of total cholesterol and

LDL-cholesterol were significantly reduced in rats fed a 1.5% taurine diet compared to those fed a 1.5% cholesterol diet, but the HDL-cholesterol level was higher in the rats fed the taurine diet than those fed the high cholesterol diet (Choi et al. 2006). Recently, several studies have reported that taurine treatment inhibits lipid peroxidation, lowers serum LDL/HDL cholesterol and elevates HDL cholesterol (Sethupathy et al. 2002;Kishida et al. 2003; Ito and Azuma 2004) Figure 39.1 shows that the adipocytes in the HF+B+TR group were different from those of the other groups. These results suggest that the size of the adipose cell may reflect the status of obesity. Shrestha et al (2007) showed that the PTP1B inhibitor and the methylenedisalicylic acid derivative significantly suppressed weight gain and adipocyte fat storage in obese rats maintained on a high-fat-diet. Similarly, Klaman et al. (2000) reported that PTP1B-deficient mice exhibited low adiposity and were protected from diet-induced obesity. Further, decreased adiposity was due to a marked reduction in fat cell mass without a decrease in adipocyte number. Collectively, the data suggest that PTP1B inhibitors and taurine treatment reduce blood lipid levels and decrease adipose cell size in the obese adolescent.

39.5 Conclusion

These results indicate that treatment with PTP1B inhibitors and taurine reduce blood triglyceride, total cholesterol, low density lipoprotein cholesterol and adipose cell size in obese adolescent rats. These results suggest that the PTP1B inhibitor and taurine act synergistically to modulate the blood lipid profile of the obese adolescent.

Acknowledgments We thank the Dong-A Pharmaceutical Co. for donating taurine.

References

Ahn JH, Cho SY, Ha JD, Kang SK, Jung SH, Kim HM, Kim SS, Kim KR, Cheon HG, Yang SD, Choi JK (2003) Protein tyrosine phosphatase 1B inhibitors: catechols. Bull Korean Chem Soc 24:1505–1508

Balkan J, Öztezcan S, Hatipoglu A, Çevikbaş U, Aykaç-Toker G, Uysal M (2004) Effect of taurine treatment on the regression of existing atherosclerotic lesions in rabbits fed a high-cholesterol diet. Biosci Biotechnol Biochem 68:1035–1039

Bays HE (2004) Current and investigational antiobesity agents and obesity therapeutic treatment targets. Obesity Research 12:384–387

Bence KK, Delidegovic M, Xue B, Gorgun CZ, Hotamisligil GS, Neel BG, Kahn BB (2006) Neuronal PTP1B regulates body weight, adiposity and leptin action. Nat Med 12:917–924

Bharatham K, Bharatham N, Leem KW (2007) Pharmacophore modeling for protein tyrosine phosphatase 1B inhibitors. Arch Pharm Res 30:533–542

Brøns C, Spohr C, Storgaard H, Dyerberg J, Vaag A (2004) Effect of taurine treatment on insulin secretion and action, and on serum lipid levels in overweight men with a genetic predisposition for type ¡ENT FONT=(normal text) VALUE=8545¿?¡/ENT¿ diabetes mellitus. Eur J Clin Nutr 58:1239–1247

Burke TR Jr, Zhang ZY (1998) Protein-tyrosine phosphatases: structure, mechanism, and inhibitor discovery. Biopolymers 47:225–241

Choi MJ, Kim JH, Chang KJ (2006) The effect of dietary taurine supplementation on plasma and liver lipid concentrations and free amino acid concentrations in rats fed a high-cholesterol diet. Adv Exp Med Biol 583:235-242

Dubé N, Tremblay ML (2005) Involvement of the small protein tyrosine phosphatase TC-PTP and PTP1B in signal transduction and diseases: from diabetes, obesity to cell cycle, and cancer. Biochim Biophys Acta 1754:108–117

Friedewald WT, Levy RI, Fredrickson DS (1972) Estimation of the concentration of low-density lipoprotein cholesterol in plasma, without use of the preparative ultracentrifuge. Clin Chem 18(6):499–502

Ito T, Azuma J (2004) Taurine is a possible anti-atherosclerotic agent. Nippon Yakurigaku Zasshi 123(5):311–317

Kishida T, Miyazato S, Ogawa H, Ebihara K (2003) Taurine prevents hypercholesterolemia in ovariectomized rats fed corn oil but not in those fed coconut oil. J Nutr 133:2616–2621

Klaman LD, Boss O, Peroni OD, Kim JK, Martino JL, Zabolotny JM, Moghal N, Lubkin M, Kim YB, Sharpe AH, Stricker-Krongrad A, Shulman GI, Neel BG, Kahn BB (2000) Increased energy expenditure, decreased adiposity, and tissue-specific insulin sensitivity in protein-tyrosine phosphatase 1B-deficient mice. Mol Cell Biol 20:5479–5489

Korner J, Aronne LJ (2003) The emerging science of body weight regulation and its impact on obesity treatment. J Clin Invest 111:565-570

Lund IK, Hansen JA, Andersen HS, Møller NPH, Billestrup N (2005) Mechanism of protein tyrosine phosphotase 1B-mediated inhibition of leptin signaling. J Mol Endocrinol 34:339-351

Obinata K, Maruyama T, Hayashi M, Watanabe T, Nittono H (1996) Effect of taurine on the fatty liver of children with simple obesity. Adv Exp Med Biol 403:607–613

Sethupathy S, Elanchezhiyan C, Vasudevan K, Rajagopal G (2002) Antiatherogenic effect of taurine in high fat diet fed rats. Indian J Exp Biol 40:1169–1172

Shrestha S, Bhattarai BR, Chang KJ, Lee KH, Cho H (2007) Methylenedisicylic acid derivatives: new PTP1B inhibitors that confer resistance to diet-induced obesity. Bioorg Med Chem Lett 17:2760–2764

Tsuboyama-Kasaoka N, Shozawa C, Sano K, Kamei Y, Kasaoka S, Hosokawa Y, Ezaki O (2007) Taurine (2-aminoethanesulfonic acid) deficiency creates a vicious circle promoting obesity. Endocrinology 147: 3276–3284

Zhang M, Bi LF, Fang JH, Su XL, Da GL, Kuwamori T, Kagamimori S (2004) Beneficial effects of taurine on serum lipids in overweight or obese non-diabetic subjects. Amino Acids 26:267-271

Chapter 40
The Effects of Dietary Taurine Supplementation on Plasma and Liver Lipid in Ovariectomized Rats

Mi-Ja Choi and Kyung Ja Chang

Abstract The purpose of this study was to investigate the effect of dietary taurine supplementation on plasma and liver lipid content in ovariectomized (OVX) rats. Forty female Sprague-Dawley rats were divided into two groups, one that was OVX and the other sham operated (Sham). They were further divided into either control or taurine supplemented dietary (2.0 g/100 g diet) subgroups. Plasma lipids (Total cholesterol, HDL-cholesterol, and Triglycerides) and liver lipids were determined by using commercial kits (Wako Pure Chemical, Osaka). LDL-cholsterol content was estimated by the equation of Fridewald et al (1972). There were no significant differences in body weight gain and food intake between the OVX or controls rats that were either control or taurine fed, but body weight gain was higher in the OVX groups. Food intake and the food efficiency ratio were not significantly different between the dietary subgroups. The concentrations of plasma total cholesterol and triglycerides were lower in the taurine, OVX group while the HDL-cholesterol and the LDL-cholesterol levels were not reduced in the two taurine fed subgroups. There were no significant differences in liver cholesterol and triglyceride content between the control, OVX and the taurine, OVX rats. There were no significant differences in the atherogenic index between control, OVX and taurine, OVX groups. These results confirmed that taurine exerts some beneficial effects on hypercholesterolemia and hypertriglyceridemia of the OVX rats.

Abbreviations *OVX*, ovariectomized; *Sham*, sham operated

40.1 Introduction

There is a proportional increase in the risk of coronary heart disease with rising serum cholesterol levels (Lloyd-Jones et al. 2003). Menopause, whether natural or surgically induced, is associated with elevated levels of circulating total cholesterol and LDL cholesterol (LDL-C), placing postmenopausal women at greater

M.-J. Choi (✉)
Department of Food and Nutrition, Keimyung University, Korea

J. Azuma et al. (eds.), *Taurine 7*, Advances in Experimental Medicine and Biology 643, DOI 10.1007/978-0-387-75681-3_40, © Springer Science+Business Media, LLC 2009

risk of coronary heart disease (CHD) (Bruschi et al. 1996; Grodstein et al. 1996). These changes are a consequence of the reduction in the levels of circulating estrogen. The mechanism in which reduced circulating levels of estrogen elevate plasma cholesterol levels is poorly understood. Reductions in the activity of the hepatic LDL receptor (LDLR) (Riedel el al. 1993) and/or cholesterol 7-hydroxylase (CYP7) activity (Kushwaha and Born 1991) are presumably involved. Numerous studies suggest that taurine, 2-amino ethane sulfonic acid, has beneficial effects on cholesterol metabolism by minimizing the effects of hypercholesterolemia (Murakami et al. 1996). Taurine upregulates LDLR activity in a human hepatoma cell line and in hamsters (Murakami et al. 2002). Therefore, upregulation of hepatic LDLR activity may be involved in the hypocholesterolemic effect of taurine. Estrogen replacement therapy (ERT) in postmenopausal women reduces the risk of CHD in part by modulating serum cholesterol. However, ERT and cholesterol-lowering pharmacologic agents may be accompanied by side effects. On the other hand, taurine is thought to be quite safe and there is little concern about side effects of excessive taurine intake (Furukawa et al. 1991).

Coronary heart disease is the leading cause of death in the U.S. and in most developed countries. Taurine, aminoethane sulfonic acid, is one of the most abundant free amino acids in animal cells and tissues and is thought to exert numerous functions, such as antioxidation, anti-inflammation, osmoregulation and nerve regulation. Since high serum cholesterol is one of the major risk factors for atherosclerosis and coronary heart diseases, taurine is thought to prevent the development of atherosclerosis. The effect of taurine on plasma lipids and coronary heart disease has been the subject of extensive research. Most studies investigating the effects of taurine on serum lipids were conducted in hypercholesterolemic subjects or animals (Yokogoshi et al. 1999), largely because of the positive relationship between baseline cholesterol and cardiovascular disease. However, the effect of taurine in an animal model of ovariecFtomy (OVX) has not been much studied. The purpose of this study was to investigate the effect of dietary taurine supplementation on plasma and hepatic lipid levels of OVX rats. Therefore, in this study we examined whether taurine benefited OVX rats from ovarian hormone deficiency-induced hypercholesterolemia.

40.2 Methods

40.2.1 Animals and Diet

Forty female Sprague-Dawley rats (Biogenomics, Seoul, Korea) weighing about 190 g were divided into two groups. One group was ovariectomized and the other group was sham operated (SHAM). They were further subdivided into control and taurine supplementation dietary subgroups.

Taurine supplementation was given at 2.0 g/100 g diet. All rats were fed an experimental diet and deionized water ad libitum for 6 weeks. The composition of each

Table 40.1 Composition of experimental diets (g/100 g diet)

Ingredients	Control	Taurine
Casein[1]	20	20
Corn starch[2]	52.9486	52.9486
Sucrose	10	10
Soybean oil	7	7
Cellulose[3]	5	5
Min-mix[3]	3.5	3.5
Vit-mix[4]	1.0	1.0
L-cystine	0.30	0.30
Choline[3]	0.25	0.25
Tert-butyl hydroquinone[3]	0.0014	0.0014
Taurine[4]	-	2.0

[1]Lactic Casein, 30 mesh, New Zealand Dairy Board, Willington, N.Z.
[2]Corn Starch, Doosan Co. 234-17 Maam-Ri, Bubal-Eup, Inchon-City, Kyunggi-Do.
[3]Mineral Mixture (AIN-93M), supplied by U.S. CORNING Laboratory Services Company. TEKLAD TEST DIETS, Madison.
[4]Vitamin mixture (AIN-93), supplied by U.S. CORNING Laboratory Services Company. TEKLAD TEST DIETS, Madison, Wisconsin.
[5]Cellulose, supplied by SIGMA Chemical Company.
[6]Choline, supplied by SIGMA Chemical Company.
[7]Soybean oil, CJ CheilJedang Co. Seoul, Korea.
[8]Taurine, Dong-A Pharm. Co. Ltd. 434-4 Moknae-dong, Ansan-City, Kyunggi-Do.

diet is shown in Table 40.1. Rats were individually housed in stainless steel cages in a room of controlled temperature (23°C) and humidity (55%). Rats were maintained on a 12 h light (07:00–19:00 h) and dark cycle. On the last day of the experimental period, a blood sample was collected from the abdominal aorta.

40.2.2 Plasma and Hepatic Lipids Analysis

The plasma was separated from the blood by centrifugation (1600 × g, 15min, 4°C) and was stored at −70°C until analysis. After blood collection, the liver was immediately removed, washed with cold saline (9 g NaCl/L), blotted dry on filter paper, weighed, and stored at −70°C until analysis.

Plasma lipids (Total cholesterol, HDL-cholesterol, and Triglycerides) were determined using commercial kits (Wako Pure Chemical, Osaka, Japan). LDL-cholesterol levels were estimated by the equation of Fridewald. About 2 g of liver were homogenized, and lipids were extracted with a chloroform:methanol mixture (2:1. v/v) as described by Folch (1957). The concentration of cholesterol and triglyceride in liver as determined enzymatically using a commercial kit (Waco, Japan).

40.2.3 Statistic Analysis

Statistical differences between the groups were evaluated by two-way ANOVA, using a computer software package (version 9.13, SAS Institute Inc, Cary, NC).

Individual comparisons were made by Duncan's multiple range test. Differences were considered significant when $p < 0.05$. Data are expressed as means ± SD.

40.3 Results

40.3.1 Weight Gain and FER

Body weight gain and food intake of rats fed the taurine diet (20 g taurine/kg diet) did not differ from those of rats fed the control diet (Table 40.2). The results of this study indicate that body weight gain was higher in OVX groups than in SHAM groups regardless of diet. Food intake and the food efficiency ratio were not significantly different between the groups. Despite taurine supplementation, body weight gain and food intake did not differ between the groups (Table 40.3).

40.3.2 Blood and Hepatic Lipids

The OVX rats fed the control diet showed higher serum total cholesterol levels than Sham rats fed the control diet, but taurine supplementation (2 g/kg) significantly reduced the concentration of serum total cholesterol. OVX rats fed the high taurine diet had lower triglyceride and total cholesterol levels than those fed the control diet (Table 40.4).

Table 40.2 The effect of diet on body weight and weight gain in rats

	Sham Control	Taurine	Ovx Control	Taurine
Initial weight(g)	204.7±9.4[1,a]	201.5±6.3[a]	211.5±6.7[a]	204.1±7.6[a]
Final weight (g)	276.6±21.9[a]	278.8±16.7[a]	317.9±26.3[b]	329.2±22.6[b]
Weight gain (g)	71.9±13.2[a]	77.30±13.9[a]	106.38±13.2[b]	119.10±11.2[b]

[1]Mean ± SD Values with different superscripts within a given row are significantly different at $p < 0.05$ by Duncan's multiple range test.

Table 40.3 The effect of diet on food intake and the food efficiency ratio (FER)

	Sham Control	Taurine	Ovx Control	Taurine
Food Intake (g)	17.60±1.09[1,a]	18.23±1.09[a]	18.89±2.26[a]	21.09±1.77[a]
FER	0.09±0.01[a]	0.10±0.01[a]	0.13±0.05[a]	0.14±0.01[a]

[1]Mean ± SD Values with different superscripts within a given row are significantly different at $p < 0.05$ by Duncan's multiple range test.

Table 40.4 The effect of diet on lipid levels

	Sham	Ovx		
	Control	Taurine	Control	Taurine
Total cholesterol (mg/dl)	76.53±6.53[1,b]	72.90±5.10[b]	83.16±5.37[a]	66.50±7.22[c]
Triglyceride (mg/dl)	85.16±7.3[ab]	77.08±12.20[b]	89.45±7.54[a]	80.00±3.59[ab]
HDL – cholesterol (mg/dl)	22.20±4.56[a]	23.48±2.75[a]	21.30±1.25[a]	21.75±6.18[a]
LDL – cholesterol (mg/dl)	35.96±4.23[a]	34.00±13.21[b]	39.60±13.43[a]	31.0±17.24[a]
Athrogenic index	2.52±0.80[a]	2.13±0.36[a]	2.51±0.42[a]	1.90±0.36[b]

[1]Mean ± SD Values with different superscripts within a given row are significantly different at $p < 0.05$ by Duncan's multiple range test.

Table 40.5 The effect of diet on liver lipid levels

	Sham	Ovx		
	Control	Taurine	Control	Taurine
Total cholesterol (mg/g)	193.8±1.5[1,a]	180.8±22.5[a]	187.2±6.6[a]	179.7±17.2[a]
Triglycerides (mg/g)	119.2±24.4[a]	93.1±46.6[a]	151.5±37.1[a]	95.1±36.6[a]

[1]Mean ± SD Values with different superscripts within a given row are significantly different at $p < 0.05$ by Duncan's multiple range test.

There were no significant differences in liver cholesterol and triglyceride content between the control, OVX and the taurine, OVX groups (Table 40.5).

40.4 Discussion

Menopause, whether natural or surgically induced, is associated with elevated levels of circulating total cholesterol and LDL cholesterol (LDL-C), placing postmenopausal women at greater risk of coronary heart disease (CHD) (Bruschi et al. 1996; Grodstein et al. 1996). These changes are a consequence of the reductions in the levels of circulating estrogen. The mechanism by which reduced levels of circulating estrogen elevate plasma cholesterol levels is poorly understood. Taurine (2-aminoethanesulfonic acid) is synthesized from the sulfur amino acid, cysteine. Numerous studies suggest that taurine has beneficial effects on cholesterol metabolism by improving the effects that hypercholesterolemia exerts. Several studies have suggested that taurine increases cholesterol elimination from the body by stimulating bile acid synthesis, thereby reducing plasma cholesterol levels (Yokogoshi et al. 1999; Kishida et al. 2001). Plasma cholesterol levels are higher in OVX rats than in Sham-operated rats (Kishida et al. 2001).

In rats fed corn oil, taurine increases fecal bile acid excretion and has a hypocholesterolemic effect. In rats fed coconut oil, however, taurine does not have a hypocholesterolemic effect even though it increases fecal bile acid excretion. The effect of taurine on ovarian hormone deficiency-induced changes in cholesterol metabolism is influenced by the types of fatty acids consumed (Kishida et al. 2003).

Supplementation of the high-cholesterol diet with taurine significantly reduces serum total cholesterol and increases HDL-cholesterol levels in rats (Nanami et al. 1996). Choi et al. (2006) also reported that taurine has a hypocholesterolemic effect in an experimentally induced hypercholesterolemic animal model. Hypercholesterolemia caused by feeding a high cholesterol diet is significantly improved in a dose-dependent manner by supplementation with taurine (Yokogoshi et al. 1999) The addition of taurine to the diet does not reduce serum levels of cholesterol in control rats.

In this study, taurine lowered plasma total cholesterol levels in OVX rats fed corn oil, but tended to increase ($P = 0.104$) it in OVX rats fed coconut oil (Kishida et al. 2003). Taurine lowered plasma total cholesterol content ($P < 0.02$) in rats fed corn oil, but not in those fed coconut oil. In rats fed both types of oils, taurine increased LDL receptor mRNA levels ($P < 0.01$), hepatic cholesterol 7-hydroxylase activity ($P < 0.01$) and fecal bile acid excretion ($P < 0.01$). Taurine increased HMG-CoA reductase mRNA levels ($P < 0.02$) in the liver of rats fed coconut oil, but not in those fed corn oil. Taurine also increased total hepatic lipid ($P < 0.05$) and triglyceride ($P < 0.05$) content in rats fed corn oil, but not in those fed coconut oil. The effect of taurine on ovarian hormone deficiency-induced changes in cholesterol metabolism is influenced by the types of fatty acids in the diet (Kishida et al. 2003). We used soybean oil rich in oleic and linoleic acids as dietary fat sources. Therefore, our results agree with the earlier study.

40.5 Conclusion

OVX rats fed taurine showed significantly lower total serum cholesterol and triglyceride levels than those fed the control diet. We confirmed that dietary supplementation with taurine decreases estrogen deficiency-induced hypercholesterolemia and hypertriglyceridemia in rats. These results indicate that taurine exerts a beneficial effect on hypercholesterolemia and hypertriglyceridemia in estrogen-deficiency induced rats.

Acknowledgments We thank the Dong-A Pharmaceutical Co. which donated taurine.

References

Bruschi F, Meschia M, Soma M, Perotti D, Paoletti R, Crosignani PG (1996) Lipoprotein(a) and other lipids after oophorectomy and estrogen replacement therapy. Obstet Gynecol 88:950–954

Choi MJ, Kim JH, Chang KJ (2006) The effect of dietary taurine supplementation on plasma and liver lipid concentrations and free amino acid concentrations in rats fed a high-cholesterol diet. Adv Exp Med Biol 583:235–242

Folch J, Lees M, Sloanestanley GH (1957) A simple method for the isolation and purification of total lipids from animal tissue. J Biol Chem 226:497–509

Fridewald WT, Lavy RI, Fredricson DS (1972) Estimation of low density lipoprotein cholesterol in plasma, without use of the preparative ultracentrifuge. Clin Chem 18:499–502

Furukawa S, Katto M, Kouyama H, Nishida I, Kikumori M, Taniguchi Y, Toda T, Araki H (1991) Repeated dose toxicity study of intravenous treatment with taurine for 13 weeks and recovery test for 5 weeks in rat. J. Jpn Pharmacol Ther 19:275–306

Grodstein F, Stampfer MJ, Manson JE, Colditz GA, Willett WC, Rosner B, Speizer FE, Hennekens, CH.(1996) Postmenopausal estrogen and progestin use and the risk of cardiovascular disease. N Engl J Med 335:453–461

Kishida T, Akazawa T, Ebihara K (2001) Influence of age and ovariectomy on the hypocholesterolemic effects of dietary taurine in rats fed a cholesterol free diet. Nutr Res 21:1025–1033

Kishida T, Miyazato S, Ogawa H, Ebihara K (2003) Taurine prevents hypercholesterolemia in ovariectomized rats fed corn oil but not in those fed coconut oil. J Nutr 133:2616–2621

Kushwaha RS, Born KM (1991) Effect of estrogen and progesterone on the hepatic cholesterol 7-alpha-hydroxylase activity in ovariectomized baboons. Biochim Biophys Act 1084:300–302

Lloyd-Jones DM, Wilson PWF, Larson MG, Leip E, Beiser A, D'Agostino RB, Cleeman JI, Levy D (2003) Lifetime risk of coronary heart disease by cholesterol levels at selected ages. Arch Intern Med 163:1966–1972

Murakami S, Kondo Y, Toda Y, Kitajima H, Kameo K, Sakono M, Fukuda N (2002) Effect of taurine on cholesterol metabolism in hamsters: up-regulation of low density lipoprotein (LDL) receptor by taurine. Life Sci 70:2355–2366

Murakami S, Yamagishi I, Asami Y, Ohta Y, Toda Y Nara Y, Yamori Y (1996) Hypolipidemic effect of taurine in stroke-prone spontaneously hypertensive rats. Pharmacolog 52:303–313

Nanami K, Oda H, Yokogoshi H (1996) Antihypercholesterolemic action of taurine on streptozotocin-diabetic rats or on rats fed a high cholesterol diet. Adv Exp Med Biol 403: 561–568

Riedel M, Rafflenbeul W, Lichtlen P (1993) Ovarian sex steroids and atherosclerosis. Clin Invest 71:406–412

Yokogoshi H, Mochizuki H, Nanami K, Hida Y, Miyachi F, Oda H (1999) Dietary taurine enhances cholesterol degradation and reduces serum and liver cholesterol concentrations in rats fed a high-cholesterol diet. J Nutr 129:1705–1712

Part VII
Potential Therapeutic Effects of Taurine

Chapter 41
Comparative Studies on 24-hour Urinary Excretion in Japanese and Chinese Adults and Children – Need for Nutritional Education

Mari Mori, Jin-Wen Xu, Hideki Mori, Cheng Feng Ling, Guo Hong Wei, and Yukio Yamori

Abstract In order to evaluate the effect of nutritional education on the risk of developing lifestyle-related diseases, we measured taurine and isoflavone content in 24-hour urine samples (24-U) of 3rd grade Chinese children (CC) and of age-matched Japanese children (JC), as well as adult Chinese and Japanese (CA, JA) according to the WHO-CARDIAC (Cardiovascular Diseases and Alimentary Comparison) Study protocol. There was a significantly higher prevalence of obesity and "thin" individuals in CC compared with JC. While K intake was not significantly different in the children, the sodium to potassium ratio (Na/ K) and the intake of sodium chloride (NaCl) were significantly higher in CC than in JC. Taurine excretion (24-U) was significantly higher in CC than in JC, but isoflavone excretion was significantly lower in CC than in JC. Taurine excretion was significantly lower in CA than in JA, while isoflavone excretion was almost the same in CA and JA. After nutritional education CC consumed more isoflavones than the control group that had been subjected to only environmental education. JC exhibited significantly higher 24-U taurine and isoflavone excretion after taking the nutritional class.

Abbreviations *24-U*, 24-hour urine samples; *CC*, Chinese children; *JC*, Japanese children; *CA*, Chinese adults; *JA*, Japanese adults

41.1 Introduction

Our world-wide cross-sectional epidemiological study coordinated by WHO (WHO-Cardiovascular Diseases and Alimentary Comparison Study= CARDIAC Study) (WHO and WHO Collaborating Centers. 1986; Yamori 1989; Yamori et al. 1990), confirmed that dietary factors related to cardiovascular disease (CVD) were markedly

M. Mori (✉)
Mukogawa Women's University Institute for World Health Development, Research Institute for Production Development, Kyoto, Japan

J. Azuma et al. (eds.), *Taurine 7*, Advances in Experimental Medicine and Biology 643,
DOI 10.1007/978-0-387-75681-3_41, © Springer Science+Business Media, LLC 2009

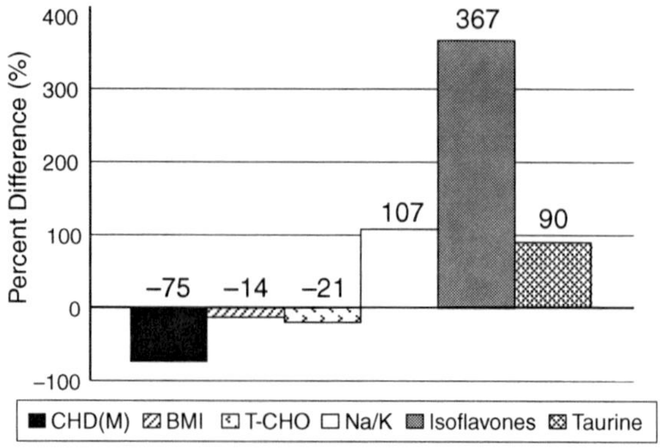

Fig. 41.1 Coronary heart disease (CHD) and related risk and dietary factors in Asian communities compared with Occidental data. CARDIAC Study sites: *Asian Communities** (Japan-Toyama, Beppu, Ohda, China-Shanghai, Shijiazhuang) *Occidental Communities** (Equador-Quito, England-Belfast, Stornoway, Newzealand-Dunedine, Portugal-Lisboa, Russia-Moscow, Spain-Madrid, Sweden-Goetheborg). CHD(M):Cardiovascular heart diesases (Male), BMI: Body mass index (weght/height^2(kg/m^2)), T-cho: Total cholesterol. *Communities, the age-adjusted CHD mortality rates which were available, were selected for data analysis

different in Asian and Occidental populations (Yamori 2006; Yamori et al. 2006). Differences between Asian communities and Occidental societies relative to cardiovascular disease and the levels of biomarkers of seafood and soy product intake, namely, the 24-U Na/ K ratio and both taurine and isoflavone excretion, are shown in Fig. 41.1.

Since traditional Asian diets, which contain large amounts of fish and soybeans, are now on a decline among the younger generation in Japan and China, we initially compared the nutritional status in both countries among adults and young people, some who have been educated to consume more fish and soybeans.

In order to evaluate the effect of nutritional education on risk reduction of lifestyle-related diseases, we measured taurine and isoflavone in 24-U from CC and age-matched JC, as well as from Chinese and Japanese adults (CA, JA) as described in the CARDIAC Study.

41.2 Methods

41.2.1 Study Protocol for CC

Four weeks before the start of the study parents were invited to an explanatory meeting where oral and written information about the study was administered and 60 children (27 boys and 33 girls) of two 3rd grade classes were recruited into the study after signing the informed consent. Prior to initiation of the study in the beginning of

September 2006, height (H) and weight (W) were measured to calculate the Rohrer index $(W(kg)/H(cm)^3) \times 10^7$).

Two classes were assigned to either the nutritional or environmental education class (E), with after school classes conducted six times over a 3 month period. The E group learned about environmental problems and did not study nutrition. In the nutritional class (N), the topic included ingredients used in cooking, the basics in how to cook, how to eat well balanced diets, and the nutritious ingredients found in fish and soybeans. Twenty four-hour urine samples (Yamori et al. 1984, 2006) were collected before and after the study period of three months. Subjects who failed to collect 24-U using an aliquot cup were excluded from data analysis of sodium (Na), potassium (K), creatinine (Cre), taurine, isoflavones and the creatinine coefficient (creatinine/ body weight). Photographs of meals were taken before and during the study period to confirm dietary habits. Morning spot urine samples were collected 3 times to confirm the content of the nutritional biomarkers.

41.2.2 Urine Samples Analysis

Urine samples were analyzed for Na, K and Cre at SRL Co., Ltd (Tokyo). Samples judged as complete were further analyzed for isoflavones and taurine by HPLC.

41.2.3 Comparison with JC

Anthropometric measurements and nutritional biomarkers of 24-U collection were obtained from the investigation of the eating habits of 172 JC, and data of 74 children of the lower grades (44 boys and 30 girls). Those values were compared with the data from the CC. JC were divided into a nutritional education group 17 boys and 12 girls) (N) and a control group lacking nutritional education (26 boys and 18 girls) (C).

Table 41.1 Schedule for Nutritional Education Class in Shanghai

	2006					
	September		October		November	
Nutrition Class	○	○	○	○	○	○
Spot Urine Collection in the Morning	○		○			○
24-hour Urine Collection	○					
Food Photos	○		○			○

41.2.4 Comparison of CA with JA in CARDIAC Study

Data from the CARDIAC Study carried out in a suburban village near Shanghai city in 1986 were compared with data collected at 7 study sites in Japan (Hirosaki, Toyama, Hiroshima, Ohda, Beppu, Kurume, Naha) in 1986-1988.

41.2.5 Statistic Analysis

Statistical analyses were conducted with Stat View 5.0 for Mac and data were expressed as means ± standard deviation (SD). The Students t-test was used to compare group difference as well as the difference from the baseline, P-values less than 0.05 were considered statistically different.

41.3 Results

41.3.1 Anthropometric Measurements

The mean ± SD of 60 CC was 132.5 ± 0.1 cm in height and 30.4 ± 6.3 kg in weight. Judged by the Rohrer index (the values below 115 were "thin" and over 145 were obese), 16 were "thin" (26.7%) and 14 were obese (23.3%). Among 74 JC there were 15 "thin" (19.2%) and 17 obese (21.8%). The prevalence of thin individuals was higher in CC than in JC.

41.3.2 Twenty-four-hour Urine

Twenty-four-hour urine samples were obtained from 40 CC and the results are shown in Fig. 41.2. While K intake was not significantly different between CC and JC, urinary levels of NaCl were significantly higher ($p < 0.05$) in CC than in JC. Therefore, the Na/ K ratio was significantly higher in CC ($p < 0.05$) than in JC.

Taurine excretion was significantly higher in CC than in JC while isoflavone excretion was significantly lower in CC than in JC. However, taurine excretion was significantly lower in CA than in JA, while isoflavone excretion was almost identical in CA and JA in the CARDIAC study (Fig. 41.3).

Nutritional education of CC was effective in changing the nutritional habits of CC, as indicated by the 24-U data, particularly isoflavone excretion after completion of the classes in nutritional education. The N group consumed more traditionally healthy foods, as isoflavone excretion tended to increase from the baseline of 8.1 ± 8.9 to 44.7 ± 82.0. Taurine was high enough at the baseline and increased from 716 ± 377 to 736 ± 368, which was not significantly different.

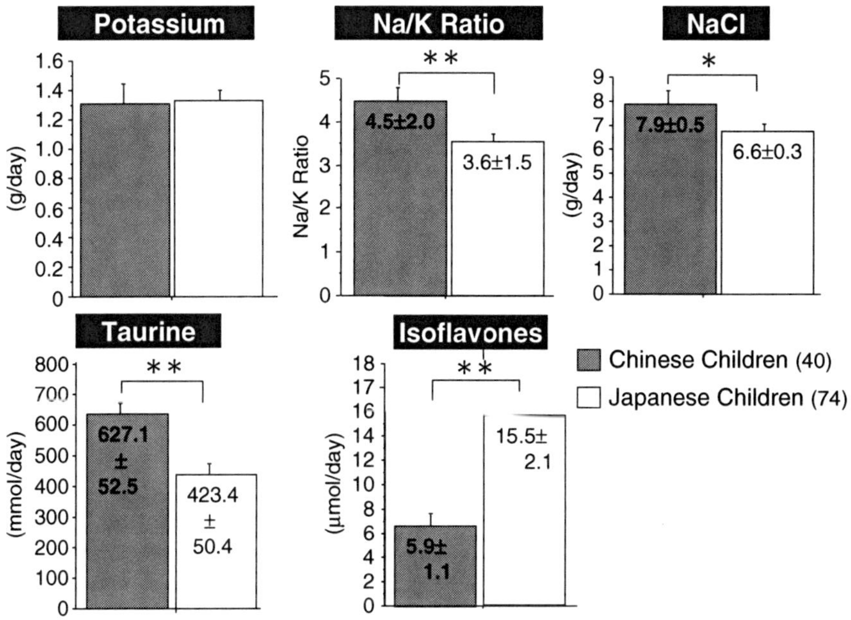

Fig. 41.2 Comparison of mineral, taurine and isoflavone excretions in 24-U among CC and JC. mean± SE, * $p < 0.05$, ** $p < 0.01$

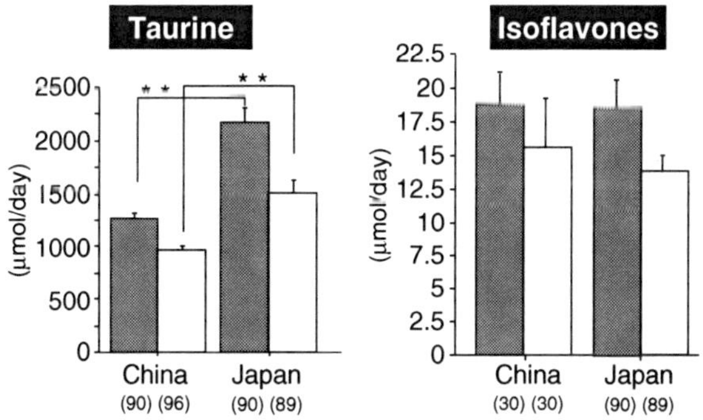

Fig. 41.3 Comparison of taurine and isoflavone excretion in 24-hour urine from adults. ** : $p < 0.001$; Gray bars: Male, White bars: Female

The 24-U taurine excretion value was 565.6±239 in the JC/ N group and 344.4± 345 in JC/ C, while isoflavone excretion was 18.4 ± 19.5 in JC/ N and 13.8 ± 17.0 in JC/ C. Taurine data among Japanese was significantly higher in the N group than in the C group while isoflavone data tended to be higher in the N group than in the C group (Fig. 41.4).

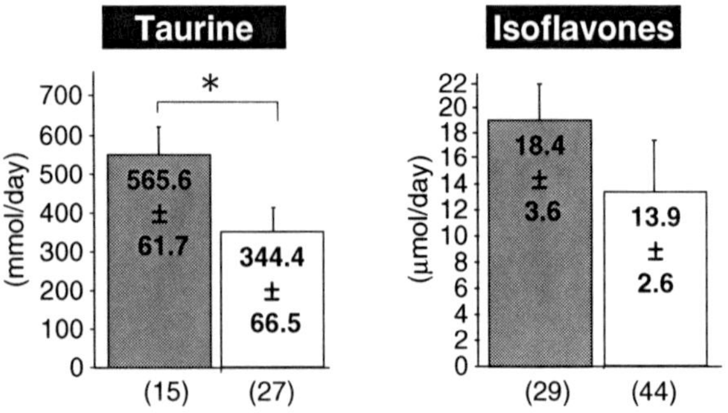

Fig. 41.4 Effect of nutrition class on 24-U taurine and isoflavone excretion from Japanese children. * $p < 0.05$; Gray bars: Nutrition Class, *White bars*: Lacking nutrition class

41.4 Discussion

The data of 24-U in CC and JC were compared with previous data from CA and JA reported in the WHO CARDIAC Study, which was carried out approximately 20 years ago. Taurine excretion in CA was significantly lower than that in JA, and isoflavone excretion in CA was nearly identical to the level in JA. The data of CC confirmed that 24-U taurine excretion in CC was significantly higher than that in JC, indicating that CC consumed more fish than JC and that Japanese dietary habits, which used to include the consumption of large amounts of fish, had changed. Based on a comparison of CA and JA, fish consumption decreased in JC, while it increased in CC in the last 20 years. The data in CC and JC were different from those in adults (CA and JA). Isoflavone excretion (24-U) was significantly lower in CC than in JC; the data indicate that the dietary habit of eating soybeans had changed, such that CC no longer consumed large amounts of soybeans.

The study also confirmed that the traditional custom of consuming fish and soybeans, while being lost in China and Japan, could recover. Both fish and soy intake increase with nutritional education of JC, while intake of soybeans increased with nutritional education of CC.

The deterioration of traditional dietary habits is one of the causes of the increase in the prevalence of lifestyle-related diseases. It is very important to convey traditional dietary habits through nutritional education of children. This should prevent the further increase in lifestyle-related disease in the adult populations of Japanese and Chinese.

41.5 Conclusion

Taurine intake was higher in CC than in JC, indicating that fish intake was increased among CC but decreased among JC when compared with taurine intake in adults, as previously reported in China and Japan. In contrast, isoflavone intake was very low in CC despite similar intake in the adult populations of both countries, indicating that traditional oriental dietary custom is being lost in China. These data represent a warning and the need for nutritional education in both countries. This study implies that proper education can prevent recent increases in the prevalence of life style related diseases through the promotion of a diet rich in soy (in CC) and in fish (in JC) (Yamori 2006; Yamori et al. 2001, 2006)

Acknowledgments We acknowledge the cooperation of Chinese and Japanese children volunteers, as well as of Chinese and Japanese scientists, who devoted themselves to the successful completion of nutritional education and the comparison study. We are grateful for research funds from the International Medical Center of Japan.

References

WHO and WHO Collaborating Centers (1986) CARDIAC (Cardiovascular Diseases and Alimentary Comparison) Study Protocol. Shimane, Geneva

Yamori Y (1989) Hypertension and biological dietary markers in urine and blood: a progress report from the CARDIAC Study group. In: Yamori Y, Strasser T (eds) New horizons in preventing cardiovascular diseases, Elsevier, Amsterdam, pp 111–126

Yamori Y (2006) Food factors or atherosclerosis prevention. Asian perspective derived from analysis of world wide dietary biomarkers. Exp Clin Cardiol 11:94–98

Yamori Y, Nara Y, Kihara M, Mano M, Horie R (1984) Simple method for sampling consecutive 24-hour urine for epidemiological and clinical studies. Clin Exp Hypertens A6:1161–1167

Yamori Y, Nara Y, Mizushima S, Mano M, Sawamura M, Kihara M, Horie R (1990) International cooperative study on the relationship between dietary factors and blood pressure: a report from the cardiovascular diseases and alimentary comparison study. J Cardiovasc Pharmacol 16; 43–47

Yamori Y, Liu L, Ikeda K, Miura A, Mizushima S, Miki T, Nara Y, WHO-Cardiovascular Disease and Alimentary Comprarison (CARDIAC) Study Group (2001) Distribution of twenty-four hour urinary taurine excretion and association with ischemic heart disease mortality in 24 populations of 16 countries: results from the WHO- CARDIAC study. Hypertens Res 24; 315–322

Yamori Y, Liu L, Mizushima S, Ikeda K, Nara Y; CARDIAC Study Group (2006) Male cardiovascular mortality and dietary markers in 25 population sampled of 16 countries. J Hypertens 24; 1499–1505

Chapter 42
Taurine Normalizes Blood Levels and Urinary Loss of Selenium, Chromium, and Manganese in Rats Chronically Consuming Alcohol

Taesun Park, Keunsook Cho, Sung-Hee Park, Dong-Hee Lee, and Ha Won Kim

Abstract The present study was undertaken to evaluate effects of dietary taurine supplementation on the homeostasis of trace elements, including Se, Cu, Mn and Cr, in rats chronically consuming alcohol. Male SD rats were fed for 8 wk a liquid form of a control diet (CD), an ethanol diet (ED), or a taurine-supplemented ethanol diet (TED). Plasma Se and Mn concentrations were significantly lower in the ED rats than in the CD rats; dietary taurine supplementation corrected alcohol-induced decreases in plasma Se and Mn levels. Chronic alcohol consumption significantly increased urinary excretion of Se (a 53% increase, $p<0.05$), Cr (a 62% increase, $p<0.05$), Mn (a 45% increase, $p<0.05$) and Cu (a 30% increase, $p<0.05$) in rats. Urinary losses of these trace elements induced by chronic alcohol consumption in rats were abolished by taurine supplementation. These results suggest that taurine supplementation in rats may protect against Se, Cr and Mn insufficiency caused by chronic alcohol-mediated loss of the trace elements in the urine.

Abbreviations *ALT*, alanine aminotransferase; *AST*, aspartate aminotransferase; *γ GTP*, γ-glutamyl transpeptidase; *CD*, control diet group; *ED*, ethanol diet group; *TED*, taurine-supplemented ethanol diet group

42.1 Introduction

Alcoholism remains one of the major causes of nutritional deficiency around the world; the interaction between nutrition and alcoholism occurs at many levels and is complex. Chronic alcoholics frequently suffer from nutritional deficiencies related to decreased intake, reduced intestinal uptake, and impaired utilization of nutrients (Ryle and Thomson 1984). Chronic alcohol-related brain damage has often

T. Park (✉)
Department of Food and Nutrition, Brain Korea 21 Project, Yonsei University, Korea

J. Azuma et al. (eds.), *Taurine 7*, Advances in Experimental Medicine and Biology 643,
DOI 10.1007/978-0-387-75681-3_42,

been attributed to nutritional depletion, particularly from reductions in thiamine, riboflavin, folic acid (Keating et al. 2007), ascorbic acid (Faizallah et al. 1986), nicotinamide, pyridoxine (Ryle and Thomson 1984), vitamin A and/or vitamin E (Marotta et al. 1994). Knowledge of the interaction between chronic alcohol consumption and trace mineral homeostasis is limited. In the serum of cirrhotic patients, Se (Burk et al. 1998) and Zn levels are significantly decreased, whereas Cu levels are significantly increased (Zarski et al. 1985). The elimination of Zn, Cu, or Fe by the urine was increased (González-Reimers et al. 1998) and that of Mn fell in patients with alcoholic liver disease (Rodríguez-Moreno et al. 1997). Chronic alcohol feeding causes Zn depletion in muscle (González-Reimers et al. 1993) and an elevated loss of Zn, Fe and Mn in the urine of rats (González-Reimers et al. 1998).

Taurine (2-amino ethane sulfonic acid) is a sulphur containing amino acid present in free form in many tissues of man and animals (Pushpakiran et al. 2004). Taurine is considered to exert protection against alcohol-induced fatty liver and hepatitis. When co-administered with alcohol, taurine not only protects against alcohol-induced hepatic steatosis but can also reverse hepatic steatosis once it has developed (Kerai 1998). The purpose of this study was to investigate the role of taurine in maintaining trace element homeostasis, including that of Mn, Cr, Cu and Se, in rats chronically loaded with alcohol.

42.2 Methods

42.2.1 Animals and Diet

The animals used in this study were treated in accordance with the Guide for the Care and Use of Laboratory Animals (Institute of Laboratory Animal Resources, Commission on Life Sciences, National Research Council, 1996), as approved by the Institutional Animal Care and Use Committee of Yonsei University. Twenty-four male SD rats (Samyuk Laboratory Animal Inc., Korea) weighing 146–191 g were randomly divided and fed one of the three experimental diets for 8 wk: the control diet (CD), the ethanol diet (ED), and the taurine-supplemented ethanol diet (TED). The rats in the ED group consumed a liquid diet formulated as described by Lieber and DeCarli (1986). The TED was identical to the ED but contained 3.75 g taurine/L diet (Table 42.1). The intake of the ED rats was ad libitum, and the pair-fed CD and TED animals received the amount that the ED rats had consumed the previous day.

At the end of the experimental period, the animals were subjected to a 12 h fast and then anesthetized with ether. Three days before sacrifice, the rats were placed in metabolic cages, and their urine was collected for three consecutive days. Blood samples were collected from the abdominal aorta with a heparinized syringe under light ether anesthesia. The livers were removed, rinsed with phosphate-buffered saline, and weighed. The plasma, urine, and liver samples were stored at −20°C and −80°C, respectively, until they were analyzed.

Table 42.1 Composition of the experimental liquid diets

	CD	ED	TED
		(g/L)	
Casein	41.4	41.4	41.4
Corn oil	2.5	8.5	8.5
Olive oil	28.4	28.4	28.4
Safflower oil	2.7	2.7	2.7
Dextrin-maltose[a]	115.2	25.6	25.6
Vitamin mix[b]	2.5	2.5	2.5
Mineral mix[c]	8.75	8.75	8.75
Choline bitartarate	0.53	0.53	0.53
Fiber	10.0	10.0	10.0
Sodium carrageenate	2.0	2.0	2.0
Ethanol	–	50.0	50.0
Taurine	–	–	3.75

[a]Starch: sucrose = 80 : 20. [b]Vitamin mixture contained (g/kg); thiamine · HCl 0.6; riboflavin 0.6; nicotinamide 25; pyridoxine · HCl 0.7; nicotinic acid 3; D-calcium pantothenate 1.6; folic acid 0.2; D-biotin 0.02; vitamin B_{12} 0.001; retinyl palmitate (250,000 IU/g) 1.6; dl-α-tocopherol acetate (250 IU/g) 20; cholecalciferol 0.25; menaquinone 0.05; sucrose, finely powdered 972.9. [c]AIN-76 mineral mixture contained (g/kg); $CaHPO_4$ 500; NaCl 74; $K_2H_6O_7H_2O$ 220; K_2SO_4 52; MgO 24; $MnCO_3$ 3.57; $Fe(C_6H_5O_7)\cdot 6H_2O_6$; $ZnCO_3$ 1.6; $CuCO_3$ 0.3; KIO_3 0.01; $Na_2SeO_3\cdot 5H_2O$ 0.01; $CrK(SO_4)_2$ 0.55; sucrose, finely powdered 118.00.

42.2.2 Biochemical Assays

For the determination of ethanol, plasma and liver homogenates were deproteinized with ice-cold trichloroacetic acid. Ethanol concentration in the protein free supernatant fraction was measured using a commercially available kit (Sigma). The activities of alanine aminotransferase (ALT), aspartate aminotransferase (AST) and γ-glutamyltranspeptidase (γ-GTP) as well as the concentration of total bilirubin in plasma were measured via an auto analyzer (Express Plus, Chiron Diagnostics). Plasma and urinary trace element concentrations were determined with an atomic absorption spectrophotometer (Duo 530, Beckman).

Liver samples were homogenized with three volumes of STE buffer (250 mM sucrose, 50 mM Tris-Cl, pH 7.5, and 1 mM EDTA with protease inhibitor cocktails), and subjected to centrifugation at 105,000 × g for 30 min at 4°C. The supernatant fractions were carefully transferred to other tubes and used for the assay of ADH activity. The concentration of cytosolic protein was determined using the BioRad protein assay kit, as described previously (Kim et al. 2006). Hepatic ADH activity was determined spectrophotometrically by measuring the rate of NADH formation from NAD^+ at 340 nm, according to the method described by Lumeng et al. (1980).

42.2.3 Statistical Analysis

All the results were expressed as means ± SEM and were analyzed by ANOVA or by two-tailed Student's t-test. ANOVA was followed by the Duncan's multiple range test. An acceptable level of significance was $p<0.05$. All statistical analyses were performed using the Statistical Analysis System (SAS/STAT Version 6, SAS Institute Inc. Cary, NC).

42.3 Results and Discussion

42.3.1 Body Weight Gain, and Plasma and Hepatic Biochemistry

ED rats exhibited significantly lower body weight gain (a 13% decrease, $p<0.05$) compared to pair-fed CD rats. Taurine administered to rats chronically loaded with ethanol significantly improved ethanol-induced suppression in body weight gain (a 11% increase, $p<0.05$). Elevated levels of hepatotoxic parameters induced by alcohol administration, such as ALT and AST activities, were reversed by taurine supplementation in rats (19% and 8% reductions, respectively, $p<0.05$) (Table 42.2). The TED group showed significantly lower plasma (a 29% reduction, $p<0.001$) and hepatic (a 72% reduction, $p<0.001$) ethanol concentrations than those of the ED group (Fig. 42.1). Cytosolic ADH activity in the liver was 11% higher in the ED rats than in rats fed the CD ($p<0.05$). Taurine co-administered with ethanol led to a significant decrease (a 23% reduction) in hepatic ADH activity compared to the value for rats administered ethanol alone (Table 42.2).

Despite strict isocaloric pair-feeding, alcohol fed animals did not gain as much weight as the pair-fed controls. The decreased rate of body weight gain of the animals administered ethanol compared to that of the CD rats is comparable to the results reported by other investigators (Lindros and Jarvelainen 1998). When alcohol is oxidized to acetaldehyde via the ADH pathway, NADH is generated. However, the oxidation of alcohol via the microsomal ethanol-oxidizing system utilizes NADPH,

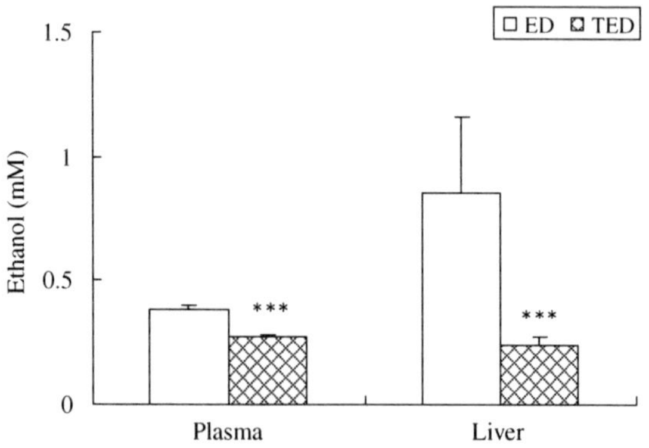

Fig. 42.1 Plasma and hepatic ethanol concentration of rats fed experimental diets. Values are means ± SEM, $n = 7$. *** Significantly different from the value for ED rats by Student's t-test, $p < 0.001$

Table 42.2 Body weight gain, and plasma and hepatic biochemistry of rats fed experimental diets

	CD	ED	TED
Body wt gain (g/ 8 wk)	310 ± 11.0[a]	270 ± 3.00[b]	300 ± 4.00[a]
ALT (IU/L)	67.0 ± 7.2[b]	88.6 ± 1.4[a]	71.8 ± 4.6[b]
AST (IU/L)	85.0 ± 3.5[c]	120 ± 1.3[a]	110 ± 2.8[b]
γ-GTP (IU/L)	12.3 ± 2.1	15.7 ± 3.2	10.8 ± 1.8
Total bilirubin (mg/L)	1.9 ± 0.4	3.0 ± 0.4	3.0 ± 0.4
Alcohol dehydrogenase (mol· min^{-1}·mg protein)	46.4 ± 3.1[ab]	51.5 ± 1.8[a]	39.6 ± 2.2[b]

Values are means ± SEM of 7 rats. [a,b,c] Values not sharing a common letter in the same row are significantly different ($p < 0.05$).

resulting in the loss of energy as heat, a mechanism that could explain the slower weight gain of rats fed the alcohol-containing diet despite similar caloric intake.

42.3.2 Plasma Levels and Urinary Excretions of Trace Elements

Rats in the ED group exhibited significantly lower plasma concentrations of Se (a 22% reduction, $p<0.05$) and Mn (a 31% reduction, $p<0.05$) than those of the CD rats. It is presumed that the lower serum Se level in alcoholics is a consequence of, rather than a cause of, liver disease (Aaseth et al. 1986). Plasma Cu and Cr concentrations were not significantly altered by the chronic consumption of alcohol in the current study. Feeding TED to rats reversed ethanol-induced decreases in plasma Se, Mn and Cu levels; the taurine-induced elevation in plasma Se was significant (a 21% increase, $p<0.05$) compared to that of the ED rats (Fig. 42.2).

Animals in the ED group exhibited significantly higher levels of urinary Se (a 53% increase, $p < 0.05$), Cr (a 62% increase, $p < 0.05$) and Mn (a 45% increase, $p < 0.05$) excretion compared to those in the CD group. Taurine administered to rats chronically loaded with ethanol significantly ameliorated the ethanol-induced increases in urinary loss of Se (a 35% reduction, $p < 0.05$), Cr (a 49% reduction, $p < 0.05$) and Mn (a 35% reduction, $p < 0.05$). Urinary Cu excretion normalized by creatinine content showed a significantly higher value in the ED rats than in the CD rats but tended to be lower in the TED rats than in the ED rats (Fig. 42.2).

The protective effect of taurine against ethanol-induced urinary loss of Se observed in the present study is in accordance with previous reports that taurine significantly lowers urinary Se excretion in rats fed alcohol (Choi et al. 2006). Se has been reported to improve immune function and ameliorate specific diseases in humans and animals (Levander 1987). More recently, convincing evidence has been presented that the consumption of Se in amounts up of 3–5 times the recommended dietary allowance may prevent certain types of cancer (Clark et al. 1996). Although there is a general agreement that liver Cu levels rise in alcoholic cirrhosis (Solomons 1980), a controversy exists about the serum Cu levels in alcoholic patients with or without liver damage (Reicks and Rader 1990). Reduced plasma concentrations of Se, Cu and Mn observed in rats chronically loaded with ethanol appears to be related to ethanol-induced losses of these trace elements in urine.

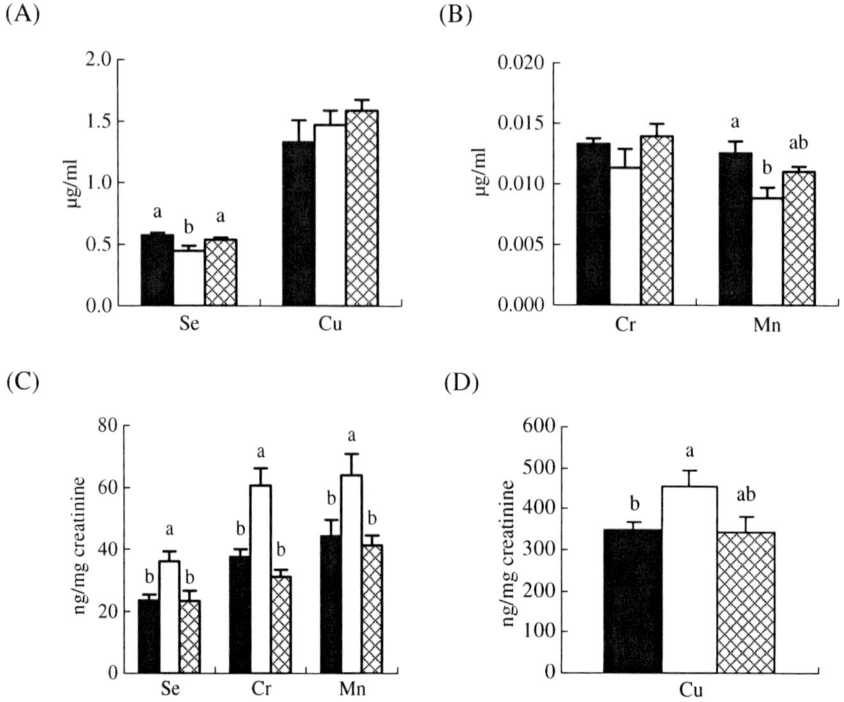

Fig. 42.2 Plasma concentration and urinary excretion of Se, Cu, Cr, and Mn in rats fed experimental diets. (**A**) Plasma Se and Cu levels; (**B**) Plasma Cr and Mn levels; (**C**) Urinary excretion of Se, Cr and Mn; (**D**) Urinary Cu excretion. Values are means ± SEM, $n = 7$. [a,b]Means not sharing a common superscript are significantly different ($p<0.05$). ■, CD; □, ED; □, TED

Se, Cu and Mn depletion caused by chronic alcohol consumption may affect the scavenging of free radicals.

Taurine is increasingly being recognized as having cytoprotective properties in numerous paradigms (Bullock et al. 1991). For example, taurine supplementation studies have documented antihypertensive (Dawson et al. 2000), antidiabetic (Trachtman et al. 1995), antioxidative (Green et al. 1991), and hepatoprotective (Dogru-Abbasoglu et al. 2001) properties of taurine. Recently it was reported that taurine may constitute an important neuroprotective mechanism during excitotoxicity and could be effective in diminishing the symptoms of alcohol-withdrawal (Bleich and Degner 2000). Taurine-induced changes in blood levels and urinary excretion of Se, Cr, Mn and Cu in rats chronically loaded with ethanol demonstrate the possible role of taurine in maintaining homeostasis of these trace elements under conditions of chronic alcohol consumption.

Acknowledgments Supported by a grant from the Korea Health 21 R&D Project, Ministry of Health and Welfare, Republic of Korea (02-PJ1-PG1-CH15-0001) and by the Brain Korea 21 Project, Yonsei University.

References

Aaseth J, Smith-Kielland A, Thomassen Y (1986) Selenium, alcohol and liver disease. Ann Clin Res 18:43–47

Bleich S, Degner D (2000) Reversal of ethanol-induced hepatic steatosis and lipid peroxidation by taurine: a study in rats. Alcohol Alcohol 35:215

Bullock R, Maxwell WL, Graham DI, Teasdale GM, Adams JH (1991) Glial swelling following human cerebral contusion: an ultrastructural study. J Neurol Neurosurg Psychiatry 54:427–434

Burk RF, Early DS, Hill KE, Palmer IS, Boeglin ME (1998) Plasma selenium in patients with cirrhosis. Hepatology 27:794–798

Choi MJ, Kim MJ, Chang KJ (2006) The effect of dietary taurine supplementation on plasma and liver lipid concentrations and mineral metabolism in rats fed alcohol. Adv Exp Med Biol 583:243–250

Clark LC, Combs GF, Turnbull BW, Slate EH, Chalker DK, Chow J, Davis LS, Glover RA, Graham GF, Gross EG, Krongrad A, Lesher JL, Park HK, Sanders BB, Smith CL, Taylor JR (1996) Effects of selenium supplementation for cancer prevention in patients with carcinoma of the skin. A randomized controlled trial. Nutritional Prevention of Cancer Study Group. JAMA 276:1957–1963

Dawson RJ, Liu S, Jung B, Messina S, Eppler B (2000) Effects of high salt diets and taurine on the development of hypertension in the strokeprone spontaneously hypertensive rats. Amino Acids 19:643

Dogru-Abbasoglu S, Kanbagli O, Balkan J, Cevikbas U, Aykac-Toker G, Uysal M (2001) The protective effect of taurine against thioacetamide hepatotoxicity of rats. Hum Exp Toxicol 20:23–27

Faizallah R, Morris AI, Krasner N, Walker RJ (1986) Alcohol enhances vitamin C excretion in the urine. Alcohol Alcohol 21:81–84

González-Reimers E, Conde-Martel A, Santolaria-Fernandez F, Martinez-Riera A, Rodriguez-Moreno F, González-Hernández T, Castro-Aleman VV (1993) Relative and combined roles of ethanol and protein malnutrition on muscle zinc, potassium, copper, iron, and magnesium. Alcohol Alcohol 28:311–318

González-Reimers E, Martinez-Riera A, Santolaria-Fernandez F, Mas-Pascual A, Rodriguez-Moreno F, Galindo-Martin L, Molina Perez, Barros-Lopez N (1998) Relative and combined effects of ethanol and protein deficiency on zinc, iron, copper, and manganese contents in different organs and urinary and fecal excretion. Alcohol 16:7–12

Green TR, Fellman JH, Eicher AL, Pratt KL (1991) Antioxidant role and subcellular location of hypotaurine and taurine in human neutrophils. Biochem Biophys Acta 1073:91

Keating E, Lemos C, Goncalves P, Martel F (2007) Acute and chronic effects of some dietary bioactive compounds on folic acid uptake and on the expression of folic acid transporters by the human trophoblast cell line BeWo. J Nutr Biochem 23:1–10

Kerai MD, Waterfield CJ, Kenyon SH, Asker DS, Timbrell JA (1998) Taurine:protective properties against ethanol-induced hepatic steatosis and lipid perowidation during chronic ethanol consumption in rats. Amino Acids 15:53–76

Kim BJ, Hood BL, Richard AA Hardwick JP, Conrads TP, Veenstra TD, Song BJ (2006) Increased oxidation and degradation of cytosolic proteins in alcohol-exposed mouse liver and hepatoma cells. Proteomics 6:1250–1260

Levander OA (1987) A global view of human selenium nutrition. Annu Rev Nutr. 7:227–250

Lieber CS, DeCarli LM (1986) The feeding of ethanol in liquid diets. Alcohol Clin Exp Res 10:550–553

Lindros KO, Jarvelainen HA (1998) A new oral low-carbohadrate alcohol liquid diet producing liver lesions: a preliminary account. Alcohol Alcohol 33:347–353

Lumeng L, Bosron WF, Li TK (1980) Rate-determining factors for ethanol metabolism in vivo during fasting. Adv Exp Med Biol 132:489–496

Marotta F, Labadarios D, Frazer L, Girdwood A, Marks IN (1994) Fat-soluble vitamin concentration in chronic alcohol-induced pancreatitis. Relationship with steatorrhea. Dig Dis Sci 39:993–998

Pushpakiran G, Mahalakshmi K, Anuradha CV (2004) Taurine restores ethanol-induced depletion of antioxidants and attenuates oxidative stress in rat tissues. Amino Acid 27:91–96

Reicks M, Rader JI (1990) Effects of dietary tin and copper on rat hepatocellular antioxidant protection. Proc Soc Exp Biol Med 195:123–128

Rodríguez-Moreno F, González-Reimers E, Santolaria-Fernández F, Galindo-Martín L, Hernandez-Torres O, Batista-López N, Molina-Perez M (1997) Zinc, copper, manganese, and iron in chronic alcoholic liver disease. Alcohol 14:39–44

Ryle PR, Thomson AD (1984) Nutrition and vitamins in alcoholism. Contemp Issues Clin Biochem 1:188–224

Solomons NW (1980) Zinc and copper in hepatobiliary and pancreatic disorders. In: Karcioglu ZA, Sarper RM et al. (eds) Zinc and copper in medicine, Springfield, IL:Charles C. Thomas, pp 317–346

Trachtman H, Futterweit S, Maesaka J, Ma C, Valderrama E, Fuchs A, Tarectecan AA, Rao PS, Sturman JA, Boles TH, Fu MX, Bayes J (1995) Taurine ameliorates chronic streptozotocin-induced diabetic nephropathy in rats. Am J Physiol 269:F429

Zarski JP, Arnaud J, Dumolard L, Favier A, Rachail M (1985) Trace elements (zinc, copper, manganese) in alcoholic cirrhosis: effect of chronic alcoholism. Gastroenterol Clin Biol 9:664–669

Chapter 43
Effect of Taurine Supplementation on Plasma Homocysteine Levels of the Middle-Aged Korean Women

Chang Soon Ahn

Abstract The purpose of this study was to evaluate the effect of taurine supplementation on plasma homocysteine (Hcy) levels, an independent risk factor of cardiovascular disease. The subjects consisted of 22 healthy middle-aged women (33 to 54 years). Serum lipids, plasma taurine and plasma Hcy levels were measured before and after supplying 3 g taurine per day for 4 weeks. The concentration of plasma taurine was significantly increased from 63.7 ± 14.2 μmol/L to 73.8 ± 16.6 μmol/L after taurine supplementation ($p<0.001$). On the other hand, the concentration of plasma Hcy was significantly decreased from 8.5 ± 1.2 μmol/L to 7.6 ± 1.1 μmol/L after taurine supplementation ($p<0.05$). The effect of taurine on the levels of plasma Hcy was assessed by regression analysis ($R^2 = 0.304$). After taurine supplementation, plasma taurine and Hcy concentration exhibited a significant negative correlation ($p<0.05$). In conclusion, taurine is an effective nutrient that antagonizes Hcy levels. Therefore, this study suggests that sufficient taurine intake might be an effective way of preventing cardiovascular diseases, such as atherosclerosis.

Abbreviations *Hcy*, homocysteine; *TC*, total cholesterol; *TG*, triglyceride; *WHR*, waist/hip ratio; *AI*; atherosclerotic index

43.1 Introduction

The prevalence of cerebrovascular and cardiovascular disease has been increasing as a result of the changing dietary life style, and has become the major cause of death. Cardiovascular diseases are particularly important factors in the health of middle-aged Korean women because of the high death rate of middle-aged Koreans. Recent studies have demonstrated that elevations in plasma homocysteine levels are related to cerebro-and cardio-vascular diseases (Yun et al. 2004). On the other hand, taurine has been reported to show opposite effects on vascular disease (Nonaka et al. 2001).

C.S. Ahn (✉)
Department of Food and Nutrition, Ansan College, Ansan, Korea

J. Azuma et al. (eds.), *Taurine 7*, Advances in Experimental Medicine and Biology 643,
DOI 10.1007/978-0-387-75681-3_43,

Taurine and homocysteine, both sulfur-containing amino acids, share the same biosynthetic pathway. However, taurine and homocysteine have been reported to exhibit opposite effects on the development of atherogenic vascular disease. Homocysteine is known to induce endothelial cell injury, proliferation of vascular smooth muscle cells and oxidative stress (Chang 1999). However, taurine is reported to attenuate homocysteine-mediated inhibition of Ca^{2+}-ATPase activity in myocardial mitochondria and regulate Ca^{2+} uptake. Moreover, taurine exerts antioxidative effects by inhibiting homocysteine-mediated generation of hydrogen peroxide and superoxide anion. Therefore, taurine exhibits significant cardiovascular protective effects (Chang et al. 2004).

In addition, taurine has been reported to lower blood lipids and cholesterol and to regulate the nervous system (Yamori et al. 1996). Taurine intake promotes the excretion of cholesterol by bile acids and decreases the absorption of cholesterol by the intestines, thereby preventing atherosclerosis. Taurine additionally suppresses the sympathetic nervous system to decrease blood pressure and attenuate stress-induced catecholamine excretion. Accordingly, taurine has been considered an effective nutrient in the prevention of hypertension, atherosclerosis and other cardiovascular diseases.

This study focused on the antioxidative and anti-atherogenic effects of taurine. Serum TBARS levels, plasma homocysteine concentration and serum lipid levels before and after taurine supplementation were determined. This study shows that taurine exerts a protective effect against various cardiovascular diseases.

43.2 Methods

The subjects were 22 healthy middle-aged women, aged 40.3 ± 4.7 (range 33 to 54 years). They were provided 3g taurine tablet per day for 4 weeks.

Taurine intake per day was calculated from a food frequency questionnaire. The subjects' weight, percent of body fat, waist/hip ratio, and body mass index were measured by a bio-impedance analysis method (Inbody 720, Biospace Co. Seoul, Korea). Blood samples were drawn from subjects after an overnight fast and collected before and after taurine supplementation.

The concentration of total serum cholesterol (TC) was analyzed by the cholestezyme-V enzyme method, and the concentration of serum triglycerides (TG) was measured by the triglyzyme-V enzyme method. The concentration of serum HDL-cholesterol was analyzed by the phosphotungstic acid-$MgCl_2$ method and the level of LDL-cholesterol was calculated by the 'Friedewald' numerical formula. The concentration of serum TBARS was measured by the 'Yagi' method, and plasma taurine levels were measured by the Dabsyl-Cl (4-dimethylamino azobenzen-4-sulfonyl-chloride) pre-column derivatization method and RP (reversed phase) HPLC and the level of plasma homocysteine was analyzed by the immunofluorimetric IMx-method that was based on a Fluorescence polarization Immunoassay (FPIA).

The data of this study were analyzed statistically using SPSSWIN 12.0. The results before and after taurine supplementation were compared by a paired t-test. Regression analysis was applied to determine the effect of taurine supplementation on the levels of plasma homocysteine. The relationship between blood components and taurine supplementation was determined by Pearson's correlation coefficient. The p-value of $\alpha = 0.05$ and $\alpha = 0.01$ were considered statistically significant.

43.3 Results and Discussion

43.3.1 Subjects

The subjects in this study were 22 healthy middle-aged women. Their mean weight and height were 57.9 ± 5.2 kg and 159.2 ± 5.2cm, respectively. Their body fat percentage and waist/hip ratio (WHR) were 26.8% and 0.84 respectively (Table 43.1).

The incidence of WHR is known as an index of abdomen obesity, which correlates with metabolic diseases, such as diabetes and arteriosclerosis. The average WHR of the subjects was 0.84, which is a little less than the normal range (0.85) but higher than the desirable value for young women (0.80).

43.3.2 Comparison of Blood Lipid Before and After Taurine Supplementation

The levels of all blood lipids before and after taurine supplementation are shown in Table 43.2. The concentration of serum total cholesterol (TC) before (154.7 ± 26.1mg/dl) and after (153.1 ± 25.1mg/dl) taurine supplementation were statistically identical and were within the normal range of middle-aged Korean women (155 ~ 219mg/dl, 40 ~ 44 yrs. Green Cross Reference Lab., Korea). The

Table 43.1 Physical characteristics of the subjects

	middle-aged women (*n*=22)
Age (yrs.)	40.27 ± 4.73
Height (cm)	159.17 ± 5.20
Weight (kg)	57.88 ± 5.17
% Bodyfat	26.77 ± 4.98
WHR	0.84 ± 0.06
BMI (kg/m^2)	22.87 ± 2.21
SBP (mmHg)	115.45 ± 8.15
DBP (mmHg)	76.82 ± 5.88
HR (beats/min)	67.00 ± 6.59
BMR (kcal)	1234.90 ± 78.00

Values are Means ±SD
WHR waist/hip ratio, *BMI* body mass Index
SBP systolic blood pressure, *DBP* diastolic blood pressure
HR heart rate, *BMR* basal metabolic rate

Table 43.2 The levels of blood lipids before and after taurine supplementation of subjects

	Subjects		
	middle-aged women (n=22)		
Variables	before	After	p-value
TC (mg/dl)	154.7 ± 26.1	153.1 ± 25.1	0.593
TG (mg/dl)	83.8 ± 41.1	81.6 ± 30.9	0.622
LDL-C (mg/dl)	81.6 ± 23.3	80.1 ± 23.9	0.682
HDL-C (mg/dl)	56.4 ± 10.58	56.7 ± 11.6	0.728
AI	1.82 ± 0.64	1.79 ± 0.65	0.700
HDL-C/TC	0.32 ± 0.34	0.37 ± 0.08	0.611
LDL-C/HDL-C	1.84 ± 0.61	1.47 ± 0.45	0.144
TBARS (nmol/ml)	5.05 ± 0.84	4.17 ± 0.64	0.001**

Values are Means ± SD
TC total cholesterol, TG triglyceride
HDL-C high density lipoprotein-cholesterol
LDL-C low density lipoprotein-cholesterol
AI atherogenic index (TC- HDL-C / HDL-C)
TBARS thiobarbituric acid reactive substances
P-values from the paired t-test between before and after taurine supplementation
**Significant differences between before and after taurine supplementation at α=0.001 by paired t-test.

concentration of serum triglycerides (TG) before and after supplying taurine were 83.8 ± 41.4mg/dl, and 81.6 ± 30.9mg/dl, respectively, values that are statistically identical. The atherogenic indexes (AI) of the subjects after taurine supplementation tended to be somewhat less than before (1.79 ± 0.65) taurine supplementation (1.82 ± 0.64). The concentration of serum HDL-C was unchanged by taurine supplementation, but the concentration of serum LDL-C tended to decrease following taurine supplementation. However, the level of serum TBARS was significantly decreased after taking taurine (p<0.001). Serum TBARS of the treated subjects was higher than the average serum TBARS of 40 ~ 49 year-old Korean women, 2.3 nmol/ml (Choi and Shin 1997). This high content of serum TBARS of the subjects is perhaps linked to their higher percentage of body fat (26.8%) and dietary fat intake (25.2%) than the recommended dietary allowance for Koreans. Chung et al. (2000) and Obrosova and Steven 1999 have reported that taurine supplementation exhibits a significant negative effect on the levels of plasma malondialdehyde (MDA). Similarly, Zhang et al. (2004) have reported that taurine supplementation (3.0 g/day) for 7 days to healthy young men reduces the levels of the serum TBARS $p < 0.05$). The significant decrease in serum TBARS levels in response to taurine supplementation is thought to be a very effective means of preventing atherosclerosis and coronary heart disease.

43.3.3 Change in Plasma Taurine and Plasma Homocysteine Levels Before and After Taurine Supplementation

The change in plasma taurine content is shown in Fig. 43.1. The concentration of plasma taurine before (63.7 ± 14.2 μmol/L) and after (73.8 ± 16.6 μmol/L)

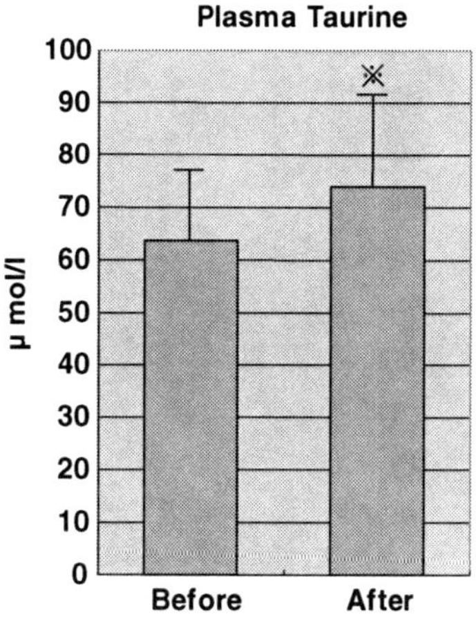

Fig. 43.1 Comparison of plasma taurine levels before and after taurine supplementation *significant difference between the before and after taurine supplementation groups (at α=0.001 by paired t-test)

taurine supplementation were significantly different ($p<0.001$). The concentration of plasma homocysteine is shown in Fig. 43.2. The levels of plasma homocysteine before and after taurine supplementation were 8.5 ± 1.2 μmol/L and 7.6 ± 1.1 μmol/L, respectively. A significant difference was detected between the before and after taurine supplementation groups [at α=0.05 by a paired t-test. Plasma homocysteine values of Korean men and women (above 40 years) are reported to be 14.4 ± 4.5 μmol/L, and 10.4 ± 4.6 μmol/L, respectively (Min 2001). It has been reported that the average plasma homocysteine level is 9.6 μmol/L in healthy Korean men ranging in age from 30 to 60 years. The report also indicated the level of plasma homocysteine increased with age (Jang et al. 1999). The subjects of this study showed comparatively low levels of plasma homocysteine

43.3.4 The Status of Dietary Taurine Intake

The average taurine-intake calculated by the food frequency questionnaire was 178.0 ± 50.4 mg/day. The amount exceeded the taurine-intake of other middle-aged Korean women (145.5 ± 64.0 mg/day) (Lim et al. 2004). The high level of taurine-intake of the subjects likely resulted from a higher intake of protein, fish and shellfish in these subjects.

43.3.5 Correlation Between Plasma Taurine and Homocysteine Levels and Serum Lipid Content

The correlation between the effect of taurine supplementation on plasma taurine and homocysteine levels and the effect of supplementation on serum lipids is shown in

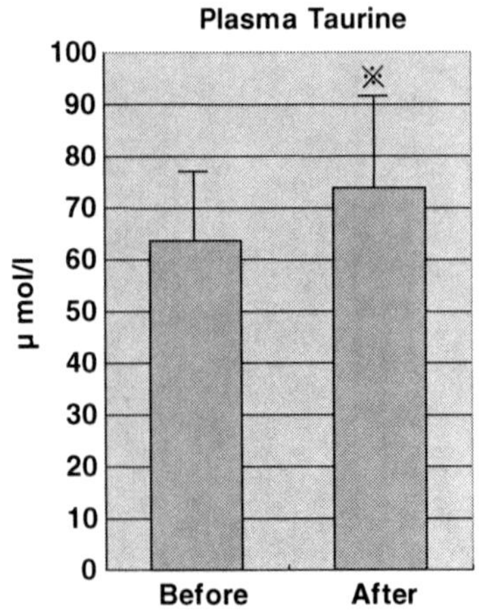

Fig. 43.2 Comparison of plasma homocysteine levels before and after taurine supplementation *significant difference between the before and after supplementation groups (at α=0.05 by paired t-test)

Table 43.3. Taurine supplementation elevated plasma taurine levels but was negatively correlated with homocysteine levels ($r = -0.551$, $p < 0.05$).

Plasma taurine content of subjects before taurine intake tended to exhibit a negative correlation with TC, TG, HDL-C, and LDL-C, an effect that became less negatively correlated after taurine intake. Matsushima et al. (2003) have reported that the levels of plasma LDL-C and VLDL-C of atherogenic mice fed a high fat diet were significantly decreased after taurine intake, while HDL-C was obviously increased. However, it was reported that the concentration of TC and HDL-C for mice fed a regular diet showed a tendency to increase after taurine intake, which is consistent with the results of our study.

The correlation between plasma homocysteine levels and serum lipid levels was positive before taurine intake. However, after taurine supplementation, the correlation between plasma homocysteine levels and serum HDL-C levels became negative. In a study focusing on coronary artery disease (CAD) patients, Liao et al. (2006) reported that the correlation between plasma homocysteine and serum HDL-C or apolipoprotein A1 (ApoA-1) was negative.

43.3.6 Effect of Taurine Supplementation on Plasma Homocysteine

The levels of plasma taurine of the subjects were significantly increased after taurine supplementation. Regression Analysis revealed a positive correlation between plasma taurine levels and taurine supplementation (significant probability = 0.008 at=0.05). However, the correlation between plasma taurine and plasma homocysteine was negative (r=–0.551, $p < 0.05$) and significant (p=0.004). A coefficient of correlation ($R^2 = 0.304$) means that 'plasma homocysteine = 8.344-0.037 × plasma

Table 43.3 Correlation of plasma taurine and homocysteine levels with serum lipid of the subjects before and after taurine supplementation

		middle-aged women (*n*=22)					
		Serum					
Plasma		TC	TG	HDL-C	LDL-C	AI	Plasma homocysteine
Taurine	before	-0.299 (0.177)	-0.080 (0.723)	-0.008 (0.970)	-0.303 (0.171)	-0.210 (0.347)	-0.403 (0.063)
	after	-0.200 (0.372)	-0.050 (0.801)	0.056 (0.803)	-0.223 (0.318)	-0.166 (0.46)	-0.551* (0.008)
Homocysteine	before	0.038 (0.867)	0.006 (0.980)	0.069 (0.762)	0.009 (0.967)	0.024 (0.915)	1.0 (0.000)
after		0.297 0.179	0.200 (0.372)	-0.179 (0.426)	0.347 (0.113)	0.350 (0.111)	

TC Total cholesterol, *TG* Triglyceride
LDL-C Low density lipoprotein-cholesterol
HDL-C High density lipoprotein-cholesterol
AI: Atherogenic Index (TC- HDL-C / HDL-C)
before, after before and after taurine supplementation *significant by Pearson's correlation coefficients at $P < 0.05$ ( ): p-value

taurine' after taurine supplementation can explain just 30.4%. Therefore, the results of this study indicate that taurine supplementation decreased the levels of plasma homocysteine.

43.4 Conclusion

This study investigated the effect of taurine on plasma homocysteine levels, an independent risk factor for atherosclerosis and other cardiovascular diseases. To determine the effect of taurine, middle-aged Korean women, who had a high risk of cardiovascular disease were recruited. After administering taurine (3 g/day) for 4 weeks, the levels of plasma homocysteine and serum TBARS were significantly decreased ($p < 0.05$, $p < 0.001$). The results by regression analysis indicated that taurine supplementation increased the levels of plasma taurine, and furthermore the increased levels of plasma taurine decreased the concentration of plasma homocysteine. In conclusion, taurine is an effective nutrient that antagonizes plasma homocysteine levels. Therefore, this study suggests that sufficient taurine intake might be an effective way of preventing cardiovascular diseases, such as atherosclerosis and coronary heart disease.

Acknowledgments This study was supported by a grant from Dong-A Pharmaceutical Co. The author would like to express her sincere appreciation for the support.

References

Chang HC (1999) Metabolism of homocysteine and cardiovascular disease. Korean Diabetes 23:621–624

Chang L, Zao Z, Xu J, Jiang W, Tang CS, Qi YF (2004) Effects of taurine and homocysteine on calcium homeostasis and hydrogen peroxide and superoxide anions in rat myocardial mitochondria. Clinical and Experimental Phamarcology and Physiology 31:237–243

Choi YS, Shin JH (1997) Age and sex related differences in serum levels of lipid peroxide, retinol and α-tocopherol in Korean adults. Korean J Nutr 30:1109–1115

Chung EJ, Um YS, Oh JY, Park TS (2000) Effect of oral taurine supplementation on blood antioxidant enzyme activities and lipid peroxidation in healthy female adults. Korean Nutrition Soc 33:745–754

Jang YS, Cho EY, Lee JH, Chung NS (1999) Relationship between plasma homocysteine levels and cardiovascular risk factors in healthy men. Korean Circulation J 29:135–145

Liao D, Tan H, Hui R, Li Z, Jiang X, Gaubatz J, Yang F, Durante W, Chan L, Schaffer AI, Powanall HJ, Yang X, Wang H (2006) Hyperhomocysteinemia decreases circulating high-density lipoprotein by inhibiting apolipoprotein A-I protein synthesis and enhancing HDL-cholesterol clearance. Circ Res 99:598–606

Lim MH, Yang HR, Chung JI, Kim ES (2004) Comparison of dietary intakes, urinary excretions and plasma taurine levels of Seoul and Kyunggi area. J East Asian Society 4:20–26

Matsushima Y, Sekine T, Kondo Y, Sakurai T, Kam K, Tachibana M, Murakami S (2003) Effects of taurine on serum cholesterol levels and development of atherosclerosis in spontaneously hyperlipidemic mice. Clinical and Experimental Pharmacology & Physiology 30:295–299

Min H (2001) Folate status and plasma homocysteine concentration of Korean adults. Korean Nutrition Society 34:393–400

Nonaka H, Tsujino T, Watari Y, Emoto N, Yokoyama M (2001) Taurine prevents the decrease in expression and secretion of extracellular superoxide dismutase induced by homocysteine. Circulation 104:1165–1170

Obrosova IG, Steven MJ (1999) Effect of dietary taurine supplementation on GSH and NAD(p)-redox status, lipid peroxidation and energy metabolism in diabetic precataractous lens. Association for research in Vision and Ophthalmology inc., etc. 40:680–688

Yamori Y, Nara Y, Ikeda K, Mizushima S (1996) Is taurine a preventive nutritional factor of cardiovasclar diseases or just a biological marker of nutrition? Taurine 2. Plenum, New York, pp 623–630

Yun SM, Kim KS, Lee HJ, Yoon HJ, Lee YS, Kim KY, Hyun DY, Han SW, Hur SH, Lee CW, Jeon DS (2004) The relationship of homocysteine, vitamin B_{12}, vitamin B_6 and folate with ischemic heart disease. Korean J Circulation 34:224–229

Zhang M, Izumi I, Kagamimori S, Sokejima S, Yamagami T, Liu Z, Qi B (2004) Role of taurine supplementation to prevent exercise-induced oxidative stress in healthy young men. Amino acids 26:203–207

Chapter 44
Correlations Between Dietary Taurine Intake and Life Stress in Korean College Students

Min Jung Sung and Kyung Ja Chang

Abstract The purpose of this study was to investigate the relationship between dietary taurine intake and life stress in Korean college students. The subjects were 320 college students (164 male and 156 female). A three day-recall method was used to assess dietary status (2 weekdays and 1 weekend). Life stress scores were determined using a self-administered life stress questionnaire. The higher stress scores indicate a high frequency and importance of the stress. Average dietary intake of taurine in male and female subjects was 124.1 ± 78.8 mg/day and 96.9 ± 71.7 mg/day, respectively. There were significant negative correlations between taurine intake and the frequency ($p<0.01$), importance ($p<0.05$) and total scores ($p<0.05$) of life stress in female subjects while there were no significant correlations between taurine intake and the frequency, importance and total scores of life stress in male subjects. In female subjects a correlation existed between taurine intake and professor problems, friend problems and future problems ($p<0.05$). These results suggest that dietary taurine intake may play an important role in reducing life stress.

44.1 Introduction

Dietary intake of taurine may play an important role in physical and psychological well-being. But standard dietary references often exclude data on taurine content of common foods (Stapleton et al. 1997). The largest amounts of taurine are present in meat and fish. Taurine is also detected in the plant kingdom, including cereals, potatoes, pulses nuts, seeds, vegetables and fruits, but the levels are much lower than those found in meat and fish (Park et al. 1998). Stress has both direct and indirect adverse effects on health, often leading to general diseases. A solution containing taurine, vitamin B_2, vitamin B_6 and caffeine increased the locomotor activity following exposure to REM sleep deprivation and immobilization stress in mice (Tadano et al. 2003). It has also been reported that taurine supplementation alleviates visual fatigue induced by VDT (visual display terminals) and work (Zhang et al. 2004). However, little information is available on the relationship between dietary taurine

M.J. Sung (✉)
Department of Food and Nutrition Inha University, Incheon, Korea

J. Azuma et al. (eds.), *Taurine 7*, Advances in Experimental Medicine and Biology 643,
DOI 10.1007/978-0-387-75681-3_44,

intake and the degree of life stress. Therefore the purpose of this study was to investigate the relationship between dietary taurine intake and the level of life stress among Korean college students.

44.2 Methods

44.2.1 Subjects

The subjects were 320 college students (164 male and 156 female) residing in the Seoul and Incheon areas and attending a non-major nutritional education class via the internet. A cross-sectional study was carried out using a self-administered questionnaire.

44.2.2 Dietary Taurine Assessment

The three day-recall method was used to assess dietary status (2 weekdays and 1 weekend). Dietary taurine intake was estimated using the computer-aided nutrition program (CAN-pro 3.0 The Korean Nutrition Society Korea).The program contained a taurine content database for 17 food groups commonly used in 310 food items (Kim et al. 1999; Park 1999, 2000).

44.2.3 Life Stress Scale

Subjects responded to each life stress event for one year using a self-administered life stress questionnaire containing fifty questions about the frequency and importance of the stress and was evaluated using a 4 point rating scale of 0 to 3 (Chon 2000). The life stress scale for college students was constructed using eight life stress areas: (interpersonal relationships for faculty, lover, friend and family) (task-related stress; grade future economy value). The total life stress score was calculated by multiplying the frequency and importance of the stress. The greater the stress scores, the greater the frequency and importance of the stress. Previously we reported a life stress level (Table 44.1) and score of life stress (Table 44.2) of the some subjects (Sung and Chang 2006). In this study we used the data from the same subjects.

Table 44.1 Life stress level of Korean college students

	Male ($n = 164$)	Female ($n = 156$)
Experience frequency**	35.5 ± 17.1	41.9 ± 17.9
Importance**	41.5 ± 20.1	49.4 ± 21.0
Total life stress scores**	68.3 ± 44.8	86.2 ± 49.2

Values represent means ± SD. Asterisks denote significant differences between groups $^{**}p < 0.01$ by Student's t-test.

Table 44.2 Life stress level by life stress category

	Male (n = 164)	Female (n = 156)
Interpersonal relationship stress		
Faculty problem	8.1 ± 11.6	10.0 ± 10.7
Lover problem	6.1 ± 13.7	7.8 ± 15.3
Friend problem*	2.7 ± 7.3	4.4 ± 8.4
Family problem***	4.5 ± 7.4	9.1 ± 13.2
Task-related stress		
Grade problem**	25.0 ± 19.8	32.8 ± 21.6
Future problem*	29.4 ± 23.1	34.7 ± 23.5
Economy problem	16.8 ± 18.7	16.9 ± 20.1
Value problem**	20.5 ± 19.9	28.7 ± 23.8

Values represent means ± SD. Asterisks denote significant differences between groups *$p < 0.05$ **$p < 0.01$ ***$p < 0.001$ by Student's t-test.

44.2.4 Statistical Analysis

The statistical analysis was conducted using the SPSS 12.0 program. Mean and standard deviations were calculated for all variables and analyzed by analysis of the Student's t-test. The correlation between life stress and dietary taurine intake were analyzed using Pearson's correlation coefficient.

44.3 Results and Discussion

44.3.1 Anthropometric Data

Anthropometric data of the subjects are shown in Table 44.3. The average age was 23.7±2.5years and 20.9±1.5years in male and female subjects, respectively.

44.3.2 Intake of Energy, Major Nutrients and Taurine

The average dietary intake of taurine is shown in Table 44.4, along with data on the intake of energy and major nutrients. The intake of energy in the male and female subjects was 1950.6 ± 472.9 kcal/day and 1495.5 ± 374.6 kcal/day, respec-

Table 44.3 Age and anthropometric data of the subjects

	Male (n = 164)	Female (n = 156)
Age (years)	23.7±2.5	20.9±1.5
Height (cm)	174.9±5.3	161.7±4.5
Weight (kg)	73.3±11.6	55.5±7.6
BMI (kg/m^2)	23.9±3.4	21.4±3.0

Values represent means ± SD.

Table 44.4 Intake of energy, major nutrients and taurine among the subjects

	Male ($n = 164$)	Female ($n = 156$)
Energy intake (kcal/day)	1950.6±472.9	1495.5±374.6
Protein intake (g/day)	74.3±24.6	56.1±19.8
Lipid intake (g/day)	61.8±28.0	47.1±21.0
Carbohydrate intake (g/day)	246.8±59.1	206.3±51.8
Taurine intake (mg/day)	124.1±78.8	96.9±71.7

Values represent means ± SD.

tively. The average intake of energy and protein of the male subjects was 75.0% and 135.1% and that of the female students was 71.2% and 124.7% of Korean dietary reference intake (%KDRIs), respectively. The average dietary intake of taurine among the male and female subjects was 124.1 ± 78.8 mg/day (range from 8.0 to 370.9 mg/day) and 96.9 ± 71.7 mg/day (range from 1.9 to 287.71 mg/day), respectively. There was a significant difference between male and female subjects relative to dietary taurine intake ($p < 0.01$).

The daily taurine intake based on the food recording method with a database on taurine (Park 1999) was 185 ± 19.1 mg per person (20–29 years) in the Seoul area. The daily taurine intake was 114.9 ± 78.7mg for women (34–54 years) living on Jeju Island in Korea (Kim et al. 2003). Also dietary taurine intake of Japanese adults (20–59 years) was 222.5 ± 210.3 mg/day for males and 162.2 ± 144.1 mg/day for females (Kibayashi et al. 2000).

44.3.3 Correlation Between Dietary Taurine Intake and the Level of Life Stress

The correlation between dietary taurine intake and the level of life stress is shown in Figs. 44.1, 44.2, 44.3. There was a difference between male and female subjects relative to the correlation between taurine intake and life stress level. There were negative correlations between taurine intake and the frequency of stress ($p < 0.01$), the importance of the stress ($p < 0.05$) and total scores of life stress ($p < 0.05$) in female subjects while there were no significant correlations between taurine intake and those stress parameters in the male subjects.

44.3.4 Correlation Between Dietary Taurine Intake and the Level of Life Stress

The correlation between dietary taurine intake and life stress is shown in Table 44.5. In female subjects correlations between dietary taurine intake and faculty problems, friend problems and future problems exhibited a significant negative correlation ($p < 0.05$). In male subjects there was no significant correlation between dietary taurine intake and life stress.

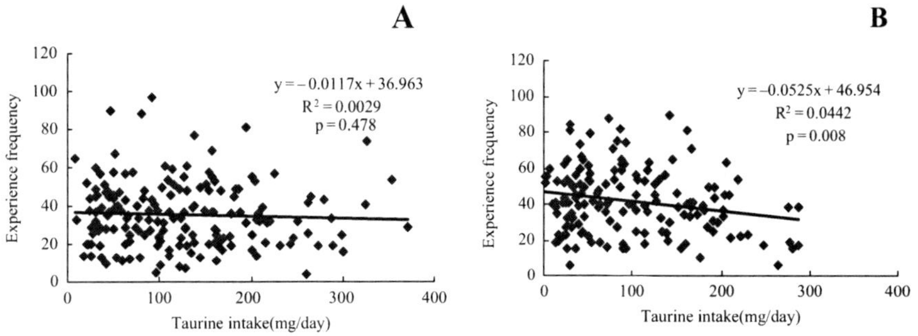

Fig. 44.1 Effect of sex on the correlation between dietary taurine intake and the frequency of stress.**A**. Male, **B**. Female

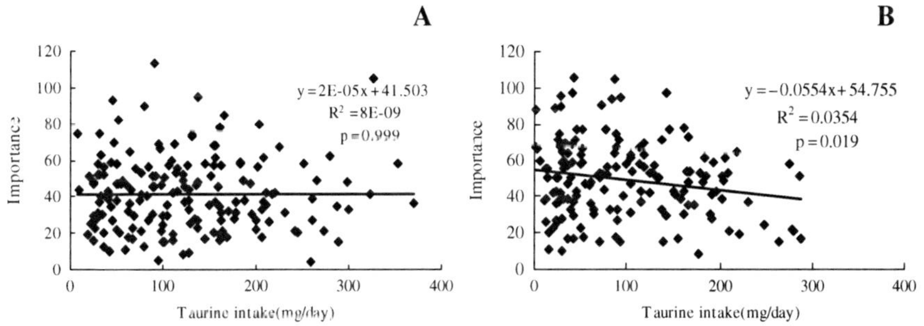

Fig. 44.2 Effect of sex on the correlation between dietary taurine intake and the importance of stress. **A**. Male, **B**. Female

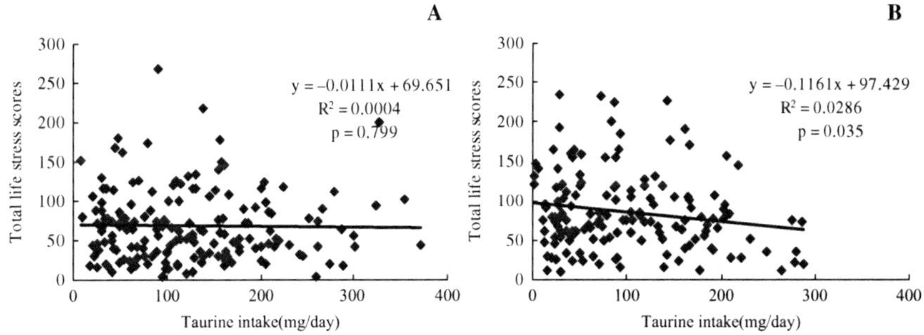

Fig. 44.3 Effect of sex on the correlation between dietary taurine intake and total life stress scores. **A**. Male, **B**. Female

Table 44.5 Correlation between dietary taurine intake and life stress category

	Taurine intake (mg/day)	
	Male (n = 164)	Female (n = 156)
Faculty problem	0.040	−0.160*
Lover problem	0.046	0.045
Friend problem	−0.020	−0.222**
Family problem	0.018	−0.098
Grade problem	−0.055	−0.063
Future problem	0.003	−0.177*
Economy problem	−0.100	−0.060
Value problem	0.046	−0.143

*, ** Significant correlations were assessed by the Pearson correlation, with significance being expressed as $p < 0.05$ and $p < 0.01$.

44.4 Conclusion

There was a difference between male and female subjects relative to the correlation between dietary taurine intake and life stress. Our results show a negative correlation between dietary taurine intake and life stress in female subjects. These results suggest that dietary taurine intake reduces the level of life stress. Therefore, further study is warranted examining the effect of seafood, which is rich in taurine, on stress.

References

Chon KK (2000) Development of the revised life stress scale for college students. Kor J Health Psychol 5:316–335

Kibayashi E, Yokogoshi H, Mizue H, Miura K, Yoshita K, Nakagawa. H, Naruse. Y, Sokejima S, Kagamimori S (2000) Daily dietary taurine intake in Japan. Adv Exp Med Biol 483:137–142

Kim ES, Kim JS, Moon HK (1999) Taurine contents in commercial milk meats and seafoods. J Kor Soc Food Sci Nutr 28:16–21

Kim ES, Kim JS, Yim MY, Jeong Y, Ko YS Watanabe T, Nakatsuka H, Nakatsuka S, Matsuda-Inouchi N (2003) Dietary taurine intake and serum taurine levels of women on jeju island. Adv Exp Med Biol 526:277–283

Park JE (1999) Taurine contents in Korean food and daily taurine intake of Korean adults. Graduate School Yonsei University Master Thesis, pp 9–43

Park TS (2000) Studies on novel activities of taurine and the development of taurine content database of foods. Final Reports of Korean Health Research and Development Project, pp 99–106, Ministry of Health and Welfare Republic of Koreas

Park TS, Park JE, Chang JS, Son MW, Sohn KH (1998) Taurine content in Korean foods of plant origin. J Kor soc Food Sci Nutr 27:801–807

Stapleton PP, Charles RP, Redmond HP, Bouchier-hayes DJ (1997) Taurine and human nutrition. Clinical Nutrition 16:103–108

Sung MJ, Chang KJ (2006) Correlations among life stress dietaryBehaviors and food Choice of college students. J East Asian Soc Dietary life 16:655–662

Tadano T, Nakagawasai O, Niijima F, Tan-no K, Hanawa M, Sakata Y, Sutoo D, Nemoto Y, Yoshiteru I, Endo Y (2003) Effect of nutritive and tonic crude drugs on physical fatigue-induced stress models in mice. Pharmacological research 47:195–199

Zhang M, Bi LF, Ai YD, Yang LP, Wang HB, Liu ZY, Sekine M (2004) Effects of taurine supplementation on VDT work induced visual stress. Amino Acids 26:59–63.

Chapter 45
Dietary Taurine and Nutrients Intake and Anthropometric and Body Composition Data by Abdominal Obesity in Korean Male College Students

Min Jung Sung and Kyung Ja Chang

Abstract The purpose of this study was to investigate the relationship between abdominal obesity and dietary taurine intake, nutrient intake, anthropometric data and body composition in Korean male college students. One hundred seventy four subjects were divided into 2 groups based on abdominal obesity as estimated by waist circumference (cm) (Lee et al. 2006): normal group (waist circumference (cm): <90 cm, $n = 141$), obese group (waist circumference (cm): $\geq$ 90 cm, $n = 33$). A three day-recall method was used to assess diet (2 weekdays and 1 weekend). Anthropometric data and body composition were measured with Inbody 3.0 (Bio-electrical Impedance Fatness Analyzer). Average dietary intake of taurine in the normal and obese groups was 123.1 ± 78.8 mg/day and 128.4 ± 79.6 mg/day, respectively. There was no significant difference in dietary taurine and nutrient intake between the normal and obese groups. However, data of anthropometric measurements and body composition in the obese group were significantly elevated compared to those of the normal group. In the normal group, dietary taurine intake was positively correlated with nutrient intake ($p < 0.01$), the exception being the intake of plant lipid and of animal calcium. In the obese group, dietary taurine intake was positively correlated with the intake of energy foods and of animal lipid ($p < 0.05$). There were positive correlations between dietary taurine intake, weight and hip circumference ($p < 0.05$) in the normal group. However, there was no significant correlation between dietary taurine intake and anthropometric and body composition data in the obese group. Therefore, the data suggest that further study is warranted to examine the relationship between dietary taurine intake and abdominal obesity.

Abbreviations *BMI*, body mass index

M.J. Sung (✉)
Department of Food and Nutrition, Inha University, Incheon, Korea

J. Azuma et al. (eds.), *Taurine 7*, Advances in Experimental Medicine and Biology 643, 429
DOI 10.1007/978-0-387-75681-3_45,

45.1 Introduction

Obesity is one of the serious health problems, with its frequency rising exponentially in the world (WHO, 2000). Using waist circumference as a measure of abdominal obesity, it is estimated that 24% of adults are abdominal obese in Korea (Ministry of Health and Welfare in Korea, 2006). An increase in waist circumference is associated with an increase in the risk of cardiovascular disease, type 2 diabetes mellitus and hypertension (Reeder et al. 1997; Janssen et al. 2002). In animal studies, taurine supplementation was found to reduce abdominal fat pads in genetically obese KKmice, but this effect was not shown in BLBL/C mice (Fujihira et al. 1970). Taurine supplementation also did not affect the weight of abdominal adipose tissue in Sprague-Dawley rats (Cheong et al. 2006). In a human study, total cholesterol and the atherogenic index were significantly decreased in the taurine supplementation group (Zhang et al. 2004). However, the effect of taurine on abdominal obesity remains unclear.

Taurine (2-aminoethanesulfonic acid) can be synthesized from cysteine in animals. However, the main source of taurine in humans is obtained from dietary intake, with seafood being an abundant source (Belli 1994). High fish and seafood intake reduce the risk of obesity and diabetes (Nkondjock and Receveur 2003). Therefore, the purpose of this study was to investigate the link between dietary taurine intake, intake of other nutrients, anthropometric and body composition data and abdominal obesity.

45.2 Methods

45.2.1 Subjects

The subjects were 174 male college students residing in Incheon and Seoul. The subjects were divided into 2 groups based on abdominal obesity using waist circumference (cm) classification (Lee et al. 2006): normal group (waist circumference (cm): < 90 cm, $n = 141$), obese group (waist circumference (cm): ≥ 90 cm, $n = 33$).

45.2.2 Anthropometric and Body Composition

Anthropometric parameters, such as height, waist circumference and hip circumference, were measured. The measurement of weight, body fat, calculation of body mass index (BMI), obesity degree, waist-hip ratio and fitness score were done by using the bioelectrical impedance body composition analyzer (Inbody 3.0, Biospace, Korea).

45.2.3 Dietary Taurine Assessment

A three day-recall method was used for dietary assessment (two weekdays and a weekend). Dietary taurine intake was estimated using the computer-aided nutrition program (CAN-pro 3.0, The Korean Nutrition Society, Korea) containing a taurine content database for 17 food groups that is commonly used for 310 food items (Kim et al. 1999; Park 1999, 2000).

45.2.4 Statistical Analysis

The statistical analysis was conducted using the SPSS 12.0 program. Mean and standard deviation were calculated for all variables. Data obtained between normal and obese groups were compared by analysis of Student's t-test. The correlation between dietary taurine intake, nutrients intake, anthropometric data and body composition data were analyzed using Pearson's correlation coefficient.

45.3 Results and Discussion

45.3.1 Anthropometric and Body Composition

Anthropometric and body composition data of subjects classified by abdominal obesity are shown in Table 45.1. In the study, 23.4% of subjects were abdominally obese and 76.6% were normal. Age was not significantly different between the 2 groups. The average height, weight and BMI of the normal and obese groups were 174.3 ± 5.5 cm, 69.2 ± 7.5 kg, 22.7 ± 2.2 kg/m^2 and 177.1 ± 4.1 cm, 90.7 ± 10.0 kg, 28.9 ± 3.0 kg/m^2, respectively. There were significant differences in height, weight, BMI, percent body fat, obesity degree, waist-hip ratio, waist circumference and hip circumference between the normal and obese groups.

Table 45.1 Age, anthropometric measurements and body composition of the subjects

	Normal ($n = 141$)	Obese ($n = 33$)
Age (years)	23.6±2.4	23.9±2.7
Height (cm)**	174.3±5.5	177.1±4.1
Weight (kg)***	69.2±7.5	90.7±10.0
BMI (kg/m^2)***	22.7±2.2	28.9±3.0
Percent body fat (%)***	17.4±3.9	26.2±4.2
Obesity degree (%)***	103.6±10.5	130.1±14.0
Waist-hip ratio***	0.82±0.0	0.91±0.0
Waist circumference (cm)***	80.3±5.4	97.5±6.5
Hip circumference (cm)***	96.3±4.3	107.7±5.2
Fitness score	82.4±5.2	83.4±3.2

Values represent means ± SD. Asterisks denote significant differences between the two groups $^{**}p < 0.01$, $^{***}p < 0.001$ by Student's t-test.

Table 45.2 Nutrients and dietary taurine intake of the subjects

	Normal ($n = 141$)	Obese ($n = 33$)
Energy (kcal)	1931.5 ± 477.3	2032.3 ± 451.7
Plant protein (g)	31.6 ± 8.6	31.7 ± 9.1
Animal protein (g)	42.6 ± 16.4	43.2 ± 14.1
Plant lipid (g)	24.8 ± 9.2	25.8 ± 8.8
Animal lipid (g)	36.8 ± 19.7	36.8 ± 15.2
Carbohydrate (g)	245.8 ± 58.1	251.4 ± 63.9
Fiber (g)	15.3 ± 5.3	14.6 ± 4.5
Plant calcium (mg)	240.8 ± 93.4	233.9 ± 90.7
Animal calcium (mg)	185.7 ± 124.6	210.5 ± 137.2
Phosphorous (mg)	944.2 ± 271.6	950.1 ± 246.6
Plant iron (mg)	8.1 ± 3.4	8.2 ± 3.5
Animal iron (mg)	3.9 ± 1.4	3.7 ± 1.0
Vitamin A (μg RE)	711.9 ± 44.4	690.7 ± 46.8
Thiamin (mg)	1.48 ± 0.6	1.37 ± 0.5
Riboflavin (mg)	1.19 ± 0.4	1.11 ± 0.3
Niacin (mg)	16.7 ± 5.5	16.9 ± 4.9
Vitamin C (mg)	66.8 ± 44.2	66.4 ± 42.1
Cholesterol (mg)	385.7 ± 160.1	381.0 ± 119.4
Taurine (mg)	123.1 ± 78.8	128.4 ± 79.6

Values represent means ± SD. No significant difference was noted between the two groups by Student's t-test.

45.3.2 Nutrients and Dietary Taurine Intake

Table 45.2 shows the average intake of nutrients and dietary taurine. The average intake of energy food, protein, lipid in the normal and obese groups were 1931.5 kcal, 74.2 g, 61.6 g and 2032.3 kcal, 74.9 g, 62.6 g, respectively. The average dietary intake of taurine in the normal and the obese groups was 123.1 mg/day and 128.4 mg/day, respectively. There was no significant difference in the intake of nutrients and dietary taurine between the normal and the obese groups.

It is relevant that the obese group might have underreported dietary intake. In studies by Lichtman et al. (1992) and Harrison et al. (2000), actual dietary intake was underreported due to insufficient knowledge of the cause of obesity or because of cultural differences regarding obesity.

45.3.3 Correlation Between Dietary Taurine Intake and Anthropometric and Body Composition Data

The correlation between dietary taurine intake and anthropometric and body composition data is shown in Table 45.3 There were several positive correlations between dietary taurine intake, weight and hip circumference ($p < 0.05$) in the normal group. However, there were no significant correlations among dietary taurine intake and

Table 45.3 Correlation between dietary taurine intake and anthropometric and body composition data

	Dietary taurine intake (mg/day)	
	Normal	Obese
Height (cm)	0.065	0.146
Weight (kg)	0.174*	−0.072
BMI (kg/m^2)	0.164	−0.138
Percent body fat (%)	0.105	−0.056
Obesity degree (%)	0.149	−0.155
Waist-hip ratio	0.137	−0.047
Waist circumference (cm)	0.081	0.008
Hip circumference (cm)	0.169*	−0.027
Fitness score	0.134	−0.173

Significant correlation according to Pearson correlation at $*p < 0.05$.

both anthropometric and body composition data in the obese group. In the normal group, animal protein intake was positively correlated with the waist-hip ratio. However, there were no significant correlations between nutrient intake and anthropometric and body composition data in the obese group (unreported data).

Table 45.4 Correlation between dietary taurine intake and intake of other nutrients

	Dietary taurine intake (mg/day)	
	Normal	Obese
Energy (kcal)	0.491**	0.409*
Plant protein (g)	0.226**	0.029
Animal protein (g)	0.506**	0.407*
Plant lipid (g)	0.131	0.127
Animal lipid (g)	0.473**	0.338
Carbohydrate (g)	0.218**	0.169
Fiber (g)	0.305**	−0.062
Plant calcium (mg)	0.294**	−0.098
Animal calcium (mg)	0.130	0.023
Phosphorous (mg)	0.471**	0.273
Plant iron (mg)	0.180*	0.066
Animal iron (mg)	0.352**	0.116
Vitamin A (μg RE)	0.244**	0.063
Thiamin (mg)	0.409**	0.328
Riboflavin (mg)	0.388**	0.239
Niacin (mg)	0.461**	0.257
Vitamin C (mg)	0.390**	−0.083
Cholesterol (mg)	0.339**	0.197

Significant correlation according to Pearson correlation at $*p<0.05$, $**p<0.01$.

45.3.4 Correlation Between Dietary Taurine Intake and the Intake of Other Nutrients

The correlation between dietary taurine intake and the intake of other nutrients is shown in Table 45.4. In normal group, dietary taurine intake was positively correlated with the intake of other nutrients, the exception being the uptake of plant lipid and animal calcium. In the obese group, dietary taurine intake was positively correlated with energy and animal protein in-take. Dietary taurine intake was positively correlated with the intake of other nutrients, the exception being riboflavin intake in adolescents and adults residing in Korea (Park et al. 2001).

45.4 Conclusion

Our results reveal a significant correlation between dietary taurine intake, anthropometrical measurement, body composition and the intake of other nutrients in the normal group. However, these correlations were not apparent in the obese group, perhaps due to underreported dietary intake. There were positive correlations between dietary taurine intake and both weight and animal protein intake, as well as with waist-hip ratio in the normal group. However, limitations of the study included its cross-sectional plan and the failure to measure plasma taurine levels. Therefore, a longitudinal study is warranted using the present preliminary data.

References

Belli DC (1994) Taurine and TPN solutions. Nutrition 10:82–84

Cheong SH, Yu CH, Choi MJ, Chang Cheong SH, Yu CH, Choi MJ, Chang KJ (2006) Effects of garlic powder and taurine supplementation on abdominal fat muscle weight and blood amino acid pattern in ovariectomized rats. Adv Exp Med Bio 583:221–226

Fujihira E, Takahashi H, Nakazawa M (1970) Effect of long-term feeding of taurine in hereditary hyperglycemic obese mice. Chem Pharm Bull 18:1636–1642

Harrison GG, Galal OM, Ibrahim N (2000) Underreporting of food intake by dietary recall is not universal: a comparison of data from Egyptian and American women. J Nutr 130:2049–2054

Janssen I, Katzmarzyk PT, Ross R (2002) Body mass index waist circumference and health risk: evidence in support of current National Institutes of Health guideline. Arch Intern Med 162:2074–2079

Kim ES, Kim JS, Moon HK (1999) Taurine contents in commercial milk meats and seafoods. J Kor soc Food Sci Nutr 28:16–21

Lee SY, Park HS, Kim SM, Kwon HS, Kim DY, Kim DJ, Cho GJ, Han JH, Kim SR, Park CY, Oh SJ, Lee CB, Lim KS, Oh SW, Kim YS, Choi WH, Yoo HJ (2006) Cut-off points of waist circumference for defining abdominal obesity in the Korean population. Kor J obesity 15:1–9

Lichtman SW, Pisarska K, Berman ER (1992) Discrepancy between self-reported and actual caloric intake and exercise in obese subjects. N Engl J Med 327:1893–1898

Ministry of Health and Welfare in Korea (2006) The third Korea national health and nutrition examination survey (KNHANES III) -Health Examination- pp 36–48

Nkondjock A, Receveur O (2003) Fish-seafood consumption obesity and risk of type 2 diabetes: an ecological study. Diabetes Metab 29:635–642

Park JE (1999) Taurine contents in Korean food and daily taurine intake of Korean adults. Graduate School Yonsei University Master Thesis, pp 9–43

Park TS (2000) Studies on novel activities of taurine and the development of taurine content database of foods. Final Reports of Korean Health Research and Development Project, pp 99–106. Ministry of Health and Welfare Republic of Koreas

Park TS, Kang HW, Park JE, Cho SH (2001) Dietary intakes plasma levels and urinary excretion of taurine in adolescents and adults residing in seoul area. Kor J Nutr 34:440–448

Reeder BA, Senthilselvan A, Despres JP, Angel A, Liu L, Wang H, Rabkin SW (1997) The association of cardiovascular disease risk factors with abdominal obesity in Canada. Canada Heart Health Surveys Research Group. CMAJ 157:S39–S45

WHO West Pacific Region (2000) The Asia-Pacific perspective: Redefining obesity and its treatment. Health Communications, Australia

Zhang M, Bi LF, Fang JH, Su XL, Da GL, Kuwamori T, Kagamimori S (2004) Beneficial effects of taurine on serum lipids in overweight or obese non-diabetic subjects. Amino Acids 26:267–271

Part VIII
Taurine as an Antioxidant: Role in Immune System and Other Tissues

Chapter 46
Taurine Haloamines and Heme Oxygenase-1 Cooperate in the Regulation of Inflammation and Attenuation of Oxidative Stress

Janusz Marcinkiewicz, Maria Walczewska, Rafał Olszanecki, Małgorzata Bobek, Rafał Biedroń, Józef Dulak, Alicja Józkowicz, Ewa Kontny, and Włodzimierz Maślinski

Abstract Taurine chloramine (TauCl) and Taurine bromamine (TauBr), products of the neutrophil myeloperoxidase halide system, exert anti-inflammatory properties. They inhibit the production of a variety of inflammatory mediators, such as prostaglandin E_2 (PGE_2), nitric oxide (NO) and proinflammatory cytokines. Heme oxygenase–1 (HO-1), a stress inducible enzyme, degrades heme to biliverdin, free iron and carbon monoxide (CO), which are involved in the anti-inflammatory and antioxidant actions of HO-1. Recently we have demonstrated that taurine haloamines induce the expression of HO-1 in inflammatory cells. In this study we examined whether HO-1 participates in taurine haloamines-mediated suppression of proinflammatory cytokine production. We have shown that TauCl/TauBr and CO inhibit the production of TNF α, IL-12 and IL-6, in a similar dose-dependent manner. However, the suppressor activity of TauCl was not altered in HO-1 deficient mice. Therefore, HO-1 and TauCl may independently regulate the production of proinflammatory cytokines. We suggest that TauCl and TauBr provide a link between the two antioxidant systems: the cysteine pathway and the heme oxygenase system.

Abbreviations *HO-1*, heme oxygenase-1; *Mϕ*, macrophages; *TauBr*, taurine bromamine; *TauCl*, taurine chloramine

46.1 Introduction

Heme oxygenase-1 (HO-1) is the rate-limiting enzyme in heme catabolism, which leads to the generation of biliverdin, free iron and carbon monoxide (CO). The products of HO-mediated heme degradation regulate important biological processes,

J. Marcinkiewicz (✉)
Chair of Immunology, Institute of Rheumatology, Poland

J. Azuma et al. (eds.), *Taurine 7*, Advances in Experimental Medicine and Biology 643,
DOI 10.1007/978-0-387-75681-3_46, © Springer Science+Business Media, LLC 2009

including oxidative stress, inflammation, apoptosis, cell proliferation and angiogenesis (Wagener et al. 2003). Over the past decade, HO-1 has been implicated in the cytoprotective defence response against oxidative stress. In addition to the antioxidant activity of bilirubin, there has been increasing interest in the anti-inflammatory effects of endogenous CO (Otterbein et al. 2000). Interestingly, it has been reported that IL-10, a well known anti-inflammatory agent, acts *via* HO-1 (Lee and Chau 2002). HO-1 expression at the site of inflammation is induced by various stimuli. Only recently we have shown that TauCl and TauBr, anti-inflammatory and anti-oxidant taurine derivatives, induce the expression of HO-1 in macrophages (Olszanecki and Marcinkiewicz 2004).

Taurine, a non-protein sulphur amino acid, is the most abundant free amino acid in the body and plays an important role in several essential biological processes (Schaffer et al. 2003; Hansen et al. 2006). It is generally accepted that taurine protects cells against oxidative injury, despite the fact that the molecule is very stable and difficult to oxidise. An indirect antioxidative role can be ascribed to taurine as it maintains mitochondrial oxidation and stabilizes the oxidative environment (Hansen et al. 2006). Moreover, taurine readily reacts with HOCl and HOBr, extremely toxic oxidants generated by activated neutrophils and eosinophils, to produce taurine haloamines (Thomas et al. 1995). It is well documented that TauCl and TauBr, the major taurine haloamines, exert anti-inflammatory properties by inhibiting the production of a variety of pro-inflammatory mediators (Marcinkiewicz 1997; Marcinkiewicz et al. 1995, 2006; Quinn et al. 1996).

In this study we examined whether the induction of HO-1 by taurine haloamines is a general phenomenon which may occur in any inflammatory cell and whether the anti-inflammatory effects of TauCl are mediated by HO-1-derived heme degradation products.

46.2 Materials and Methods

46.2.1 Synthesis of Taurine Chloramine (TauCl) and Taurine Bromamine (TauBr)

TauCl (N-chlorotaurine sodium salt) was prepared as described previously (Gottardi and Nagl 2002).

TauBr was prepared in a two-step procedure as described by Thomas et al. (1995). Stock solutions of TauCl and TauBr were kept at 4°C for a maximum period of 3 days before use.

46.2.2 Mice

Inbred Balb/c male mice, between 6 and 8 weeks of age, from the breeding unit, Department of Immunology, Jagiellonian University Medical College, Krakow, Poland, were used. Female C57BL/6xFVB, homozygous HO-1 knockout mice ($HO^{-/-}$), carrying a targeted deletion of a large portion of the HO-1 gene, and age-matched C57BL/6xFVB wild-type mice ($HO^{+/+}$) were obtained from Prof. Józef Dulak, Department of Medical Biotechnology, Faculty of Biochemistry, Biophysics and Biotechnology Jagiellonian University, Krakow. Mice were originally kindly provided by Prof. Anupam Agarwal (University of Alabama at Birmingham, USA). All mice were housed 4–5 per cage in the laboratory room with water and standard diet*ad libitum*. The authors were granted permission by the Local Ethics Committee to use mice in this study.

46.2.3 Cells

Bone marrow derived dendritic cells (DC) were propagated from bone marrow cells of Balb/c mice in the presence of granulocyte-macrophage colony stimulating factor (GM-CSF) according to the method developed by Inaba et al. (1992).

Peritoneal mouse macrophages (Mf) were induced by intraperitoneal injection of 1.0 ml of paraffin oil (Sigma, USA). Cells were collected 72 h later by washing out the peritoneal cavity with 5 ml of PBS (phosphate buffer solution) containing 5 U heparin/ml (Polfa, Poland).

The mouse macrophages cell line J774.2 was cultured in T 75 flasks in RPMI 1640 medium (JR Scientific Inc., USA) supplemented with 10% fetal calf serum (FCS), streptomycin (100 $\mu g\ ml^{-1}$) and amphotericin B (Fungizone® (0.25 $\mu g\ ml^{-1}$). Flasks were kept at 37°C in an atmosphere of humidified air containing 5% CO_2.

The cells were cultured in 24-well plates (10^6 cells per well) in 2 ml of culture medium (RPMI with 5–10% FCS). LPS (100 ng/ml) or LPS (100 ng/ml) + IFN-γ (100 U/ml) were used to activate the cells. Taurine haloamines (TauCl, TauBr) were added to the culture medium at a concentration of 100 and 300 μM. In some experiments, in order to block the activity of HO-1, the cells were cultured in the presence of chromium III mesoporphyrin chloride (CrMP 30 μM). All compounds were added to the culture medium 60 minutes prior to LPS or LPS/IFN-γ.

46.2.4 Cell Viability

Viability of the cells was routinely monitored by cellular exclusion of trypan blue. In some experiments, cell respiration, an indicator of cell viability, was assessed by mitochondrial-dependent reduction of MTT (3-(4,5-Dimethyl-2-thiazolyl)-2,5-diphenyl-2H-tetrazolium bromide) to formazan.

46.2.5 Western Blot Analysis (HO-1 Expression)

The expression of HO-1 protein in the cytosol of the cells was determined by the Western blot technique, as described previously (Olszanecki and Marcinkiewicz 2004).

46.2.6 Cytokine Determination

Cytokine concentrations (IL-6, IL-10, IL-12p40 and TNF-a) in culture supernatants were measured using capture ELISA.

46.2.7 Statistic Analysis

The data are presented as means ± S.E.M. Statistical significance was determined by the Student's *t*-test or one–way ANOVA followed by the post hoc Sheffe test, when appropriate. Differences were considered statistically significant when the P value was less than 0.05.

46.3 Results

46.3.1 TauCl and TauBr Induce Heme Oxygenase-1 (HO-1) in Various Inflammatory Cells

Previously we have shown that TauCl and TauBr, but not taurine, induced the expression of HO-1 protein in both non-activated and LPS-activated J774.2 macrophages (Olszanecki and Marcinkiewicz 2004). To examine whether taurine haloamines affect HO-1 expression to the same degree as other stress stimuli, we treated murine peritoneal macrophages (Mϕ), J774.2 cell line macrophages and dendritic cells (DC) with various well-known HO-1 stimuli and the effects were compared with those of TauCl and TauBr. Western-blot analysis (Figs. 46.1 and 46.2) indicates that TauCl and TauBr, at a concentration of 300 μM, can increase the expression of HO-1 to a similar degree as LPS, cobalt protoporphyrin [(CoPP-(Co (III) Protoporphyrin IX chloride)] and even hemin, the major physiological inducer of HO-1.

However, some differences have been observed between Mϕ (Fig. 46.1) and DC (Fig. 46.2). TauCl tends to be a more potent HO-1 inducer than TauBr in Mϕ (both peritoneal macrophages and J774.2 cells), while the opposite effect was observed in DC. Importantly, the treatment of both types of murine macrophages, J774.2 cells and peritoneal macrophages, with taurine haloamines resulted in the similar effects.

46.3.2 The Opposite Effects of HO-1 Inhibitor (CrMP) and Exogenous Carbon Monoxide (CORM-3) on Cytokine Production by Stimulated Macrophages

To demonstrate the participation of endogenous HO-1 in cytokine production by macrophages, CrMP, a competitive inhibitor of HO-1, was used. Figure 46.3 shows that stimulation of macrophages with LPS over a concentration range of 10 to

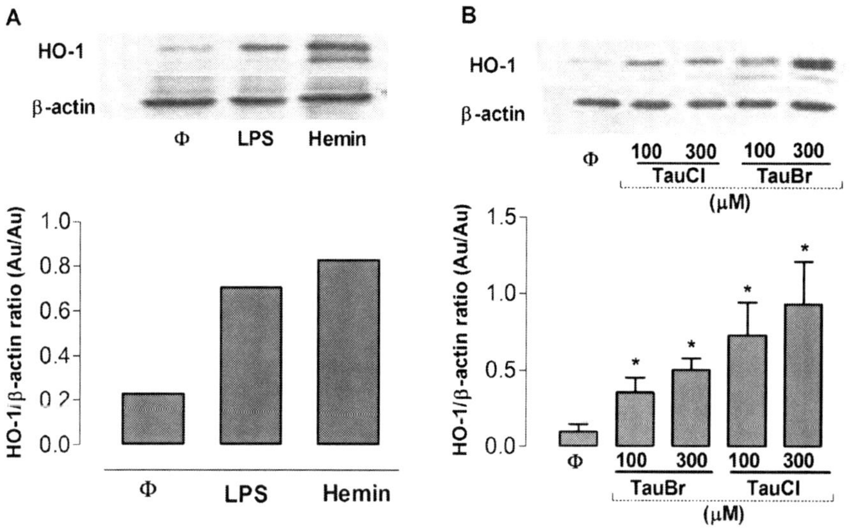

Fig. 46.1 (**A**) Western blot showing influence of LPS (100 ng/ml) and hemin (50 μM) on HO-1 expression in J774.2 cells. (**B**) Western blot showing effect of taurine haloamines (TauCl and TauBr, both at concentrations of 100 and 300 μM) on HO-1 expression in peritoneal macrophages isolated from Balb/c mice. $^*p < 0,05$ vs. ϕ group, ($n = 3$)

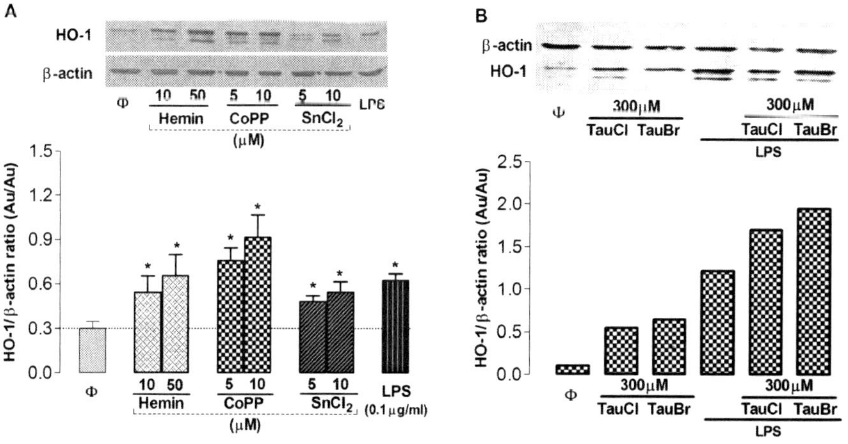

Fig. 46.2 (**A**) Western blot showing effect of hemin (10, 50 μM), CoPP (5 μM, 10 μM), $SnCl_2$ (5, 10 μM) and LPS (100 ng/ml) on HO-1 expression in dendritic cells (DC) isolated from Balb/c mice. (**B**) Western blot showing effect of taurine haloamines (TauCl and TauBr, both at concentrations of 100 and 300 μM) on HO-1 expression in resting and LPS-stimulated dendritic cells (DC) isolated from Balb/c mice. $*p < 0,05$ vs. ϕ group, ($n = 3$)

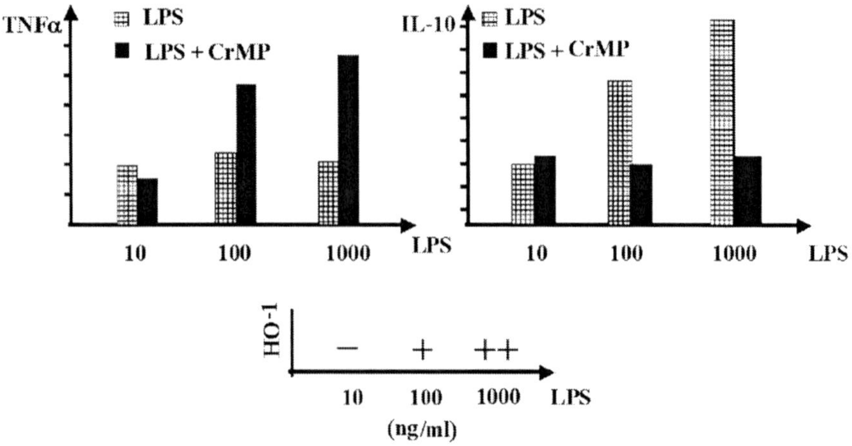

Fig. 46.3 The effect of CrMP, a competitive inhibitor of HO-1, on TNF-α and IL-10 synthesis by LPS-stimulated J774.2 macrophages. To block HO-1 activity, the cells were cultured in the presence of 30 μM CrMP. The capacity of different concentrations of LPS to induce the expression of HO-1 (determined by Western blot) was depicted in the lower panel

1000 ng/ml led to a dose-dependent increase in the production of IL-10 while the production of TNF-α was maintained at the same level. The increase in the production of IL-10 correlated with the expression of HO-1 protein. The inhibition of HO-1 activity by CrMP resulted in an enhancement of TNF-α production in a dose dependent manner. By contrast, the effect of LPS stimulation on IL-10 production was inhibited in the presence of the inhibitor of HO-1 activity.

Inhibition of HO-1 activity results in a reduction of carbon monoxide (CO) generation, a major product of heme degradation, which is responsible for HO-1's anti-inflammatory properties (Motterlini 2007). To mimic the effect of HO-1-derived CO on cytokine production we stimulated macrophages *in vitro* in the presence of CORM-3, a donor of CO. As shown in Fig. 46.4, CORM-3 inhibited in a dose-dependent manner the production of all cytokines tested, namely TNF-α, IL-6 and IL-12p40.

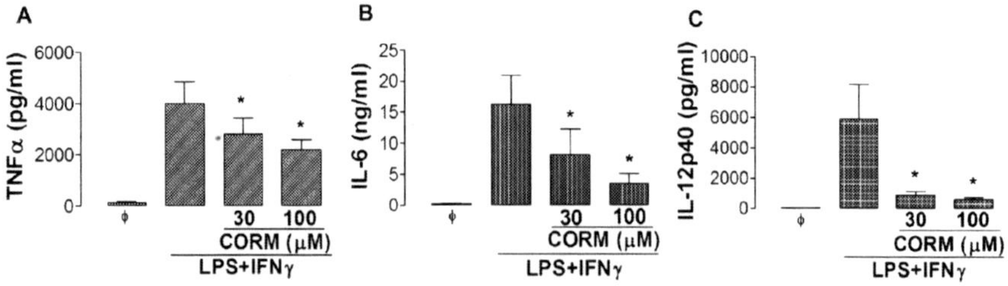

Fig. 46.4 The effect of the CO donor, CORM-3, on the production of TNF-α (**A**), IL 6 (**B**) and IL-12p40 (**C**) by LPS/IFN-γ stimulated peritoneal macrophages isolated from Balb/c mice. The macrophages were stimulated with LPS (100 ng/ml) and IFN-γ (100 U/ml) and incubated in the presence of CORM-3 (30, 100 μM) for 24 hours. $*p<0{,}05$ vs. LPS+IFN-γ group, ($n = 3$)

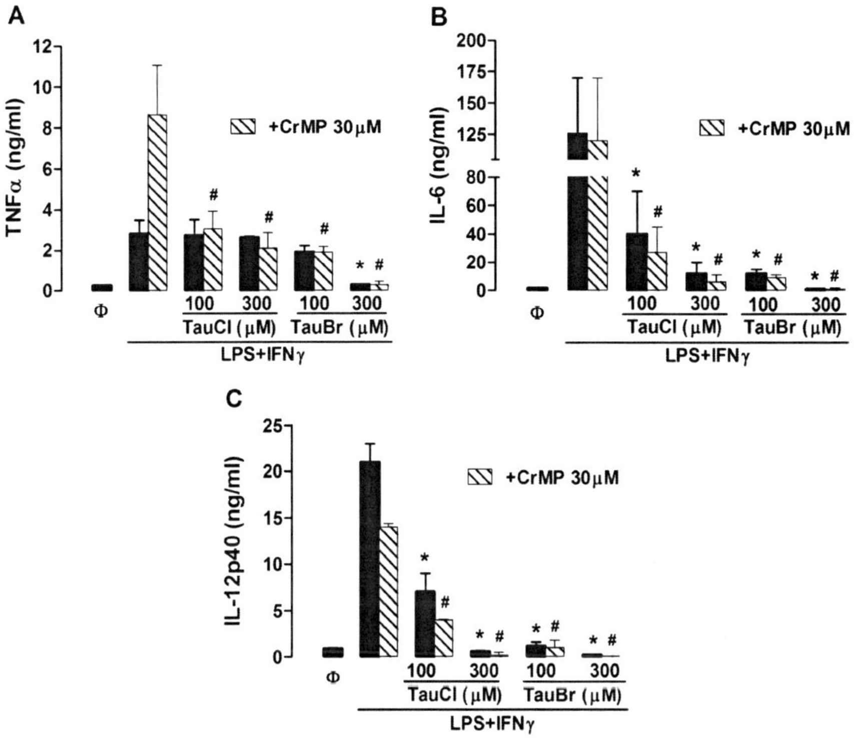

Fig. 46.5 The Effect of taurine haloamines (TauCl and TauBr, both at concentrations of 100 and 300 μM) on the production of TNF-α **(A)**, IL-6 **(B)** and IL 12p40 **(C)** by LPS/IFN-γ stimulated peritoneal macrophages isolated from Balb/c mice. The macrophages were stimulated with LPS (100 ng/ml) and IFN γ (100 U/ml) and incubated for 24 hours in the presence of CrMP, an inhibitor of HO-1 (30 μM). $*p<0{,}05$ vs. LPS+IFN-γgroup, $\#p < 0, 05$ vs. [(LPS+IFN-γ) + CrMP group], ($n = 5$)

46.3.3 Role of HO-1 in TauCl/TauBr-Mediated Inhibition of Cytokine Production by Stimulated Macrophages. Evidence from the HO-1 Knockout Mice

To estimate whether HO-1 mediates or enhances/supports the anti-inflammatory activities of the taurine haloamines, the effect of TauCl on cytokine production by activated macrophages was analyzed in two experimental systems: (i) *In vitro* activated peritoneal macrophages taken from Balb/c mice were cultured with different concentrations of TauCl/TauBr in the presence of CrMP, an inhibitor of HO-1 activity. As shown in Fig. 46.5 TauCl and TauBr, in a dose dependent manner, inhibited the production of all cytokines tested. The effect was not reversed by the inhibition of endogenous HO-1 activity. (ii) The effect of TauCl on cytokine production was demonstrated on peritoneal macrophages taken from HO-1 deficient mice and compared with control macrophages of wild type mice. HO-1 deficient

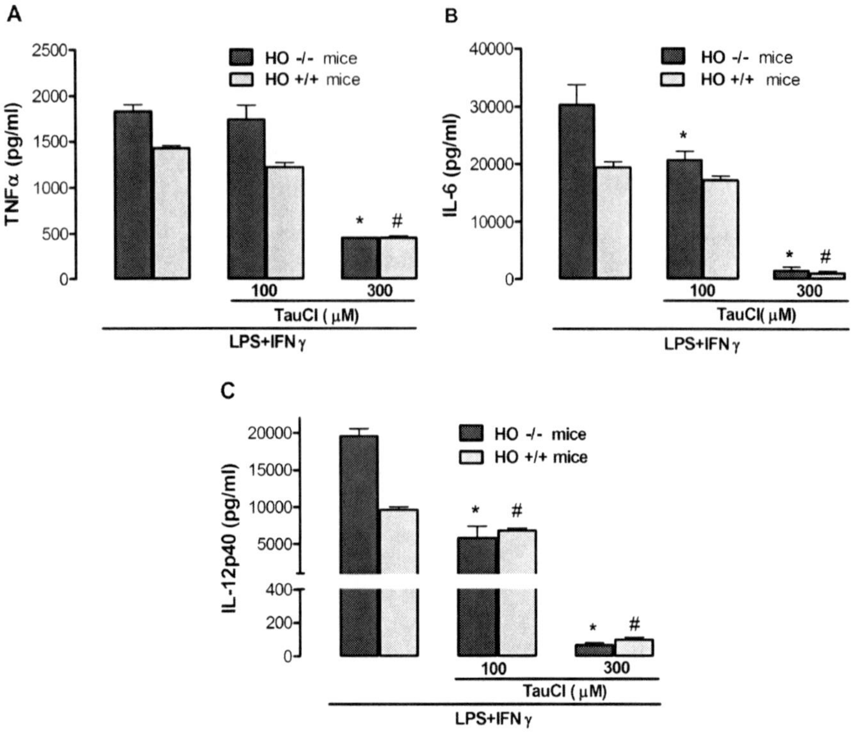

Fig. 46.6 The effect of TauCl on the production of TNF-α (**A**), IL-6 (**B**) and IL 12p40 (**C**), by LPS/IFN-γ-treated (100 ng/ml and 100 U/ml) peritoneal macrophages isolated from HO−/− and HO+/+ mice. $*p < 0, 05$ vs. LPS+IFN-γ group (HO-/-mice), $\#p < 0, 05$ vs. LPS+ IFN-γ group (HO+/+mice), ($n = 1$)

macrophages, spontaneously released approximately 2-fold more proinflammatory cytokines (TNF-α, IL-6, IL-12p40) than wild-type macrophages (data not shown). Stimulation of both types of macrophages with LPS and IFN-γ resulted in a massive release of these cytokines. As shown in Fig. 46.6, 300 μM TauCl strongly inhibited the production of all cytokines tested. The effect was similar in HO-1 $^{-/-}$ and HO-1 $^{+/+}$ macrophages, suggesting that TauCl exerts its anti-inflammatory activities in the absence of HO-1.

46.4 Discussion

In this paper we examined whether HO-1 participates in taurine haloamines-mediated immunoregulatory effects, especially in the anti-inflammatory activity of TauCl. It was reasonable to expect such a scenario for the following reasons: (i) TauCl and TauBr, but not taurine, induce the expression of HO-1 in macrophages (Olszanecki and Marcinkiewicz 2004); (ii) TauCl and carbon monoxide (CO), the product of HO-1 activity, share many anti-inflammatory properties (Marcinkiewicz et al. 1995; Quinn et al. 1996; Otterbein et al. 2000); (iii) HO-1, via CO, mediates the suppressor activity of IL-10, a major anti-inflammatory agent (Lee and Chau 2002).

Over the last few years, numerous studies have demonstrated that HO-1 is able to modulate the inflammatory process and protect tissue from oxidative damage (Wagener et al. 2003). Interestingly, IL-10, a major suppressor cytokine of inflammatory cells, induces the expression of HO-1, which in turn, *via* CO, mediates the inhibitory effect of IL-10, namely it down-regulates the production of a variety of pro-inflammatory mediators, such as TNF-α, PGE_2 and NO (Lee and Chau 2002; Lee et al. 2003). Moreover, an interplay between the heme-HO-1 system and the synthesis of heme proteins (e.g. NOS-2, COX-2) has been recently demonstrated (Alcaraz et al. 2001).

Our previous studies have demonstrated that TauCl and TauBr, the products of the MPO-halide system, are potent inducers of HO-1 protein expression in murine primary macrophages and J774.2 cells, the murine macrophage cell line (Olszanecki and Marcinkiewicz 2004). We have also shown that TauCl, but not TauBr, similarly to IL-10, TGF-β and oxidative stress mediators, stimulate HO-1 *via* p38 mitogen-activated kinase (MAPKp38) (Olszanecki et al. 2007; Ning et al. 2002). In this paper we have extended these observations into dendritic cells, showing that taurine haloamines can induce HO-1 with an effectiveness comparable to that of hemin, the primary physiological inducer of HO-1 expression. Moreover, the up-regulation of HO-1 protein by TauCl was observed in human endothelial cells (HUVEC) (data not shown) and human fibroblast-like synoviocytes (FLS) (Muz et al. 2007), suggesting that the effect of TauCl is not limited to rodents.

To determine the role of HO-1 in the suppressor (anti-inflammatory) activity of taurine haloamines, we used two experimental models: (i) Murine macrophages, stimulated *in vitro* with LPS or LPS + IFN-γ (Olszanecki et al. 2007). (ii) Human rheumatoid fibroblast-like synoviocytes (FLS), stimulated *in vitro* with IL-1β (Muz et al. 2007). In the murine model we found that TauCl/TauBr–mediated suppression of PGE_2 production by macrophages was reversed by chromium III mesoporphyrin (CrMP), an inhibitor of HO-1 (Olszanecki et al. 2007). We concluded that in activated macrophages *in vitro* TauCl and TauBr decreased the production of PGE_2 due to the induction of HO-1, with the subsequent inhibition of the COX-2 pathway. Indeed, an increase in HO-1 activity by non-heme inducers is accompanied by a reduction of cellular heme to suboptimal levels and subsequent dysfunction of heme containing enzymes (COX-2) (Alcaraz et al. 2001). We have shown that in human rheumatoid FLS, TauCl significantly up-regulates HO-1 expression at the mRNA and protein levels and simultaneously inhibits the production of the proinflammatory cytokines, IL-6 and IL-8. However, pretreatment of FLS with the HO-1 inhibitor fully restored IL-8 secretion but had a weaker effect on the IL-6 response. These results together with our previous reports suggest that in rheumatoid FLS TauCl inhibits the production of pro-inflammatory cytokines by two mechanisms: (i) the reduction in the activity of crucial transcription factors (NFkB, AP-1) (Kontny et al. 2000) and (ii) the up-regulation of HO-1 (Muz et al. 2007). The present findings confirm that HO-1 activation is not crucial for the inhibition of pro-inflammatory cytokine production by taurine haloamines. Both TauCl and

TauBr inhibited the production of TNF-α, IL-6 and IL-12 by activated macrophages in a dose-dependent manner. A similar effect was observed when macrophages were treated with a CO-donor (CORM-3). Importantly, the effects of both CO and TauCl/TauBr on cytokine production were not altered in the presence of CrMP, which suggests that endogenous HO-1 was not involved.

Therefore, to summarize the above data one can conclude that HO-1 participates in TauCl-mediated suppression of inflammatory heme-containing proteins whereas in the regulation of cytokine production the relationship between HO-1 and TauCl is still unclear and seems to be cytokine specific (e.g. IL-8 vs. IL-6). In this study, to estimate the role of HO-1 in the regulation of inflammation by TauCl, we performed a decisive experiment using HO-1 deficient mice. It is known that these mice are characterized by a Th1-weighted shift in the cytokine response suggesting a general pro-inflammatory tendency associated with HO-1 deficiency (Kapturczak et al. 2004). The present study confirms these observations. The results show a pro-inflammatory polarization of cytokine production by LPS activated macrophages taken from HO-1 knockout mice. Importantly, TauCl was equally effective in inhibiting the production of all cytokines tested (TNF-α, IL-6 and IL-12) in HO-1 deficient macrophages and wild type macrophages. These results clearly indicate that, at least in this system, the TauCl mediated anti-inflammatory properties are HO-1 independent.

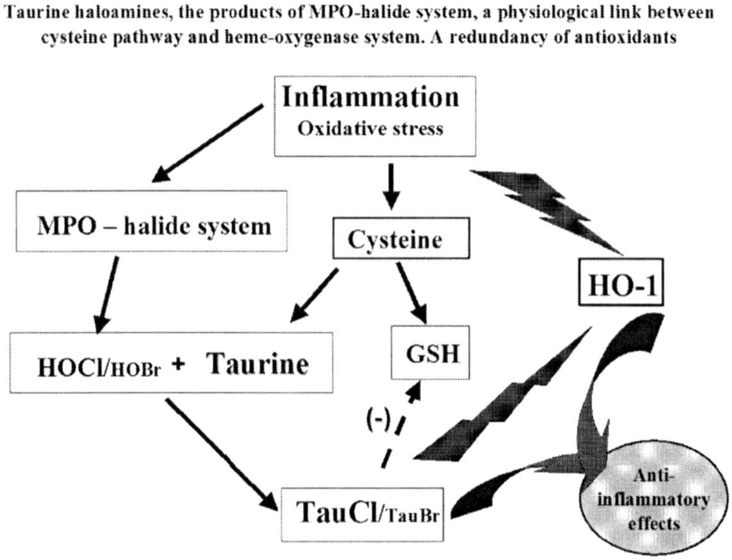

Fig. 46.7 The presumed pathway of taurine haloamines' interaction with antioxidants generated at the site of inflammation. TauCl and TauBr, physiological products of the MPO-halide system, provide a physiological link between the sulphur-containing amino acid system (cysteine – taurine - glutathione) and the heme oxygenase-1 system (HO-1 – heme). The final anti-inflammatory effects depend on a common but independent action of TauCl/TauBr and the products of heme degradation, namely CO and bilirubin

Finally, we propose the following scenario for the interaction between taurine haloamines and the HO-1 system *in vivo* (Fig. 46.7). At the site of inflammation activated phagocytes generate a number of ROS species leading to oxidative stress. Simultaneously, a variety of antioxidants are generated (taurine, glutathione, HO-1) which exhibit a high redundancy in their action. The best known example of such action is super-induction of HO-1 by prior depletion of glutathione (Andre and Felley-Bosco 2003). We suggest that the generation of taurine haloamines results in an induction of HO-1, with the subsequent generation of biologically active heme degradation products, such as CO. Both, TauCl and CO display similar anti-inflammatory properties and act in concert with enhanced antioxidant activity to promote cytoprotection. Therefore, HO-1 induction enhances the effect of TauCl, however TauCl can act directly on the target cell to inhibit their pro-inflammatory activity.

Acknowledgments We want to thank Prof. Marcus Nagl from the Institute of Hygiene and Social Medicine, Leopold Franzens University of Innsbruck, Austria for giving us N-chlorotaurine sodium salt and Dr Roberto Motterlini from Northwick Park Institute for Medical Research, United Kingdom for giving us CORM–3. This work was supported by grants from the State Committee for Scientific Research of Poland (No 2 PO5A 091 29) and partly by Jagiellonian University Medical College (grant number WŁ/291/P/L).

References

Alcaraz MJ, Habib A, Creminon C, Vicente AM, Lebret M, Levy-Toledano S, Maclouf J (2001) Heme oxygenase-1 induction by nitric oxide in RAW 264.7 macrophages is upregulated by a cyclo-oxygenase-2 inhibitor. Biochim Biophys Acta 1526:13–16

Andre M, Felley-Bosco E (2003) Heme oxygenase-1 induction by endogenous nitric oxide: intracellular glutathione. FEBS Lett 546:223–227

Gottardi W, Nagl M (2002) Chemical properties of N-chlorotaurine sodium, a key compound in the human defence system . Arch Pharm (Weinheim) 335:411–421

Hansen SH, Andersen ML, Birkedal H, Cornett C, Wibrand F (2006) The important role of taurine in oxidative metabolism. Adv Exp Med Biol 583:129–135

Inaba K, Inaba M, Romani N, Aya H, Deguchi M, Ikehara S, Muramatsu S, Steinman RM. (1992) Generation of large numbers of dendritic cells from bone marrow cultures supplemented with granulocyte/macrophage colony-stimulating factor. J Exp Med 176:1693–1702

Kapturczak MH, Wasserfall C, Brusko T, Campbell-Thompson M, Ellis TM, Atkinson MA, Agarwal A (2004) Heme oxygenase-1 modulates early inflammatory responses: evidence from the heme oxygenase-1-deficient mouse. Am J Pathol 165:1045–1053

Kontny E, Szczepańska K, Kowalczewski J, Kurowska M, Janicka I, Marcinkiewicz J, Maśliński W (2000) The mechanism of taurine chloramine inhibition of cytokine (interleukin-6, interleukin-8) production by rheumatoid arthritis fibroblast-like synoviocytes. Arthritis Rheum 43: 2169–2177

Lee TS, Chau LY (2002) Heme oxygenase-1 mediates the anti-inflammatory effect of interleukin-10 in mice. Nat Med 8(3):240–246

Lee TS, Tsai HL, Chau LY (2003) Induction of heme oxygenase-1 expression in murine macrophages is essential for the anti-inflammatory effect of low dose 15-deoxy-Delta 12, 14-prostaglandin J2. J Biol Chem 278(21):19325–30

Marcinkiewicz J (1997) Neutrophil chloramines: missing links between innate and acquired immunity. Immunol Today 18:577–580
Marcinkiewicz J, Grabowska A, Bereta J, Stelmaszynska T (1995) Taurine chloramine, a product of activated neutrophils, inhibits in vitro the generation of nitric oxide and other macrophage inflammatory mediators. J Leukoc Biol 58:667–674
Marcinkiewicz J, Kurnyta M, Biedroń R, Bobek M, Kontny E, Maśliński W (2006) Anti- inflammatory effects of taurine derivatives (Taurine chloramine, taurine bromamine, and taurolidine) are mediated by different mechanisms. Adv Exp Med Biol 583:481–492
Motterlini R (2007) Carbon monoxide-releasing molecules (CO-RMs): vasodilatory, anti-ischaemic and anti-inflammatory activities. Biochem Soc Trans 35:1142–1146
Mu B, Kontny E, Marcinkiewicz J, Maśliński W (2007) Heme oxygenase-1 participates in the anti-inflammatory activity of taurine chloramine. Amino Acids doi:10.1007/s00726-007-0605-1
Ning W, Song R, Li C, Park E, Mohsenin A, Choi AM, Choi ME (2002) TGF-beta1 stimulates HO-1 via the p38 mitogen-activated protein kinase in A549 pulmonary epithelial cells. Am J Physiol Lung Cell Mol Physiol 283:L1094–L1102
Olszanecki R, Marcinkiewicz J (2004) Taurine chloramine and taurine bromamine induce heme oxygenase –1 in resting and LPS-stimulated J774.2 macrophages. Amino Acids 27:29–35
Olszanecki R, Kurnyta M, Biedroń R, Chorobik P, Bereta M, Marcinkiewicz J (2007) The role of heme oxygenase-1 in down regulation of PGE(2) production by taurine chloramine and taurine bromamine in J774.2 macrophages. Amino Acids doi:10.1007/s00726-007-0609-x
Otterbein LE, Bach FH, Alam J, Soares M, Tao LH, Wysk M, Davis RJ, Flavell RA, Choi AM (2000) Carbon monoxide has anti-inflammatory effects involving the mitogen-activated protein kinase pathway. Nat Med 6:422–428
Quinn MR, Park E, Schuller-Levis G (1996) Taurine chloramine inhibits prostaglandin E2 production in activated RAW 264.7 cells by post-transcriptional effects on inducible cyclooxygenase expression. Immunol Lett 50:185–188
Schaffer S, Azuma J, Takahashi K, Mozaffari M (2003) Why is taurine cytoprotective? Adv Exp Med Biol 526:307–321
Thomas EL, Bozeman PM, Jefferson MM, King CC (1995) Oxidation of bromide by the human leukocyte enzymes myeloperoxidase and eosinophil peroxidase. Formation of bromamines. J Biol Chem 270:2906–2913
Wagener FA, Volk HD, Willis D, Abraham NG, Soares MP, Adema GJ, Figdor CG (2003) Different faces of the heme-heme oxygenase system in inflammation. Pharmacol Rev 55:551–571

Chapter 47
Taurine Chloramine: A Possible Oxidant Reservoir

Tetsuya Ogino, Tin Aung Than, Mutsumi Hosako, Michitaka Ozaki, Masako Omori, and Shigeru Okada

Abstract Taurine is abundant in polymorphonuclear leukocytes (PMNs) where it reacts with PMN-derived hypochlorous acid to form taurine chloramine (Tau-NHCl), a substance that does not readily cross the cell membrane. When PMNs were stimulated in PBS lacking taurine, extracellular oxidant concentration was low, but the concentration increased 3–4 fold when 15 mM taurine was added, indicating that taurine lowers oxidant levels inside the cell. When Tau-NHCl was added to Jurkat cells in suspension, its half life was about 75 min. In contrast, membrane-permeable ammonia mono-chloramine (NH_2Cl) has a half life of only 6 min. Accordingly, NH_2Cl oxidizes cytosolic proteins, such as IκB, and inhibits NF-κB activation, whereas Tau-NHCl exhibits no comparable effect. However, when NH_4^+ was added to the medium, Tau-NHCl oxidizes IκB and inhibits NF-κB activation, probably through oxidant transfer to NH_4^+ leading to NH_2Cl formation. These results indicate that Tau-NHCl can serve as an oxidant reservoir, exhibiting either delayed oxidant effects or acting as an oxidant at a distant site.

Abbreviations *PMN*, polymorphonuclear leukocytes; *Tau-NHCl*, taurine chloramine; *NH_2Cl*, ammonia monochloramine

47.1 Introduction

Reactive oxygen species are constantly produced from various sources, such as the mitochondrial respiratory chain, activated polymorphonuclear leukocytes (PMN) and metal-catalyzed reactions. Recent studies reveal that these reactive oxygen species are not simply tissue-damaging molecules but affect various cellular responses, such as cell proliferation, apoptosis, gene expression and signal transduction. Activated PMNs produce HOCl, a reaction catalyzed by NADPH oxidase and myeloperoxidase. HOCl reacts with various amines to form chloramines. Although

T. Ogino (✉)
Pathology and Experimental Medicine, Okayama University Graduate School of Medicine, Dentistry and Pharmaceutical Sciences, Okayama, Japan

J. Azuma et al. (eds.), *Taurine 7*, Advances in Experimental Medicine and Biology 643,
DOI 10.1007/978-0-387-75681-3_47,

chloramines are generally less reactive than HOCl, they often show more definitive biological effects than HOCl, probably because chloramine's oxidation targets are limited to thiols and thioethers.

Chloramines can be produced in substantial amounts by activated PMNs and can modify cell response (Test et al. 1984; Pero et al. 1996). PMNs possess abundant levels of taurine (Learn et al. 1990). Taurine chloramine (Tau-NHCl) is considered to be one of the major chloramines products formed during inflammation. The various chloramines differ in membrane permeability and biological effects. Ammonia monochloramine (NH_2Cl) is readily membrane-permeable and has various biological effects (Omori et al. 2002; Hosako et al. 2004; Ogino et al. 2005). Tau-NHCl, on the other hand, does not effectively enter the cell staying outside the cell for a prolonged period of time. However, Tau-NHCl may transfer its oxidant activity to other compounds, such as NH_4^+, generating lipophilic chloramine, NH_2Cl, which can enter the cell and mediate biological effects. Thus, Tau-NHCl may work as an oxidant reservoir and mediate oxidant effects at some distant time or site.

In this chapter we compared the membrane permeability and biological effects of NH_2Cl and Tau-NHCl, focusing on their ability to transfer oxidant capacity among themselves.

47.2 Methods

47.2.1 Respiratory Burst and Chloramine Accumulation

Rat peritoneal PMNs were collected 15–18 h after injection of 2% casein in PBS. To induce the respiratory burst, PMNs were stimulated with 200 ng/ml of PMA, and H_2O_2 production was measured after the conversion of superoxide anion to H_2O_2 by addition of superoxide dismutase, horseradish peroxidase and p-hydroxyphenylacetate. The production of H_2O_2 was monitored by the fluorescent oxidation product of p-hydroxyphenylacetate (Hyslop and Sklar 1984).

To determine the degree of chloramine accumulation, PMNs were suspended in PBS with or without 15 mM taurine, and stimulated with PMA (200 ng/ml). Chloramines in the suspension were measured by the oxidation of 5-thio 2-nitrobenzoic acid (Thomas et al. 1986). Oxidant levels, of which most are chloramines (Weiss et al. 1983), are expressed as nmol H_2O_2 equivalent / million cells.

47.2.2 Membrane Permeability of Chloramines and Thiol/Protein Oxidation

Jurkat T cells were obtained from Hayashibara Biochemical Laboratories Inc. (Okayama, Japan), and cultured in RPMI1640 supplemented with 10% FBS, 2 mM L-Gln and 110 mg/l sodium pyruvate. Tau-NHCl and NH_2Cl were prepared immediately prior to the onset of the experiment and their concentrations were determined by absorption spectra (Thomas et al. 1986).

Jurkat cells or PMNs were suspended in D-PBS(-) and mixed with the indicated doses of chloramine. Chloramine concentration in the D-PBS(-) was measured by the oxidation of 5-thio 2-nitrobenzoic acid. Cellular glutathione was measured as described previously (Tietze 1969). IκB oxidation was detected by its band-shift on SDS-PAGE (Kanayama et al. 2002).

47.2.3 *Cell Cycle Arrest by PMN-Derived Oxidants and the Effect of Taurine*

Jurkat cells were co-cultured with rat PMNs using cell culture inserts with membrane filter bottom (0.4 μm pores). The PMNs were stimulated by opsonized zymosan. Where indicated, the PMN suspension was supplemented with 5 mM taurine. After the co-culture, the Jurkat cells were collected and cultured for an additional 24 h before cell cycle analysis.

47.2.4 *Exogenous NH_4^+ and the Oxidant Effects of Tau-NHCl*

Jurkat cells were treated with Tau-NHCl in the presence or absence of 5 mM ammonium ion. IκB oxidation was detected by Western blot. Tau-NHCl-treated cells were stimulated with TNFα, and NF-κB activation was studied by a gel-shift assay.

47.2.5 *Statistic Analysis*

Group means were compared using the Student's t-test. Analysis of variance was also used for multiple comparisons using Statcel QC software (OMS publishing Inc., Saitama, Japan). P values less than 0.05 were considered significant.

47.3 Results

47.3.1 *Taurine Captures PMN-Derived Oxidants*

Figure 47.1 shows the rate of oxidant generation by activated PMNs. In this experiment, all superoxide produced outside the cell was converted to H_2O_2 by superoxide dismutase, and the levels of H_2O_2 was measured by the fluorescent oxidation product of p-hydroxyphenylacetate. The result showed that cumulative H_2O_2 production was about 25 nmol / million cells after 60 min of stimulation.

Figure 47.2 shows the levels of chloramine produced by neutrophils suspended in buffer. When the PMNs were suspended in PBS without taurine, oxidant levels

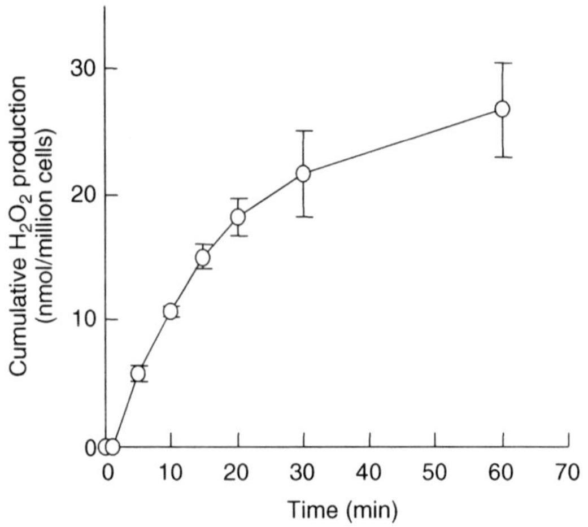

Fig. 47.1 Respiratory burst activity measured as H_2O_2 production in the presence of SOD. The assay mixture contained 1×10^6 cells/ml, 1.1 mM p-hydroxy phenylacetate, 50 μg/ml SOD, 50μg/ml HRP, in PBS. The respiratory burst was initiated by the addition of 200 ng/ml of PMA. H_2O_2 production at 37°C was monitored as the fluorescent oxidation product of p-hydroxy phenylacetate at an excitation 323 nm and an emission 400 nm. (Each point is the mean ± S.D. for 3 determinations.) Reprinted from Ogino et al (1997b) with permission from Elsevier

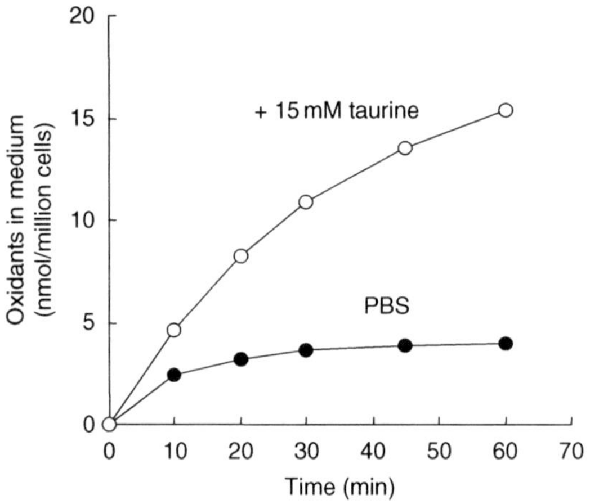

Fig. 47.2 Measurable oxidants in the neutrophil suspension. Neutrophils were suspended in PBS with or without 15 mM taurine, and stimulated with PMA (200 ng/ml). Oxidants in PBS were measured by the oxidation of 5-thio 2-nitrobenzoic acid, and expressed as nmol H_2O_2 equivalent/million cells

outside the cell were only 3–4 nmol/million cells. However, when the PMNs were supplemented with 15 mM taurine, oxidant levels significantly increased, rising to about 15 nmol/million cells. Taurine addition did not change oxygen consumption, as measured by an O_2 electrode (data not shown). These results indicate that a substantial portion of the oxidants produced outside the cell may enter the cell and react with intracellular components, but taurine traps oxidants in the extracellular space as Tau-NHCl.

47.3.2 *Tau-NHCl is not Readily Membrane-Permeable*

When oxidative equivalents are trapped by taurine as Tau-NHCl, they stay in the medium for a prolonged period of time. Figure 47.3 shows the rate of chloramine decline following addition to a Jurkat cell suspension. NH_2Cl disappeared quickly from the buffer with a $T_{1/2}$ of approximately 6 min. In contrast, Tau-NHCl remained in the buffer for a long time, with a $T_{1/2}$ of approximately 75 min.

The decline in medium NH_2Cl content was associated with an increase in cellular NH_2Cl, which resulted in the oxidation of cytosolic glutathione and proteins. Table 47.1 shows that when NH_2Cl was added to the medium containing PMNs in which de novo GSH synthesis was inhibited by DL-buthionine-(S,R)-sulfoximine, cellular glutathione significantly decreased, indicating that it was oxidized as a non-glutathione form. Tau-NHCl was a less effective oxidant. Indeed, cytosolic proteins, such as IκB, were efficiently oxidized by NH_2Cl, but not by Tau-NHCl (Ogino et al. 2005).

Thus, it is likely that exogenous taurine can attenuate the cellular effects of PMN-derived oxidants by sequestering oxidants in the extracellular space. We reported that NH_2Cl oxidizes retinoblastoma protein, resulting in the inhibition of cell cycle

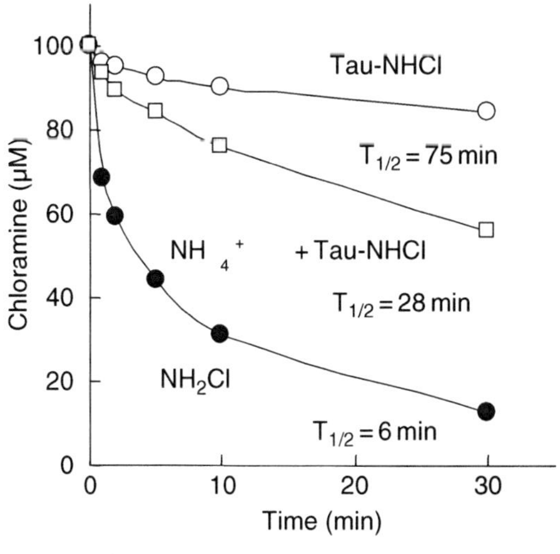

Fig. 47.3 Jurkat cells (1×10^6 cells/ml) were suspended in D-PBS(-), and Tau-NHCl or NH_2Cl was added at 100 μM. Where indicated, 5 mM NH_4Cl was added to the buffer prior to the addition of Tau-NHCl. Samples were collected at the indicated times and the chloramine concentration was measured by the oxidation of 5-thio 2-nitrobenzoic acid

Table 47.1 Glutathione loss in resting neutrophils by chemically produced chloramines in the presence of BSO

Addition	GSH + GSSG (nmol/10^6 cells)
None	1.15 ± 0.17
Taurine-chloramine 30 μM	1.04 ± 0.15
Taurine-chloramine 100 μM	0.93 ± 0.14
Taurine-chloramine 300 μM	0.61 ± 0.15*
NH_2Cl 30 μM	1.10 ± 0.16
NH_2Cl 100 μM	0.52 ± 0.16*
NH_2Cl 300 μM	0.09 ± 0.05*

Resting cells (5×10^6/ml) were incubated at 37°C for 60 min with membrane-permeable monochloramine (NH_2Cl) or membrane-impermeable taurine-chloramine at the indicated concentrations and the cellular glutathione was measured. (Results represent the means ± S.D. for 3 determinations.) Significant difference between "*" and "None" ($p < 0.01$). Reprinted from Ogino et al (1997b) with permission from Elsevier.

Table 47.2 Effect of neutrophil-derived oxidants on the cell cycle

	S phase cells (%)
No neutrophils	24.1 ± 1.0
Resting neutrophils	24.0 ± 1.1
Activated neutrophils	21.4 ± 1.3*
Activated neutrophils + 5 mM taurine	23.6 ± 1.2

Jurkat cells were co-cultured with rat peritoneal neutrophils using a chamber with a membrane filter bottom. The chamber contained a Jurkat cell suspension (1×10^6 cells/ml in HBSS, 0.6 ml). It was placed in a six-well culture plate that contained a neutrophil suspension (5×10^6 cells/ml in HBSS + 2 mM NH_4Cl, 2.5 ml). Where indicated, the neutrophil suspension was supplemented with 5 mM taurine. The neutrophils were stimulated with opsonized zymosan (400 μg/ml). Neutrophil suspensions were replaced 3 times with fresh suspensions at 30 min intervals. The Jurkat cells were collected, the medium was replaced with the cell culture medium, and the incubation allowed to proceed for 24 h before cell cycle analysis. *Significantly decreased from the other samples ($p < 0.05$). Reprinted from Hosako et al. (2004) with permission from Elsevier.

progression, because oxidized retinoblastoma protein was hardly phosphorylated by cyclin-CDK complex (Hosako et al. 2004). Indeed, activated PMNs inhibit cell cycle progression, but this phenomenon was not observed when taurine was added to the medium (Table 47.2).

47.3.3 Tau-NHCl Reacts with NH_4^+ to Form Membrane-Permeable Chloramine

It is noteworthy that Tau-NHCl can function as an oxidant. The oxidizing activity of Tau-NHCl is transferred to other molecules, such as NH_4^+. As mentioned above, Tau-NHCl in Jurkat cell suspension remains in the PBS buffer with a $T_{1/2}$ of approximately 75 min. The consumption of Tau-NHCl was significantly accelerated by the addition of NH_4^+, which reduced the $T_{1/2}$ to 28 min (Fig. 47.3). When chemically prepared NH_2Cl or Tau-NHCl was added to Jurkat cells, NH_2Cl oxidized IκB and attenuated TNFα-induced NF-κB activation, while Tau-NHCl at the same concentration had almost no effect (Fig. 47.4). However, when the incubation medium was supplemented with 5 mM NH_4Cl, Tau-NHCl did oxidize IκB and inhibit NF-κB activation. This effect was probably through the conversion of Tau-NHCl to NH_2Cl (Fig. 47.5).

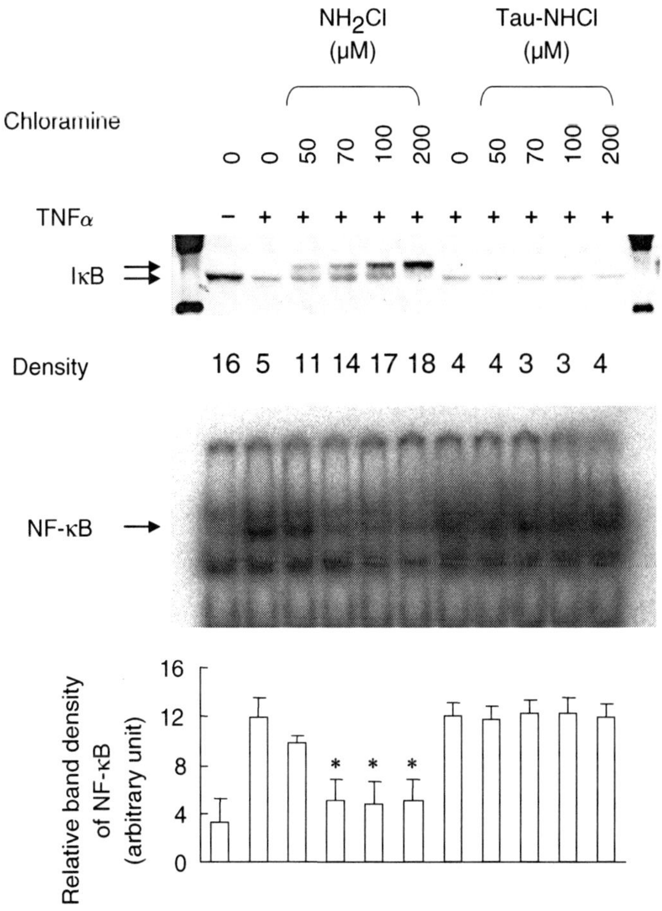

Fig. 47.4 NH_2Cl, but not Tau-NHCl, attenuated TNFα-induced NF-κB activation. Jurkat cells were treated with the indicated concentration of NH_2Cl or Tau-NHCl for 10 min at 37°C, at which point the medium was changed and the cells were stimulated with TNFα (20 ng/ml) for 10 min. Cellular protein was extracted using the lysis buffer supplemented with 0.42 M NaCl and 10% (v/v) glycerol. The band densities (arbitrary unit) were indicated for IκB Western blot. The two IκB bands were measured together. The bottom bar diagram shows the relative band densities of NF-κB (mean ± S.D.) from 3 independent experiments. *Significantly decreased from chloramine (0 μM), TNFα (+) samples ($P < 0.05$). Reprinted from Ogino et al. (2005) with permission from Elsevier

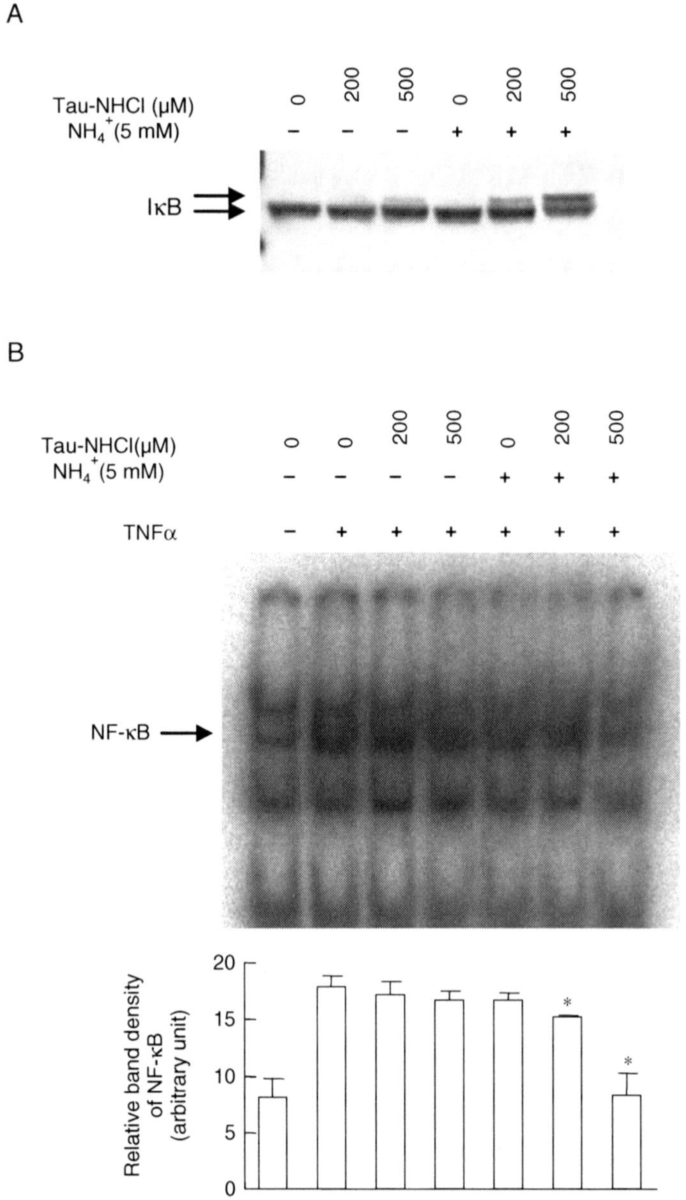

Fig. 47.5 Ammonium ion enhanced Tau-NHCl-induced IκB band-shift and attenuated the degree of NF-κB activation. Jurkat cells were treated with the indicated concentrations of Tau-NHCl for 10 min at 37°C in the presence or absence of 5 mM ammonium ion. **A**: Protein samples were analyzed by Western blot using anti-IκB antibody. **B**: After Tau-NHCl treatment, the medium was changed and the cells were stimulated with TNFα (20 ng/ml) for 10 min. NF-κB activation was studied by electrophoretic mobility-shift assay. The bar diagram shows the relative band densities of NK-κB (mean ± S.D.) from 3 independent experiments. *Significantly decreased from Tau-NHCl (0 μM), NH_4^+ (+), TNFα (+) samples ($P < 0.05$). Reprinted from Ogino et al. (2005) with permission from Elsevier

47.4 Discussion

The possible role of taurine / Tau-NHCl in inflammation is summarized in Fig. 47.6. Activated PMNs produce hypochlorous acid, which reacts with various amines to form chloramines. Some chloramines, such as α-amino acid chloramines, are not stable and decompose spontaneously. NH_2Cl is chemically stable but rapidly enters nearby cells and induces biological effects. Tau-NHCl stays outside the cell for much longer times. Although it has been reported that RAW 264.7 cells have a Tau-NHCl transport system, the efficiency appears to be far less than that of NH_2Cl (Park et al. 1993). Thus, most of the Tau-NHCl cannot readily enter the cell. Rather, it diffuses to a distant site, where it transfers oxidative equivalents to other molecules generating cell permeable oxidants. The generated oxidants may enter the distant cell and induce biological effects.

As an oxidant reservoir, Tau-NHCl is likely to play a unique role in inflammation. When inflammatory stimuli such as bacteria or damaged tissue exist, oxidants are required for bacterial killing or inactivation of toxic substances. Under these conditions, it is desirable that taurine can trap the oxidants outside the cells as Tau-NHCl, protecting the cells from extracellular and intracellular oxidative damage. When such pathogens are oxidized and digested, then the inflammatory response needs to be suppressed. At that time, Tau-NHCl can gradually transfer its oxidant activity to other molecules, forming membrane-permeable oxidants, such as NH_2Cl. These chloramines can inhibit cytokine production (Marcinkiewicz et al. 1995), cell proliferation (Than et al. 2003), PMN respiratory burst (Ogino et al. 1997a), and stimulate lymphocyte apoptosis. These reactions contribute to the termination of the

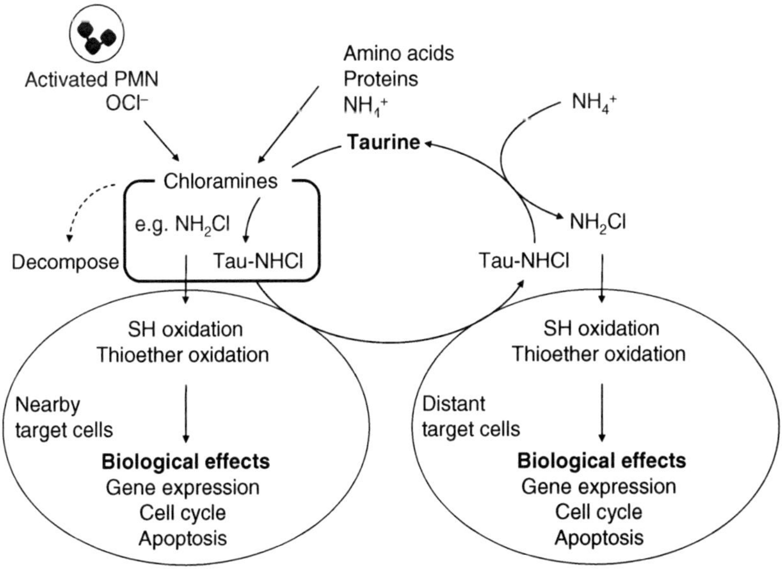

Fig. 47.6 Changes in taurine and Tau-NHCl during inflammation, see text for details

inflammatory reaction. Thus, the hypothesis that Tau-NHCl may actually modify the inflammatory response through its oxidation-reduction cycle warrants further study.

47.5 Conclusion

Taurine/Tau-NHCl may play an interesting role in inflammation. Taurine is rich in PMNs and traps PMN-derived HOCl forming Tau-NHCl. As Tau-NHCl is not readily membrane-permeable, taurine works as an antioxidant by keeping oxidants outside the cells and protecting cellular components from oxidative damage. Nevertheless, Tau-NHCl still works as an oxidant. Tau-NHCl transfers its oxidant capacity to other molecules, such as NH_4^+ generating in the process NH_2Cl, which is readily membrane-permeable and affects various cellular functions through the oxidation of intracellular components. Thus, Tau-NHCl can work as an oxidant reservoir, and may mediate its oxidant effects at distant times or sites.

Acknowledgments This work was supported in part by grants from National Institutes of Health Grant GM27345, WESCO research foundation, and the Program for Promotion of Fundamental Studies in Health Sciences of the National Institute of Biomedical Innovation (NIBIO).

References

Hosako M, Ogino T, Omori M, Okada S (2004) Cell cycle arrest by monochloramine through the oxidation of retinoblastoma protein. Free Radic Biol Med 36:112–122

Hyslop PA, Sklar LA (1984) A quantitative fluorimetric assay for the determination of oxidant production by polymorphonuclear leukocytes:its use in the simultaneous fluorimetric assay of cellular activation processes. Anal Biochem 141:280–286

Kanayama A, Inoue J, Sugita-Konishi Y, Shimizu M, Miyamoto Y (2002) Oxidation of IκBα at methionine 45 is one cause of taurine chloramine-induced inhibition of NF-κB activation. J Biol Chem 277:24049–24056

Learn DB, Fried VA, Thomas EL (1990) Taurine and hypotaurine content of human leukocytes. J Leukoc Biol 48:174–182

Marcinkiewicz J, Grabowska A, Bereta J, Stelmaszynska T (1995) Taurine chloramine, a product of activated neutrophils, inhibits in vitro the generation of nitric oxide and other macrophage inflammatory mediators. J Leukoc Biol 58:667–674

Ogino T, Hosako M, Hiramatsu K, Omori M, Ozaki M, Okada S (2005) Oxidative modification of IκB by monochloramine inhibits tumor necrosis factor α-induced NF-κB activation. Biochim Biophys Acta 1746:135–142

Ogino T, Kobuchi H, Sen CK, Roy S, Packer L, Maguire JJ (1997a) Monochloramine inhibits phorbol ester-inducible neutrophil respiratory burst activation and T cell interleukin-2 receptor expression by inhibiting inducible protein kinase C activity. J Biol Chem 272:26247–26252

Ogino T, Packer L, Maguire JJ (1997b) Neutrophil antioxidant capacity during the respiratory burst: loss of glutathione induced by chloramines. Free Radic Biol Med 23:445–452

Omori M, Ogino T, Than TA, Okada S (2002) Monochloramine inhibits the expression of E-selectin and intercellular adhesion molecule-1 induced by TNF-α through the suppression of NF-κB activation in human endothelial cells. Free Radic Res 36:845–852

Park E, Quinn MR, Wright CE, Schuller-Levis G (1993) Taurine chloramine inhibits the synthesis of nitric oxide and the release of tumor necrosis factor in activated RAW 264.7 cells. J Leukoc Biol 54:119–124

Pero RW, Sheng Y, Olsson A, Bryngelsson C, Lund-Pero M (1996) Hypochlorous acid/N-chloramines are naturally produced DNA repair inhibitors. Carcinogenesis 17:13–18

Test ST, Lampert MB, Ossanna PJ, Thoene JG, Weiss SJ (1984) Generation of nitrogen-chlorine oxidants by human phagocytes. J Clin Invest 74:1341–1349

Than TA, Ogino T, Hosako M, Omori M, Tsuchiyama J, Okada S (2003) Physiological oxidants induce apoptosis and cell cycle arrest in a multidrug-resistant natural killer cell line, NK-YS. Leuk Lymphoma 44:2109–2116

Thomas EL, Grisham MB, Jefferson MM (1986) Preparation and characterization of chloramines. Methods Enzymol 132:569–585

Tietze F (1969) Enzymic method for quantitative determination of nanogram amounts of total and oxidized glutathione: applications to mammalian blood and other tissues. Anal Biochem 27:502–522

Weiss SJ, Lampert MB, Test ST (1983) Long-lived oxidants generated by human neutrophils: characterization and bioactivity. Science 222:625–628

Chapter 48
Production of Reactive Oxygen and Nitrogen Species in Phagocytes is Regulated by Taurine Chloramine

Chaekyun Kim and Young-Nam Cha

Abstract Taurine is abundantly present in phagocytic cells and provides protection against cytotoxicity caused by reactive oxygen species (ROS). The reaction between taurine and HOCl, a toxic product of the myeloperoxidase (MPO) system, generates a more stable and less toxic product, taurine chloramine (TauCl). TauCl has also been shown to inhibit the production of superoxide anion (O_2^-) and nitric oxide (NO). In this review, we compare the effect of taurine and TauCl on the production of these reactive species in phagocytes. First, TauCl inhibit PMA-derived O_2^- production and this is associated with inhibition of p47phox phosphorylation and of p47phox and p67phox translocation. Second, TauCl inhibits LPS-induced iNOS expression and NO production. This occurs by direct inhibition of Ras activation, ERK1/2 phosphorylation and NF-κB activation. Third, TauCl by itself increases the expression of heme oxygenase-1 (HO-1) and enhances HO activity. Carbon monoxide (CO), a product of HO activity, is able to inhibit both O_2^- and NO production. Combined, these effects of TauCl appear to provide cytoprotection against the inadvertent cytotoxicity caused by overproduction of O_2^- and NO.

Abbreviations *TauCl*, taurine chloramine; *ROS*, reactive oxygen species; *MPO*, myeloperoxidase; *iNOS*, nitric oxide synthase; *HO-1*, heme oxygenase-1

48.1 Introduction

TauCl is produced by the interaction of taurine and HOCl, a reaction catalyzed by the MPO system, which is abundant in neutrophils. Large amounts of HOCl are produced in neutrophils to kill phagocytosed bacteria. The microbicidal oxidant HOCl is generated by the reaction between the chloride ion present abundantly within cells and H_2O_2 that is overproduced from O_2^- by superoxide dismutase (Thomas et al. 1985). In activated leukocytes, up to 110 mM of highly toxic HOCl is produced

C. Kim (✉)
Laboratory for Leukocyte Signaling Research and Center for Advanced Medical Education by BK21 Project, Incheon, Korea

J. Azuma et al. (eds.), *Taurine 7*, Advances in Experimental Medicine and Biology 643,
DOI 10.1007/978-0-387-75681-3_48,

by the MPO system (Baron 1969). Taurine, the decarboxylation product of cysteine, is one of the most abundant free amino acids not incorporated into proteins and is present in high levels in most mammalian phagocytic cells and blood cells (Learn et al. 1990; Vinton et al. 1986). The intracellular concentration of taurine in most mammalian phagocytic cells is 10–70 mM and in human extracellular fluid is present at a concentration of 20–100 μM (Fukuda et al. 1982). Taurine reacts rapidly with HOCl on a one to one molar ratio and generates TauCl. It is estimated that neutrophils equal to the concentration found normally in blood ($2–3 \times 10^6$/ml) can produce up to 100 μM TauCl due to the high content of MPO and taurine (Pero et al. 1996). Because TauCl is produced only after phagocytosis and activation of phagocytes, the physiological concentration of TauCl produced remains unclear. Reported rate constants for the formation of TauCl by MPO is $(4.4\pm0.2) \times 10^5\ M^{-1}S^{-1}$ at pH 7.4 (Marquez and Dunford 1994). HOCl is readily permeable across the cell membrane and is highly toxic even at low concentrations, not only to phagocytes themselves but also to surrounding cells (Grisham et al. 1984). However, TauCl is easily ionized and charged, making it less permeable (Midwinter et al. 2004). However, TauCl can be transported into cells via a Na^+- and Cl^-- dependent transport mechanism and modulate several intracellular events (Kim et al. 1998; Tallan et al. 1983).

Taurine is reported to provide protection against inflammatory cytotoxicity and is also needed for many other cellular functions, such as osmoregulation, membrane stabilization, detoxification, anti-oxidation and anti-inflammation. Taurine modulates other physiological functions, such as reproduction, calcium mobilization and neurotransmission ((Huxtable 1992; Schuller-Levis and Park 2003). The protection of phagocytic cells against oxidative injury caused by inflammatory stress provided by taurine results primarily from the elimination of highly toxic HOCl and generation of non-toxic TauCl, the latter which can produce anti-inflammatory effects on its own. Although the underlying mechanisms involved in the anti-inflammatory actions of TauCl remain unclear, TauCl has been reported to suppress the production of many inflammatory mediators, such as O_2^-, NO, tumor necrosis factor (TNF-α), interleukin (IL)-1β, -2, -6, -8, and -10, prostaglandin-2 (PGE_2), macrophage inflammatory protein-2 (MIP-2), monocyte chemo-attractant protein-1 and -2 (MCP-1 and -2) in phagocytic cells (Kim et al. 1996; Marcinkiewicz et al. 1995; Park et al. 2002; Park et al. 1993; Park et al. 1995).

Upon inflammatory stimulation, O_2^-, NO and carbon monoxide (CO) are produced in sequence in phagocytic cells and each of these gaseous signal molecules are known to play important roles in modulating subsequent inflammatory processes, as well as associated cell signaling pathways. In addition, TauCl produced in neutrophils by inflammatory stimulation inhibits the production of many of these inflammatory mediators, stopping the continued and chronic inflammatory processes that are responsible for many pathophysiological events. Thus, in the present report, we review the underlying mechanisms involved in the anti-inflammatory effect of TauCl by examining the effects of TauCl on the sequential overproduction of O_2^-, NO and CO that occur in activated phagocytic cells like macrophages and neutrophils.

48.2 TauCl Inhibits Overproduction of O_2^- in PMA-Stimulated Neutrophils by Inhibiting Phosphorylation of p47phox and Assembly of Functional NADPH-Oxidase Complex

Upon stimulation of neutrophils, the membrane associated NADPH-oxidase (phox) generates O_2^- by catalyzing the transfer of electrons from NADPH to molecular oxygen. The functional NADPH-oxidase complex ,which can catalyze such overproduction of O_2^-, is comprised of cytosolic subunits such, as p47phox, p67phox, p40phox and Rac GTPase, and membrane-bound subunits, such as gp91phox and p22phox. These membrane bound complexes combine with flavocytochrome b_{558} and catalyze the actual production of O_2^- via a heme-dependent transfer of electrons from NADPH to molecular oxygen (Kuribayashi et al. 2002; Lopes et al. 2004). Upon stimulation, multiple serine residues of cytosolic p47phox are phosphorylated. This signals other cytosolic components of NADPH-oxidase, namely, p67phox, p40phox and Rac, to move either to the plasma membrane or to the phagosomal membrane. Subsequently, they become associated with flavocytochrome b_{558}, the component embedded in the membrane, forming a functional complex that catalyzes O_2^- production (Babior 1999; Dinauer 2003; Groemping and Rittinger 2005; Nauseef 2004). Although adequate production of O_2^- is essential for the killing of phagocytosed bacteria, excessive overproduction of O_2^- and H_2O_2 is known to cause oxidative stress and cytotoxicity in neutrophils. In response to the oxidative stress caused by overproduction of H_2O_2, redox-sensitive transcription factor NF-κB is activated and iNOS expression is enhanced. This leads to overproduction of NO that can rapidly scavenge O_2^-. The rapid interaction between NO and O_2^- generates a highly reactive product, peroxynitrite ($ONOO^-$), which is readily reduced to nitrite by GSH, yielding in the process GSSG. The resulting depletion of intracellular GSH can lead to further oxidative stress. The increase in the GSSG/GSH redox ratio activates another redox-sensitive transcription factor Nrf2 and induces the expression of heme oxygenase-1 (HO-1), enhancing HO activity (Huie and Padmaja 1993; Motterlini et al. 2002).

Several studies have demonstrated that TauCl inhibits the overproduction O_2^- in murine and human neutrophils in a dose-dependent and reversible manner (Kim et al. 1996; Park et al. 1998; Witko et al. 1992). After the initial report on TauCl-mediated inhibition of O_2^- overproduction in PMA-stimulated neutrophils obtained from murine peritoneum (Kim et al. 1996), Park et al. (1998) found a similar effect with neutrophils obtained from mouse, rat and human. TauCl inhibited PMA-stimulated phosphorylation of p47phox in neutrophils, the initial event that is required for the assembly of a functional NADPH-oxidase complex (Choi et al. 2006). TauCl also inhibited the next step, namely the translocation of assembled p47phox and p67phox to flavocytochrome b_{558} that is localized in the plasma membrane (Choi et al. 2006). Inhibition of this translocation occurred without affecting the translocation of Rac. In any case, TauCl inhibited the assembly of functional NADPH-oxidase by impairing the translocation of cytosolic subunits to the membrane subunits. This suggested that TauCl stops the overproduction of O_2^- and protects cells from cytotoxicity caused by further oxidative stress. In addition to TauCl, other monochloramines (NH_2Cl) arising from the interaction of HOCl with other free amines not incorporated into proteins can also inhibit the overproduction of O_2^- in PMA activated neutrophils (Ogino et al. 1997). Combined, the results suggest that inhibition of O_2^- overproduction by the product of HOCl and taurine, namely, TauCl, via inhibition of p47phox phosphorylation and assembly of a functional NADPH-oxidase complex, could serve as a self-limiting protective mechanism in stimulated phagocytic cells.

48.3 TauCl Inhibits Overproduction of NO in Macrophages by Suppressing the Activation of Ras, Phosphorylation of ERK1/2, Nuclear Translocation of NF-κB and Induction of iNOS Expression

NO is produced from L-arginine by a family of nitric oxide synthases (NOSs) and serves as a signal molecule involved in modulating a variety of cell functions. Although small amounts of NO are produced by the constitutively expressed NOSs (i.e., eNOS and nNOS) and serves as a physiological signal molecule, the large amount of NO produced by inducible NOS (iNOS), whose expression is triggered initially by oxidative stress, causes further oxidative stress and cytotoxicity (Ignarro 1996). The NO overproduced by iNOS reacts rapidly with O_2^- and scavenges mildly toxic O_2^- to generate the highly toxic $ONOO^-$ and thus, plays a pivotal role in many patho-physiological processes mediated by inflammatory cells, particularly those of macrophages, in which iNOS is induced readily by stimuli causing oxidative stress. The reaction product of NO with O_2^-, namely $ONOO^-$, which is overproduced by macrophages in association with the immune response, is cytotoxic not only to the invading pathogens but also to the macrophages themselves, as well as to the surrounding host cells.

It has been previously shown that in macrophages TauCl inhibits the overproduction of NO, which is stimulated by LPS or interferon-γ (IFN-γ) (Barua et al. 2001; Marcinkiewicz et al. 1995; Park et al. 1993; Park et al. 1995). TauCl-mediated inhibition of NO production results from the inhibition of iNOS mRNA transcription, iNOS protein induction, and NO-producing enzyme activity (Park et al. 1993; Park et al. 1995). However, the effect of TauCl on the signal transduction pathways leading to activation of NF-κB and induction of iNOS mRNA transcription remain to be determined. The overproduction of O_2^- and H_2O_2 causes oxidative stress, activation of NF-κB, induction of iNOS and overproduction of NO in RAW 264.7 macrophages that had been stimulated by LPS through the formation of the Ras-GTP complex. In LPS-stimulated BAC-1.2F5 macrophage cells, however, it has been reported that formation of the Ras-GTP complex does not occur (Buscher et al. 1995). In any case, we have hypothesized that TauCl may interfere with LPS-triggered activation of GTPase Ras and the phosphorylation of its downstream signaling molecule, ERK1/2 MAPK. In support of our hypothesis, LPS appears to be a rather potent activator of Ras and TauCl inhibits LPS-derived activation of Ras in RAW 264.7 macrophages (Kim and Kim 2005).

The binding of GTP with Ras in RAW 264.7 macrophages is mediated by LPS. The activation of Ras in turn promotes the phosphorylation of MAPKs, like ERK and p38. The phosphorylation of both ERK and p38 MAPK supports the activation of NF-κB, which leads to the induction of iNOS expression. However, it is controversial as to whether the activation of both ERK and p38 MAPK is necessary for LPS-induced upregulation of iNOS expression (Ajizian et al. 1999; Bhat et al. 1998; Watters et al. 2002). In RAW 264.7 macrophages, LPS-induced upregulation of iNOS expression and NO overproduction appear to require the activation of both ERK and p38 MAPK (Kim and Kim 2005). TauCl mediates a selective inhibition of LPS-induced ERK 1/2 phosphorylation both in RAW 264.7 and peritoneal macrophages without affecting the phosphorylation of p38 MAPK. However, TauCl has no effect on TNF-α or IL-1β stimulated ERK phosphorylation in fibroblast-like synoviocytes isolated from rheumatoid arthritis (Kim et al. 2007; Kontny et al. 2003a; Kontny et al. 2003b). This suggests that TauCl-derived inhibition of ERK phosphorylation is cell and stimuli specific.

The transcription factors that are activated by phosphorylated MAPKs, namely NF-κB and AP-1, are involved in the induction of iNOS and also the expression of other pro-inflammatory gene products. Recently, it was reported that TauCl inhibits LPS-stimulated activation of NF-κB and induction of iNOS in various cell types, such as macrophages, Jurkat T-cells, and synoviocytes (Barua et al. 2001; Kanayama et al. 2002; Kontny et al. 2003a). We found that TauCl inhibits LPS-induced NF-κB activation in RAW 264.7 macrophages, however, TauCl does not reduce LPS-induced activation of AP-1; instead, LPS-derived activation of AP-1 is facilitated by TauCl (Kim and Kim 2005).

48.4 TauCl Overproduces Carbon Monoxide (CO) by Induction of HO-1 Expression

Heme oxygenase (HO) is an NADPH- and oxygen-dependent enzyme and catalyzes the rate-limiting step in oxidative degradation of free-heme to yield ferrous iron, CO and biliverdin. Biliverdin is converted to bilirubin by the enzyme, biliverdin reductase. Both biliverdin and bilirubin serve as potent antioxidants (Tenhunen et al. 1968). Two isoforms of HO exist in all mammalian cells; the non-inducible HO-2, which is expressed constitutively, and inducible HO-1, which is expressed upon exposure to either free-heme or a wide variety of stimuli that cause oxidative stress. An increase in HO activity enhances the rate of removal of free-heme, which can catalyze the production of highly reactive and toxic hydroxyl radical (HO$*$) from H_2O_2 via Fenton chemistry. The induction of HO-1 has been demonstrated to decrease the inflammatory response, as well as apoptotic death of phagocytic cells via rapid removal of potentially toxic free-heme that is released from diverse heme-containing proteins by oxidative stress. Increased HO activity is also involved in the rapid production of not only the strong antioxidants, biliverdin/bilirubin, but also the anti-inflammatory agent, CO. CO binds avidly to reduced iron (Fe^{++}) contained in heme-containing enzymes, such as NADPH-oxidase and NOS, to prevent the overproduction of O_2^- and NO. In addition, CO is known to modulate many other cellular functions, such as the leukocyte adhesion, apoptosis, and production of cytokines, like TNF-α, IL-1β, MIP-1 and IL-10 (Otterbein et al. 2000; Petrache et al. 2000). Therefore, CO and bilirubin can protect cells against the toxicity of O_2^- and NO by inhibiting their production and by scavenging them. In our previous study, we reported that ROS overproduced in response to LPS stimulation causes secondary induction of iNOS and overproduction of NO (Srisook and Cha 2004). Furthermore, $ONOO^-$ causes tertiary induction of HO-1 and overproduction of CO. This CO then inhibits additional overproduction of O_2^- and NO in macrophages stimulated with LPS, by blocking NADPH-oxidase and iNOS activity (Srisook et al. 2006).

Based on the previous report that TauCl can induce HO-1 expression in J774.2 macrophages in a dose-dependent manner (Olszanecki and Marcinkiewicz 2004), we studied whether TauCl can induce HO-1 expression and elevate HO activity in RAW 264.7 macrophage cells. TauCl by itself or in combination with LPS elevated HO-1 mRNA, protein and HO activity in RAW 264.7 cells (MS in preparation). CO overproduced as a result of TauCl-derived induction of HO-1 suppresses LPS-induced overproduction of O_2^- and NO in phagocytes (Choi et al. 2006). Combined, these results suggest that in stimulated phagocytes the upregulation of HO-1 expression and attendant increase of HO activity induced by TauCl, the product arising from the reaction of taurine with HOCl, appears to protect activated phagocytes from the cytotoxicity of ROS and RNS overproduction in response to LPS stimulation. This may occur both by the increased rate of elimination of potentially toxic free-heme and the coupled overproduction of the antioxidant bilirubin and the anti-inflammatory agent, CO, which result from the tertiary induction of HO-1 expression and enhancement of HO activity.

48.5 Summary

In phagocytes, the production of ROS and RNS markedly increases upon phagocytosis of pathogens and inflammatory stimulation that cause oxidative stress. Overproduction of ROS and RNS can lead to inadvertent cytotoxicity not only to the phagocytes themselves but also to neighboring cells. TauCl, produced by the reaction of taurine with HOCl, the latter a highly toxic oxidant generated by the MPO system in activated phagocytic cells to kill invading pathogens, induces HO-1 expression and elevates HO activity. Elevated HO activity is involved in the rapid degradation and elimination of intracellular free-heme, which is released from diverse heme-containing enzymes as a result of oxidative stress. Because free-heme catalyzes the production of highly toxic HO* via Fenton chemistry and is responsible for oxidative cytotoxicity, rapid removal of free-heme is essential for the survival of oxidatively stressed cells. In addition, the rapid removal of free-heme and elevated HO activity enhances the production of CO, which can in turn binds to reduced hemes involved in electron transfer, causing reduced production of O_2^-. CO also inhibits heme-containing enzymes, such as NADPH-oxidase and iNOS, to further block overproduction of O_2^- and NO.

The inflammatory response is initiated by overproduction of O_2^-, which is catalyzed by NADPH-oxidase. This initial overproduction of O_2^- is followed by induction of iNOS, another heme-containing enzyme that is involved in the overproduction of NO. Together NO and O_2^- combine to produce the potent oxidant and highly toxic $ONOO^-$. Considerable evidence suggests that chronic oxidative toxicity caused by overproduction of $ONOO^-$ can trigger pathobiology and contribute to the severity of various diseases. As these three gaseous signal molecules, namely O_2^-, NO and CO, can regulate the production of each other via alterations in intracellular redoxes, timely induction of HO-1 expression and elevation of HO activity is essential for the protection of cells from oxidative toxicity by $ONOO^-$. In previous studies, we demonstrated that LPS-driven O_2^- overproduction triggers the expression of iNOS and overproduction of NO. This in turn induces the expression of HO-1 and causes overproduction of CO (Srisook and Cha 2004). Overproduced CO, resulting from LPS-mediated overproduction of O_2^- and NO, inhibits PMA-induced overproduction of O_2^- in neutrophils and also LPS-mediated iNOS expression and NO overproduction in macrophages (Srisook et al. 2006). Thus, induction of HO-1 and overproduction of CO protects activated phagocytes against the cytotoxicity caused by oxidative stress. In this connection, the ability of TauCl to induce HO-1 in macrophages (Olszanecki and Marcinkiewicz 2004) provides the basis for exploring the molecular mechanisms involved in the cytoprotective effect of TauCl against chronic oxidative toxicity. Because TauCl is produced endogenously in activated phagocytes, future studies dealing with adaptive survival mechanisms by HO-1 induction may provide new insights into our understanding of cytoprotection against inflammatory cytotoxicity. In this connection, results obtained in this laboratory showing that macrophages exposed to TauCl in combination with LPS elevate HO activity over and above that caused separately by either LPS or TauCl (MS in preparation). Furthermore, TauCl induces HO-1, elevates HO activity, suppresses LPS-mediated iNOS expression and prevents NO overproduction. TauCl also inhibits PMA-elicited overproduction of O_2^- via induction of HO-1, elevation of HO activity and overproduction of CO (Choi et al. 2006). This indicates that TauCl provides cytoprotection against the cytotoxicity of ROS and RNS in stimulated phagocytes. Accordingly, these results suggest that upregulation of HO-1 expression and increased HO activity induced by TauCl is essential for the adaptive cytoprotection of activated phagocytic cells through the suppression of severe overproduction of NO and O_2^- that can occur in inflammatory lesions.

Acknowledgments This work was supported by an intramural research grant of Inha University.

References

Ajizian SJ, English BK, Meals EA (1999) Specific inhibitors of p38 and extracellular signal-regulated kinase mitogen-activated protein kinase pathways block inducible nitric oxide synthase and tumor necrosis factor accumulation in murine macrophages stimulated with lipopolysaccharide and interferon-gamma. J Infect Dis 179(4):939–944

Babior BM (1999) NADPH oxidase: an update. Blood 93(5):1464–1476
Baron DN (1969) Down with plasma. Intracellular chemical pathology studied by analysis of cells of solid tissues, erythrocytes, and leukocytes. Proc R Soc Med 62(9):945–953
Barua M, Liu Y, Quinn MR (2001) Taurine chloramine inhibits inducible nitric oxide synthase and TNF-alpha gene expression in activated alveolar macrophages: decreased NF-kappaB activation and IkappaB kinase activity. J Immunol 167(4):2275–2281
Bhat NR, Zhang P, Lee JC, Hogan EL (1998) Extracellular signal-regulated kinase and p38 subgroups of mitogen-activated protein kinases regulate inducible nitric oxide synthase and tumor necrosis factor-alpha gene expression in endotoxin-stimulated primary glial cultures. J Neurosci 18(5):1633–1641
Buscher D, Hipskind RA, Krautwald S, Reimann T, Baccarini M (1995) Ras-dependent and -independent pathways target the mitogen-activated protein kinase network in macrophages. Mol Cell Biol 15(1):466–475
Choi HS, Cha YN, Kim C (2006) Taurine chloramine inhibits PMA-stimulated superoxide production in human neutrophils perhaps by inhibiting phosphorylation and translocation of p47(phox). Int Immunopharmacol 6(9):1431–1440
Dinauer MC (2003) Regulation of neutrophil function by Rac GTPases. Curr Opin Hematol 10(1):8–15
Fukuda K, Hirai Y, Yoshida H, Nakajima T, Usui T (1982) Free amino acid content of lymphocytes and granulocytes compared. Clin Chem 28(8):1758–1761
Grisham MB, Jefferson MM, Thomas EL (1984) Role of monochloramine in the oxidation of erythrocyte hemoglobin by stimulated neutrophils. J Biol Chem 259(11):6757–6765
Groemping Y, Rittinger K (2005) Activation and assembly of the NADPH oxidase: a structural perspective. Biochem J 386(Pt 3):401–416
Huie RE, Padmaja S (1993) The reaction of no with superoxide. Free Radic Res Commun 18(4):195–199
Huxtable RJ (1992) Physiological actions of taurine. Physiol Rev 72(1):101–163.
Ignarro LJ (1996) Physiology and pathophysiology of nitric oxide. Kidney Int Suppl 55:S2–S5
Kanayama A, Inoue J, Sugita-Konishi Y, Shimizu M, Miyamoto Y (2002) Oxidation of Ikappa Balpha at methionine 45 is one cause of taurine chloramine-induced inhibition of NF-kappa B activation. J Biol Chem 277(27):24049–24056
Kim C, Chung JK, Jeong JM, Chang YS, Lee YJ, Kim YJ, Lee MC, Koh CS, Kim BK (1998) Uptake of taurine and taurine chloramine in murine macrophages and their distribution in mice with experimental inflammation. Adv Exp Med Biol 442:169–176
Kim C, Park E, Quinn MR, Schuller-Levis G (1996) The production of superoxide anion and nitric oxide by cultured murine leukocytes and the accumulation of TNF-alpha in the conditioned media is inhibited by taurine chloramine. Immunopharmacology 34(2–3):89–95
Kim JW, Kim C (2005) Inhibition of LPS-induced NO production by taurine chloramine in macrophages is mediated though Ras-ERK-NF-kappaB. Biochem Pharmacol 70(9):1352–1360
Kim KS, Park EK, Ju SM, Jung HS, Bang JS, Kim C, Lee YA, Hong SJ, Lee SH, Yang HI, Yoo MC (2007) Taurine chloramine differentially inhibits matrix metalloproteinase 1 and 13 synthesis in interleukin-1beta stimulated fibroblast-like synoviocytes. Arthritis Res Ther 9(4):R80
Kontny E, Maslinski W, Marcinkiewicz J (2003a) Anti-inflammatory activities of taurine chloramine: implication for immunoregulation and pathogenesis of rheumatoid arthritis. Adv Exp Med Biol 526:329–340
Kontny E, Rudnicka W, Kowalczewski J, Marcinkiewicz J, Maslinski W (2003b) Selective inhibition of cyclooxygenase 2-generated prostaglandin E2 synthesis in rheumatoid arthritis synoviocytes by taurine chloramine. Arthritis Rheum 48(6):1551–1555
Kuribayashi F, Nunoi H, Wakamatsu K, Tsunawaki S, Sato K, Ito T, Sumimoto H (2002) The adaptor protein p40(phox) as a positive regulator of the superoxide-producing phagocyte oxidase. Embo J 21(23):6312–6320
Learn DB, Fried VA, Thomas EL (1990) Taurine and hypotaurine content of human leukocytes. J Leukoc Biol 48(2):174–182

Lopes LR, Dagher MC, Gutierrez A, Young B, Bouin AP, Fuchs A, Babior BM (2004) Phosphorylated p40PHOX as a negative regulator of NADPH oxidase. Biochemistry 43(12):3723–3730
Marcinkiewicz J, Grabowska A, Bereta J, Stelmaszynska T (1995) Taurine chloramine, a product of activated neutrophils, inhibits in vitro the generation of nitric oxide and other macrophage inflammatory mediators. J Leukoc Biol 58(6):667–674
Marquez LA, Dunford HB (1994) Chlorination of taurine by myeloperoxidase. Kinetic evidence for an enzyme-bound intermediate. J Biol Chem 269(11):7950–7956
Midwinter RG, Peskin AV, Vissers MC, Winterbourn CC (2004) Extracellular oxidation by taurine chloramine activates ERK via the epidermal growth factor receptor. J Biol Chem 279(31):32205–32211
Motterlini R, Green CJ, Foresti R (2002) Regulation of heme oxygenase-1 by redox signals involving nitric oxide. Antioxid Redox Signal 4(4):615–624
Nauseef WM (2004) Assembly of the phagocyte NADPH oxidase. Histochem Cell Biol 122(4):277–291
Ogino T, Kobuchi H, Sen CK, Roy S, Packer L, Maguire JJ (1997) Monochloramine inhibits phorbol ester-inducible neutrophil respiratory burst activation and T cell interleukin-2 receptor expression by inhibiting inducible protein kinase C activity. J Biol Chem 272(42):26247–26252
Olszanecki R, Marcinkiewicz J (2004) Taurine chloramine and taurine bromamine induce heme oxygenase-1 in resting and LPS-stimulated J774.2 macrophages. Amino Acids 27(1):29–35
Otterbein LE, Bach FH, Alam J, Soares M, Tao Lu H, Wysk M, Davis RJ, Flavell RA, Choi AM (2000) Carbon monoxide has anti-inflammatory effects involving the mitogen-activated protein kinase pathway. Nat Med 6(4):422–428
Park E, Alberti J, Quinn MR, Schuller-Levis G (1998) Taurine chloramine inhibits the production of superoxide anion, IL-6 and IL-8 in activated human polymorphonuclear leukocytes. Adv Exp Med Biol 442:177–182
Park E, Jia J, Quinn MR, Schuller-Levis G (2002) Taurine chloramine inhibits lymphocyte proliferation and decreases cytokine production in activated human leukocytes. Clin Immunol 102(2):179–184
Park E, Quinn MR, Wright CE, Schuller-Levis G (1993) Taurine chloramine inhibits the synthesis of nitric oxide and the release of tumor necrosis factor in activated RAW 264.7 cells. J Leukoc Biol 54(2):119–124
Park E, Schuller-Levis G, Quinn MR (1995) Taurine chloramine inhibits production of nitric oxide and TNF-alpha in activated RAW 264.7 cells by mechanisms that involve transcriptional and translational events. J Immunol 154(9):4778–4784
Pero RW, Sheng Y, Olsson A, Bryngelsson C, Lund-Pero M (1996) Hypochlorous acid/N-chloramines are naturally produced DNA repair inhibitors. Carcinogenesis 17(1):13–18
Petrache I, Otterbein LE, Alam J, Wiegand GW, Choi AM (2000) Heme oxygenase-1 inhibits TNF-alpha-induced apoptosis in cultured fibroblasts. Am J Physiol Lung Cell Mol Physiol 278(2):L312–L319
Schuller-Levis GB, Park E (2003) Taurine: new implications for an old amino acid. FEMS Microbiol Lett 226(2):195–202
Srisook K, Cha YN (2004) Biphasic induction of heme oxygenase-1 expression in macrophages stimulated with lipopolysaccharide. Biochem Pharmacol 68(9):1709–1720
Srisook K, Han SS, Choi HS, Li MH, Ueda H, Kim C, Cha YN (2006) CO from enhanced HO activity or from CORM-2 inhibits both O2- and NO production and downregulates HO-1 expression in LPS-stimulated macrophages. Biochem Pharmacol 71(3):307–318
Tallan HH, Jacobson E, Wright CE, Schneidman K, Gaull GE (1983) Taurine uptake by cultured human lymphoblastoid cells. Life Sci 33(19):1853–1860
Tenhunen R, Marver HS, Schmid R (1968) The enzymatic conversion of heme to bilirubin by microsomal heme oxygenase. Proc Natl Acad Sci USA 61(2):748–755
Thomas EL, Grisham MB, Melton DF, Jefferson MM (1985) Evidence for a role of taurine in the in vitro oxidative toxicity of neutrophils toward erythrocytes. J Biol Chem 260(6):3321–3329

Vinton NE, Laidlaw SA, Ament ME, Kopple JD (1986) Taurine concentrations in plasma and blood cells of patients undergoing long-term parenteral nutrition. Am J Clin Nutr 44(3):398–404
Watters JJ, Sommer JA, Pfeiffer ZA, Prabhu U, Guerra AN, Bertics PJ (2002) A differential role for the mitogen-activated protein kinases in lipopolysaccharide signaling: the MEK/ERK pathway is not essential for nitric oxide and interleukin 1beta production. J Biol Chem 277(11): 9077–9087
Witko V, Nguyen AT, Descamps-Latscha B (1992) Microtiter plate assay for phagocyte-derived taurine-chloramines. J Clin Lab Anal 6(1):47–53

Chapter 49
Taurine Chloramine Inhibits LPS-Induced Glucose Uptake and Glucose Transporter 1 Expression in RAW 264.7 Macropages

Chaekyun Kim and Seongtag Kim

Abstract Inflammatory cells use glucose as a primary source of metabolic energy, and thus increased uptake of glucose and high rates of glycolysis are characteristics of inflamed cells. Taurine chloramine (TauCl) is the product of a reaction between cellular taurine and hypochlorous acid ($HOCl/OCl^-$), the latter produced by the halide-dependent myeloperoxidase (MPO) system in inflammatory cells. Taurine, a major metabolite of cysteine, protects cells from inflammatory injury by removing toxic hypochlorous acid formed by the MPO system, and also by inhibiting the production of inflammatory mediators. In the present study, we examined the effect of TauCl on glucose uptake and the expression of the glucose transporter 1 (GLUT1) in RAW 264.7 murine macrophages stimulated with lipopolysaccharide (LPS). Glucose uptake was measured by employing labeled glucose analogue [18F]-2-fluoro-2-deoxy-D-glucose (FDG). Stimulation RAW 264.7 cells with LPS increased glucose uptake and led to an upregulation in GLUT1 expression, effects that were abrogated in macrophages treated with TauCl. These data suggest that TauCl can inhibit LPS-mediated enhancement of glucose uptake through inhibition of the upregulation of glucose transporter expression in activated macrophages. This represents one of the mechanisms by which TauCl modulates inflammatory cell function.

Abbreviations *TauCl*, taurine chloramine; *GLUT*, glucose transporter; *LPS*, lipopolysaccharide; *FDG*, [^{18}F]-2-fluoro-2-deoxy-D-glucose

49.1 Introduction

Many mammalian cells utilize glucose as the primary source of energy and reducing equivalents needed for their metabolism and growth. The phagocytic inflammatory macrophages need glucose more acutely to replenish the reducing

C. Kim (✉)
Laboratory for Leukocyte Signaling Research, and Center for Advanced Medical Education by BK21 Project, Inha University School of Medicine, Incheon, Korea

J. Azuma et al. (eds.), *Taurine 7*, Advances in Experimental Medicine and Biology 643,
DOI 10.1007/978-0-387-75681-3_49, © Springer Science+Business Media, LLC 2009

equivalents utilized in the oxidative burst. Glucose uptake is markedly increased in macrophages that accumulate at sites of injury and infection. Animals injected with LPS experience increased glucose consumption particularly in macrophage-rich organs (Fukuzumi et al. 1996; Gamelli et al. 1996). Increased uptake of glucose is closely associated with increased expression of glucose transporters. There are two classes of glucose transporters that regulate glucose transport in mammalian cells, the facilitated glucose transporter (GLUT) mediates passive diffusion of glucose and the sodium-dependent glucose transporter (SGLT) mediates active transport against the glucose concentration gradient. In activated lymphocytes and macrophages, both the uptake of glucose and the expression of GLUT1 are increased (Chakrabarti et al. 1994; Fu et al. 2004; Malide et al. 1998). [^{18}F]-2-fluoro-2-deoxy-D-glucose (FDG) is employed frequently to measure glucose uptake, as it is taken up by cells and then phosphorylated by hexokinase to FDG-6-phosphate. However, as it is not further metabolized and accumulates in the cell, where it can be easily measured.

Taurine chloramine (TauCl) is the major chloramine produced by the reaction between taurine and hypochlorous acid (HOCl/OCl^-), the cytotoxic metabolites arising from the neutrophil myeloperoxidase (MPO) system (Babior, 1999; Thomas et al. 1985). Elimination of cytotoxic HOCl by formation of TauCl attenuates the toxicity of HOCl and serves as a protective mechanism in neutrophils. Recent studies indicate that TauCl is not simply the inactivate taurine conjugate of HOCl but TauCl can also reduce the production of many pro-inflammatory mediators, such as nitric oxide (NO), superoxide anion (O_2^-), tumor necrosis factor (TNF)-α, interleukin (IL)-6, IL-8 and prostaglandins (Kim et al. 1996; Marcinkiewicz et al. 1995; Marcinkiewicz et al. 1998; Park et al. 1995). Although TauCl is charged and not readily membrane-permeable, it can be transported into cells in a Na^+- and Cl-dependent manner similar to taurine (Kim et al. 1998; Tallan et al. 1983).

Uptake of glucose into cells mediated by GLUT and SGLT is modulated by numerous cellular and molecular factors. Taurine has been reported to lower blood sugar content and cellular glucose uptake by partial inhibition of SGLT-1 activity (Arany et al. 2004; Diebold and Bokoch 2001; Hansen, 2001; Kim et al. 2006). However, it has not been determined whether TauCl has any effect on glucose uptake. Thus, in this study we examined the effect of TauCl on glucose uptake and the expression of GLUT1 using the murine macrophage cell line, RAW 264.7 cell.

49.2 Materials and Methods

49.2.1 Antibodies and Reagents

Rabbit polyclonal antibodies against GLUT1 was purchased from Chemicon (Temecula, CA), the mouse monoclonal antibody against beta-actin was obtained from Sigma (St. Louis, MO), and the peroxidase-linked secondary antibodies were purchased from Upstate Biotechnology (Lake Placid, NY). Fetal bovine serum (FBS), phosphate buffered saline (PBS), Hanks' balanced salt solution (HBSS),

RPMI 1640 medium, penicillin and streptomycin were from HyClone (Logan, UT). Other routinely used chemicals were purchased from Sigma unless stated otherwise. TauCl was synthesized freshly on the day of use by adding equimolar amounts of NaOCl (Aldrich Chemical, Milwaukee, MI) to taurine. The authenticity of TauCl formation was monitored by UV absorption (200–400 nm) (Thomas et al. 1986). Endotoxin-free or low endotoxin grade water and buffers were used.

49.2.2 Preparation of Human Peripheral Blood Monocytes and Cell Culture

Human peripheral blood monocytes (PBMC) were isolated from the blood of healthy donors. The blood was layered on top of Histopaque 1077 and 1119 gradients, and centrifuged at 700 × g for 30 min at room temperature. The mono nuclear cell layer was washed three times in 10 ml of PBS and collected by centrifugation at 200 × g for 10 min. Red blood cells were lysed with Tris-buffered ammonium chloride (pH 7.2) and washed in PBS. The PBMC used for experiments was obtained by adhering them onto plates via incubation in RPMI 1640 medium for 4 h at 37°C under 5% CO_2. RAW 264.7 cells, murine macrophage cell line obtained from ATCC (Manassas, VA) were grown in RPMI 1640 medium supplemented with 10% FBS, 100 units/ml of penicillin and 100 μg/ml of streptomycin and maintained at 37°C in 5% CO_2incubator.

49.2.3 Measurement of Glucose (FDG) Uptake

PBMC in RPMI 1640 medium containing various concentrations of glucose (0, 5.5 and 16.7 mM) was placed 48-well plates for either 4 h or 24 h at 37°C in a 5% CO_2 incubator. For activation, PBMC and RAW 264.7 cells were treated with 10 μg/ml LPS in the presence and absence of taurine and TauCl. Cells were washed with HEPES-buffered saline (HBS; 140 mM NaCl, 1 mM $CaCl_2$, 5 mM KCl, 2.5 mM $MgSO_4$, and 20 mM HEPES, pH 7.4) and then incubated with 500 μl of 1 μCi/ml FDG in HBS for 20 min at 37°C. The uptake was terminated by washing the cells three times with HBS. Cells were lysed with 1% sodium dodecyl sulfate (SDS) and the radioactivity in the lysates was determined using a gamma counter (Cobra II, Packard, Downers Grove, IL). Cellular protein was quantified by the bicinchoninic acid (BCA) protein assay kit (Pierce, Rockford, IL) using bovine serum albumin (BSA) as the standard.

49.2.4 Immunoblot Analysis of GLUT1

RAW 264.7 cells were treated with LPS for 12 h in the presence and absence of taurine and TauCl, washed with ice-cold HBSS, and then harvested by gentle scraping. Cells were then lysed in a lysis buffer containing 20 mM Tris-Cl (pH 8.0), 150 mM NaCl, 1 mM EDTA, 1% Triton X-100, 20 μg/ml chymostatin, 2 mM phenylmethylsulfonyl fluoride (PMSF), 10 μM leupeptin and 1 mM 4-(2-aminoethyl)-benzenesulfonyl fluoride (AEBSF). Protein was quantified using the BCA protein assay kit. Whole cell lysates were mixed with Laemmli sample buffer, heated at 95°C for 5 min, and subjected to 10% SDS-PAGE. Separated proteins were transferred onto nitrocellulose membrane (MSI, Westborough, MA), and non-specific binding was blocked with 6% non-fat milk dissolved in TBST buffer (10 mM Tris-Cl, pH 8.0, 150 mM NaCl, 0.05% Tween 20). Blots were probed with GLUT1 antibody (1/1000 dilution) and developed using the ECL method (Amersham, Arlington Heights, IL).

49.2.5 Measurement of Cell Viability Using MTT Reduction

To determine whether 10 μg/ml LPS and 0.7 mM taurine or TauCl cause cell death each by themselves, we determined cell viability using the conventional MTT (3-(4,5-dimethylthiazol-2yl)2,5-dipheyltetrazolium bromide) reduction assay. Briefly, cells (2×10^5) were incubated with LPS, taurine or TauCl for 4 h at 37°C in 5% CO_2. Thereafter, 10 μl of 5 mg/ml MTT solution (final concentration 500 μg/ml) was added to each well. At 2 h after incubation with MTT at 37°C, the culture media was removed and cells were dissolved in DMSO. The absorbance was measured at 550 nm using spectrophotometer (Bio-Tek Instrument EL900; Highland Park, VT).

49.2.6 Statistical Analysis

The two-tailed Student's t-test (paired) was performed using Microsoft Excel software (Redmond, WA). Data are expressed as means ± SEM and a p value < 0.05 was considered significant.

49.3 Results and Discussion

49.3.1 Glucose Uptake is Increased in Activated Cells and Hyperglycemia Attenuates LPS-Stimulated Glucose Uptake

To examine the effect of the cell's activation state on glucose uptake, glucose transport was determined using unstimulated human PBMC and LPS-stimulated PBMC (A-PBMC) at various glucose concentrations. Glucose uptake was increased in LPS-stimulated PBMC after 4 h and 24 h incubation (Fig. 49.1), confirming the reports that glucose uptake is increased in activated macrophages, and supporting the notion that glucose consumption in macrophage-rich organs is increased in animals having LPS-induced endotoxemia. (Fukuzumi et al. 1996; Gamelli et al. 1996).

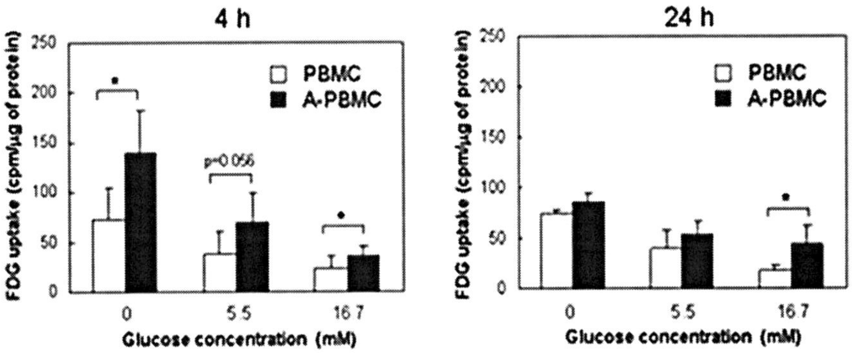

Fig. 49.1 Activation of human peripheral blood monocytes (PBMC) increases glucose uptake. FDG uptake in PBMC and LPS-activated PBMC (A PBMC) was measured after 4 h (A) and 24 h (B) of incubation in media containing various concentrations of glucose. The result represents means ± SEM of 3–4 independent experiments, * $p < 0.05$ PBMC vs. A-PBMC (paired t test)

Both unstimulated and LPS-stimulated PBMCs, FDG uptake was higher when cells were incubated in glucose-free media and FDG uptake decreased with an increase in glucose concentration (Fig. 49.1).

49.3.2 TauCl Inhibits LPS-Stimulated Glucose Uptake in Macrophages

To examine the effect of TauCl on the LPS-stimulated enhancement in glucose uptake, RAW 264.7 cells were treated with LPS in the presence and absence of 0.7 mM TauCl. Although the rate of glucose uptake in RAW 264.7 cells was increased by ~200% following 10 μg/ml LPS treatment, when 0.7 mM TauCl was added, LPS-induced increase in glucose uptake was abolished and was similar to that of the control (Fig. 49.2A).

In some experiments, RAW 264.7 cells were treated with 5 μM cytochalasin B, which inhibited glucose uptake even in the unstimulated state (data not shown). LPS (10 μg/ml) induced around 20% cell death after a 24 h incubation, however, addition of 0.7 mM taurine and TauCl had no cytoprotective effect against LPS-induced cell death (Fig. 49.2B).

49.3.3 TauCl Inhibits GLUT1 Expression in Macrophages

To uncover the mechanism by which TauCl inhibits glucose uptake in RAW 264.7 cells, we measured the effect of TauCl on GLUT1 expression. Enhanced glucose uptake is closely related with increased expression of GLUTs. Among the seven known GLUTs, GLUT1 is the most widely expressed GLUT, and it is the prominent glucose transporter in macrophages and cancer cells (Chung et al. 1999; Flier et al. 1987; Gamelli et al. 1996). Whole cell lysates were extracted after 12 h

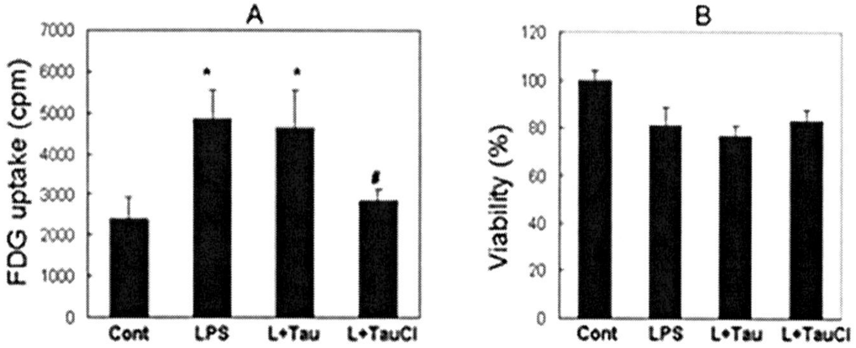

Fig. 49.2 (**A**) TauCl (0.7 mM0) inhibits FDG uptake in RAW 264.7 macrophages after 4 h incubation in medium containing 5.5 mM glucose. The result represents means ± SEM of three independent experiments, $*p < 0.05$ compared to control, $\#p < 0.05$ compared to LPS. (**B**) Effect of TauCl on RAW 264.7 cell viability. Cell viability was measured using the MTT assay after 24 h incubation with TauCl. The result represents means ± SEM of three independent experiments

incubation with LPS, taurine and TauCl, and the expression of GLUT1 was detected by Western blotting. LPS increased GLUT1 expression and TauCl inhibited LPS-stimulated GLUT1 expression in a dose dependent manner, while taurine had no significant effect (Fig. 49.3). However, addition of taurine and TauCl in the absence of LPS had no effect on GLUT1 expression.

The present study shows that glucose uptake is increased in activated macrophages, and TauCl may attenuate LPS-mediated stimulation of glucose uptake through partial prevention of the upregulation of the glucose transporter in activated macrophages.

49.3.4 Mathematical Analysis

In this part, we interpret experimental data using mathematics. The mathematical correlation between glucose uptake and glucose concentration in the presence

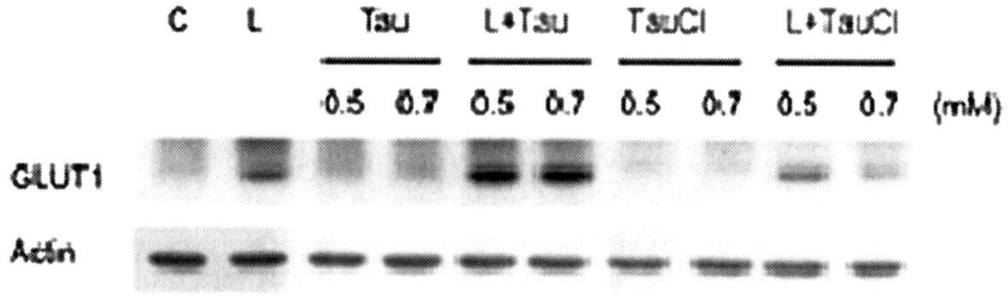

Fig. 49.3 TauCl inhibits LPS-stimulated GLUT1 expression in RAW 264.7 macrophages. Cells were incubated with TauCl for 12 h and whole cell lysates were separated on 10% SDS-PAGE, transferred to PVDF membrane. Immunoblotting was performed with antibodies against GLUT1 and beta-actin. Immunoblots are representative of 3–4 independent experiments

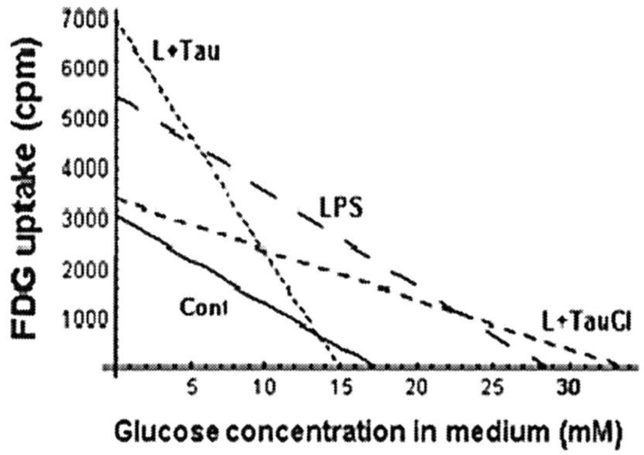

Fig. 49.4 Mathematical analysis of glucose uptake in RAW 264.7 macrophages. FDG uptake was measured after incubation with taurine and TauCl in the presence of LPS and the relative comparison is illustrated. This figure is drawn using program Mathematica (Wolfram, Illinois)

or absence of stimulus is not known. For mathematical analysis, we assume that glucose uptake depends linearly on glucose concentration in the medium (Fig. 49.4). With this assumption, we determined equations expressing the relation between glucose uptake and glucose concentration in the medium. Here we provide detailed mathematical analysis of the experimental data.

Let y be the amount of glucose uptake and x the glucose concentration in the media. The calculated equations are:

control: $y = -178x + 3078$
LPS: $y = -192.9x + 5498$
L+ Tau: $y = -469.6x + 7021$
L+TauCl: $y = -100.9x + 3394$, respectively.

We also calculated the effect of 0.7 mM TauCl on glucose uptake after LPS treatment at various glucose concentrations. TauCl reduced glucose uptake by 38.3% and 36.0% with respect to LPS in glucose free medium and in 5.5 mM glucose containing medium, respectively.

Acknowledgments We are grateful to Dr. Young-Nam Cha for critical review on this manuscript and Mi-Ran Cho for valuable technical support. This work was supported by an INHA University Research Grant (INHA-33909).

References

Arany E, Strutt B, Romanus P, Remacle C, Reusens B, Hill DJ (2004) Taurine supplement in early life altered islet morphology, decreased insulitis and delayed the onset of diabetes in non-obese diabetic mice. Diabetologia 47:1831–1837

Babior BM (1999) NADPH oxidase: an update. Blood 93:1464–1476

Chakrabarti R, Jung CY, Lee TP, Liu H, Mookerjee BK (1994) Changes in glucose transport and transporter isoforms during the activation of human peripheral blood lymphocytes by phytohemagglutinin. J Immunol 152:2660–2668

Chung JK, Lee YJ, Kim C, Choi SR, Kim M, Lee K, Jeong JM, Lee DS, Jang JJ, Lee MC (1999) Mechanisms related to [18F]fluorodeoxyglucose uptake of human colon cancers transplanted in nude mice. J Nucl Med 40:339–346

Diebold BA, Bokoch GM (2001) Molecular basis for Rac2 regulation of phagocyte NADPH oxidase. Nat Immunol 2:211–215

Flier JS, Mueckler MM, Usher P, Lodish HF (1987) Elevated levels of glucose transport and transporter messenger RNA are induced by ras or src oncogenes. Science 235:1492–1495

Fu Y, Maianu L, Melbert BR, Garvey WT (2004) Facilitative glucose transporter gene expression in human lymphocytes, monocytes, and macrophages: a role for GLUT isoforms 1, 3, and 5 in the immune response and foam cell formation. Blood Cells Mol Dis 32:182–190

Fukuzumi M, Shinomiya H, Shimizu Y, Ohishi K, Utsumi S (1996) Endotoxin-induced enhancement of glucose influx into murine peritoneal macrophages via GLUT1. Infect Immun 64:108–112

Gamelli RL, Liu H, He LK, Hofmann CA (1996) Augmentations of glucose uptake and glucose transporter-1 in macrophages following thermal injury and sepsis in mice. J Leukoc Biol 59:639–647

Hansen SH (2001) The role of taurine in diabetes and the development of diabetic complications. Diabetes Metab Res Rev 17:330–346

Kim C, Chung JK, Jeong JM, Chang YS, Lee YJ, Kim YJ, Lee MC, Koh CS, Kim BK (1998) Uptake of taurine and taurine chloramine in murine macrophages and their distribution in mice with experimental inflammation. Adv Exp Med Biol 442:169–176

Kim C, Park E, Quinn MR, Schuller-Levis G (1996) The production of superoxide anion and nitric oxide by cultured murine leukocytes and the accumulation of TNF-alpha in the conditioned media is inhibited by taurine chloramine. Immunopharmacology 34:89–95

Kim HW, Lee AJ, You S, Park T, Lee DH (2006) Characterization of taurine as inhibitor of sodium glucose transporter. Adv Exp Med Biol 583:137–145

Malide D, Davies-Hill TM, Levine M, Simpson IA (1998) Distinct localization of GLUT-1, -3, and -5 in human monocyte-derived macrophages: effects of cell activation. Am J Physiol 274(3 Pt 1):E516–E526

Marcinkiewicz J, Grabowska A, Bereta J, Stelmaszynska T (1995) Taurine chloramine, a product of activated neutrophils, inhibits in vitro the generation of nitric oxide and other macrophage inflammatory mediators. J Leukoc Biol 58:667–674

Marcinkiewicz J, Grabowska A, Chain BM (1998) Modulation of antigen-specific T-cell activation in vitro by taurine chloramine. Immunology 94:325–330

Park E, Schuller-Levis G, Quinn MR (1995) Taurine chloramine inhibits production of nitric oxide and TNF-alpha in activated RAW 264.7 cells by mechanisms that involve transcriptional and translational events. J Immunol 154:4778–4784

Tallan HH, Jacobson E, Wright CE, Schneidman K, Gaull GE (1983) Taurine uptake by cultured human lymphoblastoid cells. Life Sci 33:1853–1860

Thomas EL, Grisham MB, Jefferson MM (1986) Preparation and characterization of chloramines. Methods Enzymol 132:569–585

Thomas EL, Grisham MB, Melton DF, Jefferson MM (1985) Evidence for a role of taurine in the in vitro oxidative toxicity of neutrophils toward erythrocytes. J Biol Chem 260(6):3321–3329

Chapter 50
Inhibition of Apoptosis by Taurine in Macrophages Treated with Sodium Nitroprusside

So Young Kim, Taesun Park, and Ha Won Kim

Abstract Nitric oxide (NO) induces apoptotic cell death in murine RAW264.7 macrophages. To elucidate the mechanism underlying the inhibitory effect of taurine on NO-induced apoptosis, a cell was exposed to sodium nitroprusside (SNP), an NO donor, in the absence and presence of taurine. Taurine treatment prevented SNP-mediated cellular apoptosis in a concentration dependent manner. The exposure of the cell to taurine prior to SNP treatment inhibited DNA fragmentation more than addition of taurine to the medium after SNP treatment. Agarose gel electrophoresis data revealed that taurine reduced the intensity of SNP-induced DNA laddering. The taurine-mediated reduction in the number of apoptotic cells was also observed using the Hoechst 33258 stain. These results support the idea that taurine has the potential to function as an inhibitory modulator of NO-mediated cell injury.

Abbreviations *DPA*, diphenylamine; *MPO*, myeloperoxidase; *NO*, nitric oxide; *NOS*, nitric oxide synthase; *SNP*, sodium nitroprusside

50.1 Introduction

Taurine is a beta-amino acid found in high concentration (20–50 mM) in mammarian cells, such as muscle (Park et al. 1995). It is biosynthesized from cysteine using pathways that proceed either through hypotaurine or cysteic acid. The biological functions of taurine are anti-neurodegeneration (Dawson 2003), anti-oxidation and anti-apoptosis (Yu et al. 2007), membrane stabilization (You and Chang 1998), and antagonism of the Ca^{++} paradox (Kramer et al. 1981). These functions of taurine are reported from various organs such as the brain, retina, liver, muscle and bones (Huxtable 1992).

Nitric oxide (NO) is a kind of intracellular free radical gas that is involved in important functions in diverse organs. It is synthesized in endothelial cells as a transcellular signal, with the biosynthesis occurring when L-arginine is acted upon

S.Y. Kim (✉)
College of Medicine, Ewha Women's University, Seoul, Korea

J. Azuma et al. (eds.), *Taurine 7*, Advances in Experimental Medicine and Biology 643,
DOI 10.1007/978-0-387-75681-3_50,

by the enzyme, NO synthase (NOS) (Solomonson et al. 2003). It also serves as a biological messenger during the inflammatory processes in macrophages (Alvarez and Evelson 2007) or as a toxic effector molecule in autoimmune diseases, such as spontaneous glomerulonephritis and arthritis (Weinberg et al. 1994). Moreover, it activates or inhibits many enzyme reactions (Brune and Lapetina 1990), causing genotoxicity via DNA damage (Felley-Bosco 1998), and mediating oxidative and nitrosative reactions (Stamler et al. 1992a, 1992b). Recently, induction of apoptosis or programmed cell death by NO was reported in various cells. Typical characteristics of apoptosis include DNA fragmentation, membrane blebbing, and the formation of apoptotic bodies (Higuchi 2003).

During inflammation or phagocytosis, neutrophils or macrophages secrete myeloperoxidase (MPO), which catalyzes the formation of hypochlorous acid (HOCl) from H_2O_2 and Cl^- (Weiss et al. 1982). Due to the very strong oxidative properties of HOCl, it can damage various cellular organelles and organs. However, HOCl is partially neutralized by antioxidants, such as taurine, which convert HOCl to a chloramine, a subtancewhich exhibits less intrinsic toxicity and has the potential to suppress neutrophil activity. Thus, taurine protects cells and organs from toxicity mediated by $HOCl/OCl^-$ (Marquez and Dunford 1994).

In this report, the toxic effect of sodium nitroprusside, an NO donor, was assessed in RAW264.7 cells, a macrophage cell line. The study examines the hypothesis that taurine reduces or prevents cellular apoptosis resulting from NO generated by an NO donor.

50.2 Methods

50.2.1 Cell and Reagents

RAW 264.7 is an Abelson leukemia virus-transformed murine macrophage cell line. It was obtained from the Korea Cell Line Bank (KCLB) of the Seoul National University Hospital (SNUH).

Dulbecco's modified Eagle's medium (DMEM), heat-inactivated fetal bovine serum (FBS), penicillin and streptomycin were from Gibco-BRL (Grand Island, NY). Taurine, sodium nitroprusside (SNP), 3-[4,5-dimethylthiazol-2-yl]-2, 5-diphenyltetrazolium bromide (MTT), diphenylamine, Hoechst 33258, agarose, ethidium bromide, and trichloroacetic acid were from Sigma Chem. Co. (St. Louis, MO).

50.2.2 Cell Culture

The RAW 264.7 cell line was maintained in medium that contained 3.7 g/l sodium bicarbonate, 10% FBS, 100 U/ml penicillin, and 100 μg/ml streptomycin, which were passed through with a 0.22-μm membrane filter. The cells were cultured in 100-mm-diameter dishes in 7 ml medium, with the medium changed every other day. The cells were harvested by digesting them with trypsin and scraping them with a cell scraper. Cell viability was checked via a Trypan blue exclusion test. Only the cells that had more than 95% viability were used in this experiment.

50.2.3 DNA Fragmentation

The TTE cell lysis buffer contained 0.5% Triton X-100, 5-mM Tris-buffer (pH 7.6), and 20-mM EDTA (pH 7.6). The diphenylamine (DPA) solution was prepared by dissolving 10 mg of diphenylamine in 10 ml of glacial acetic acid and adding to it 150 μl of conc-H_2SO_4 and 50 μl of acetaldehyde solution. For the DNA fragmentation analysis, the cells grown in 100-mm culture dishes were harvested and lysed with 0.5 ml TTE buffer for 30 min at 4°C. The supernatant was incubated overnight with 25% TCA, and the pellet was obtained through centrifugation for 10 min at 13,000 rpm. The obtained pellet was re-suspended in 80 μl of 5% TCA and heated for 14 min at 90°C, after which it was incubated with 160 μl of a DPA solution for 4 hrs. The degree of DNA fragmentation was determined by measuring the optical density at 610 nm.

50.2.4 DNA Gel Electrophoresis and Nuclear Staining

RAW264.7 cells, which were grown in 60-mm culture plates, were treated with taurine and SNP, followed by lysis, for 10 mins at 4°C by adding the TTE lysis buffer. The DNA was extracted with phenol : chloroform : isoamylalcohol (25:24:1) from the DNA-containing supernatant solution and then purified again with chloroform:isoamylalcohol (24:1). The resulting DNA was precipitated with 3 M of sodium acetate and absolute alcohol, and the precipitant was treated with 5 μl of 1-mg/ml RNase A and re-suspended with 20 μl of the TE buffer. The DNA in the TE buffer was subjected to electrophoresis (1.8% agarose gel), after which the DNA fragmentation patterns were analyzed. The RAW264.7 cells treated with taurine and SNP were fixed with 4% formaldehyde for 5 mins at room temperature and washed twice with PBS buffer. The fixed cells were stained with 5 μg/ml of Hoechst 33258 for 5 mins, washed with PBS, and analyzed under a fluorescent microscope in the dark room.

50.3 Results

50.3.1 SNP Cytotoxicity and Recovery by Taurine

To analyze the cytotoxicity of the NO donor, SNP, RAW264.7 cells were cultured with various concentrations of SNP ranging from 200 to 1,000 μM. Cell viability was measured via the MTT assay after 24 hrs of SNP exposure. Cell viability was rapidly reduced below 50% at SNP concentrations greater than 200 μM (Fig. 50.1).

The antioxidant activity of taurine was examined by pretreating the cells for 24 hrs with various concentrations of taurine prior to exposure of the cells to SNP

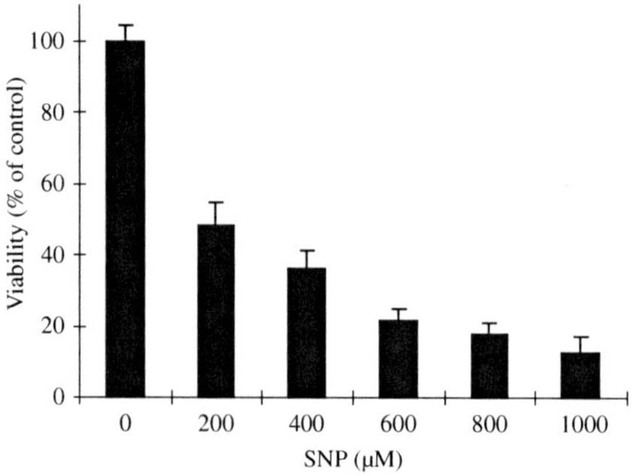

Fig. 50.1 Effect of SNP on viability of RAW264.7 cells, a macrophage cell line. Viability was measured by the MTT assay after a 24 hr-incubation with SNP

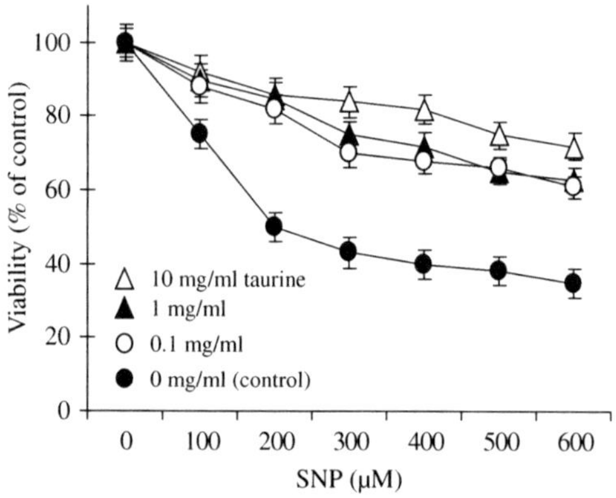

Fig. 50.2 Recovery of viability by taurine in SNP treated macrophages. RAW264.7 cells were pretreated with taurine for 24 hrs and then cultured with SNP for an additional 8 hrs

for 8 hrs. Taurine (0.1-mg/ml) mediated a similar degree of cell protection at most SNP concentrations. At 300 μM SNP, taurine (10 mg/ml) increased cell viability two fold with respect to the untreated control group (Fig. 50.2).

50.3.2 Inhibition of DNA Fragmentation by Taurine Pretreatment

DNA fragmentation was analyzed after addition of SNP to RAW264.7 cells in culture. Taurine treatment (0.1 to 10 mg/ml) reduced DNA fragmentation 5.7% in the absence of SNP. However, when 300 μM SNP was added to untreated cells, 29% of the cells exhibited DNA fragmentation. However, pretreatment with 0.1, 1.0 or 10 mg/ml taurine reduced SNP-mediated DNA fragmentation to 18.3, 12.4 and 9.9%, respectively (Table 50.1).

Table 50.1 Inhibition of DNA fragmentation by taurine-pretreated macrophages in the presence of SNP

	DNA Fragmentation (%)	
Taurine (mg/ml)	Control	SNP (300 μM)
0	5.8 ± 1.0	29.0 ± 1.0
0.1	5.7 ± 0.6	18.3 ± 1.3
1	5.7 ± 1.3	12.4 ± 2.9
10	5.6 ± 2.1	9.9 ± 1.0

The RAW cells (1×10^6 cells/assay) were treated with taurine for 24 hrs followed by 8 hrs with SNP. The data represent the mean values ± SDs of three separate experiments.

50.3.3 Effect of the Taurine Treatment Order

As taurine prevented DNA fragmentation in a concentration-dependent manner, the effect of the taurine treatment order was analyzed. When the cells were pretreated with 10 mg/ml of taurine for 24 hrs followed by 300-μM SNP treatment for 8 hrs, DNA fragmentation was inhibited by 85%, inhibition of DNA fragmentation was observed, which was almost similar to that of the control. However, when taurine (10 mg/ml) was added either simultaneously with 300 μM SNP (simultaneous group) or for 8 hrs following SNP exposure (post-treatment group) taurine-mediated inhibition of DNA fragmentation was reduced to 60% and 25%, respectively (Fig. 50.3).

50.3.4 Effect of SNP Exposure Time on DNA Fragmentation

To investigate the effect of taurine on extended exposure to SNP, RAW264.7 cells were exposed for a maximum of 12 hrs to 300 μM SNP following pretreatment with 10 mg/ml of taurine for 24 hrs. When the cells were exposed to SNP in the absence

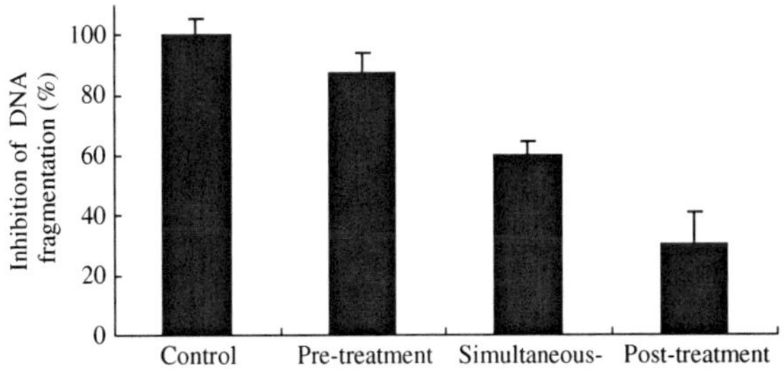

Fig. 50.3 Time-dependent effect of treatment on SNP-mediated DNA fragmentation of RAW264.7 macrophage cells. Some cells were pretreated with 10 mg/ml taurine for 24 hrs and then 300 μM SNP for 8 hrs. Other cells were co-treated with 10 mg/ml taurine and 300 μM SNP for 8 hrs. Finally, some cells were pretreated with SNP for 8 hrs and then exposed to taurine for 8 hrs

Table 50.2 Effect of the time course of SNP treatment on DNA fragmentation in macrophages

	DNA Fragmentation (%)				
Exposure Time (hrs)	1	3	6	8	12
Control	5.8 ± 1.8	5.8 ± 2.0	5.8 ± 1.7	5.8 ± 1.1	5.8 ± 1.4
SNP	12.2 ± 3.1	18.7 ± 2.0	25.0 ± 1.3	29.9 ± 2.4	31.5 ± 1.1
SNP+Taurine	7.8 ± 2.2	9.6 ± 1.0	9.7 ± 3.8	10.5 ± 1.8	14.1 ± 2.7

RAW cells (1×10^6 cells/assay) were exposed to 10 mg/ml of taurine for 24 hrs and then to 300 μM SNP for various duration. The data represent the mean values ± SDs of three separate experiments.

of taurine, the degree of DNA fragmentation increased in a time-dependent fashion. However, after pretreatment of the cells with taurine for 24 hrs, SNP exposure led to little DNA fragmentation. The degree of DNA fragmentation of the groups that were not treated with taurine but were exposed to SNP for 6 and 12 hrs was 25.0% and 31.5%, respectively. However, cells that were pretreated with taurine prior to exposure to SNP for either 6 or 12 hrs experienced only 9.7% and 14.1% DNA fragmentation, respectively (Table 50.2). Therefore, pretreatment appears to have elevated intracellular taurine levels, which in turn neutralized NO derived from SNP.

These protective effects of taurine on DNA fragmentation were confirmed by agarose gel electrophoresis. When the cells were untreated or treated with 10 mg of taurine for 24 hrs, no DNA fragmentation was observed (Fig. 50.4, columns 2 and 3). Treatment with 300 μM SNP for 8 hrs, however, led to DNA fragmentation, as reflected by the formation of a DNA ladder on agarose gel electrophoresis (Fig. 50.4, column 4). As shown in column 5 of Fig. 50.4, cells pretreated with 10 mg of taurine for 24 hrs before exposure to SNP, exhibited very weak DNA ladders. Furthermore, after extended pretreatment of the cells with taurine alone for 24 hrs followed by

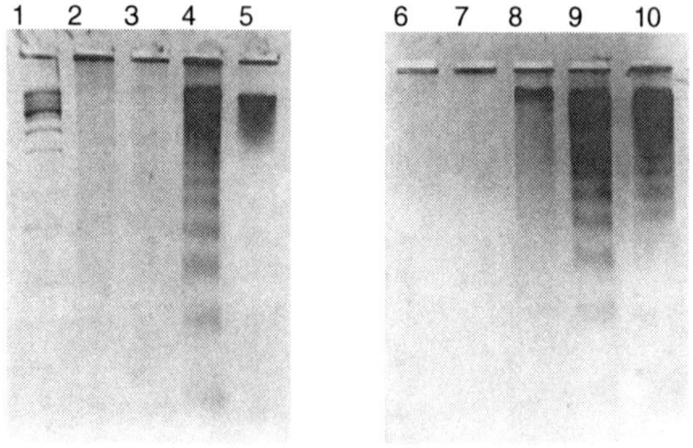

Fig. 50.4 Effect of taurine on SNP-induced DNA fragmentation in macrophages. Lane 1: 1 Kb DNA ladder; lane 2: control; lane 3: 10 mg/ml taurine (24 hr); lane 4: 300 μM SNP; lane 5: 10 mg/ml taurine (24 hrs) pretreatment followed by 300 μM SNP (8 hrs) in the presence of taurine; lane 6: control; lane 7: 10 mg/ml taurine (24 hrs); lane 8: 10 mg/ml taurine (32 hrs) + 300 μM SNP (8 hrs); lane 9: 300 μM SNP (8 hrs); lane 10: 10 mg/ml taurine (24 hrs) and then 300 μM SNP (8 hrs) in the absence of taurine

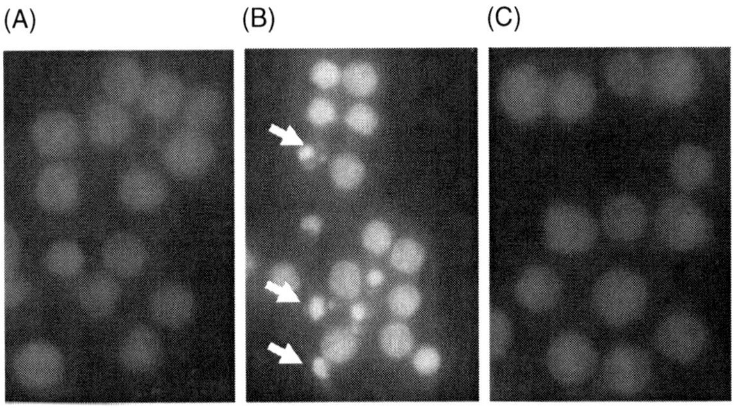

Fig. 50.5 Staining of fragmented DNA in macrophages with Hoechst 33258. Apoptotic cells are indicated with arrows. (**A**) untreated control, (**B**) treated with 300 μM SNP for 8 hrs; arrows indicate DNA fragmented cells, (**C**) pretreatment with 10 mg/ml taurine for 24 hrs, followed by exposure to 300 μM SNP for 8 hrs

an additional 8 hrs of taurine treatment in the presence of SNP, DNA fragmentation caused by SNP exposure was significantly reduced compared with DNA fragmentation in cells pretreated for 24 hrs with taurine followed by 8 hrs exposure to SNP without taurine (Fig. 50.4, column 8).

The Hoechst 33258 fluorescent dye also revealed that SNP treatment induces DNA fragmentation in many cells. After pretreatment with taurine for 24 hrs, however, SNP-exposed cells were protected against DNA fragmentation (Fig. 50.5).

50.4 Discussion

Taurine is an amino acid that is abundant in many tissues. One of the important actions of taurine is its antioxidant activity, which protects tissues from oxidative and nitrosative stress. During exposure to cytokines, such as IFN-γ, TNF-α and LPS, the halide-dependent myeloperoxidase of immune cells protects cells by promoting the conversion of taurine to taurine chloramines, in the process neutralizing the toxic metabolite, HOCl. Taurine (Giri et al. 2000) and taurine chloramine (Barua et al. 2001; Kim et al. 2006) prevent cell or tissue damage arising from NO toxicity by suppressing NO synthase. Taurine can also protect cells from apoptosis and DNA fragmentation induced by NO generated outside the cell. In this paper, taurine is reported as capable of protecting cell viability from oxidative stress due to NO-induced injury of RAW264.7 cells, which are a murine macrophage cell line.

The addition of SNP to the cell culture system in this experiment led to a decrease in viability of RAW264.7 cells, which depends on SNP concentration over a range of 200~1,000 μM. SNP is a well known NO donor which induces cell death or apoptosis in smooth muscle cells (Lau 2003), chondrocytes (Kuhn and Lotz 2003), and the retina (Ju et al. 2001). Pretreatment of the cells with taurine for 24 hrs prevented cell death caused by SNP-mediated nitrosative stress, possibly by directly neutralizing NO. Among the various concentrations of taurine administered to the cells, the lowest concentration, 0.1 mg/ml, showed a degree of protection equivalent to that of the highest concentration, 10 mg/ml. This implies that taurine is a very effective, potent antioxidant in macrophage cells.

Cell death from SNP treatment may be due to the induction of apoptosis. To confirm this phenomenon, it was found that addition of 300 μM SNP to the cells induced 29.0% DNA fragmentation compared to only 5.8% in the control group. Taurine reduced the extent of DNA fragmentation in a concentration-dependent manner. Attenuation of DNA fragmentation by taurine has been previously reported in human vascular endothelial cells subjected to oxidative stress mediated by hyperglycemia (Wu et al. 1999). In the present study, the order of taurine addition was very important. Pretreatment of cells with taurine decreased DNA fragmentation more than simultaneous or post-treatment with taurine. This implies that the antioxidant actions of taurine depend upon a rise in intracellular taurine levels prior to the increase in intracellular NO. When the cells were exposed to taurine prior to the addition of SNP, the DNA ladder was blunted compared to cells exposed to SNP in the absence of taurine treatment. However, maximal protection against DNA fragmentation was noted in cells incubated for 24 hrs with taurine containing medium prior to SNP addition and for 8 hrs in the presence of SNP. Therefore, taurine should be added to the medium to prevent apoptosis from NO toxicity.

These data show that taurine protects the cell against NO generated extracellularly. However, further research studying the effect of taurine on intracellularly generated NO or ROS is warranted.

50.5 Conclusion

This work demonstrates that taurine inhibits apoptosis in a macrophage cell line exposed to SNP, a source of extracellular NO. The inhibition of SNP-induced DNA fragmentation was measured by DPA assay, agarose gel electrophoresis and nuclear staining using Hoechst 33258. Taurine can function as a protective reagent against NO-mediated cell injury.

Acknowledgments This work (HWKim) was supported by Dong-A Pharmaceutical Company. The authors thank the company for its financial support of this project.

References

Alvarez S,Evelson PA (2007) Nitric oxide and oxygen metabolism in inflammatory conditions: sepsis and exposition to polluted ambients. Front Biosci 12:964–974

Barua M, Liu Y, Quinn MR (2001) Taurine chloramine inhibits inducible nitric oxide synthase and TNF-alpha gene expression in activated alveolar macrophages: decreased NF-kappaB activation and IkappaB kinase activity. J Immunol 167:2275–2281

Brune B, Lapetina EG (1990) Properties of a novel nitric oxide-stimulated ADP-ribosyltransferase. Arch Biochem Biophys 279:286–290

Dawson R Jr (2003) Taurine in aging and models of neurodegeneration. Adv Exp Med Biol 526:537–545

Felley-Bosco E (1998) Role of nitric oxide in genotoxicity:implication for carcinogenesis. Cancer Metastasis Rev 17:25–37

Giri SN, Gurujeyalakshmi G, Wang Y (2000) Suppression of bleomycin-induced increased production of nitric oxide and NF-kB activation by treatment with taurine and niacin. Adv Exp Med Biol 483:545–561

Higuchi Y (2003) Chromosomal DNA fragmentation in apoptosis and necrosis induced by oxidative stress. Biochem Pharmacol 66:1527–1535

Huxtable RJ (1992) Physiological actions of taurine. Physiol Rev 72:101–163

Ju WK, Chung IW, Kim KY, Gwon JS, Lee MY, Oh SJ, Chun MH (2001) Sodium nitroprusside selectively induces apoptotic cell death in the outer retina of the rat. Neuroreport 12:4075–4079

Kim C, Choi HS, Kim JW (2006) Taurine chloramine inhibits the production of nitric oxide and superoxide anion by modulating specific mitogen-activated protein kinases. Adv Exp Med Biol 583:493–498

Kramer JH, Chovan JP, Schaffer SW (1981). Effect of taurine on calcium paradox and ischemic heart failure. Am J Physiol 240:H238–H246

Kuhn K, Lotz M (2003) Mechanisms of sodium nitroprusside-induced death in human chondrocytes. Rheumatol Int 23:241–247

Lau HK (2003) Cytotoxicity of nitric oxide donors in smooth muscle cells is dependent on phenotype, and mainly due to apoptosis. Atherosclerosis 166:223–232

Marquez LA, Dunford HB (1994) Chlorination of taurine by myeloperoxidase. Kinetic evidence for an enzyme-bound intermediate. J Biol Chem 269:7950–7956

Park E, Schuller-Levis G, Quinn MR (1995) Taurine chloramine inhibits production of nitric oxide and TNF-alpha in activated RAW 264.7 cells by mechanisms that involve transcriptional and translational events. J Immunol 154:4778–4784

Solomonson LP, Flam BR, Pendleton LC, Goodwin BL, Eichler DC (2003) The caveolar nitric oxide synthase/arginine regeneration system for NO production in endothelial cells. J Exp Biol 206(Pt 12):2083–2087

Stamler JS, Simon DI, Osborne JA, Mullins ME, Jaraki O, Michel T, Singel DJ, Loscalzo J (1992a) S-nitrosylation of proteins with nitric oxide:synthesis and characterization of biologically active compounds. Proc Natl Acad Sci USA 89:444–448

Stamler JS, Singel DJ, Loscalzo J (1992b) Biochemistry of nitric oxide and its redox-activated forms. Science 258:1898–1902

Weinberg JB, Granger DL, Pisetsky DS, Seldin MF, Misukonis MA, Mason SN, Pippen AM, Ruiz P, Wood ER, Gilkeson GS (1994) The role of nitric oxide in the pathogenesis of spontaneous murine autoimmune disease:increased nitric oxide production and nitric oxide synthase expression in MRL-lpr/lpr mice, and reduction of spontaneous glomerulonephritis and arthritis by orally administered NG-monomethyl-L-arginine. J Exp Med 179:651–660

Weiss SJ, Klein R, Slivka A, Wei M (1982) Chlorination of taurine by human neutrophils. Evidence for hypochlorous acid generation. J Clin Invest 70:598–607

Wu QD, Wang JH, Fennessy F, Redmond HP, Bouchier-Hayes D (1999) Taurine prevents high-glucose-induced human vascular endothelial cell apoptosis. Am J Physiol 277(6 Pt 1): C1229–1238

You JS, Chang KJ (1998) Taurine protects the liver against lipid peroxidation and membrane disintegration during rat hepatocarcinogenesis. Adv Exp Med Biol 442:105–112

Yu X, Chen K, Wei N, Zhang Q, Liu J, Mi M (2007) Dietary taurine reduces retinal damage produced by photochemical stress via antioxidant and anti-apoptotic mechanisms in Sprague-Dawley rats. Br J Nutr 98:711–719

Chapter 51
Effect of Taurine on Antioxidant Enzyme System in B16F10 Melanoma Cells

Jisun Yu and An Keun Kim

Abstract There is now increasing evidence that free radicals and reactive oxygen species (ROS) are involved in a variety of pathological events. Reactive oxygen species are produced during normal cellular function and lead to lipid peroxidation, massive protein oxidation and degradation. Taurine is an abundant free amino acid in inflammatory cells, where it is thought to be cytoprotective. The aim of the present study was to examine whether taurine enhances endogenous antioxidant enzyme activity and/ or regulates ROS generation in B16F10 mouse melanoma cells. B16F10 cells were exposed to medium containing taurine for a period of 24 h. Cell viability, measured by the MTT assay, exhibited a dose-dose dependent inhibition. Taurine increased the activities of superoxide dismutase, glutathione peroxidase and CAT compared to those of the control, an effect paralleling an increase in gene expression. Taurine also reduced ROS content in a dose-dependent manner. Taken together, our results suggest that taurine decreases ROS levels by increasing the levels of the antioxidant enzymes.

Abbreviations *SOD*, superoxide dismutase; *GPx*, glutathione peroxidase; *CAT*, catalase

51.1 Introduction

Free radicals are chemical species containing one or more unpaired electrons. The unpaired electrons of oxygen react to form partially reduced highly reactive species that are classified as ROS, such as hydroxyl radicals ($\cdot OH$), superoxide anion ($\cdot O_2^-$) and hydrogen peroxide (H_2O_2) (Halliwell 1991; Fruehauf and Meyskens 2007; Matés 2001) ROS are normal products of aerobic metabolism, including the mitochondrial electron transport chain, cytochrome P450, the NADPH oxidase complex and the peroxisomes (Inoue et al. 2003). They are transient species due to their high chemical reactivity, which leads to lipid peroxidation, oxidation of some enzymes,

J. Yu (✉)
College Pharmacy, Sookmyung Women's University, Seoul, Korea

J. Azuma et al. (eds.), *Taurine 7*, Advances in Experimental Medicine and Biology 643,
DOI 10.1007/978-0-387-75681-3_51,

protein oxidation and DNA mutation (Matés et al. 1999). They can consequently initiate or further the development of many diseases, such as cancer, liver injury, cardiovascular disease and neurodegenerative disease (Slater 1984; Halliwell 1994; Yagi 1987). To prevent injury from oxidative stress, aerobic organisms have evolved a system of chemical and enzymatic antioxidants. Among the antioxidant enzymes are superoxide dismutase (SOD), glutathione peroxidase (GPx) and catalase (CAT). SOD catalyzes the dismutation of the superoxide radical anion to hydrogen peroxide and oxygen. CAT and GPx convert H_2O_2 to H_2O (Beatty et al. 2000).

Taurine, a free amino acid, is found naturally in food, especially in seafood and meat. It is found in millimolar concentrations in most mammalian tissues. Mammals are able to endogenously synthesize taurine, but some species such as humans are more dependent on dietary sources of taurine. (Bouckenooghe et al. 2006) Taurine can act as a direct antioxidant by lowering ROS and/ or as an indirect antioxidant by preventing changes in membrane permeability due to oxidant injury (Schuller-Levis and Park 2004). Clinically, taurine has been used with varying degrees of success in the treatment of a wide variety of conditions, including cardiovascular diseases, hypercholesterolemia, epilepsy and other seizure disorders, macular degeneration, Alzheimer's disease, hepatic disorders, alcoholism, and cystic fibrosis (Birdsall 1998)

The aim of this study was to evaluate the possibility of using taurine as a chemopreventive agent in mouse melanoma cells. For this purpose, we examined whether taurine affects the antioxidant enzymes and ROS content.

51.2 Methods

51.2.1 Cell Culture and Experiments

The murine melanoma cell line B16F10 was obtained from the Korean Cell Line Bank (KCLB, Seoul, South Korea). This cell line was used while in the log phase of growth. Cells were maintained in DMEM (Wellgene Inc.) supplemented with 10% fetal bovine serum (Wellgene Inc.), 100 units/ml penicillin and 100 μg/ml streptomycin (Wellgene Inc.), at 37°C in a humidified atmosphere of 5% CO_2.

51.2.2 Cell Viability

Cell viability was assessed by 3-(4, 5-dimethylthiazol-2-yl)-2, 5-diphenyl tetrazolium bromide (MTT) staining as described by Vistica (Vistica et al., 1991). Briefly, B16F10 cells were plated at a density of 10^4 cells/ well into 96-well plates and treated with different concentrations of taurine for a further 24 h. At the end of treatment, 50 μl of MTT was added, and cells were incubated for a further 4 h. Cell viability was determined by scanning with an ELISA reader using a 570 nm filter.

51.2.3 The Measurement of Antioxidant Activities

51.2.3.1 Measurement of Superoxide Dismutase Activity

We used the Martin method with hematoxylin (Martin et al. 1987). Hematoxylin is autoxidized to the red colored hematin. SOD inhibits this autoxidation. After adding the sample to 50 mM potassium phosphate buffer (pH 7.5) with 0.1 mM EDTA, the samples were preincubated for 5 min. Inhibition of the conversion of hematoxylin to hematin was monitored at 560 nm by UV/ visible spectrophotometer.

51.2.3.2 Measurement of Catalase Activity

Catalase converts H_2O_2 to H_2O and $1/2\ O_2$. Catalase activity was measured by the Aebi method (Aebi 1984). The principle of this method is based on the hydrolyzation of H_2O_2 with a decrease in absorbance at 240 nm. The conversion of H_2O_2 into H_2O and $1/2\ O_2$ in 1 min under standard conditions was considered to be the enzyme reaction velocity. Optical density was measured by UV/ visible spectrophotometer using an extinct coefficient of 43.6 $M^{-1}\ cm^{-1}$.

51.2.3.3 Measurement of Glutathione Peroxidase Activity

We measured the activity of GPx using the method of Paglia and Valentine (Paglia and Valentine 1967). Glutathione reductase (GR) reduces glutathione (GSSG), with the oxidation of reduced nicotinamide adenine dinucleotide phosphate (NADPH) measured. The reaction medium consisted of 0.4 M Tris-HCl (pH 7.2), 0.04 M GSH, 0.075 mM H_2O_2 and 6 mM NADPH. After addition of each sample to the medium, optical density was measured at 340 nm using a UV/ visible spectrophotometer.

51.2.4 Measurement of Intracellular ROS Levels

ROS production was monitored using DCHF-DA. A total of 10^6 cells were incubated with 50 μM DCFH-DA (2′, 7′-dichlorofluorescein diacetate) for 5 minutes at 37°C and subsequently washed twice in cold phosphate buffer saline (PBS) before analysis using a Cytofluor 2350 plate reader at an excitation of 485 nm and an emission of 530 nm. All experiments were performed at least three times with three separate samples.

51.2.5 RNA Extraction, cDNA Synthesis and RT-PCR

Total RNA was extracted from B16F10 cells by using the easy-BLUE total RNA extraction kit (Intron Biotechnology, Korea) according to the manufacture' s instructions. RNA was precipitated with ethanol, washed and dissolved in diethylpyrocarbonated treated water. After quantification by spectrophotometry, 2 μg RNA was

Table 51.1 Sequence of primers and RT-PCR conditions

Gene	Primer sequence	Annealing (°C)	Product size (bp)
β-actin		52.5	220
Forward	CATCCATCATGAAGTGTGACG		
Reverse	CATACTCCTGCTTGCTGATCC		
Cu/ Zn SOD			
Forward	TTAACTGAAGGCCAGCATGGG	60	335
Reverse	ATCACTCCACAGGCCAAGCGG		
Mn SOD			
Forward	TGCACCACAGCAAGCACCATG	55	413
Reverse	CTCCCACACGCAATCCCCAG		
GPx			
Forward	CTCGGTTTCCCGTGCAATCAG	65	431
Reverse	GTGCAGCCAGTAATCACCAAG		
CAT			
Forward	TCTGCAGATACCTGTGAACTG	55	357
Reverse	TAGTCAGGGTGGACGTCAGTG		

reverse-transcribed into cDNA at 42°C for 1 h using an Power cDNA synthesis kit (Intron Biotechnology, Korea). The cycling condition and primer sequences are presented on Table 51.1. The products of RT-PCR were subjected to 1.5% agarose gel eletrophoresis .

51.2.6 Statistic Analysis

Each experiment was performed at least in triplicate. Statistical significance was determined by the Student's t-test. The results are expressed as the means ± S.D. Results with P value < 0.05 were considered statistically significant from the control group.

51.3 Results

51.3.1 Effect of Taurine on the Cell Viability

The effect of taurine on cell viability was assessed in B16 F10 cells using the MTT assay. As shown in Fig. 51.1, viability was slightly decreased in a dose-dependent manner. In subsequent experiments, the highest concentration used was 40 mM, as there was no significant difference between a taurine concentration of 40 mM and 80 mM (data not shown).

51.3.2 Effect of Taurine on the Activity of the Antioxidant Enzymes

The antioxidant enzymes of cells play an important role in the defense against oxidative stress. SOD, GPx and CAT are the main detoxifying enzymes. To examine the

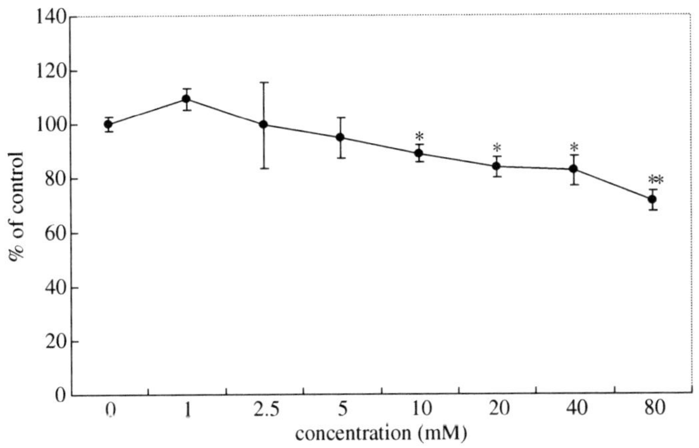

Fig. 51.1 Cell viability of B16 melanoma cells after treatment with taurine. The cells were exposed to various concentrations of taurine for 24 h. The percentage of variable cells was determined using the MTT assay. Results are expressed as percentage of control. Values represent means ± S.E and were obtained from three different experiments. $^{*}P < 0.05$, $^{**}P < 0.01$ vs. controls

effect of taurine on the activity of these antioxidant enzymes, melanoma cells were treated with taurine at concentrations ranging from 5 to ~ 40 mM for 24 h, at which time the activity of SOD, GPx and CAT was measured. Compared with the control, the activities of SOD, GPx and CAT increased in a dose-dependent manner. Taurine (40 mM) increased the activities of SOD, GPx and CAT 1.4 fold, 1.4 fold and 1.9 fold, respectively (Fig. 51.2).

51.3.3 *Effect of Taurine on Expression of Antioxidant Enzymes*

To determine whether the changes in SOD, GPx and CAT activity came from alterations in gene expression, RT-PCR was performed. B16F10 cells were treated with different concentrations of taurine for 24 h. Figure 51.3A shows the RT-PCR and Fig. 51.3B the densitometric data. Taurine treatment (40 mM) led to a dose-dependent increase in the mRNA levels of Cu/ Zn SOD, GPx, and CAT, reaching levels that were 2 fold, 1.58 fold, 2.4 fold elevated, respectively. However MnSOD was not statistically affected by taurine treatment.control (means ± SEM from three different experiments). $^{**}P < 0.01$, $^{***}P < 0.001$ compared with control.

51.3.4 *ROS*

In order to demonstrate the role that ROS plays in taurine-treated cells, the intracellular ROS level was determined by the fluorescence intensity of DCF. The result shows that taurine significantly reduced the intracellular level of ROS in a dose-dependent manner (Fig. 51.4). It suggests that taurine inhibits the production of ROS.

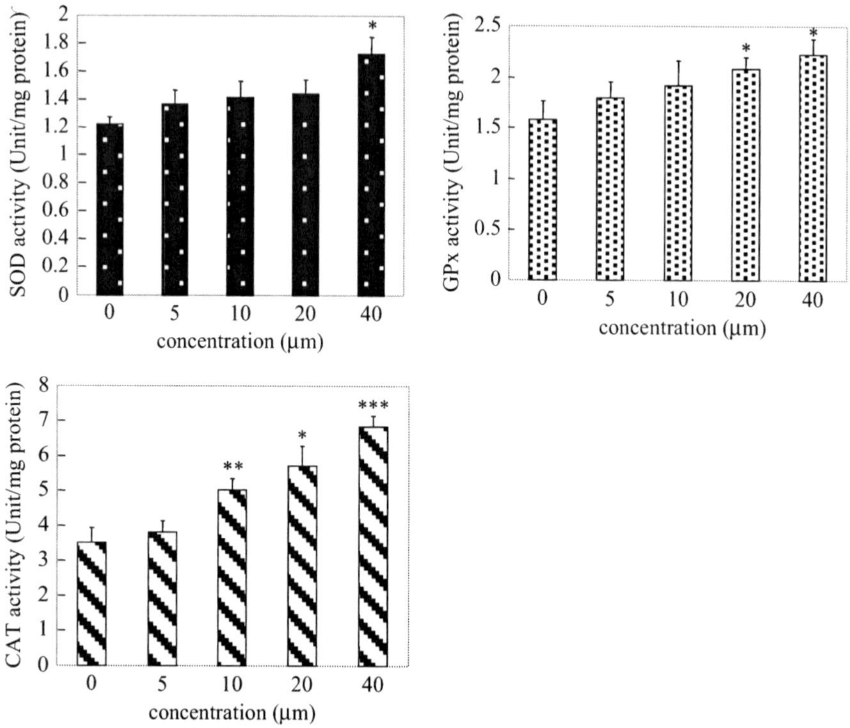

Fig. 51.2 Effect of taurine on the activities of the antioxidant enzymes of B16F10 murine melanoma cells. The cells were exposed to various concentrations of taurine for 24 h. The taurine-treated and control cells were harvested and protein was isolated. Results are expressed as the means ± S.E for $n = 3$. $^{*}P < 0.05$, $^{**}P < 0.01$, $^{***}P < 0.001$ compared with the control

51.4 Discussion

Recently, the cytoprotective and antioxidant activity of naturally occurring compounds has attracted a great deal of attention. Cytoprotective agents are substances that either directly or indirectly protect cells against the adverse effects of xenobiotics, drugs, carcinogens and toxic radical reactions (Matés 2001). Antioxidants function by either scavenging reactive species or mediating the upregulation of members of the antioxidant defense system. Vitamin E, vitamin C, melatonin, and flavonoids are biologically important compounds that have been reported to exert antioxidant action. Among other factors, the cellular antioxidant system contains different free radical scavenging antioxidant enzymes. SOD, GPx and CAT represent first line antioxidants, which act to reduce the levels of substances containing unpaired electrons. When excess free radicals are produced, oxidative stress ensues and the antioxidant defenses can be overwhelmed.

Chemoprevention, namely, the use of either synthetic or naturally occurring agents to inhibit pre-cancerous events, has become recognized as a credible,

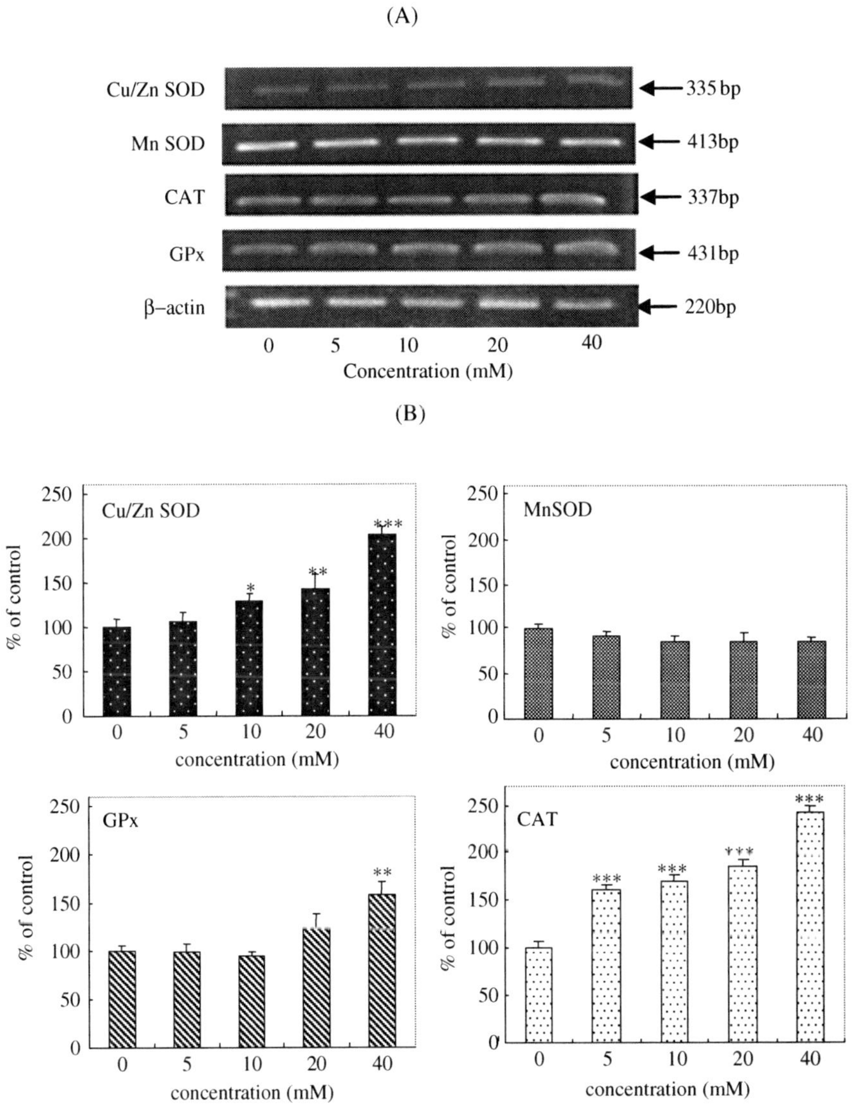

Fig. 51.3 Cu/ ZnSOD, MnSOD, GPx and CAT mRNA expression in B16F10 melanoma cells after taurine exposure. (**A**) PCR blot of each antioxidant enzyme obtained by RT-PCR after 24 h exposure of B16F10 melanoma cells at the indicated concentration of taurine. (**B**) Relative quotient of ethidium-stained bands representing the integrated area under the curve by densitometric tracing are reported as the ratio of each antioxidant to β-actin. Values are expressed as percentage of the control (means ± SEM from three different experiments). $^{**}P < 0.01$, $^{***}P < 0.001$ compared with control

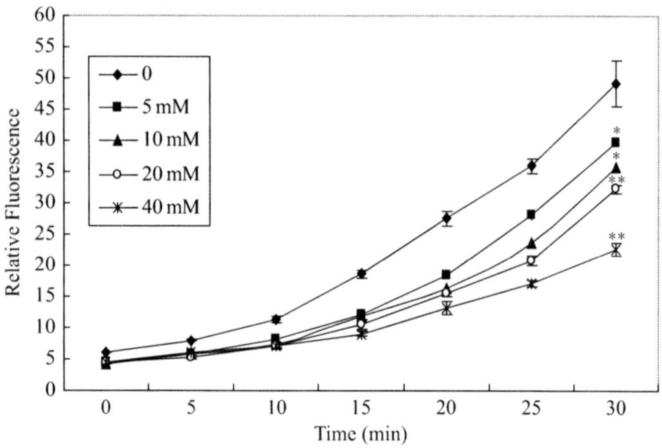

Fig. 51.4 Measurement of the level of reactive oxygen species (ROS) after taurine treatment. The melanoma cells were incubated with taurine at various concentrations and then 200 μlDCFH – DA(5 μM) was added as a substrate for ROS. After incubation for 30 min, ROS levels were measured by spectrofluoremetry (excitation: 485 nm, emission: 530 nm). Results represent the means±S.E.M. of three experiments. $^{*}P < 0.05$, $^{**}P < 0.01$ compared with control

cost-effective and essential approach of reducing cancer mortality (Sporn and Suh 2002; Hong and Spron 1997).

Taurine exerts cytoprotection against ROS generation in various cell lines (Chang et al. 2004; Pokhrel and Lau-Cam 2000; Sener et al. 2005). Accordingly, we evaluated the potential antioxidant properties of taurine as a chemopreventive agent in mouse melanoma cells. After taurine (40 mM) treatment for 24h, the activities of SOD, GPx and CAT increased, an effect correlated with the increase in the expression of the antioxidant enzymes, the exception being MnSOD. These results show that taurine affects the activities and gene expression of the antioxidant enzymes in melanoma cells. To further confirm the role that ROS plays in taurine-treated cells, we observed intracellular ROS levels by DCF fluorescence; using this method taurine was found to decrease ROS content. Thus, taurine reduces ROS and stimulates detoxification within cells. In conclusion, taurine may serve as a chemopreventive agent against cancer through a reduction in ROS.

References

Aebi Hugo (1984) Catalase in vitro. Method Enzymol 105:93–127

Beatty S, Koh H, Phil M, Henson D, Boulton M (2000) The role of oxidative stress in the pathogenesis of age-related macular degeneration. Surv Ophthalmol 45:115–34

Birdsall TC (1998) Therapeutic applications of taurine. Altern Med Rev 3:128–136

Bouckenooghe T, Remacle C, Reusens B (2006) Is taurine a functional nutrient? Curr Opin Clin Nutr Metab Care 9:728–33

Chang L, Xu JX, Zhao J, Pang YZ, Tang CS, Qi YF (2004) Taurine antagonized oxidative stress injury induced by homocysteine in rat vascular smooth muscle cells. Acta Pharmacol Sin 25:341–346

Fruehauf J, Meyskens F (2007) Reactive Oxygen Species:A Breath of Life or Death? Clin Cancer Res 13:789–794
Halliwell B (1991) Reactive oxygen species in living systems: source, biochemistry, and role in human disease. Am J Med 91:14–22
Halliwell B (1994) Free radicals, antioxidants and human diseases: curiosity, cause or consequently? Lancet 334:721–724
Hong WK, Spron MB (1997) Recent advances in chemoprevention of cancer. Science 278: 1073–1077
Inoue M, Sato EF, Nishikawa M (2003) Mitochondrial generation of reactive oxygen species and its role in aerobic life. Curr Med Chem 10:2495–505
Martin JP, Dailey M, Sugarman E (1987) Negative and positive assays superoxide dismutase based on hematoxylin autoxidation. Arch Biochem Biophys 255:329–336
Matés, JM., Pe´rez-Go´mez, C., Nu´n˜ez de Castro, I., (1999) Antioxidant enzymes and human diseases. Clin Biochem 32:595–603
Matés JM (2001) Effect of antioxidant enzymes in the molecular control of reactive oxygen species toxicology. Toxicology 153:83–10.
Paglia DE, Valentine WN (1967) Studies on the quantitative and qualitative characterization of erythrocyte glutathione peroxidase. J Lab Clin Med 70:158–169
Pokhrel PK, Lau-Cam CA (2000) In vitro and in vivo effects of taurine and structurally related sulfur-containing compounds against phenylhydrazine-induced oxidative damage to erythrocytes. Adv Exp Med Biol 483:503–522
Schuller-Levis GB, Park E (2004) Taurine and its chloramine: modulators of immunity. Neurochem Res 29(1):117–26
Sener G, Ozer Sehirli A, Ipçi Y, Cetinel S, Cikler E, Gedik N, Alican I (2005) Taurine treatment protects against chronic nicotine-induced oxidative changes. Fundam Clin Pharmacol 19:155–164
Slater T (1984) Free radical mechanisms in tissue injury. Biochem J 222:1–15
Sporn MB, Suh N (2002) Chemoprevention:an essential approach to controlling cacner. Nat Rev Cancer 2:537–554
Vistica DT, Skehan P, Scudiero D, Monks A, Pittman A, Boyd MR (1991) Tetrasodium-based assays for cellular viability: a critical examination of selected parameters affecting formazan production. Cancer Res 51:2515–2520
Yagi K (1987) Lipid peroxidation and human diseases. Chem. Phys Lipids 45:337–351

Part IX
Regulation of the Taurine Transporter

Chapter 52
Inhibition of Taurine Transport by Cyclosporin A is Due to Altered Surface Abundance of the Taurine Transporter and is Reversible

Thanh T. Tran, Xiaojun Qian, Cynthia Edwards, and Hemanta K. Sarkar

Abstract We have investigated the underlying mechanism of the CsA-induced inhibition of taurine transport using a cell line permanently expressing the mouse taurine transporter (mTauT) tagged with the green-fluorescence protein (GFP). CsA inhibited the uptake activity of the expressed mTauT.GFP fusion protein in both dose and time dependent manner. Surface biotinylation assay revealed that the CsA-treatment reduced the relative surface abundance of the taurine transporter without affecting its total expression level. CsA treatment reduced both the taurine uptake and the relative surface abundance of the transporter by similar magnitudes. Conversely, when the CsA was washed off, both the uptake and the relative surface abundance of the transporter recovered fully to the control level. Remarkably, the recovery process was insensitive to the protein synthesis inhibitor cycloheximide. These results suggested that the CsA inhibited taurine transport by altering the surface abundance, possibly by internalization of the expressed taurine transporters.

Abbreviations *CsA*, cyclosporine A; *mTauT*, mouse taurine transporter; *GFP*, green fluorescence protein

52.1 Introduction

The β-amino acid taurine, which is abundantly present in its free form in several mammalian tissues and cells (Huxtable, 1992), is believed to play an important role in the fetal and neonatal development (for reviews, see Huxtable 1992; Chesney et al. 1998; Redmond et al. 1998; Aerts and Van Assche 2002). Despite its importance, both human fetus and newborn virtually lack the endogenous capacity to synthesize taurine even though the tissue concentrations of taurine are known to be highest in the fetal stage and lowest in adults (for reviews, see Huxtable 1992; Chesney et al. 1998; Redmond et al. 1998; Aerts and Van Assche 2002). Taurine is

T.T. Tran (✉)
Department of Molecular Physiology and Biophysics, Baylor College of Medicine, Houston, TX, USA

J. Azuma et al. (eds.), *Taurine 7*, Advances in Experimental Medicine and Biology 643, DOI 10.1007/978-0-387-75681-3_52, © Springer Science+Business Media, LLC 2009

therefore considered a conditionally essential nutrient for the fetus and newborn. To meet their demands, fetuses depend solely on the transplacental transfer of taurine from their mothers, whereas the newborns acquire taurine from the mother's milk through intestinal absorption; in both cases, taurine transfer occurs via a mediated process (for reviews, see Chesney et al. 1998; Redmond et al. 1998). Accordingly, a Na^+-dependent taurine transport activity was found in the vesicles prepared from the microvillous (Karl and Fisher 1990) and brush border (Miyamoto et al. 1988) membranes of the human placenta, as well as in a human placental cell line, JAR, (Kulanthaivel et al. 1991).

Consistent with taurine's importance in the fetal development, several studies suggested a close link between the lower placental taurine concentration to the intrauterine growth retardation (IUGR) and low birth weight of infants (Ghisolfi et al. 1989; Ejiri et al. 1987). Incidentally, cases of low birth weight and severe intrauterine growth retardation are commonly reported in renal transplant recipient mothers undergoing treatment with the immunosuppressive drug cyclosporin A (Pickrell et al. 1988; Willis et al. 2000). Based on the observation that cyclosporin A (CsA) adversely affected the taurine transport in the human placental JAR cells, Ramamoorthy et al. (1992) first proposed that impairment of the placental taurine transporter function might be a causative factor in the development of the CsA-induced intrauterine growth retardation. Any understanding of the mechanism of the CsA-induced inhibition of the taurine transporter function, however, is lacking.

Molecular cloning and functional expression of the taurine transporter encoding cDNAs from a variety of mammalian species and tissues, including the human placental cells, revealed that the primary structure and the biochemical properties of the mammalian taurine transporter proteins are highly conserved (see Qian et al. 2000 and references therein).

In this communication, we describe creating a cell line that permanently expresses the mouse taurine transporter (mTauT), cloned in our laboratory (Vinnakota et al. 1997), as a GFP-fusion protein. Subsequently, we used these mTauT.GFP fusion protein expressing stable cells as a model system to investigate the mechanism of the CsA-induced inhibition of taurine transport. Our results suggest that CsA inhibits taurine transport by altering the surface abundance of the taurine transporter and the inhibition is reversible.

52.2 Methods

52.2.1 Materials

Dulbecco's modified Eagle's medium (DMEM), fetal bovine serum, G418, penicillin and streptomycin were purchased from Gibco BRL/Life Technologies, Inc. (Gaithersburg, MD). 1,2-[^{3}H]-Taurine (1.18 TBq/mmol) was purchased from Amersham Pharmacia Biotech (Arlington Heights, IL). The green fluorescence protein (GFP) fusion plasmid pEGFP and the Living Colors A.v. peptide antibody, a polyclonal anti-GFP antibody, were purchased from Clontech (Palo Alto, CA). Cyclosporin A was a generous gift from Dr. J. Clifford Waldrep (Baylor College of Medicine, Houston, TX) and cycloheximide was purchased from Sigma Chemicals (St. Louis, MO).

52.2.2 Recombinant Plasmids

The mouse taurine transporter (mTauT) cDNA (Vinnakota et al. 1997) was cloned into the pEGFP plasmid (Clontech, Palo Alto, CA) to create the plasmid pmTauT.GFP in which the GFP was fused in frame to the C-terminal end of the mouse taurine transporter (mTauT.GFP).

52.2.3 Stable Cells

HEK293 cells (3.5×10^5 cells/60-mm dish) were transfected with 5.6 μg of pmTauT.GFP or pEGFP (GFP expressing plasmid) using Fugene 6 (Boehringer Mannheim, IL). Stable cells were selected with G418 (250 μg/ml final concentration). Several selected clones were subsequently screened for taurine transporter (taurine uptake) and GFP expression (GFP-fluorescence using a Zeiss fluorescence microscope). Expression of the mTauT.GFP and the GFP proteins were further confirmed by immunoblot assay using the anti-GFP polyclonal antibody. One clone, HEK/mTauT.GFP, expressing the taurine transporter-GFP fusion protein and high taurine transport activity, was chosen for the subsequent studies. For negative control, one clone, HEK/GFP, expressing the GFP protein and low taurine transport activity (i.e., same as HEK 293 cells) was selected. Another stable cell line expressing the HA-tagged mTauT (HEK/HA.mTauT) was used as a control in some experiments. These three stable cells were maintained in DMEM containing 100 μg/ml G418. For uptake studies, the cells were seeded in poly-D-lysine coated 12-well plates at a density of $\sim 1–2 \times 10^5$ cells/well and were grown for 3–4 days before using for uptake assays.

52.2.4 Cyclosporin A Treatment

Cyclosporin A (CsA) or vehicle solvent (ethanol) was added directly to the growth medium and the cells were maintained in the absence or presence of indicated concentration of CsA for a given period at 37°C in the 95% O_2/5 % CO_2 incubator as described earlier (Tran et al. 2000).

52.2.5 Taurine Uptake

Unless otherwise stated, taurine transport was measured by performing taurine uptake assay as described (Qian et al. 2000) using 5 μM of [^{3}H]-taurine as a substrate.

52.2.6 *Immunoblot Assay*

Immunoblot assays were performed using commercially available polyclonal anti-GFP antibody (Living Colors A.v. peptide antibody, Clontech, CA) as described (Tran et al. 2000). Briefly, extracts of total cellular proteins prepared from the experimental cells were separated on 10% SDS-PAGE and then electrophoretically transferred onto a Hybond-P polyvinylidene difluoride (PDVF) membrane. The blotted membrane was first incubated with 3% BSA and 2% non-fat dry milk to block any non-specific binding, and then incubated with 1:1000 dilution of the anti-GFP polyclonal antibody for 1 h at room temperature. Next, the blot was washed several times with TBST (137 mM NaCl, 2.6 mM KCl, 25 mM Tris-HCl, pH 7.4, and 0.1% Tween 20) to remove the unbound antibody and then incubated for 1 h at room temperature with a 1:10,000 dilution of the horseradish peroxidase-conjugated goat anti-rabbit IgG (Pierce, IL) for detecting the bound anti-GFP. The bound antibody-antigen complex was then detected using the ECL Plus Chemiluminescent Detection kit (Amersham Pharmacia Biotech, IL). Subsequently, the immunoreactive bands were quantified by densitometric scanning of the autoradiogram using AlphaImager 2000 (Alpha Innotech Co., San Leandro, CA).

52.2.7 *Cell Surface Labeling Assay*

Cell surface labeling assays were performed to determine the relative surface abundance of the taurine transporter in control (solvent-treated) and CsA-treated cells as described (Tran et al. 2000) using the membrane impermeable biotinylation reagent NHS-SS-biotin (Pierce, IL). The labeled cells were solubilized in detergent, and the biotinylated surface proteins were captured from the cell extracts (containing 100 μg of protein) using streptavidin-agarose beads (Sigma Chemicals, MO). The captured proteins were separated on 10% SDS-PAGE and then electrophoretically transferred onto a PDVF membrane. The membrane was subsequently subjected to immunoblot assay using the anti-GFP antibody. Finally, the immunoreactive bands were quantified by densitometric scanning as described above.

52.3 Results

52.3.1 *Expression of Taurine Transporter-GFP Fusion Protein*

The taurine uptake in the HEK/mTauT.GFP stable cell clones were between 7 and 10-fold higher than that in the HEK/GFP cells. We chose one clone that expressed the highest level of uptake ($\sim$ 10-fold over the control cell uptake) for our studies. Although we have not fully examined the biochemical properties of the permanently expressed mTauT.GFP transporter, we found that the induced taurine uptake in the stable cells was Na^+- and Cl^--dependent, and was inhibited by known taurine transport inhibitors β-alanine, guanidinoethane sulfonic acid, hypotaurine, and

3-amino-propionic acid (results not shown). Thus, fusion of GFP at the C-terminal end of the mTauT did not seem to affect the basic transport properties of the transporter.

Immunoblot assay using the polyclonal anti-GFP antibody detected a single protein band of ~ 70 kDa, representing the mTauT.GFP fusion protein, only in the total extract prepared from the HEK/mTauT.GFP cells (Fig. 52.1, lane 1) but not in the total extracts prepared from the HEK/GFP (Fig. 52.1, lane 2) or the HEK/HA.mTauT cells (Fig. 52.1, lane 3). In contrast, the anti-GFP antibody recognized the expressed GFP protein in the HEK/GFP cell extract as the only major immunoreactive protein band of ~ 27 kDa (Fig. 52.1, lane 2).

52.3.2 *Effect of CsA on Taurine Uptake in Stable Cells*

The taurine uptake in HEK/GFP and HEK/mTauT.GFP cells were inhibited by CsA in both dose and time dependent manner (Fig. 52.2). In both cells, a 50% reduction in uptake was observed between 25 and 28 h post-treatment with 1–3 μM of CsA. These results were very similar to those reported earlier in human placental JAR cells (Ramamoorthy et al. 1992).

Remarkably, the inhibition was reversed when the CsA was removed. Thus, when the CsA-treated HEK/mTauT.GFP cells were washed three times with the normal growth medium and were subsequently allowed to recover for 24 h, the uptake level in the CsA-treated and washed cells was virtually identical to that of the of the solvent-treated control cells processed in parallel (Fig. 52.3). As shown, addition of 10 μM of the protein synthesis inhibitor cycloheximide during the recovery period did not affect the uptake reversal process (Fig. 52.3).

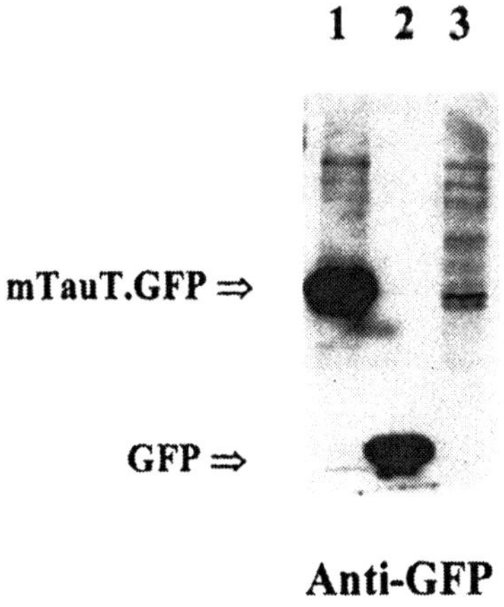

Fig. 52.1 Immunoblot analysis of cellular proteins. Cell extracts (each containing 50 μg of total protein) prepared from the HEK/mTauT.GFP (lane 1), HEK/GFP (lane 2) and HEK/HA.mTauT (lane 3) cells were analyzed by immunoblot assay using the anti-GFP antibody as described in Methods

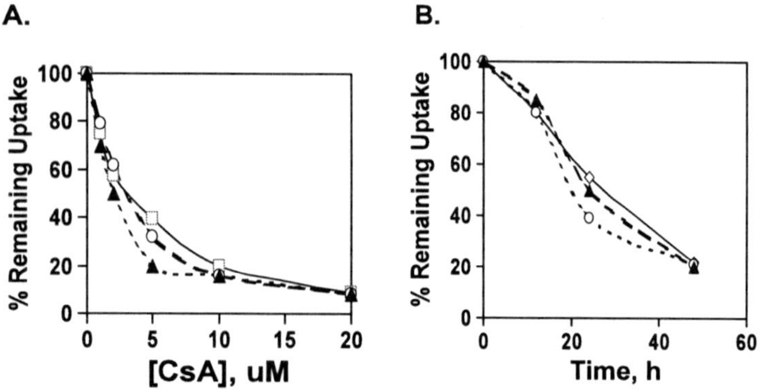

Fig. 52.2 CsA inhibited taurine uptake in dose (**A**) and time (**B**) dependent manner in HEK/mTauT.GFP (*closed triangle*), HEK/HA.mTauT (*open circle*) and HEK/GFP (*open square*) cells. A 30-min uptake assay was performed using 5 μM of [^{3}H]-taurine as described in Methods. In (A), cells were treated with the indicated concentrations of CsA for ~ 48 h at 37°C; and in (B), cells were treated with 2 μM CsA for 12, 24 and 48 h at 37°C. Results are average of two independent experiments

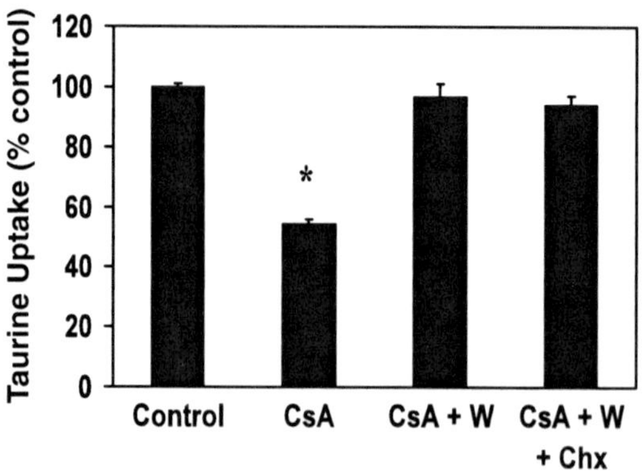

Fig. 52.3 Effect of CsA on taurine transporter activity is reversible. Taurine uptake was measured using the following cells (in triplicate): Control, solvent-treated cells (grown in parallel with the CsA-treated cells); CsA, cells treated with 2 μM CsA for 24 h at 37°C; CsA + Wash, cells treated with 2 μM CsA for 24 h at 37°C, washed three times and then allowed to recover for 24 h at 37°C; CsA + Wash + Chx: cells treated with 2 μM CsA for 24 h, washed three times, and then allowed to recover for 24 h in the presence of 10 μM Chx at 37°C. Results are expressed as mean ± SE ($n = 3$)

52.3.3 Effect of CsA on Taurine Transporter Expression

We next analyzed the total and cell surface expression levels of the mTauT.GFP protein in solvent-treated and CsA-treated HEK/mTauT.GFP cells by using the immunoblot and surface biotinylation assays, respectively. Figure 52.4 (Total; lanes 1 and 2) shows a typical immunoblot of the total cellular extracts probed with the anti-GFP antibody. As shown, the relative intensity of the 70 kDa protein band,

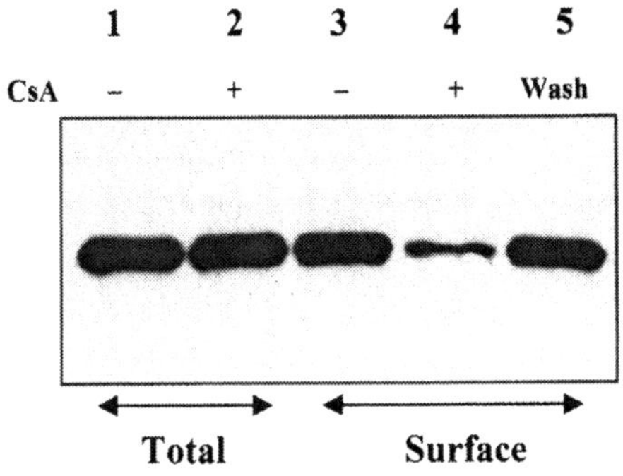

Fig. 52.4 Effect of CsA on taurine transporter expression. Total cellular (Total) and cell surface expression levels (Surface) of mTauT.GFP were analyzed by immunoblot assay and cell surface biotinylation assay, respectively, as described in Methods. CsA-treatment of HEK/mTauT.GFP cells were carried out as described in Fig. 52.3. Total: Immunoblot of total cell extracts (each containing 40 μg of protein) prepared from the solvent-treated control (lane 1; – symbol) and the CsA-treated (lane 2; + symbol) HEK/mTauT.GFP cells. Surface: Immunoblot of biotin-labeled and streptavidin-bead captured surface proteins from the solvent-treated control (lane 3; – symbol), CsA-treated (lane 4; + symbol), and CsA-treated and washed (lane 5; Wash) HEK/mTauT.GFP cells

representing the mTauT.GFP protein, was very similar in detergent extracts of both solvent-treated and CsA treated cells (Fig. 52.4; lanes 1 and 2), suggesting that CsA treatment did not alter the total expression level of the mTauT.GFP protein.

Results of the cell surface biotinylation assay are also shown in Fig. 52.4 (Surface; lanes 3, 4, and 5). The anti-GFP antibody detected the surface biotinylated and streptavidin-agarose bead-captured 70-kDa protein band in both solvent-treated and CsA-treated HEK/mTauT.GFP cells (Fig. 52.4; lanes 3 and 4), but not in the HEK/GFP cells (result not shown). As shown, the intensity of the 70-kDa protein band in the solvent-treated cells is significantly stronger than that in the CsA-treated cells (Fig. 52.4; compare lanes 3 and 4). In three independent experiments, the relative intensity of the 70-kDa protein band in CsA-treated cells, as determined by densitometric scanning, was 50%, 61%, and 70%, respectively, of that in the control cells. Remarkably, in about 24 h after the CsA was washed off, the relative surface expression level of the transporter in the CsA-treated and washed cells was found to be virtually similar to that in the solvent-treated control cells (Fig. 52.4, lane 5). Similar increase in the taurine transporter surface expression level was found even in the presence of cycloheximide (result not shown), suggesting that the increase in activity after the CsA wash was not due to new protein synthesis.

52.4 Discussion

Our results show that CsA inhibited the uptake activity of the permanently expressed mouse taurine transporter in a dose and time dependent manner. These results are consistent with the earlier report by Ramamoorthy et al. (1992), who initially demonstrated that CsA inhibited the taurine uptake in human placental cells. Our studies for the first time show that the CsA-induced inhibition of taurine uptake is reversed when the CsA is removed, and the reversal in uptake is insensitive to the protein synthesis inhibitor cycloheximide. Taken together, these results indicate that the reversal of uptake inhibition is not due to the synthesis of new transporter molecules.

Surface labeling assays clearly demonstrate that the CsA-treatment reduces the relative surface abundance of the taurine transporter without affecting its total expression level. Thus, the decrease in taurine uptake is effectively due to a decrease in the relative abundance of the surface expressed taurine transporters, and not due to an altered transporter expression. Our results also demonstrate that after the CsA is removed, the relative abundance of the surface expressed taurine transporters increases to that of the control level, and this increase in the surface abundance is insensitive to the protein synthesis inhibitor cycloheximide. These results further confirm our earlier contention that the reversal of the CsA-induced inhibition in taurine uptake is not due to any newly synthesized taurine transporter.

One possible explanation for the observed decrease in the surface abundance is that the CsA-treatment triggers internalization of the surface expressed transporters into an internal membrane compartment via a yet to be determined mechanism. After the CsA is washed off, the internalized transporters recycle back to the cell surface, thus reversing the inhibitory action of CsA. Further experiments are, however, needed to gain better insight into the inhibitory mechanism.

Previously, Ramamoorthy et al. (1992) suggested that CsA treatment lowered the fetal plasma taurine concentration due to the impaired taurine transporter function, which might be a causative factor for the development of intrauterine growth retardation. Consistent with their suggestion, Norberg et al. (1998) found low plasma concentrations of taurine in intrauterine growth retardation fetuses, and subsequently demonstrated that the lower plasma taurine concentration was due to the reduced activity of the placental taurine transporter. More recently, Roos et al. (2004) demonstrated that the reduced taurine transport associated with the intrauterine growth retardation was in fact not due to a decrease in the transporter expression, but due to a decrease in the transporter activity. Although our results using the model system described here do not emulate the clinical conditions of the intrauterine growth retardation, our observation that CsA inhibits the taurine transporter activity by altering the surface expression of the transporter without affecting its expression level has remarkable similarity to the results of the clinical studies summarized above.

As demonstrated above, the inhibition of taurine uptake, induced by short-term CsA treatment, is clearly reversible in our cell culture system. Whether or not the same will hold true in transplant patients undergoing prolonged CsA treatment, however, remains to be a matter of further investigation. Nonetheless, we suggest that in the transplant recipient mothers undergoing CsA treatment, the transplacental taurine transfer, mediated by the placental taurine transporter, is effectively reduced by a mechanism similar to the one proposed above.

Acknowledgments This work was supported by a grant from USDA/ARS under Cooperative Agreement 58-6255-6001 to H.K.S. and was carried out at the Baylor College of Medicine, Houston, Texas. Authors also thank Dr. J. Clifford Waldrep for the use of Zeiss fluorescence microscope and Dr. Susan Hamilton for the use of AlphaImager 2000.

References

Aerts L, Van Assche FA (2002) Taurine and taurine-deficiency in perinatal period. J Perinat Med 30:281–286

Chesney RW, Holms RA, Christensen M, Burdreau AM, Han X, Sturman JA (1998) The role of taurine in infant nutrition. Adv Exp Med Biol 442:463–476

Ejiri K, Akahori S, Kudo K, Sekiba K, Ubuka T (1987) Effect of guanidinoethyl sulfonate on tissue concentrations and fetal growth in pregnant rats. Biol Neonate 51:234–240

Ghisolfi J, Berrebi A, Nguyen VB, Thouvenot JP, Rolland M, Putet G, Dabadie A, Pontonnier G (1989) Placental taurine and low birth weight infants. Biol Neonate 56:181–185

Huxtable RJ (1992) Physiological actions of taurine. Physiol Rev 72:101–163

Karl PI, Fisher SE (1990) Taurine transport by microvillous membrane vesicles and the perfused cotyledon of the human placenta. Am J Physiol 285:C443–C451

Kulanthaivel P, Cool DR, Ramamoorthy S, Mahesh VB, Leibach FH, Ganapathy V (1991) Transport of taurine and its regulation by protein kinase C in JAR human placental choriocarcinoma cell line. Biochem J 277:53–58

Miyamoto Y, Balkovetz DF, Leibach FH, Mahesh VB, Ganapathy V (1988) $Na^{+} + Cl^{-}$ -gradient-driven, high-affinity, uphill transport of taurine in human placental brush-border membrane vesicles. FEBS Lett 231(1):263-7

Norberg S, Powell TL, Jansson T (1998) Intrauterine growth restriction is associated with a reduced activity of placental taurine transporters. Pediatr Res 44:233–238

Pickrell MD, Sawers R, Michael J (1988) Pregnancy after renal transplantation:severe intrauterine growth retardation during treatment with cyclosporin A. Br Med J (Clin Res) 296:825

Qian X, Vinnakota S, Edwards E, Sarkar HK (2000) Molecular characterization of taurine transport in bovine aortic endothelial cells. Biochim Biophys Acta 1509:324–334

Ramamoorthy S, Leibach FH, Mahesh VB, Ganapathy V (1992) Selective impairment of taurine transport by cyclosporin A in a human placental cell line. Pediatr Res 32:125–127

Redmond HP, Stapleton PP, Neary P, Bouchier-Hayes D (1998) Immunonutrition:The role of taurine. Nutrition 14:599–604

Roos S, Powell TL, Jansson T (2004) Human placental taurine transporter in uncomplicated and IUGR pregnancies: cellular localization, protein expression, and regulation. Am J Physiol Regul Integr Comp Physiol 287:R886–R893

Tran TT, Dai W, Sarkar HK (2000) Cyclosporin A inhibits creatine uptake by altering the surface expression of the creatine transporter. J Biol Chem 275:35708–35714

Vinnakota S, Qian X, Egal H, Sarthy V, Sarkar HK (1997) Molecular characterization and in situ localization of a mouse retinal taurine transporter. J Neurochem 69:2238–2250

Willis FR, Findlay CA, Gorrie MJ, Watson MA, Wilkinson AG, Beattie TJ (2000) Children of renal transplant recipient mothers. J Paediatr Child Health 36:230–235

Chapter 53
Downregulation of Taurine Transport by Calcium Blockers in Osteoblast Cells

Young-Sook Kang

Abstract Taurine is found in a high concentration in bone cells and is thought to help enhance bone tissue formation and inhibit bone loss. It is mainly transported by a sodium and chloride ion dependent taurine transporter (TauT), which is expressed in a variety of tissues, such as brain, retina, and placenta, but in bone the transporter has not yet been identified. The purpose of this study is to clarify the uptake mechanism of taurine in osteoblasts using mouse osteoblast cell lines. Mouse stromal ST2 cells and mouse osteoblast-like MC3T3-E1 cells were used as osteoblast cell lines. Detection of TauT mRNA expression in these cells was performed by RT-PCR. The activity of the taurine transporter was assessed by measuring the uptake of [^{3}H]taurine in cell lines in the presence and absence of inhibitors. TauT mRNA was detected in ST2 and MC3T3-E1 cells. [^{3}H]Taurine uptake by these cells exhibited a time dependent increase that was linear for at least 10 min. [^{3}H]Taurine uptake was dependent on the presence of extracellular sodium and chloride ions, and was inhibited by unlabeled taurine, β-alanine and γ-amino-n-butyric acid. Moreover, uptake of [^{3}H]taurine by these cells was dependent on the presence of extracellular calcium. The uptake of [^{3}H]taurine in ST2 cells treated with 4 mM calcium was increased 1.7-fold. The initial rate of [^{3}H]taurine uptake was significantly inhibited by 100 μM nifedipine and 100 μM verapamil. These results suggest that in mouse osteoblast cell lines taurine transport is controlled by extracellular calcium.

Abbreviations *TauT*, taurine transporter; *mRNA,* messenger ribonucleic acid; *RT-PCR*, reverse transcriptase polymerase chain reaction; *BGT-1*, betaine transporter

53.1 Introduction

Bone formation and resorption in adults are carried out by osteoblasts (bone-forming cells) and osteoclasts (bone-resorption cells), respectively. These two kinds of cells cooperate with each other to maintain homeostasis of bony tissue (Manolagas 2000).

Y.-S. Kang (✉)
College of Pharmacy, Sookmyung Women's University, Seoul, Korea

J. Azuma et al. (eds.), *Taurine 7*, Advances in Experimental Medicine and Biology 643,
DOI 10.1007/978-0-387-75681-3_53, © Springer Science+Business Media, LLC 2009

Taurine, 2-aminoethanesulfonic acid, is present in a variety of tissues and exhibits many important physiological functions, such as stabilization of membranes, osmoregulation, antioxidation, and detoxification (Huxtable 1992). In bone cells, taurine is found in high concentration (Lubec et al. 1997), and is said to help in the enhancement of bone tissue formation, as evidenced by increased matrix formation and collagen synthesis (Park et al. 2001). Besides stimulating bone tissue formation, it also inhibits bone loss through the inhibition of bone resorption and osteoclast formation. It has been suggested that taurine, by entering bone tissue, plays an important role in bone formation, however, the mechanism remains unclear.

Taurine is mainly transported by the Na^+/Cl^- ion dependent taurine transporter (TauT) (Uchida et al. 1992; Takeuchi et al. 2000). TauT is expressed in a variety of tissues, such as brain, kidney, liver, retina, and placenta (Ramamoorthy et al. 1994; Chen et al. 2004), where its physiological function is to transport taurine into the cell. In bone tissue, it is possible that taurine is transported by TauT, however the expression of TauT in bone tissue remains to be established.

Calcium is involved in bone formation, possibly in part through the regulation of taurine uptake. However, the mechanism by which calcium modulates the uptake of taurine in bone cells has not been determined. How these two factors affect bone formation in osteoblast cell lines is the focus of this study. In this study, we used mouse stromal ST2 cells and mouse osteoblast-like MC3T3-E1 cells. The stromal cell line ST2, which is derived from mouse bone marrow, differentiates into osteoblast-like cells in response to ascorbic acid (Otsuka et al. 1999). It is important to understand the mechanism of taurine uptake in these two types of cells, which differ in maturation level. We investigated the expression of TauT and the effect of various factors on taurine uptake in osteoblastic cells.

53.2 Methods

53.2.1 Reagent

[2-^{3}H(N)]Taurine ([^{3}H], 30.3 Ci/mmol) was purchased from NEN Life Science (Boston, MA, USA). All other chemicals were of reagent grade and were commercially available.

53.2.2 Cell Culture

ST2 cells and MC3T3-E1 cells were grown routinely in tissue culture dishes at 37°C under 5% CO_2 and 95% air. The cells were cultured in Dulbecco's modified Eagle's medium (DMEM; Gibco, Grand Island, NY) 100 U/mL benzylpenicillin,

100 mg/mL streptomycin sulfate (Gibco, Grand Island, NY), 10% fetal bovine serum (Gibco, Grand Island, NY).

53.2.3 [^{3}H]Taurine Uptake Study by Culture ST2 and MC3T3-E1 Cells

[^{3}H]Taurine uptake was performed as described in a previous report (Kang et al. 2002). ST2 cells (1×10^5 cells/well) and MC3T3-E1 cells (1.5×10^5 cells/well) were cultured at 37°C for 2 days on rat tail collagen type-I coated 24-well plates (Becton Dickinson) and washed with 1 mL extracellular fluid (ECF) buffer consisting of 122 mM NaCl, 25 mM $NaHCO_3$, 3 mM KCl, 1.4 mM $CaCl_2$, 1.2 mM $MgSO_4$, 0.4 mM K_2HPO_4, 10 mM D-glucose and 10 mM Hepes (pH 7.4) at 37°C. Uptake was initiated by applying 200 μL ECF buffer containing 1.0 mCi [^{3}H]taurine for water adhesion at 37°C in the presence or absence of inhibitors. Na^+ or Cl^- free conditions were obtained by replacement with equimolar choline or gluconate. After appropriate time periods, uptake was terminated by removing the applied solution and cells, which were immersed in ice-cold ECF buffer. The cells were then solubilized in 750 μL 1 M NaOH. An aliquot (50 mL) was taken for protein assay using a DC protein assay kit (Bio-Rad, Hercules, CA, USA) with bovine serum albumin serving as a standard. The remaining solution (500 μL) was mixed with 5 mL scintillation cocktail (Hionic-fluor, Packard, Meriden, CT, USA) for measurement of radioactivity in a liquid scintillation counter (LS6500, Beckman, Fullerton, CA, USA).

53.2.4 Reverse Transcription-Polymerase Chain Reaction (RT-PCR) Analysis

Expression of rat taurine transporter, rat glyceraldehyde 3-phosphate dehydrogenase (G3PDH), by osteoblast cell lines was analyzed by RT-PCR. The sequences of sense and antisense primers and experimental conditions were given in our previous paper. Total RNA was isolated by the acid phenol procedure using ISOGEN (Wako Pure Chemical Industries, Ltd., Tokyo, Japan) according to the manufacturer's protocol. The first standard cDNA reaction was performed using ReverTra Ace M-MLV reverse Transcriptase (Rever Tra Ace, Toyobo Co., Ltd., Osaka, Japan). RT-PCR was performed with TAKARA Ex TaqTM (Takara Shuzo Co., Ltd., Shiga, Japan) according to the manufacturer's instructions. The PCR was carried out using the following protocol. After an initial melting temperature of 85°C for 5 min, primers were added according to the hot start method. After 1 min of denaturation at 94°C, 1 min of annealing at 57°C or 59°C, and 1 min of extension at 72°C for repeated cycles of amplification, a final extension at 72°C was allowed to proceed for 10 min. The PCR products were analyzed on a 5% acrylamide gel, stained with ethidium bromide, and visualized under ultraviolet light.

53.3 Results

53.3.1 mRNA Expression of Taurine Transporter (TauT) in Two Osteoblast Cell Lines

The expression of TauT mRNA in ST2 and MC3T3-E1 cells was analyzed by RT-PCR (Fig. 53.1). The bands corresponding to a 459 bp of mTauT from ST2 cells and MC3T3-E1 cells were amplified, with mouse brain serving as a positive control. The DNA sequences of the bands in these cells were identical to mouse TauT; the homology was 100%.

53.3.2 Characterization of Taurine Transport

To characterize taurine transport in the osteoblast cell lines, [^{3}H]taurine uptake was performed using ST2 and MC3T3-E1 cells. As shown in Fig. 53.2, [^{3}H]taurine cells exhibited a time-dependent increase and was linear for at least 10 min.

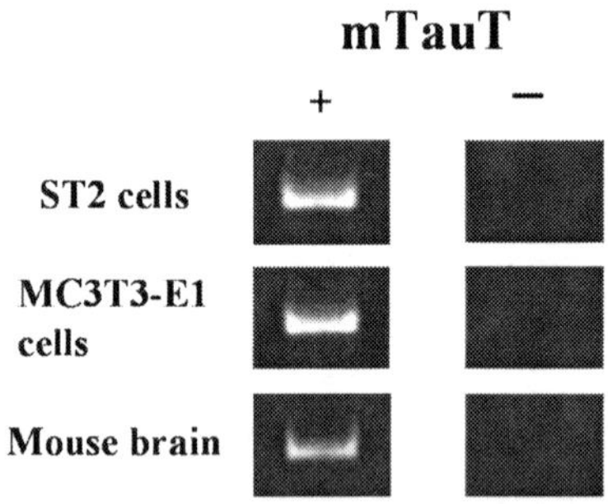

Fig. 53.1 Detection of mTauT mRNA in ST2 cells, MC3T3-E1 cells and mouse brain. (+) and (−) represent the presence or absence of reverse transcriptase, respectively

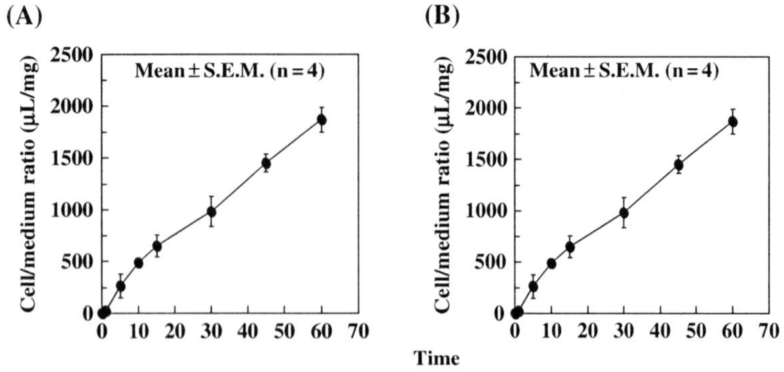

Fig. 53.2 Time-course of [^{3}H]taurine uptake by ST2 cells (**A**) and MC3T3-E1 cells (**B**)

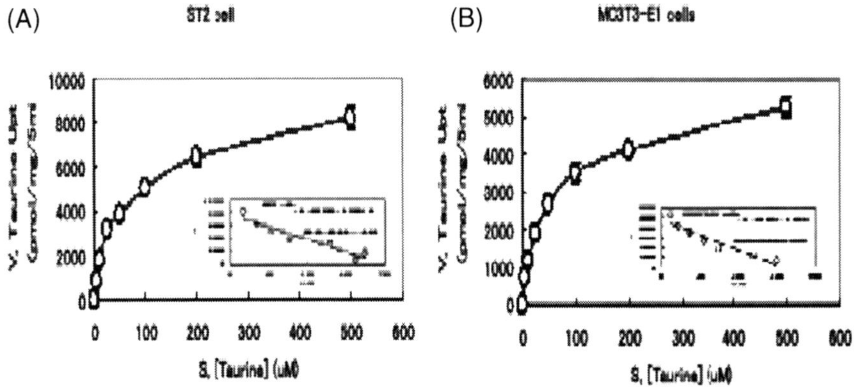

Fig. 53.3 The kinetic parameters of [^{3}H]taurine uptake by ST2 and MC3T3-E1 cells

53.3.3 Determination of Kinetic Parameters of Taurine Uptake

The concentration dependence of taurine uptake was examined at 5 min (initial uptake). Taurine concentrations up to 500 mmol/L were examined (Fig. 53.3). A simple Michaelis-Menten equation did not adequately fit the data, thus requiring the addition of a linear component to the equation. The kinetic parameters for [^{3}H]taurine uptake revealed saturation kinetics with a Michaelis constant (K_m) of 38.8 ± 0.71 mM and a maximum rate of uptake (V_{max}) of 7.73 ± 0.42 nmol/5 min/mg of protein in ST2, and K_m of 29.8 ±1.95 mM and V_{max} of 4.75 ± 0.97 nmol/5 min/mg of protein in MC3T3-E1.

53.3.4 Ion Dependency of Taurine Uptake

The effect of ions on the initial rate of taurine uptake into osteoblast cells was examined in medium lacking sodium, chloride or calcium (Figs. 53.4, 53.5). In sodium-free medium, [^{3}H]taurine uptake was reduced by 94.1% and 96.3% in ST2 cells and MC3T3-E1 cells, respectively, while in chloride-free medium, [^{3}H]taurine uptake was reduced by 93.6% and 96.0% in ST2 cells and MC3T3-E1 cells, respectively. Moreover, in ST2 and MC3T3-E1 cells incubated in medium lacking calcium, [^{3}H]taurine uptake was reduced by 57.7% and 30.4%, respectively. On the other hand, the uptake of [^{3}H] taurine by ST2 cells increased in the presence of 4 mM calcium ion, but higher extracellular calcium concentrations had no effect on taurine uptake by MC3T3-E1 cells, indicating that taurine uptake was also independent of calcium ion. Taurine uptake is regulated by Ca^{++} in progenitor osteoblast cells but, less so by osteoblast cells.

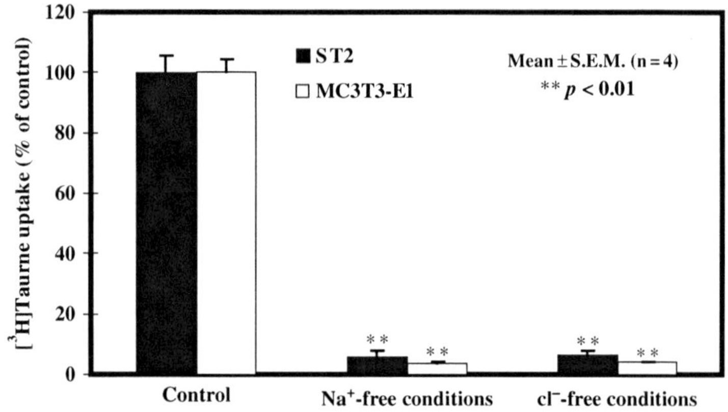

Fig. 53.4 Na^+-and Cl^- dependence of [^{3}H]taurine uptake by ST2 and MC3T3-E1 cells

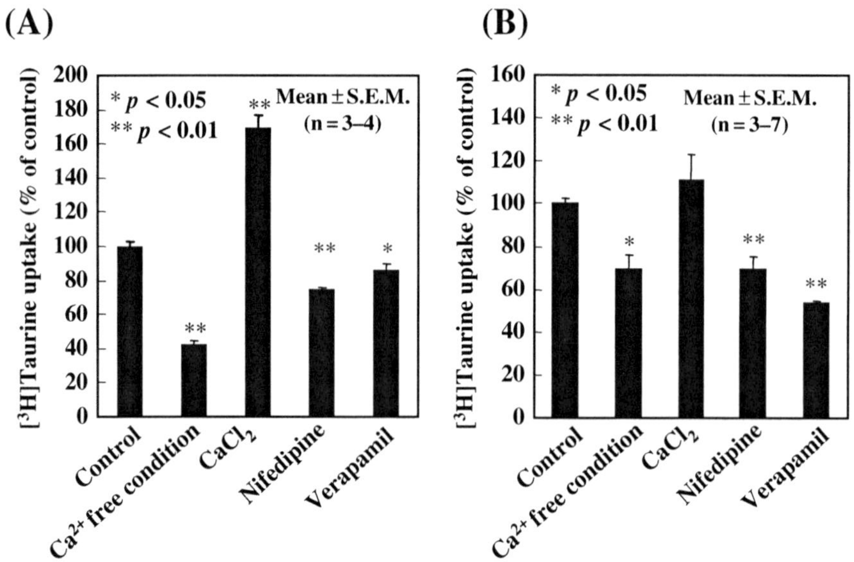

Fig. 53.5 Effect of Ca^{2+} and calcium channel blockers on the uptake of [^{3}H]taurine by ST2 (**A**) and MC3T3-E1 (**B**) cells

53.3.5 Effect of Various Compounds on Taurine Uptake

To determine the specificity of the TauT the effect of other compounds on taurine uptake was examined in ST2 and MC3T3-E1 cells (Table 53.1). The initial rate of [^{3}H]taurine uptake was significantly inhibited by cold taurine, β-alanine and γ-amino-n-butyric acid. Betaine did not affect [^{3}H]taurine uptake in ST2 and MC3T3-E1 cells.

Table 53.1 Inhibitory effect of various compounds on [^{3}H]taurine uptake by ST2 and MC3T3-E1 cells

Compounds	Uptake ratio	
	ST2 cells	MC3T3-E1 cells
Control	100 ± 21.03	100 ± 22.25
50 mM Taurine	2.12 ± 0.57**	6.42 ± 4.01**
0.5 mM β-alanine	15.81 ± 1.03**	12.08 ± 1.71**
0.5 mM GABA	33.89 ± 5.25**	54.22 ± 3.74**
0.5 mM Betaine	77.18 ± 18.22	92.72 ± 12.47
10 mM Betaine	103.38 ± 5.89	93.12 ± 3.72
0.5 mM Glycine	78.29 ± 11.43#	102.00 ± 10.60**
10 mM Choline	92.059 ± 1.66	89.90 ± 6.51

# $P < 0.05$ * $P < 0.01$ ** $P < 0.001$
The uptake ratio is expressed as % control and represents means ± S.E.

53.3.6 *Effect of Calcium Channel Antagonists on [^{3}H]taurine Uptake*

To examine the effect of calcium inhibitors on [^{3}H]taurine uptake, we used the L-type calcium channel blockers, nifedipine, and verapamil (Fig. 53.5). The initial rate of [^{3}H]taurine uptake was significantly reduced by 100 μM nifedipine and 100 μM verapamil.

53.4 Discussion

The expression of mTauT in ST2 and MC3T3-E1 cells was analyzed by RT-PCR. Both ST2 and MC3T3-E1 cells expressed mTauT mRNA (Fig. 53.1). Mouse brain was used as a positive control. To characterize taurine transport by osteoblast cells, [^{3}H]taurine uptake was performed using ST2 and MC3T3-E1 cells. [^{3}H]Taurine uptake exhibited a time-dependent increase and was linear for at least 10 min in ST2 cells and more than 60 min in MC3T3-E1 cells (Fig. 53.2). The effect of $Na^+$and Cl^- on taurine uptake into osteoblastic cell lines was examined in medium lacking $Na^+$and Cl^- (Fig. 53.4). In the absence of Na^+, [^{3}H]taurine uptake was reduced in ST2 and MC3T3-E1 cells by 94.1% and 96.3%, respectively. Removal of Cl^- from the medium reduced [^{3}H]taurine uptake in ST2 cells and MC3T3-E1 cells by 93.6% and 96.0%, respectively. Since taurine transport by TauT is dependent on $Na^+$and Cl^-, taurine transport in ST2 and MC3T3-E1 cells likely involves TauT. In ST2 cell, the K_m value for taurine transport was 38.8 ± 0.71 μM and the V_{max} value was 7.73 ± 0.42 nmol/5min/mg of protein (Fig. 53.3A) while in MC3T3-E1 cells the K_m and V_{max} values were 29.8 ± 1.95 μM and 4.75 ± 0.97 nmol/5min/mg of protein, respectively (Fig. 53.3B). A similar K_m value has been reported by Lubec et al. (1997) and Uchida et al. (1992).

To determine the substrate specificity of taurine uptake by osteoblastic cells, taurine uptake was examined in the presence of other compounds (Table 53.1). The initial rate of [^{3}H]taurine uptake was significantly inhibited by cold taurine, β-alanine and γ-amino-n-butyric acid (GABA). Betaine, which is the betaine transporter substrate, did not affect [^{3}H]taurine uptake in ST2 and MC3T3-E1 cells, despite the view that the betaine transporter (BGT-1) is capable of transporting taurine. One would predict that taurine transport by BGT-1 would be blocked by betaine in cells containing BGT-1. However, the BGT-1 system does not appear to function in ST2 and MC3T3-E1 cells. Moreover, choline does not affect taurine transport, suggesting that taurine is transported by TauT in osteoblast cells.

Taurine also affects Ca^{++} homeostasis, activating Ca^{++} uptake in many tissues, such as heart, brain and retina. It is also known that calcium plays an important role in bone formation. Because taurine conductance is inhibited by L-type calcium channel blockers in rat astrocytes, it is possible that a change in extracellular Ca^{++} concentration also affects taurine uptake in osteoblasts. Therefore, we examined the effect of Ca^{++} and L-type calcium channel blockers on [^{3}H]taurine uptake by osteoblastic cells (Fig. 53.5). In the absence of extracellular Ca^{++}, [^{3}H]taurine uptake was reduced in ST2 and MC3T3-E1 cells by 57.7% and 30.4%, respectively. On the other hand, the uptake of [^{3}H]taurine increased in the presence of 4 mM Ca^{++} in ST2 cells, but was unaffected by Ca^{++} in MC3T3-E1 cells. As previously reported (Ramamoorthy et al. 1994), taurine uptake in rat retina cells is not affected at 0.1mM Ca^{++}. However, in this study, taurine uptake by ST2 cells is reduced at lower Ca^{++} concentrations but increased at 4 mM Ca^{++} (2.8 times the normal condition). This reveals the need to further review the relationship between taurine and Ca^{++} concentration.

The uptake of [^{3}H]taurine in ST2 and MC3T3-E1 cells was significantly inhibited by the L-type calcium channel blockers, nifedipine and verapamil. It has been previously reported that taurine excretion in rat astrocytes is inhibited by Ca^{++} and a calmodulin inhibitor. Based on this study, it is assumed that inhibition of taurine uptake by calcium channel antagonists involves the inhibition of TauT expression. That is supported by the finding that calcium channel inhibitors block the interaction of Ca^{++} with calmodulin. In this study, the decrease in taurine uptake by Ca^{++} channel blockers supports the view that channel blockers inhibit the effects of Ca^{++} on calmodulin, as both Ca^{++}-dependent taurine uptake and TauT expression are decreased. The effect of Ca^{++} and calcium channel blockers on taurine uptake is similar in ST2 and MC3T3-E1 cells. Therefore, our results suggest that Ca^{++} regulates taurine uptake in osteoblast cells.

These results suggest that taurine uptake into osteoblast cells contribute to bone formation and is regulated by various factors. Although there are some differences between ST2 and MC3T3-E1 cells, many factors, such as calcium, modulate bone formation and bone absorption.

Acknowledgments The author thanks to N. Ishido for her excellent technical assistance.

References

Chen NH, Reith ME, Quick MW (2004) Synaptic uptake and beyond: the sodium- and chloride-dependent neurotransmitter transporter family SLC6. Pflugers Arch 447:519–531

Huxtable RJ (1992) Physiological actions of taurine. Physiol Rev 72:101–163

Kang YS, Ohtsuki S, Takanaga H, Tomi M, Hosoya K, Terasaki T (2002) Regulation of taurine transport at the blood-brain barrier by tumor necrosis factor-alpha, taurine and hypertonicity. J Neurochem 83:1188–1195

Lubec B, Ya-hua Z, Pertti S, Pentti T, Kitzmuller E, Lubec G (1997) Distribution and disappearance of the radiolabeled carbon derived from L-arginine and taurine in the mouse. Life Sci 60: 2373–2381

Manolagas SC (2000) Birth and death of bone cells: basic regulatory mechanisms and implications for the pathogenesis and treatment of osteoporosis. Endocr Rev 21:115–137

Otsuka E, Yamaguchi A, Hirose S, Hagiwara H (1999) Characterization of osteoblastic differentiation of stromal cell line ST2 that is induced by ascorbic acid. Am J Physiol 277:C132–C138

Park S, Kim H, Kim SJ (2001) Stimulation of ERK2 by taurine with enhanced alkaline phosphatase activity and collagen synthesis in osteoblast-like UMR-106 cells. Biochem Pharmacol 62:1107–1111

Ramamoorthy S, Del Monte MA, Leibach FH, Ganapathy V (1994) Molecular identity and calmodulin-mediated regulation of the taurine transporter in a human retinal pigment epithelial cell line. Curr Eye Res 13:523–529

Takeuchi K, Toyohara H, Sakaguchi M (2000) A hyperosmotic stress-induced mRNA of carp cell encodes Na(+)- and Cl(-)-dependent high affinity taurine transporter. Biochim Biophys Acta 1464(2):219–30

Uchida S, Kwon HM, Yamauchi A, Preston AS, Marumo F, Handler JS (1992) Molecular cloning of the cDNA for an MDCK cell Na(+)- and Cl(-)-dependent taurine transporter that is regulated by hypertonicity. Proc Natl Acad Sci USA 89:8230

Chapter 54
Involvement of Transcriptional Factor TonEBP in the Regulation of the Taurine Transporter in the Cardiomyocyte

Takashi Ito, Yasushi Fujio, Stephen W. Schaffer, and Junichi Azuma

Abstract Taurine is found in high concentrations in heart where it exerts several actions that could potentially benefit the diseased heart. The taurine transporter (TauT) is crucial for the maintenance of high taurine levels in the heart. Although cardiac taurine content is altered in various pathological conditions, little is known about the regulatory mechanisms governing TauT expression in cardiac myocytes. In the present study, we found that treatment with the antineoplastic drug doxorubicin (DOX), which is also known as a cardiotoxic agent, decreases the expression of the TauT gene in cultured cardiomyocytes isolated from the neonatal rat heart. Based on data obtained using a luciferase assay, DOX significantly reduced transcriptional activity driven by the TauT promoter, while deletion or mutation of a tonicity-response element (TonE) in this promoter eliminated the change of promoter activity. The protein level of the TonE-binding protein (TonEBP) was reduced by DOX treatment. In addition, the reduction in TonEBP protein content was suppressed by proteasome inhibitors. In conclusion, the DOX-enhanced degradation of TonEBP resulting in reduced TauT expression in the cardiomyocyte.

Abbreviations *TauT*, taurine transporter; *DOX*, doxorubicin; *TonE*, tonicity-response element; *TonEBP*, tonicity-response element-binding protein

54.1 Introduction

Taurine (2-aminoethylsulfonic acid), a sulfur-containing amino acid, is widely distributed in the body and exerts many physiological functions (Huxtable 1992; Schaffer et al. 2003). Taurine is found in millimolar concentrations in most mammalian tissues (i.e. 10 mM in heart) but is found at a concentration of 20–100 μM in plasma (Chapman et al. 1993; Chesney 1985). Taurine has been reported to protect the heart against ischemia and the development of a cardiomyopathy and heart failure. Furthermore, it has been shown that taurine deficiency in cats, foxes and dogs

T. Ito (✉)
Department of Clinical Pharmacology and Pharmacogenomics, Graduate School of Pharmaceutical Sciences, Osaka University, Japan; Department of Pharmacology, University of South Alabama, College of Medicine, AL, USA

J. Azuma et al. (eds.), *Taurine 7*, Advances in Experimental Medicine and Biology 643, DOI 10.1007/978-0-387-75681-3_54,

leads to the development of a dilated cardiomyopathy (Moise et al. 1991; Pion et al. 1987). These cardioprotective effects may be caused by multiple biological and physiological actions, such as a direct modulation of contractile function, an anti-apoptotic action, a membrane stabilizing action, an osmoregulatory action and an antioxidative effect (Huxtable 1992; Oudit et al. 2004; Satoh and Sperelakis 1998; Schaffer et al. 2000a, 2000b; Takatani et al. 2004).

The Na^+-, Cl^--dependent taurine transporter (TauT) principally transports taurine from plasma into the cells, thereby playing a role in the maintenance of the large taurine gradient between the tissues and plasma (Uchida et al. 1992). The expression of TauT is susceptible to several modulators, such as osmolality (Ito et al. 2004; Uchida et al. 1992). The 5'-flanking promoter region of TauT gene was recently identified and has been shown to bind a tonicity response element binding protein (TonEBP), MEF, p53 and the WT1 transcription factor (Han and Chesney 2003; Han et al. 2002; Ito et al. 2004; Uozumi et al. 2006). Importantly, disruption of the TauT gene in mice leads to a dramatic decrease in myocardial taurine content compared with wild type mice (Heller-Stilb et al. 2002). Collectively, TauT may be a crucial factor in the regulation of cellular taurine levels, although the molecular mechanisms involved in the control of cardiac taurine content remain to be elucidated. This would imply that the TauT plays a role under pathological conditions, which lead to a marked elevation in myocardial taurine content during congestive heart failure in humans and rabbits (Huxtable and Bressler 1974; Takihara et al. 1986), but large decreases in taurine content during ischemia and hypoxia (Schaffer et al. 2000b). However, the role of TauT in the regulation of cardiac taurine concentration has never been clarified.

The antineoplastic agent doxorubicin (DOX) is used against a wide variety of tumors (Minotti et al. 2004; Singal and Iliskovic 1998). However, the clinical use of the drug is limited because of its cardiac toxicity that leads to the development of a cardiomyopathy and heart failure. Although the mechanisms of doxorubicin-induced irreversible myocardial damage remain unclear, several factors, including the generation of reactive oxygen species (ROS), myocardial gene inhibition, mitochondrial impairment, disorder of calcium handling and apoptosis, have been implicated in the toxicity (Arai et al. 1998; Ito et al. 1990; Minotti et al. 2004). We previously reported that DOX-induced mortality and the severity of the DOX cardiomyopathy were diminished by taurine treatment (Hamaguchi et al. 1989; Harada et al. 1990). In the present study, we investigated the effect of DOX exposure on TauT expression in the compromised heart.

54.2 Methods

54.2.1 Animals and Treatment

The investigation conformed with the Institutional Animal Care and Use Committee of Osaka University. Six-week-old male C57BL6 mice (SLC) were given doxorubicin (Kyowa Hakko; Tokyo, Japan) with 15 mg/kg of single injection. After 2 days, the hearts were isolated from mice, frozen and stored at −80°C.

54.2.2 Cell Culture

Cardiac ventricles of 1-day-old Wister Kyoto rat were minced and dissociated with 0.1% trypsin (GIFCO) and 0.1% collagenase type IV (Sigma). Dispersed cells were plated and incubated for 1 h at 37°C. Nonattached myocytes were collected and cultured in Dulbecco's modified essential medium / Ham's F-12 containing 5% neonatal bovine serum and 100 μM bromodeoxyuridine (BrdU) at 37°C for 2 days. After being washed twice with serum-free DMEM, cells were treated with doxorubicin in serum- and BrdU- free DMEM.

54.2.3 Western Blot

Tissues were homogenized in SDS/PAGE buffer and boiled immediately. After electrophoresis, Western blotting was performed as previously described (Ito et al. 2004). Anti-GAPDH antibody (CHEMICON) and anti-TonEBP antibody (SantaCruz) were used.

54.2.4 Northern Blot

Total RNA was isolated from mice heart and Northern blot was performed as previously described (Ito et al. 2004).

54.2.5 Luciferase Assay

Promoter-reporter constructs pTauT/-862-Luc, pTauT/-124-Luc, pTauT/-99-Luc and pTauT/-124m-Luc were previously generated (Fig. 54.2A) (Ito et al. 2004). Transient transfection into cardiac myocytes was performed by using Fugene 6 transfection reagent (Roche). The luciferase assay was performed as previously described (Ito et al. 2004).

54.2.6 Static Analysis

Each value was expressed as the mean ± SE. Statistical significance was determined by the Student's t-test. Differences were considered statistically significant when the calculated P value was less than 0.05.

54.3 Results

54.3.1 The Effect of DOX on Tissue Taurine Level and TauT Gene Expression in Heart

To determine the influence of DOX on the level of TauT, TauT mRNA content of the heart of mice treated with DOX was analyzed by Northern blot analysis. Figure 54.1A reveals that TauT protein was decrease by DOX injection. Moreover, DOX treated cardiomyocytes exhibited reduced TauT expression (Fig. 54.1B).

54.3.2 TonEBP Activity of Cardiac Myocyte is Reduced by DOX Treatment

Next, we determined the regulatory mechanism responsible for the DOX-mediated reduction in the TauT gene. To evaluate whether the downregulation of the TonEBP protein was accompanied by a reduction in its transcriptional activity, the promoter-reporter assay was performed by using reporter plasmids consisting of the 5′- flanking region of the TauT gene (Fig. 54.2A). Luciferase assays demonstrated that DOX exposure reduced the promoter activity of pTauT/-862-Luc in cardiac myocytes. DOX-dependent downregulation of the promoter activity was also observed in myocytes transfected with pTauT/-124-Luc. However, the reporter plasmid pTauT/-99-Luc did not exhibit a downregulation of TauT activity in response to DOX. Consistently, the reporter activity of the plasmid pTauT/-124mut-Luc, which is mutated in the TonEBP-binding sequence, was not affected by DOX exposure (Fig. 54.2B).

54.3.3 The Change of TonEBP Expression by DOX

Previously, we identified the presence of the TonE motif (TonEBP binding sequence) in the -110 to -100 region of the 5′-flanking region of *TauT* gene (Ito et al. 2004).

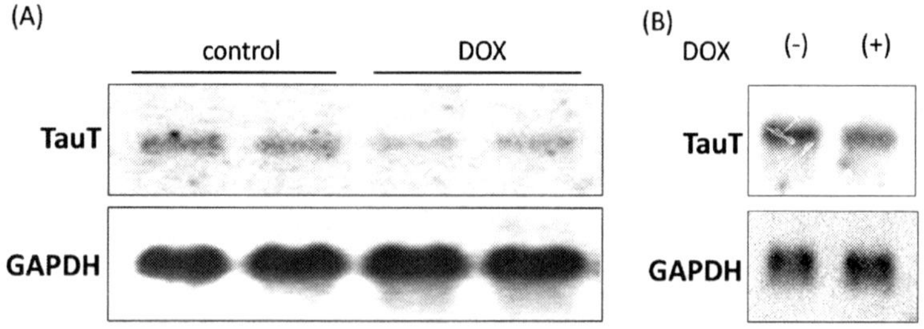

Fig. 54.1 Northern blot analyses of total RNA extracted from heart of mice injected with DOX (15 mg/g body weight) (**A**) from cardiomyocytes cultured with DOX (0.3 μM) for 24 h (**B**). Representative bands from three independent cell preparations are shown

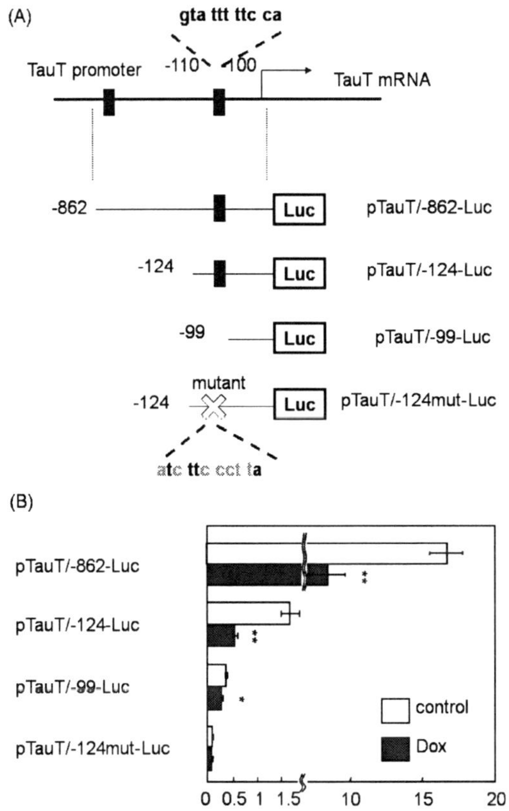

Fig. 54.2 (**A**) The luciferase (Luc) constructs containing different lengths of the 5′-flanking region of the TauT gene were generated. (**B**) Cells transfected with promoter-reporter constructs containing various lengths of the TauT promoter region were exposed to DOX (0.3 μM) for 24 h. Promoter activity was normalized using the luciferase activity of pTK-RL. Data represent means ± SE, $n = 3$. $^{**}p < 0.01$ vs. DOX (−)

To elucidate the influence of DOX on TonEBP expression, we analyzed the protein level of TonEBP in cultured cardiomyocytes treated with DOX. Consistent with the reduction in promoter activity, Western blot analyses revealed that the levels of the TonEBP protein were significantly reduced in the DOX treated cells compared with non-treated myocytes (Fig. 54.3). However, GAPDH protein content was not changed. On the other hand, the mRNA level of TonEBP was not altered by DOX treatment (Data not shown), indicating that reductions in TonEBP protein content are not dependent on the transcriptional rate but are caused by a reduction in translation or in selective degradation.

It has been reported that exposure of isolated cardiomyocytes to DOX resulted in the activation of the ubiquitin-proteasome pathway. To determine the mechanism underlying the reduction in TonEBP, we analyzed the role of proteasomes in the DOX-mediated decrease in TonEBP. Isolated cardiomyocytes in culture were

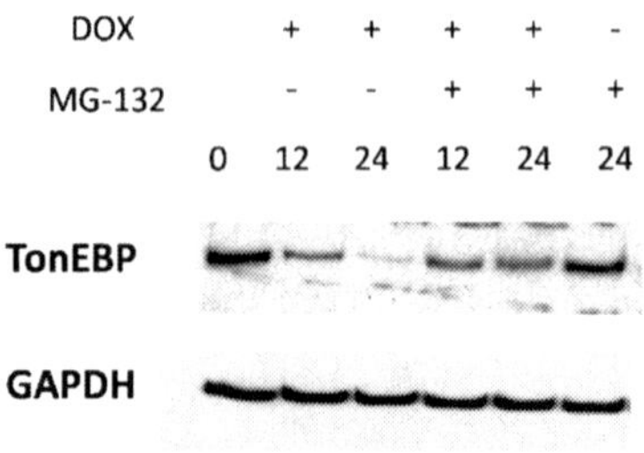

Fig. 54.3 Western blot analyses of whole cell lysates from cardiomyocytes that were pretreated with either DMSO or proteasome inhibitor MG-132 (5 μM) for 1 h and then exposed to DOX (0.3 μM) for 12–24 h

incubated with DOX in the presence or absence of a proteasome inhibitor MG-132 (5 μM) for 12–24 h before examining the expression of TonEBP by Western blot analysis (Fig. 54.3). While TonEBP expression was decreased by DOX alone, treatment with the proteasome inhibitor prevented the downregulation of TonEBP. These data suggest that degradation of TonEBP caused by DOX exposure is mediated by the activation of the UPS pathway. However, we failed to detect ubiquitinated TonEBP in cardiac myocytes exposed to DOX (Data not shown). Therefore, it remains unclear whether proteasome-mediated degradation of TonEBP in the DOX treated cardiomyocyte is dependent on ubiquitination.

54.4 Discussion

We have previously reported that TonEBP is involved in the regulation of TauT expression in response to hypertonicity (Ito et al. 2004). However, the role of TonEBP in the regulation of TauT mRNA in cells exposed to other types of stress has not been addressed. In the present study, we demonstrate that DOX mediates the downregulation of the TauT gene through a reduction in TonEBP protein of the cardiomyocyte.

TonEBP, a member of the rel/NFkappaB/NFAT family of transcription factors, was originally identified as a transcriptional factor involved in the cellular response to hypertonic stress (Ho 2003; Woo et al. 2002). While NFAT 1-4 are activated by Ca^{2+}/calcineurin, TonEBP activity is regulated in a calcineurin-independent manner, largely because it lacks the amino-terminal NFAT homology region containing a calcineurin regulatory motif (Ho 2003; Woo et al. 2002). TonEBP is activated by hyperosmotic stress (Ho 2003; Woo et al. 2002), and in turn transcriptionally regulates the expression of target genes responsible for the metabolism of organic osmolytes, such as aldose reductase (Ko et al. 2000), betaine/GABA transporter (BGT-1) (Miyakawa et al. 1998) and sodium/myo-inositol transporter (Rim et al. 1998). Although TonEBP plays a central role in the transcriptional response to extracellular hypertonic stress, TonEBP is widely expressed in tissues, such as the heart, which are not usually exposed to a hypertonic environment

(Maouyo et al. 2002; Trama et al. 2000). The present study shows that TonEBP is localized not only in the cytosol but also in the nucleus of cardiomyocytes cultured in isotonic medium (Ko et al. 2000), indicating that TonEBP is also regulated by factors that are not associated with osmotic stress. Furthermore, we identified that TonEBP is degraded through a proteasome pathway that is enhanced by DOX. To our knowledge, this study is the first demonstration showing that the process regulating TonEBP protein levels responds to cytotoxic stress even in the absence of an osmotic imbalance.

Although Kumarapeli et al. (2005) demonstrated that DOX activates the UPS-mediated proteolysis pathway in cardiomyocytes, we could not detect ubiquitinated TonEBP in cardiac myocytes exposed to DOX. However, previous reports demonstrate that the proteasome degrades a wide range of proteins without ubiquitinization. Recently, p300 has been reported to be degraded by DOX exposure via an ubiquitin-independent and proteasome-dependent pathway in cultured cardiomyocytes (Poizat et al. 2000), suggesting that DOX may activate the ubiquitin-independent proteasome pathway of proteolysis. Taken together, it is likely that DOX-induced degradation of TonEBP is mediated by the proteasome-linked proteolytic process in an ubiquitin-independent manner.

Accumulating evidence supports a role for TonEBP in cell viability and in regulating cell proliferation. Many of the mice that lack the *TonEBP* gene undergo late gestational lethality while others develop atrophy of the kidney medulla (Lopez-Rodriguez et al. 2004). Furthermore, suppression of TonEBP transcription by expression of the dominant-negative form of TonEBP leads to an increase in cell death in T-cell and lens fiber cells (Trama et al. 2002, Wang et al. 2005). Additionally, the transcriptional targets of TonEBFP are cytoprotective. In particular, taurine is effective against Dox-induced cardiotoxicity and hypoxia-induced apoptosis (Hamaguchi et al. 1989; Harada et al. 1990; Takahashi et al. 2003). Therefore, this study

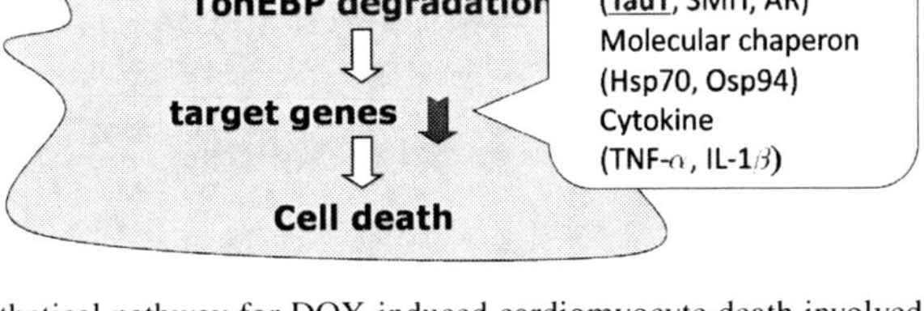

Fig. 54.4 Hypothetical pathway for DOX-induced cardiomyocyte death involved with the degradation of TonEBP. *SMIT*: sodium/myo-inositol transporter, *AR*: Aldose reductase, *Hsp70*: heat shock protein 70, *Osp94*: Osmotic stress protein 94 kDa, *TNF*-α, Tumor necrosis factor-α, *IL-1*β: Interleukin-1β

implicates accelerated degradation of TonEBP protein in the type of cardiotoxicity caused by DOX administration. Indeed, inhibition of TonEBP by either siRNA or overexpressing the dominant-negative form of TonEBP improves cell viability of isolated cardiomyocytes in culture (Ito et al. 2007). Taken together, the regulation of TauT expression via TonEBP is a critical pathway of cardioprotection (Fig. 54.4).

54.5 Conclusion

This study indicates that TonEBP regulates the expression of the TauT gene in the cardiomyocyte, an effect that may result in an alteration in cardiac taurine content. Furthermore, degradation of TonEBP may contribute to the development of DOX-linked cardiotoxicity through a decrease in tissue taurine content.

Acknowledgments This study was partly granted by Taisho Pharmaceutical Ltd. We acknowledge to Kyowa Hakko for the donation of doxorubicin. We thank Ms. Yasuko Murao for her secretary work.

References

Arai M, Tomaru K, Takizawa T, Sekiguchi K, Yokoyama T, Suzuki T, Nagai R (1998) Sarcoplasmic reticulum genes are selectively down-regulated in cardiomyopathy produced by doxorubicin in rabbits. J Mol Cell Cardiol 30:243–254

Chapman RA, Suleiman MS, Earm YE (1993) Taurine and the heart. Cardiovasc Res 27:358–363

Chesney RW (1985) Taurine: its biological role and clinical implications. Adv Pediatr 32:1–42

Hamaguchi T, Azuma J, Harada H, Takahashi K, Kishimoto S, Schaffer SW (1989) Protective effect of taurine against doxorubicin-induced cardiotoxicity in perfused chick hearts. Pharmacol Res 21:729–734

Han X, Chesney RW (2003) Regulation of taurine transporter gene (TauT) by WT1. FEBS Lett 540:71–76

Han X, Patters AB, Chesney RW (2002) Transcriptional repression of taurine transporter gene (TauT) by p53 in renal cells. J Biol Chem 277:39266–39273

Harada H, Cusack BJ, Olson RD, Stroo W, Azuma J, Hamaguchi T, Schaffer SW (1990) Taurine deficiency and doxorubicin: interaction with the cardiac sarcolemmal calcium pump. Biochem Pharmacol 39:745–751

Heller-Stilb B, van Roeyen C, Rascher K, Hartwig HG, Huth A, Seeliger MW, Warskulat U, Haussinger D (2002) Disruption of the taurine transporter gene (taut) leads to retinal degeneration in mice. Faseb J 16:231–233

Ho SN (2003) The role of NFAT5/TonEBP in establishing an optimal intracellular environment. Arch Biochem Biophys 413:151–157

Huxtable R, Bressler R (1974) Taurine concentrations in congestive heart failure. Science 184:1187–1188

Huxtable RJ (1992) Physiological actions of taurine. Physiol Rev 72:101–163

Ito H, Miller SC, Billingham ME, Akimoto H, Torti SV, Wade R, Gahlmann R, Lyons G, Kedes L, Torti FM (1990) Doxorubicin selectively inhibits muscle gene expression in cardiac muscle cells in vivo and in vitro. Proc Natl Acad Sci USA 87:4275–4279

Ito T, Fujio Y, Hirata M, Takatani T, Matsuda T, Muraoka S, Takahashi K, Azuma J (2004) Expression of taurine transporter is regulated through the TonE (tonicity-responsive element)/TonEBP (TonE-binding protein) pathway and contributes to cytoprotection in HepG2 cells. Biochem J 382:177–182

Ito T, Fujio Y, Takahashi K, Azuma J (2007) Degradation of NFAT5, a transcriptional regulator of osmotic stress-related genes, is a critical event for doxorubicin-induced cytotoxicity in cardiac myocytes. J Biol Chem 282:1152–1160

Ko BC, Turck CW, Lee KW, Yang Y, Chung SS (2000) Purification, identification, and characterization of an osmotic response element binding protein. Biochem Biophys Res Commun 270:52–61

Kumarapeli AR, Horak KM, Glasford JW, Li J, Chen Q, Liu J, Zheng H, Wang X (2005) A novel transgenic mouse model reveals deregulation of the ubiquitin-proteasome system in the heart by doxorubicin. Faseb J 19:2051–2053

Lopez-Rodriguez C, Antos CL, Shelton JM, Richardson JA, Lin F, Novobrantseva TI, Bronson RT, Igarashi P, Rao A, Olson EN (2004) Loss of NFAT5 results in renal atrophy and lack of tonicity-responsive gene expression. Proc Natl Acad Sci USA 101:2392–2397

Maouyo D, Kim JY, Lee SD, Wu Y, Woo SK, Kwon HM (2002) Mouse TonEBP-NFAT5: expression in early development and alternative splicing. Am J Physiol Renal Physiol 282:F802–F809

Minotti G, Menna P, Salvatorelli E, Cairo G, Gianni L (2004) Anthracyclines: molecular advances and pharmacologic developments in antitumor activity and cardiotoxicity. Pharmacol Rev 56:185–229

Miyakawa H, Woo SK, Chen CP, Dahl SC, Handler JS, Kwon HM (1998) Cis- and trans-acting factors regulating transcription of the BGT1 gene in response to hypertonicity. Am J Physiol 274:F753–F761

Moise NS, Pacioretty LM, Kallfelz FA, Stipanuk MH, King JM, Gilmour RF, Jr. (1991) Dietary taurine deficiency and dilated cardiomyopathy in the fox. Am Heart J 121:541–547

Oudit GY, Trivieri MG, Khaper N, Husain T, Wilson GJ, Liu P, Sole MJ, Backx PH (2004) Taurine supplementation reduces oxidative stress and improves cardiovascular function in an iron-overload murine model. Circulation 109:1877–1885

Pion PD, Kittleson MD, Rogers QR, Morris JG (1987) Myocardial failure in cats associated with low plasma taurine: a reversible cardiomyopathy. Science 237:764–768

Poizat C, Sartorelli V, Chung G, Kloner RA, Kedes L (2000) Proteasome-mediated degradation of the coactivator p300 impairs cardiac transcription. Mol Cell Biol 20:8643–8654

Rim JS, Atta MG, Dahl SC, Berry GT, Handler JS, Kwon HM (1998) Transcription of the sodium/myo-inositol cotransporter gene is regulated by multiple tonicity-responsive enhancers spread over 50 kilobase pairs in the 5'-flanking region. J Biol Chem 273:20615–20621

Satoh H, Sperelakis N (1998) Review of some actions of taurine on ion channels of cardiac muscle cells and others. Gen Pharmacol 30:451–463

Schaffer SW, Azuma J, Takahashi K, Mozaffari M (2003) Why is taurine cytoprotective? Adv Exp Med Biol 526:307–321

Schaffer SW, Solodushko V, Azuma J (2000a) Taurine-deficient cardiomyopathy: role of phospholipids, calcium and osmotic stress. Adv Exp Med Biol 483:57–69

Schaffer SW, Takahashi K, Azuma J (2000b) Role of osmoregulation in the actions of taurine. Amino Acids 19:527–546

Singal PK, Iliskovic N (1998) Doxorubicin-induced cardiomyopathy. N Engl J Med 339:900–905

Takahashi K, Ohyabu Y, Solodushko V, Takatani T, Itoh T, Schaffer SW,Azuma J (2003) Taurine renders the cell resistant to ischemia-induced injury in cultured neonatal rat cardiomyocytes. J Cardiovasc Pharmacol 41:726–733

Takatani T, Takahashi K, Uozumi Y, Matsuda T, Ito T, Schaffer SW, Fujio Y, Azuma J (2004) Taurine prevents the ischemia-induced apoptosis in cultured neonatal rat cardiomyocytes through Akt/caspase-9 pathway. Biochem Biophys Res Commun 316:484–489

Takihara K, Azuma J, Awata N, Ohta H, Hamaguchi T, Sawamura A, Tanaka Y, Kishimoto S, Sperelakis N (1986) Beneficial effect of taurine in rabbits with chronic congestive heart failure. Am Heart J 112:1278–1284

Trama J, Go WY, Ho SN (2002) The osmoprotective function of the NFAT5 transcription factor in T cell development and activation. J Immunol 169:5477–5488

Trama J, Lu Q, Hawley RG, Ho SN (2000) The NFAT-related protein NFATL1 (TonEBP/NFAT5) is induced upon T cell activation in a calcineurin-dependent manner. J Immunol 165:4884–4894

Uchida S, Kwon HM, Yamauchi A, Preston AS, Marumo F, Handler JS (1992) Molecular cloning of the cDNA for an MDCK cell Na(+)- and Cl(-)-dependent taurine transporter that is regulated by hypertonicity. Proc Natl Acad Sci USA 89:8230–8234

Uozumi Y, Ito T, Hoshino Y, Mohri T, Maeda M, Takahashi K, Fujio Y, Azuma J (2006) Myogenic differentiation induces taurine transporter in association with taurine-mediated cytoprotection in skeletal muscles. Biochem J 394:699–706

Wang Y, Ko BC, Yang JY, Lam TT, Jiang Z, Zhang J, Chung SK, Chung SS (2005) Transgenic mice expressing dominant-negative osmotic-response element-binding protein (OREBP) in lens exhibit fiber cell elongation defect associated with increased DNA breaks. J Biol Chem 280:19986–19991

Woo SK, Lee SD, Kwon HM (2002) TonEBP transcriptional activator in the cellular response to increased osmolality. Pflugers Arch 444:579–585

Chapter 55
Effects of Taurine on Cardiovascular and Autonomic Nervous Functions in Cold Exposed rats

Masayoshi Kuwahara, Tomohiro Kawaguchi, Koichi Ito, and Hirokazu Tsubone

Abstract Exposure to cold temperature might affect on cardiovascular and autonomic nervous function. Although there are a lot of studies on physiological and pathophysiological responses of taurine, it was poorly understood the effects of taurine on cardiovascular and autonomic nervous function during cold circumstances. Therefore, the purpose of this study was to clarify the possible role of taurine on cardiovascular and autonomic nervous function in rats exposed to cold temperature. For this purpose, heart rate, blood pressure and locomotive activity were recorded from conscious and unrestrained rats using a telemetry system. Moreover, the autonomic nervous function was investigated by power spectral analysis of heart rate variability. After the recovery period of implantation of transmitter, 1% taurine was supplied during experimental period ad libitum. After the 7 days control period, both taurine administrated and control groups of rats were exposed a cold temperature. There were no differences in heart rate, blood pressure and locomotive activity between taurine and control groups before cold exposure. However, parasympathetic nervous function was somewhat predominant in taurine group. Heart rate and blood pressure in both groups increased greatly by cold exposure. Heart rate in taurine group was much higher than that in control group. LF and HF powers were decreased by cold exposure in both groups. Although no differences were observed in LF, decrease of HF in taurine group was greater than that in control group. These results suggested that taurine might provide some reservoir for cardiovascular and autonomic nervous function to cold stress in rats.

M. Kuwahara (✉)
Department of Comparative Pathophysiology, Graduate School of Agricultural and Life Sciences, The University of Tokyo, Tokyo, Japan

J. Azuma et al. (eds.), *Taurine 7*, Advances in Experimental Medicine and Biology 643,
DOI 10.1007/978-0-387-75681-3_55,

55.1 Introduction

Taurine is one of the most abundant free amino acids in animal tissues (Jacobsen and Smith 1968). Taurine possesses many important physiological roles such as osmoregulator, diabetes modulator, antioxidant, heparoprotector, neurotransmitter, antiatherosclerotic, and cardioprotective actions. Moreover, antihypertensive action of taurine by suppression of sympathetic overactivity was also reported (Sato et al. 1987).

Cold-induced hypertension represents a prototypical model of environmentally induced hypertension. It is a "naturally occurring" form of experimental hypertension that is induced without genetic manipulation, administration of excessive doses of drugs or hormones, or surgical intervention (Sun et al. 1997). Numerous studies have shown that cold exposure activates the sympathetic nervous system, and increased heart rate (HR) and blood pressure (BP) are found in cold-exposed rats (Fregly et al. 1989; Papanek et al. 1991). However, the time course of the cold-induced responses on cardiovascular and autonomic nervous functions is poorly understood.

Therefore, we have examined (1) the time course of the cold-induced responses on cardiovascular and autonomic nervous functions and (2) the possible role of taurine on these functions during cold exposure.

55.2 Materials and Methods

55.2.1 Animals and Housing

Twelve male Wistar rats were purchased from Saitama Experimental Animals Supply Co., Ltd. (Saitama, Japan). They were housed in individual cages within a light-proof chamber (MIR-553, Sanyo, Tokyo, Japan). In the chamber, a light–dark cycle (LD 12:12, lights on at 08:00 h) was maintained under constant temperature (24°C). Standard diet (MF, Oriental Yeast, Co., Ltd., Tokyo, Japan) and water were supplied ad libitum.

55.2.2 Surgical Procedure

Animals were anesthetized (pentobarbital sodium, 50 mg/kg) and instrumented with a catheter in the descending aorta coupled with a sensor and transmitter (TA11PA-C40; Data Sciences, St. Paul, MN, USA) for telemetric monitoring of cardiovascular status.

55.2.3 Experimental Protocol

After the recovery period (7 days) from surgery, rats in taurine group were supplied 1% taurine solution throughout the experiment. Temperature in chamber was decreased from 24 to 6°C at 13:00 h after the 7-day supplementation of taurine. After the 21-day exposure to 6°C, temperature was returned to 24°C at 13:00 h for a 3-day recovery period.

55.2.4 Data Recording and Analysis

Each rat cage was placed on a signal receiving board (RLA1020, Data Sciences) in the chamber. The signals of HR, BP, and locomotor activity (LA) were recorded every 5 min by a Data Quest analyzing system (Data Sciences) throughout the experiment. The BP signal was continuously recorded by an ECG processor analyzing system (Softron, Tokyo, Japan). The recorded signals were sampled at 1 msec intervals and stored on a compact disket using the ECG processor.

Power spectral analysis of HR variability was made on an ECG processor analyzing system using BP data recorded on a compact disket. The computer program first detected and calculated the RR interval tachogram from BP wave as the raw HR variability in sequence order, and following time domain parameters were obtained: the mean of all normal RR intervals (mean RR); the SD of all normal RR intervals (SDRR); and the coefficient of variation of all normal RR intervals (CV). From this tachogram, data sets of 512 points were re-sampled at 80 msec. Any RR intervals before and after artifacts were excluded from analysis. The Hanning window and the fast Fourier transform were applied to each set of data to obtain the power spectrum of the fluctuation. Squared magnitudes and the products of the computed discrete Fourier transforms were averaged to obtain spectral estimates. We used frequency bands of low frequency (LF) and high frequency (HF) according to a previous study (Kuwahara et al. 1994): LF 0.04–1.0 Hz, HF 1.0–3.0 Hz. The ratio of LF and HF power (LF/HF) was also calculated. For analysis of BP variability, continuous BP signals were analyzed in the same way with RR interval time series. These sets of data underwent the fast Fourier transform directly.

55.2.5 Statistical Analysis

All values are expressed as mean ± SD. An analysis of variance (ANOVA) was used to compare the values in control and taurine groups. A paired Student's t-test was used to compare the effects of cold exposure in same group. A value of $P < 0.05$ was considered significant.

55.3 Results

55.3.1 Heart Rate and Blood Pressure

The HR and systolic BP obtained from baseline were 358.5 ± 14.8 beats/min and 116.3 ± 14.1 mmHg in control and 361.8 ± 11.3 beats/min and 117.9 ± 17.5 mmHg in taurine groups. Typical recordings of HR and BP throughout of experimental period in control rat are shown in Fig. 55.1. HR and BP changes in averaged 24 h in control and taurine groups are shown in Fig. 55.2. Exposure to cold resulted in elevated HR and BP in both groups. The levels of HR and BP in taurine group were slightly higher than those in control group during cold exposure, but there were no significant differences in HR and BP between control and taurine groups before, during and after cold exposure.

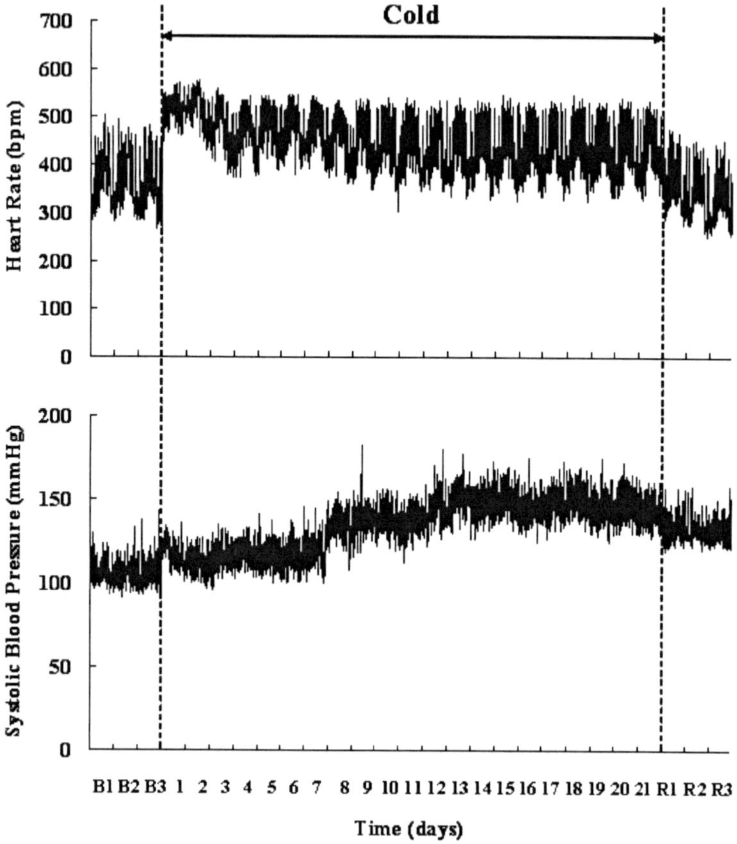

Fig. 55.1 Representative records of HR and BP throughout experiment in control rat

55.3.2 Autonomic Nervous Function

Changes in averaged 24 h in parameters obtained from power spectral analysis of HR variability are shown in Fig. 55.3. The changes in plot of the SDRR, CV, and LF power in both groups had the same patterns as did mean RR. Although the changes in HF power had resemble patterns, control group was higher than taurine group. Cold exposure significantly lowered these parameters in both groups. LF power

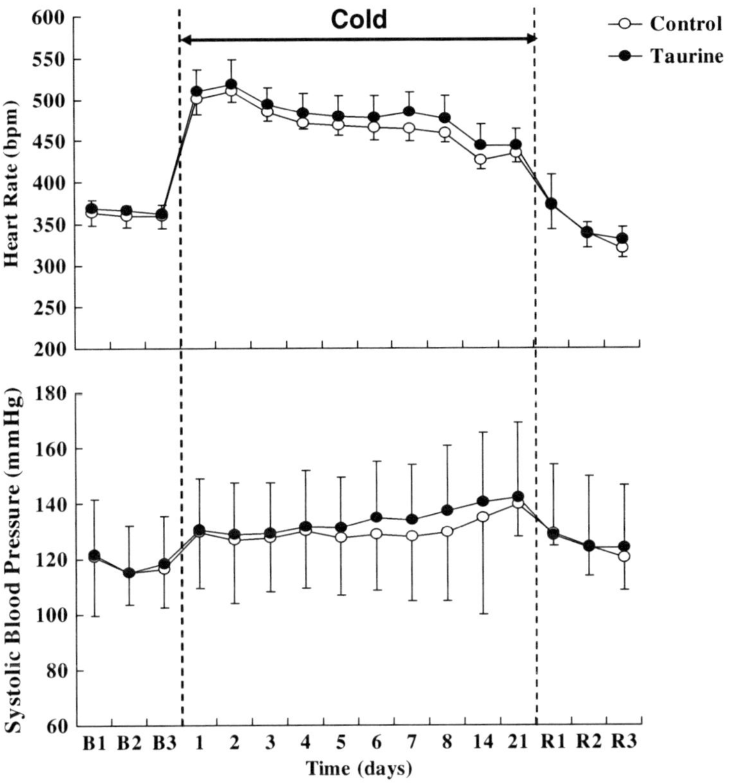

Fig. 55.2 Time course changes in HR and BP

of BP variability (BP-LF) and LF/HF changes were relatively similar patterns and no significant changes were observed by cold exposure in both groups. BP-LF in taurine group was slightly higher than that in control group and the values of this parameter were increased from 7- to 21-day cold exposure (Fig. 55.4).

55.4 Discussion

The present experiment clearly shows the time course changes of HR and BP to cold exposure, and these are promptly elevated by cold onset and decreased by cold offset in rats. Withdrawal of parasympathetic nervous activity might be mainly relevance to the increase of HR to onset and during cold exposure. BP elevation in chronic phase of cold exposure might be dependent with increased sympathetic nervous activity.

Moreover, enhancement of parasympathetic and diminished sympathetic nervous activities might decrease HR and BP to cold offset. There was no significant difference in these responses between control and taurine supplemented rats. These

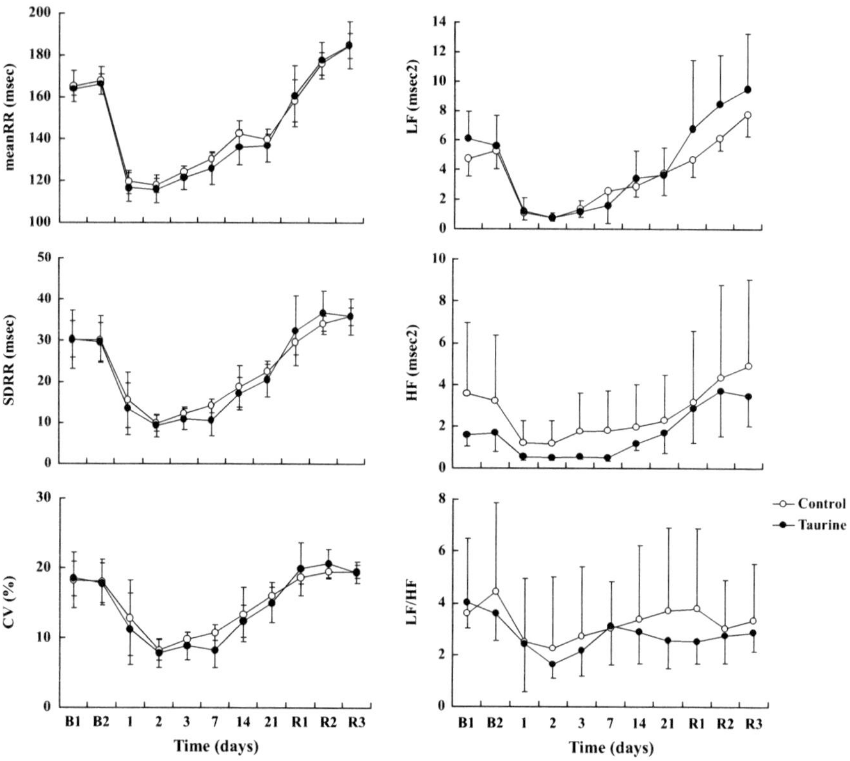

Fig. 55.3 Time course changes in time-domain and frequency-domain analysis of HR variability

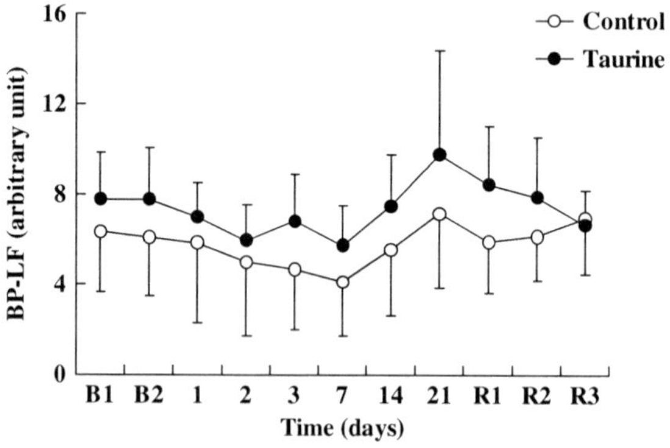

Fig. 55.4 Time course changes in BP-LF

results suggest that alteration of autonomic nervous activity might closely relate to the HR and BP responses to cold exposure.

It was reported that taurine supplementation attenuates the development of hypertension as well as reduces BP in DOCA-salt rats (Sato et al. 1987). We could not detect the antihypertensive effect of taurine in cold-induced hypertension model of rats. There is no obvious explanation for this discrepancy, and this may simply reflect the different model of hypertension. However, it should be noted the difference of measurement technique for BP. Although they used a tail cuff measurement for BP, nowadays a telemetry measurement is a much popular and reliable to evaluate many physiological phenomenon of animals. Even if the time course of cold-induced increase in BP may depend on measurement technique for BP (Chambers et al. 2000). Therefore, the tail cuff technique may not be the best approach to assess the influence of taurine on cardiovascular function.

HR variability has provided increasing interest as a noninvasive index of autonomic nervous activity (Task Force of the European Society of Cardiology and the North American Society of Pacing and Electrophysiology 1996). It is generally thought that LF reflects both sympathetic and parasympathetic nervous activity, HF reflects parasympathetic nervous activity, and LF/HF ratio reflects balance of autonomic nervous activity. Moreover, LF in BP variability reflects sympathetic nervous activity. It seems that changes in autonomic nervous activity obtained by HR and BP variabilities to cold exposure might reasonably explain the cardiovascular responses. One of the important findings is that withdrawal of parasympathetic nervous activity might be mainly relevance to the increase of HR to cold exposure.

Although there was no significant difference in autonomic nervous activity between control and taurine groups, sympathetic nervous activity became somewhat predominant by taurine supplementation throughout the experiment. Because effects of taurine on neuronal system are still poorly understood, the possible mechanisms of taurine on changes in autonomic balance are unknown. Suppression of parasympathetic nervous activity may be relevance to these changes, since Hue and Chanelet (1981) have shown that taurine decreases the release of acetylcholine from adrenals and superior cervical ganglia. However, in vitro addition of taurine significantly attenuated the Ca^{2+}-dependent, K^{+}-evoked release of [^{3}H]NE from a variety of neuronal tissues without affecting uptake or unstimulated release (Kuriyama et al. 1978; Muramatsu et al. 1978). Therefore, further studies will be necessary to clarify the effect of taunine on neuronal system.

In conclusion, the present study suggests that withdrawal of parasympathetic nervous function might be relevance to HR increase to cold exposure, and predominance of sympathetic nervous function might respond to BP increase during chronic cold exposure. Because we could not clearly show effects of taurine supplementation on cardiovascular and autonomic nervous functions, further studies will be needed to clarify mechanisms of taurine on these systems.

References

Chambers JB, Williams TD, Nakamura A, Henderson RP, Overton JM, Rashotte ME (2000) Cardiovascular and metabolic responses of hypertensive and normotensive rats tone week of cold exposure. Am J Physiol Regulatory Integrative Comp Physiol 279:R1486–R1494

Fregly MJ, Kikta DC, Threatte RM, Torres JL, Barney CC (1989) Development of hypertension in rats during chronic exposure to cold. J Appl Physiol 66:741–749

Hue B, Chanelet T (1981) Modulation of acetylcholine release by taurine in the central nervous system of the cockroack. J Physiol (Paris) 15:65–78

Jacobsen JG, Smith LH (1968) Biochemistry and physiology of taurine and taurine derivatives. Physiol Rev 48:424–511

Kuriyama K, Muramatsu K, Nakagawa K, Kakita K (1978) Modulating role of taurine on release of neurotransmitters and calcium transport in excitable tissues. In: Barbeau A, Huxtable RJ (eds) Taurine and neurological disorders. Raaven Press, New York, pp 201–216

Kuwahara M, Yayou K, Ishii K, Hashimoto S, Tsubone H, Sugano S (1994) Power spectral analysis of heart rate variability as a new method for assessing autonomic activity in the rat. J Electrocardiol 27:333–337

Muramatsu K, Kakita K, Nakagawa K, Kuriyama K (1978) A modulating role of taurine on release of acetylcholine and norepinephrine from neuronal tissues. Jpn J Pharmacol 28:259–268

Papanek PE, Wood CE, Fregly MJ (1991) Role of the sympathetic nervous system in cold-induced hypertension in rats. J Appl Physiol 71:300–306

Sato Y, Anodo, K, Fujita T (1987) Role of sympathetic nervous system in hypotensive action of taurine in DOCA-salt rats. Hypertension 9:81–87

Sun Z, Cade JR, Fregly MJ (1997) Cold-induced hypertension: a model of mineralocorticoid-induced hypertension. Ann NY Acad Sci 813:682–688

Index

Printed in the United Kingdom
by Lightning Source UK Ltd.
134982UK00001B/70-90/P

9 780387 756806